BIOCHEMISTRY AND MOLECULAR BIOLOGY IN THE POST GENOMIC ERA

TRENDS IN BIOCHEMISTRY AND MOLECULAR BIOLOGY

Biochemistry and Molecular Biology in the Post Genomic Era

Additional books and e-books in this series can be found on Nova's website under the Series tab.

BIOCHEMISTRY AND MOLECULAR BIOLOGY IN THE POST GENOMIC ERA

TRENDS IN BIOCHEMISTRY AND MOLECULAR BIOLOGY

HOSSAIN UDDIN SHEKHAR
AND
M. M. TOWHIDUL ISLAM
EDITORS

nova science publishers
New York

Library of Congress Cataloging-in-Publication Data

Names: Shekhar, Hossain Uddin, 1965- editor.
Title: Trends in biochemistry and molecular biology / Hossain Uddin Shekhar, M.M. Towhidul Islam.
Description: New York : Nova Science Publishers, Inc., [2019] | Series: Biochemistry and molecular biology in the post genomic era | Includes bibliographical references and index. |
Identifiers: LCCN 2019042878 (print) | LCCN 2019042879 (ebook) | ISBN 9781536164343 (paperback) | ISBN 9781536164350 (adobe pdf)
Subjects: LCSH: Biochemistry. | Molecular biology.
Classification: LCC QH345 .T766 2019 (print) | LCC QH345 (ebook) | DDC 572--dc23
LC record available at https://lccn.loc.gov/2019042878
LC ebook record available at https://lccn.loc.gov/2019042879

Published by Nova Science Publishers, Inc. † New York

CONTENTS

vi *Contents*

Chapter 9 The Molecular Determinants of Preeclampsia Accountable
 for Multifactorial Disorder in Pregnant Women **211**
 Md. Alauddin and Yearul Kabir

Chapter 10 Role of Autophagy in Cancer: Mechanistic and Therapeutic
 Understanding from the Cellular and Molecular Point of View **233**
 Omar Hamza Bin Manjur, Akib Mahmud Khan,
 Hamida Nooreen Mahmood, Sohidul Islam
 and Mahmud Hossain

Chapter 11 Insights into the Bifidobacterium:
 Integral Members of the Gut Microbial Community **259**
 Parag Palit and Farhana Tasnim Chowdhury

Chapter 12 Coordination of Molecular Techniques and Advanced
 Bioinformatics Tools for Analyzing Viral Evolutionary
 Distance with an Emphasis on Foot-and-Mouth Disease Virus
 (FMDV) Serotype O **273**
 Salma Akter, Mohammad Anwar Siddique,
 A. S. M. Rubayet Ul Alam, Munawar Sultana
 and M. Anwar Hossain

Chapter 13 Endophytes: A Diverse World of Microorganisms **295**
 Farhana Tasnim Chowdhury and Tonny Tabassum

Chapter 14 Morphological and Molecular Detection of Fungi in Bangladesh **315**
 Shamim Shamsi

Chapter 15 Application of Molecular Biology Techniques in Aquatic
 Animal Health Research **337**
 Mohammad Shamsur Rahman, Nusrat Jahan Punom,
 Md. Mostavi Enan Eshik and Mst. Khadiza Begum

Chapter 16 Utilities of the Concept of Statistical Analysis
 in the Biological Sciences **383**
 Murshida Khanam

Editors' Contact Information **397**

Index **399**

Related Nova Publications **413**

PREFACE

"Trends in Biochemistry and Molecular Biology" provides the essential information necessary for students in the life and health sciences. The book adopts a readable, student-friendly style that helps introduce students to this fascinating and often-times daunting subject. Each chapter begins with a summary of essential facts followed by descriptions of the subjects that focus on core information with clear, simple diagrams that are easy for students to understand and recall in essays. The extensive use of cross-referencing makes it possible for students to return to individual sections for review purposes without difficulty. Whether students' interests lie in biological, chemical, or medical aspects of biochemistry and molecular biology, "Trends Biochemistry and Molecular Biology" will help make students able, excited, and eager to read more widely and more deeply on this engaging subject. This important new book not only covers an extensive set of topics of current and special interest, but includes more traditional areas in biochemistry as well. Covering a wide range of topics, from classical biochemistry to proteomics and genomics, it also details the properties of commonly used biochemicals, laboratory solvents, and reagents. Coverage is expanded to include a section on stem cells, chapters on immunochemical techniques and spectroscopy techniques, and additional chapters on drug discovery and development, and clinical biochemistry. Moreover, a number of techniques used in molecular biology, for example, molecular cloning, gel electrophoresis, polymerase chain reaction, microarrays, etc. are also explained with practical examples. It also includes some of the vital pieces of work being conducted across the world, on various topics related to molecular biology. Through it, we attempt to further enlighten the readers about the new concepts in this field. Altogether, presented in an organized, concise, and simple-to-use format, "Trends in Biochemistry and Molecular Biology" allows quick access to the most frequently used data. There is an emphasis on biological aspects of biochemistry and new topics are introduced in their biological context wherever possible. Experimental design and the statistical analysis of data are emphasized at the end to ensure students are equipped to successfully plan their own experiments and examine the results obtained.

Hossain Uddin Shekhar and M. M. Towhidul Islam

In: Trends in Biochemistry and Molecular Biology
Editors: Hossain Uddin Shekhar et al.
ISBN: 978-1-53616-434-3

Chapter 1

STEM CELLS: HISTORY, APPLICATIONS AND THE FUTURE

Hamida Nooreen Mahmood[1], Talita Zahin Choudhury[1], Ar-Rafi Md. Faisal[1], Hussain Md. Shahjalal[2], PhD and Mahmud Hossain[1,*], PhD

[1]Department of Biochemistry and Molecular Biology, University of Dhaka, Dhaka, Bangladesh
[2]Department of Biochemistry and Molecular Biology, Jahangirnagar University, Savar, Dhaka, Bangladesh

ABSTRACT

The discovery of stem cells (SC) in the late 1970s has introduced a paradigm shift in the way scientists study developmental and cellular biology. Extensive studies on clonogenic and self-renewing capabilities have sensationalized stem cells' role in regenerative medicine and their applications are now expanding into therapies for various chronic diseases. They are the class of cells possessing the unlimited replicative ability and are capable to be differentiated into all types of specialized cells under certain conditions. Based on the developmental stages, stem cells can be categorized into two major groups: embryonic stem cells (ESC), and adult stem cells. In this chapter, the roles of these cells in every stage of development and the complex regulatory processes that govern these cells are discussed. The discovery of the Nobel-prize winning induced pluripotent stem cells (iPSC) and their various regenerative applications are also explained. Finally, the recent successes of stem cell-based therapy and the future scopes are mentioned.

* Corresponding Author's E-mail: mahmudbio1480@du.ac.bd.

Keywords: stem cells (SC), induced pluripotent stem cells (iPSC), embryonic stem cells (ESC)

1. INTRODUCTION

Most of the cells that constitute a multicellular organism have very specific functions and morphologies; they are said to be differentiated. However, each cellular unit has been derived from fertilized eggs. With the epitome of totipotency and the gold standard of a stem cell, the fertilized egg will divide countless times and undergo many complex processes to generate a complete set of specialized diploid and haploid cells, and become ready to transmit the genetic code to the next generation. Stem cells, with their self-renewal capacity as well as to differentiate into specialized cell types, have established themselves on the top of the developmental hierarchy of the cells.

The stem cells can largely be classified into three types of cells according to their differentiation potential: totipotent, pluripotent, and multipotent stem cells. Totipotent stem cells possess the ability of self-renewal by dividing and can develop into all the three primary germ layer cells of the early embryo. As such, the fertilized egg cells are totipotent in nature those retain the capability to develop into any specialized cell type of an organism. On the other hand, pluripotent stem cells have the ability to divide and develop into all the three germ layers– ectoderm, endoderm, mesoderm; but its inability to form extra-embryonic tissues is a limitation. In case of multipotent stem cells, the major limitation is the restricted developmental ability into specific cell types from the same lineage, e.g., hematopoietic stem cells, which ultimately differentiated into all blood cell types, but they are unable to differentiate into brain cells. They persist into adulthood and serve to replace or replenish cells of a specific tissue. Alternatively, depending on their origin stem cells can further classify. In mammals, stem cells are widely classified into two major categories: 1. Embryonic stem cells (ESC), derived from the inner mass cells of an embryo that have been fertilized *in vitro*. ESCs are obtained by the transfer of pre-implantation stage embryo cells into a nutrient rich culture medium. These cells persist as undifferentiated as long as these cells are grown within a maintenance medium. However, by altering the surface medium or by changing the chemical composition of the medium or even by manipulation of the genes in the ESCs, they can be induced to develop cells of specific types. Due to their being plastic in nature and containing unlimited self-renewal potential, ESCs are a reliable source of transplantable cells and thus can efficiently serve in regenerative medicine. Additionally, they can be used to generate immunocompatible tissues/organs, toxicology study and as a model to study early development. 2. Adult stem cells, are the undifferentiated kind of cells that resides in a specific region of an organ or tissue (for example, gut, brain, blood vessels, peripheral blood, bone marrow, liver, heart, pancreas etc.) called the cell niche, among the cells that are differentiated. The natural function of adult stem cells includes the conservation of tissue homeostasis and damaged tissue repair that is incurred all along the lifespan of an organism. These cells are multipotent and have limited capacity to divide. Various somatic cells can also be induced to generate embryonic stem cell-like cells, known as induced pluripotent stem cells (iPSC), through the introduction of certain pluripotency genes. iPSC can also be used to generate tissues and organs for regenerative medicine and overrides the ethical limitations of using ESCs. This has introduced a new arena of stem cell research by providing a source of donor specific cell, therefore solving the compatibility issues in case of tissue regeneration purposes. Although several treatment strategies have been reported in the recent years based on stem cells, the methods for generating, propagating and directing stem cells to specific cell lineages and their effective applications are still under investigation.

2. Characteristics of Stem Cells

According to the classical definition, stem cells must have two salient features: Self-renewal, and Potency.

2.1. Self-Renewal

Self-renewal is the process of a stem cell's symmetrical or asymmetrical division with a view to produce daughter stem cells having developmental potency alike the mother cell. This self-renewal capability is a prerequisite for stem cells for the purpose of (i) propagation in course of development; (ii) their maintenance within adult tissue; and (iii) reformation of stem cell repertoire after a damage. Any defect in this process may bring about developmental anomalies, may cause cancer and untimely aging (Shenghui et al., 2009). Two different mechanisms mainly come into play for maintaining stem cell population:

2.1.1. Obligatory Asymmetric Replication

This replication process is also known as asymmetric cell division, in which stem cells undergo cell division to produce two different types of daughter cells where one cell is indistinguishable to the mother cell but the other one is different from the initial stem cell.

2.1.2. Stochastic Differentiation

In this differentiation process along with one stem cell differentiate into two daughter cells, another undifferentiated produces two copies of stem cells alike the original one through mitosis and thus an existing population of the undifferentiated stem cells is maintained.

2.2. Potency

Stem cell potency refers to its potential to produce a range of new cell types. More precisely, potency is the capacity of stem cells to differentiate into particular cell types. To be more specific, stem cells are required to be of totipotent or pluripotent nature, though in several cases progenitor cells' multipotent or unipotent in nature are attributed to stem cells as well. Stem cells of different developmental stages and from different tissue types produce different numbers of cells with different character. Thus, function of stem cell in its originating tissue is largely suggested by its potency.

3. Classification of Stem Cells

After division, some of the progeny cells persist as stem cells responsible for repairing damages promptly while other cells continue to become mature into specific cell types like heart, blood, muscle, brain, etc. In this way stem cells attain the capability to reproduce themselves continually in order to promote tissue renewal throughout the life span of an individual. Throughout different developmental stages, stem cells can be collected. Moreover, adult somatic cells can be induced to develop cells alike embryonic stem cell using reprogramming technology (Figure 1).

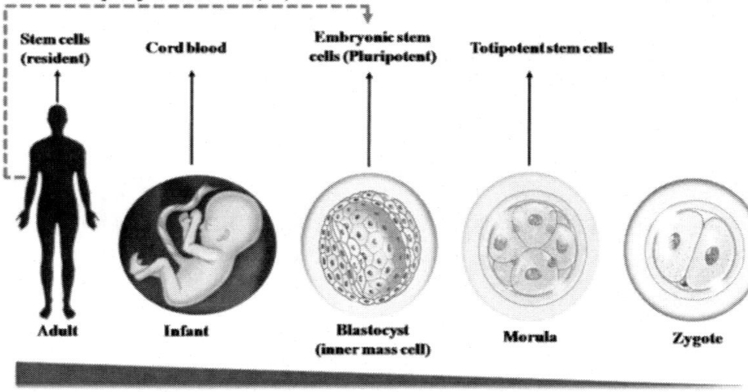

Figure 1. Sources of stem cells. Different developmental stages of life provide rich stem cell sources. Moreover, with the help of reprogramming technology, specific inductions can subject adult somatic cells to develop embryonic "stem cell–like" cells called Induced Pluripotent Stem Cell (iPSC) (modified from *Plast. Reconstr. Surg.* 126: 1163, 2010).

3.1. Classification Based on Differentiation Potential

Based differentiation potency, stem cells can be classified into: totipotent, pluripotent, and multipotent, these three types of cells. A totipotent stem cell can be differentiated into an entire organism while a pluripotent stem cell lacks this characteristic. But they have the potency to be differentiated into any cell type. In contrast, multipotent stem cells' differentiation capacity are limited to only a few cell types like bone, blood etc. (Schöler, 2016).

3.1.1 Totipotent Stem Cells

Totipotent (also known as omnipotent) stem cells have the ability to develop into embryonic and extra-embryonic cells inevitable for human development. Such cells have the competence to manufacture an entire viable organism (Condic, 2013). These cells are the end products of an egg cell and a sperm cell fusion that ultimately formulates zygote which further endures cell division to structure a two-cell stage. In successive cell divisions there occurs a boost in cell numbers for several days until the morula (consisted of eight cells) stage. The zygote is the eventual stem cell for its potential to be developed into all three embryonic germ layer cells, i.e., ectoderm, mesoderm and endoderm as well as extra-embryonic cell types responsible for generating the placenta and supporting tissues.

3.1.2. Pluripotent Stem Cells

Pluripotent stem cells are the progenies of totipotent cells that can develop into almost all the three germ layer cells produced in the initial differentiation stages of embryonic stem cells (Shenghui et al., 2009). Nearly all cell types can be produced from Pluripotent stem cells *in vitro*, but despite this characteristic upon implantation in a uterus, they can't structure a complete living organism due to their inability to give rise to extra-embryonic tissues needed for fetal development. Embryonic germ cells (EG cells) are the pluripotent stem cells collected from the fetal primordial germ cells, whereas embryonic stem cells (ESC) are the stem cells collected from the inner mass cell of an embryo blastocyst. However, ESC-like cells can also be produced through reprogramming of various somatic cells, which are recognized as induced pluripotent stem cells (iPSCs) (Wilson and Wu, 2015). This ground breaking nuclear reprogramming technology of converting adult somatic cells into pluripotent stem cells, was unlocked by two scientists, Kazutoshi Takahashi and Shinya Yamanaka, who reprogrammed skin fibroblast cells into ESC-like cells (i.e., iPSCs). They introduced four different pluripotency genes, *KLF4, SOX2, c-MYC* and *Oct-3/4* to bring both self-renewal and pluripotency characteristics in iPSCs (Mahla, 2016). As a result, these cells can be differentiated into different functional cells, such as- cardiomyocytes, adipocytes, neural cells, hematopoietic progenitor cells, pancreatic beta-like cells, hepatocytes etc. (Figure2). For this outstanding discovery, Shinya Yamanaka was awarded the Nobel

Prize in Physiology in 2012. The discovery of iPSCs offers the researchers a sublime way to create and study diseased cells and to develop the treatments for various diseases (Yamanaka, 2012).

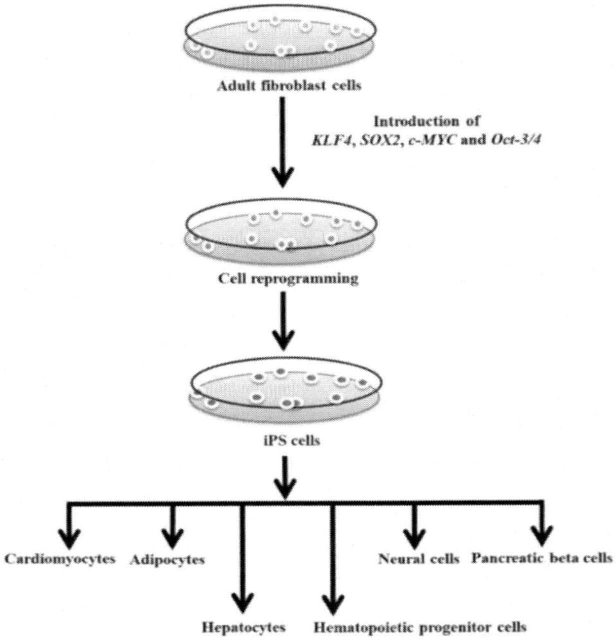

Figure 2. Generation of Induced Pluripotent Stem Cells (iPSCs) and their differentiation potential into major somatic cell lineages.

3.1.3. Multipotent Stem Cells

Multipotent stem cells are of the characteristic to differentiate into a number of cells that are closely associated (Liang et al., 2013). They have the aptitude to produce different terminally differentiated cell types working as the constituents of an organ or tissue and thus being liable for the maintenance of cell, tissue and organ homeostasis. Hair follicles, mesenchymal stem cells, hematopoietic (adult) stem cells, etc. are the common examples of multipotent stem cells. Different blood cell lineages, such as white blood cells (WBC), red blood cells (RBC) or platelets are differentiated from hematopoietic stem cells. The hematopoietic stem cells can readily be collected from blood, bone marrow and umbilical cords (immediately after birth) like sources. On the other hand, neural stem cells are one of the best characterized multipotent stem cells that can develop into all cell types of the nervous system. Multipotent stromal cells or mesenchymal stem cells are examples of some best characterized stem cells and collected conveniently from adipose tissue and bone marrow. Having an extensive differentiation repertoire, mesenchymal stem cells can develop into numerous cell types including tissues like fat cells, bone/bone marrow, tendon, cartilage, smooth muscle cells etc.

In addition, two other types of stem cells are also found: Oligopotent stem cells and unipotent stem cells.

3.1.4. Oligopotent Stem Cells

They can differentiate into only a limited number of cell types. For example, myeloid stem cells or lymphoid stem cells (Kondo et al., 1997).

3.1.5. Unipotent Stem Cells

Their ability is restricted to producing only one type of differentiated cells. Examples include- skin cells, (adult) muscle cells, etc. (Slack, 2000).

3.2. Classification of Stem Cells Based on Their Origin

Stem cells can be classified according to their origin of derivation. In mammals, stem cells are widely classified into two types: (i) embryonic stem cells, collected from the inner mass cell of blastocysts of embryo and (ii) adult stem cells, obtained from several tissues. Mostly stem cells are extracted from either a dividing zygote or adult tissues and then are placed in a controlled culture that has the ingredients to restrict their differentiation but allows replication. Both types of stem cells are normally represented by their differentiation potency into different types of cells for example- muscle, bone, skin, heart, pancreas, hepatocytes, etc.

3.2.1. Embryonic Stem Cells

Embryonic stem cells are the ancient type of cells that can persuade themselves to differentiate into most of the 220 type cells of the human body. Embryonic stem cells come from inner mass cells of 5-6 days old embryos or blastocysts, which are like a ball of cells (50-150 cells) obtained from *in vitro* fertilization (IVF) for infertility treatment or from 5-9 week old fetuses provided by facultative abortion. Somatic cell nuclear transfer or cloning can also work as a source of embryo (Holland et al., 2001). However, embryonic stem cells can no longer possess totipotency, when they are disassociated. Rather, embryonic stem cells are pluripotent in character, for their being able to generate any cell types (Marshak et al., 2001). Despite not being a part of the placenta or extra-embryonic membranes during development, embryonic stem cells can derive all of the three primary germ layers, i.e., ectoderm (derives skin, nervous system), endoderm (derives lung and the total gut tube) and mesoderm (derives bone, muscle, blood and all the connecting constituents between endoderm and ectoderm). The ability of embryonic stem cells to develop into all three germ layer cell types emphasizes their propitious role in regenerative medicine despite the existence of multiple limitations. For example, embryonic stem cells' unlimited self-renewal capability mostly in the undifferentiated states makes them vulnerable to recalcitrant differentiation, which ultimately paves the way of teratocarcinoma formation (Wu et al., 2007). Therefore, elimination of this possibility is the prerequisite for the embryonic stem cells' to be used in clinical purposes successfully. On the other hand, potential immunogenic responses towards a tissue graft which is derived from embryonic stem cell have been a concern (Martin et al., 2005). Moreover, there are significant ethical drawbacks hindering further inspection of embryonic stem cells in human. The principal interest of scientists to research and get the best out of embryonic stem cells is especially based on three objectives: (i) excel the knowledge for human health and development, (ii) cultivating several types of cells, tissues and even organs (iii) testing and exploring new and existing compounds or drugs for effective treatments (Garfinkel, 1999).

3.2.2. Adult Stem Cells

Adult stem cells are the special type of cells those are undifferentiated in nature, but endowed among the other differentiated cells of tissues or organs. They are accessible in many parts of our body, for example brain, bone marrow, skin, fat and in several other locations. Adult stem cells, also known as progenitor cells possess two indisputable characteristics: (i) dividing and producing a cell which is self-identical, and (ii) dividing and producing a more differentiated cell than itself. Adult stem cells (germline or somatic) generally reside in the bone marrow and their other common sources are spinal cord, gum tissue, skin epithelia, digestive system, brain, cornea and retina, liver, teeth, pancreas, etc. Adult stem cells are usually multipotent in character and are fundamental for homeostasis maintenance of most cell system, tissues and organs throughout mammalian lifespan. In some tissues, these cells have sustainable turnover to carry out repair mechanism during the whole life. For instance, skin stem cells keep on producing new skin cells that ensure replenishment of aged or damaged skin. They are a very small number of stem cells residing in the distinct place of every tissue where they may maintain dormancy for several years before being induced by tissue injuries and diseases (Rodriguez et al., 2005). The unique advantage that adult stem cells provide is certainly the ability to be harvested from the patient himself. This advantage overcomes the probability of immune rejection after transplantation process. Moreover, their limited proliferation capacity provides them with an advantage of reducing the risk of malignancy. Due to these advantages in bone marrow transplantation and other therapeutic processes, adult stem cells are being used regularly.

4. Methods of Obtaining Stem Cells

In case of regenerative medicine, shortage for organ donor has increased the need of seeking help from stem cell research for its having the potential to be an effective source. An overview of the methods for obtaining stem cells is summarized below.

4.1. Stem Cell Derived from Amniotic Fluid

Amniotic fluid consists of heterogeneous cell populations originated from the developing fetus (Polgar et al., 1989). These heterogeneous cell populations are comprised of adipocytes, osteocytes, neurogenic and endothelial cells, and are called amniotic fluid-derived stem cells (Schmidt et al., 2007). It has been reported that both mouse and human amniotic fluid derived-stem cells have extensive self-renewal capability which allow them to generate cells of osteogenic, adipogenic, myogenic, hepatogenic, neurogenic, and endothelial lineages (De Coppi et al., 2007).

4.2. Transfer of Somatic Nucleus

Somatic nuclear transfer process involves the elimination of oocyte in culture and then replacing its original nucleus with a nucleus of the patient's somatic cell (Figure 3).

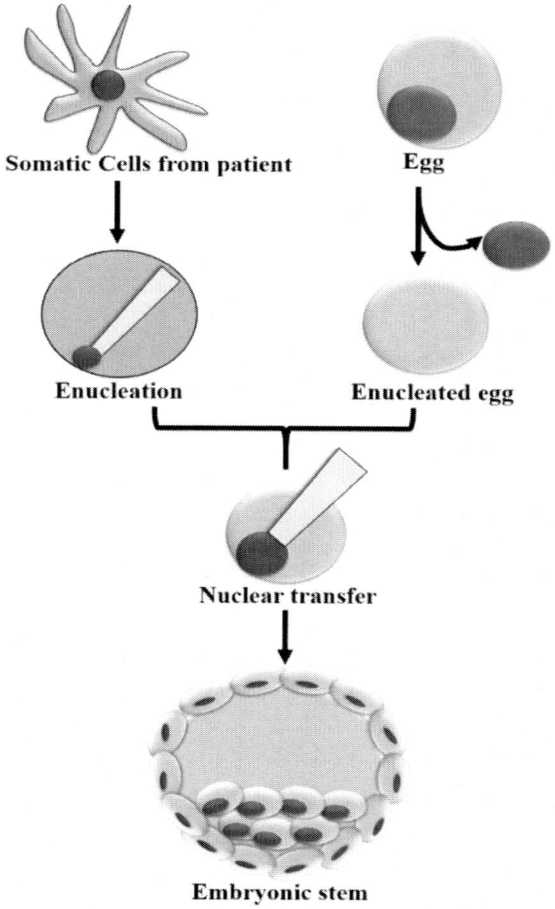

Figure 3. Postnatal somatic nucleus transfer into an enucleated ovum during somatic nuclear cell transfer technology (modified from *Plast. Reconstr. Surg.* 126: 1163, 2010).

Cells are then induced to divide up to blastocyst stage through chemical or electrical stimuli. In subsequent steps inner mass cells are then isolated and cultured in order to produce genetically identical (to patient) embryonic stem cells. Being genetically identical these embryonic stem cells are completely adapted to the immune system of the patient that excludes immune rejection and thus use of immune suppressant is not needed (Ying et al., 2002). The major breakthrough in this field was obtained by sheep Dolly (July 5, 1996 - February 14, 2003), the first mammal in the history of Bioscience to be cloned from adult stem cell. In case of other mammals for example, mice, cattle, cats, pigs and dogs this technique was used as well (Ying et al., 2002). Human embryonic stem cells formulated in this manner can be further differentiated into cells important for therapeutic purposes and transplanted into patients in order to cure degenerative diseases. However, somatic nuclear cell transfer is ethically challenged by a complex argument about obtaining human unfertilized eggs and the embryo produced in this process. Moreover, low efficiency (about 10%) of this process enhances the controversy (Shin et al., 2002). Nevertheless, somatic nuclear transfer is still considered to be an effective approach.

4.3. Using Spare Embryos from Fertility Treatment

Production of 'Spare' embryos is normally induced by the use of fertility hormones. All of the multiple eggs that women produce are fertilized. The only egg that is not fertilized and thus re-implanted due to the risks associated with multiple births is called 'spare' embryo. Dr. Edwards and other workers have produced many 'spare' embryos for fertilization techniques' testing and development (Choudhary et al., 2004).

4.4. Using Adipose Tissue as a Stem Cell Source

Stem cells derived from adipose tissue can be obtained under local anesthesia through minor liposuction surgeries. This is a convenient option for patients going through iliac crest cell harvest or bone marrow transplantation, since this process eliminates significant sensations of pain. In USA, 478,251 cases of elective surgeries of liposuction were reported in 2004 (Swami et al., 2009). Moreover, an outpatient cosmetic liposuction (1994-2000) study carried out by the American Society for Dermatologic Surgery reported 0% death rate in case of66,570 procedures and the rate of serious adverse event was only 0.68 out of 1000 cases, bolstering on this technique's being highly safe (Housman et al., 2002). Stem cells derived from adipose tissue retain the ability to be differentiated into bone, muscle, cartilage, adipose tissue, nerve, etc., and have been proved to be a progressive option to be used in the applications of regenerative medicine in several cases of animal model (Figure 4). For example, osteogenic potential and myogenic capability of this type of stem cells have been manifested in mouse calvarial (skull cap) bone defective model and in a mouse model of Duchenne muscular dystrophy, respectively (William et al., 2003).

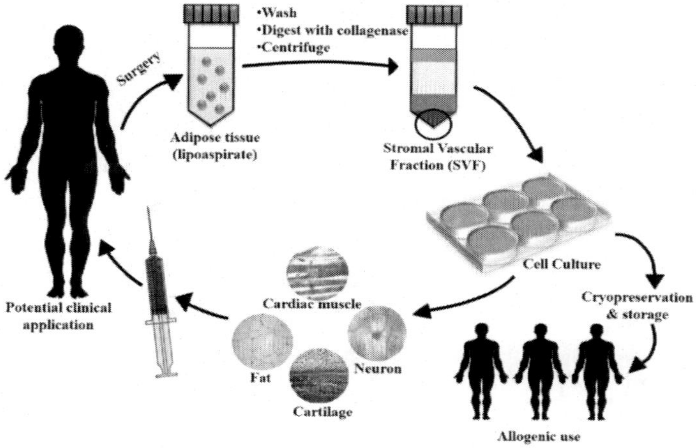

Figure 4. Stem cells derived from adipose tissue in a reconstructive cycle. Liposuction is a rich source of harvesting adipose tissue derived stem cells.

To be used in regenerative application purposes stem cells derived from adipose tissues need to fulfill the following prerequisites (Sengenes et al., 2005):

a) Should be found amply (up to billions of cells)
b) Their harvest procedure should be minimal invasive
c) Their differentiation pattern should be reproducible and at the same time highly controllable along with multiple cell lineage pathways
d) Their transplantation into allogenic or autologous hosts should be effective and safe
e) Should be able to be manufactured according to the of good manufacturing practice guidelines

4.5. Single Cell Embryo Biopsy

The primary and the most objectionable issue of embryonic stem cell research in human is its association with the living embryo extermination. Therefore, scientists thought of a method that would isolate embryonic without any destruction in 2006. Chung et al. group was the pioneer to differentiate blastomere-derived embryonic stem cells into teratomas and all the three germ layer cell types (Klimanskaya et al. 2006). It was reported that no developmental decrement occurred in the biopsied embryo of mouse if developed full term. In order to figure out the possibility of this process to be translated into human, scientists are performing further experiments.

5. Applications of Stem Cells

A stem cell is a non-specialized, generic cell that has self-renewal capability and has the capacity of differentiating into specialized cell types (Mahla, 2016). It is a "Hot Cake" of 21st century medical research as it offers scientists and researchers to have a better understanding of cellular differentiation which further helps to understand and treat various diseases including birth defects, whole tissue development by tissue engineering, genetic manipulations in gene therapy, and to provide resources for testing new drugs and what not. Ongoing research on stem cells gives hope to treat patients who would normally receive treatment just to alleviate the symptoms of their chronic illness (Yamanaka, 2012). Stem cell therapy involves more than transplanting cells into the body and allowing them to grow healthy tissues. It also provides the ample opportunities for screening pharmaceuticals and to develop personalized-precision medicine (Wilson and Wu, 2015).

5.1. Stem Cells Can Be Used for Better Understanding of Diseases

The major step while developing precise therapy for a particular disease is to understand the exact molecular mechanism of the disease, how it works, and what effect it does to the body. For this, researchers need to study the affected cells or tissues from the early stages of disease progression, but it is practically almost impossible to obtain cells or tissues in the early stages of a disease when an individual is not aware of any symptoms. Reprogramming of diseased cells to pluripotent stem cells (i.e. IPSCs) allows the opportunity to track a large number of diseases from the early stages as these cells have the same genetic background (Mimeault et al., 2007). In turn, this will help to acquire more knowledge about the molecular mechanism of a disease and what goes wrong inside the cells and why.

5.2. Stem Cells Have the Potential of Damaged Cell Repair and Disease Treatment

The attribute of stem cells to replace damaged cells or tissues, has already been exploited in case of extensive burn treatments (skin replacement) and to treat leukemia and other blood disorders. Moreover, stem cells also possess the potentials to replace damaged cells or cells lost due to diseases for which no well-grounded medications are currently available. Most often donated organs and tissues are the main sources to replace damaged tissues, but they have the chances of potential immune rejections. Furthermore, the supply of transplantable organ or tissues can't meet the current demand. Stem cells having the characteristics of directed differentiation into specific cell types offer a great opportunity to adopt them as a renewable source of replacing tissue disease treatment including- stroke, Parkinson's, diabetes, heart disease, etc. This amazing approach needs more significant research to overcome the technical hurdles to ground it as an effective treatment strategy in the future.

5.3. Stem Cells in Autologous Cell Therapies

Embryonic stem cells are isolated from the inner mass cells of a blastocyst and are the only source of studying pluripotency, especially in human systems until the discovery of iPSCs (Thomson et al., 1998). However, studying pluripotency from human ESCs faced much controversy and ethical challenges as loss of human embryos was unethical in terms of public acceptance. Instead of ESCs, iPSCs are easier to derive and are free of ethical issues. Regenerative therapy involving iPSCs differentiated cells is thought as self if transplanted into the original donor patient while excluding the major concern of longstanding use of immunosuppressants. This strategy opens up a new window for autologous cell therapy where disease affected cells are collected from the patients and are subjected to reprogramming with *KLF4, SOX2, c-MYC* and *Oct-3/4* for the generation of iPSCs possessing pluripotency. Subsequently, genetic modification of these cells can also be done followed by differentiation and transplantation of the reformed cells or tissues into the donor patients (Figure 5) (Wilson and Wu, 2015). However, iPSC approach needs some level of immunosuppression when subjected to study in some animal models (Zhao et al., 2011). The first-in-human clinical trial of treating neovascular age-related macular degeneration by transplantation of a sheet of retinal pigment epithelial cells using iPSC approach (i.e., iPSC derived retinal pigment epithelium cells) demonstrated that autologous iPSC-derived transplanted retinal pigment epithelial cell sheet remained intact along with a brighten vision after one year of transplantation (Mandai et al., 2017). As a result, the autologous iPSC approach remains a long term goal for many diseases because of the concern of safety, immunosuppression, potential tumorigenicity and various advantages over other stem cell therapies, as well as currently available treatments.

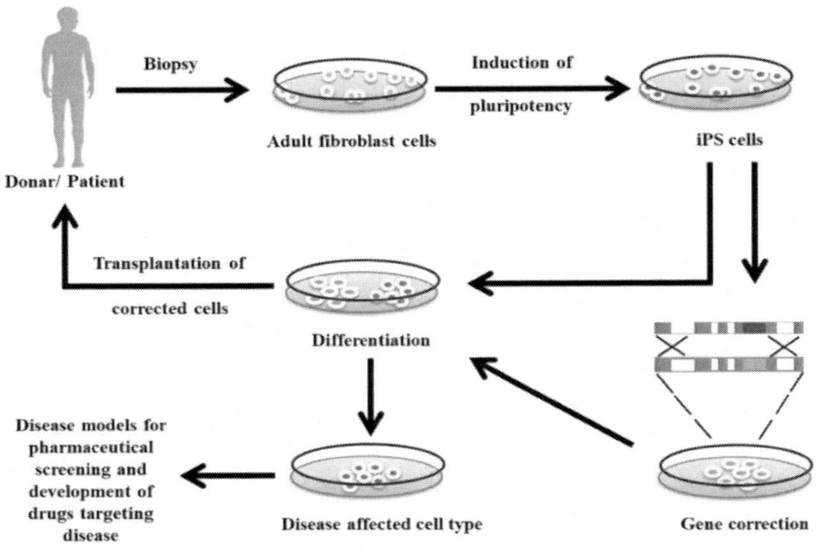

Figure 5. Therapeutic potential and application of induced pluripotent stem cells (iPSCs).

5.4. Stem Cells in Pharmaceutical Screening and Drug Development

Patient specific iPSCs can be differentiated into the cells, which possess the same genetic background as the diseased cells. Therefore, they provide an alternative option for screening pharmaceuticals (Figure 5). High throughput screening of pharmaceuticals on cells iPSCs derived cells would allow finding out specific drugs targeting disease along with the accelerated assessment of their efficacy, toxicity and off-target effects (Liang et al., 2013). Thus, iPSC approach is considered an important tool to design and develop personalized precision medicine.

CONCLUSION

Stem cells are the undifferentiated cells that can be classified according to their origin and their potential to differentiate. Stem cells formed early in zygote development are totipotent, which means they can develop into both embryonic and extra-embryonic cell types required for human development. Pluripotent stem cells can develop into almost all of the three germ layers derived cells formed during embryogenesis. In contrast, multipotent stem cells, which retain in adults, differentiate only into closely related cells serving to replenish the tissue or organ and thus maintain homeostasis. There are a number of established methods for obtaining both adult stem cells and embryonic stem cells, ranging from extraction of amniotic fluid, somatic nuclear transfer, and IVF to the use of adipose tissue. Besides, fibroblast cells can be reprogrammed into “Induced

Pluripotent Stem Cells" (iPSC) through the introduction of genes involved in promoting pluripotency, namely *SOX2, Oct-3/4, c-MYC and KLF4*. This breakthrough discovery allows scientists to generate different types of functional cells from iPSCs, mitigating the organ donor shortage and thus endorsing regenerative medicine. In addition, stem cells are frequently used to study the pathogenesis of various diseases, and can be used to develop personalized therapy and drug screening.

References

Aguilar-Gallardo, C. and C. Simón (2013). *Cells, stem cells, and cancer stem cells. Seminars in reproductive medicine*, Thieme Medical Publishers.

Baker, Christopher L., and Martin F. Pera. Capturing totipotent stem cells. *Cell stem cell* 22.1 (2018): 25-34.

Choudhary, M., Haimes, E., Herbert, M., Stojkovic, M., and Murdoch, A. (2004). Demographic, medical and treatment characteristics associated with couples' decisions to donate fresh spare embryos for research. *Human Reproduction* 19, 2091-2096.

Condic, M.L. (2013). Totipotency: what it is and what it is not. *Stem cells and development* 23, 796-812.

Daley, G. Q. (2015). Stem cells and the evolving notion of cellular identity. *Phil. Trans. R. Soc. B* 370(1680): 20140376.

De Coppi, P., Bartsch, Jr, G., Siddiqui, M.M., Xu, T., Santos, C.C., Perin, L., Mostoslavsky, G., Serre, A.C., Snyder, E.Y., and Yoo, J.J. (2007). Isolation of amniotic stem cell lines with potential for therapy. *Nature biotechnology* 25, 100.

Dulak, J., Szade, K., Szade, A., Nowak, W., & Józkowicz, A. (2015). Adult stem cells: hopes and hypes of regenerative medicine. *Acta Biochimica Polonica* 62(3).

Garfinkel, M. (1999). *Stem Cell Controversy*. Professional Ethics Report 12.

Holland, S., Lebacqz, K., and Zoloth, L. (2001). The human embryonic stem cell debate: *Science, ethics, and public policy*, Vol 3 (MIT Press).

Jenson, G. (2017). *Histology Utilization in Biomedical Research*.

Kondo, M., Weissman, I.L., and Akashi, K. (1997). Identification of clonogenic common lymphoid progenitors in mouse bone marrow. *Cell* 91, 661-672.

Liang, P., Lan, F., Lee, A.S., Gong, T., Sanchez-Freire, V., Wang, Y., Diecke, S., Sallam, K., Knowles, J.W., and Nguyen, P.K. (2013). Drug screening using a library of human induced pluripotent stem cell-derived cardiomyocytes reveals disease specific patterns of cardiotoxicity. *Circulation*, CIRCULATIONAHA. 113.001883.

Liu, Y.W., Chen, B., Yang, X., Fugate, J.A., Kalucki, F.A., Futakuchi-Tsuchida, A., Couture, L., Vogel, K.W., Astley, C.A., Baldessari, A. and Ogle, J. (2018). Human embryonic stem cell–derived cardiomyocytes restore function in infarcted hearts of non-human primates. *Nature Biotechnology* 36.7 (2018): 597.

Mahla, R.S. (2016). Stem cells applications in regenerative medicine and disease therapeutics. *International Journal of Cell Biology* 2016.

Mandai, M., Watanabe, A., Kurimoto, Y., Hirami, Y., Morinaga, C., Daimon, T., Fujihara, M., Akimaru, H., Sakai, N., and Shibata, Y. (2017). Autologous induced stem-cell–derived retinal cells for macular degeneration. *New England Journal of Medicine* 376, 1038-1046.

Marshak, D.R., Gardner, R.L., and Gottlieb, D. (2001). *Stem Cell Biology* (Cold Spring Harbor Monograph Series, 40).

Martin, M.J., Muotri, A., Gage, F., and Varki, A. (2005). Human embryonic stem cells express an immunogenic nonhuman sialic acid. *Nature Medicine* 11, 228.

Mimeault, M., Hauke, R., and Batra, S. (2007). Stem cells: a revolution in therapeutics—recent advances in stem cell biology and their therapeutic applications in regenerative medicine and cancer therapies. *Clinical Pharmacology & Therapeutics* 82, 252-264.

Polgar, K., Adany, R., Abel, G., Kappelmayer, J., Muszbek, L., and Papp, Z. (1989). Characterization of rapidly adhering amniotic fluid cells by combined immunofluorescence and phagocytosis assays. *American Journal of Human Genetics* 45, 786.

Rodriguez, A. M., Pisani, D., Dechesne, C.A., Turc-Carel, C., Kurzenne, J. Y., Wdziekonski, B., Villageois, A., Bagnis, C., Breittmayer, J. P., and Groux, H. (2005). Transplantation of a multipotent cell population from human adipose tissue induces dystrophin expression in the immunocompetent mdx mouse. *Journal of Experimental Medicine* 201, 1397-1405.

Schmidt, D., Achermann, J., Odermatt, B., Breymann, C., Mol, A., Genoni, M., Zund, G., and Hoerstrup, S.P. (2007). Prenatally fabricated autologous human living heart valves based on amniotic fluid–derived progenitor cells as single cell source. *Circulation* 116, I-64-I-70.

Schöler, H.R. (2016). The potential of stem cells: An inventory. In *Human Biotechnology as Social Challenge* (Routledge), pp. 45-72.

Shenghui, H., Nakada, D., and Morrison, S.J. (2009). Mechanisms of stem cell self-renewal. *Annual Review of Cell and Developmental* 25, 377-406.

Shin, T., Kraemer, D., Pryor, J., Liu, L., Rugila, J., Howe, L., Buck, S., Murphy, K., Lyons, L., and Westhusin, M. (2002). Cell biology: a cat cloned by nuclear transplantation. *Nature* 415, 859.

Slack, J.M. (2000). Stem cells in epithelial tissues. *Science* 287, 1431-1433.

Thomson, J.A., Itskovitz-Eldor, J., Shapiro, S.S., Waknitz, M.A., Swiergiel, J.J., Marshall, V.S., and Jones, J.M. (1998). Embryonic stem cell lines derived from human blastocysts. *Science* 282, 1145-1147.

Tolar, Jakub, et al. Sarcoma derived from cultured mesenchymal stem cells. *Stem cells* 25.2 (2007): 371-379.

Vairagade, Amishi, Julia Michelle Smith, and Thai Thanh Trinh. *Developing a Multiplexing Assay System for the Quality Control of Cell Therapy Products*. (2017).

Wilson, K.D., and Wu, J.C. (2015). Induced pluripotent stem cells. *JAMA* 313, 1613-1614.

Wu, D.C., Boyd, A.S., and Wood, K.J. (2007). Embryonic stem cell transplantation: potential applicability in cell replacement therapy and regenerative medicine. *Front Biosci* 12, 4525-4535.

Wyles, Saranya, Emma Brandt, and Timothy Nelson. Stem cells: the pursuit of genomic stability. *International Journal of Molecular Sciences* 15.11 (2014): 20948-20967.

Yamanaka, S. (2012). Induced pluripotent stem cells: past, present, and future. *Cell Stem Cell* 10, 678-684.

Ying, Q. L., Nichols, J., Evans, E.P., and Smith, A.G. (2002). Changing potency by spontaneous fusion. *Nature* 416, 545.

Zhao, T., Zhang, Z. N., Rong, Z., and Xu, Y. (2011). Immunogenicity of induced pluripotent stem cells. *Nature* 474, 212.

In: Trends in Biochemistry and Molecular Biology
Editors: Hossain Uddin Shekhar et al.
ISBN: 978-1-53616-434-3

Chapter 2

L1: SHAPING HUMAN GENOME

***Tania Sultana*[*], PhD**
Department of Biochemistry and Molecular Biology,
University of Dhaka, Dhaka, Bangladesh

ABSTRACT

The blueprint of life, DNA, is changing itself; transposons are one of the mechanisms of how DNA does it. Transposons are bohemian genes, inserts in new locations by cut- or copy-paste manner. Such insertion events are mutagenic and may lead to rearrangements at the insertion sites. The consequences of an insertion thus depend largely on the feature which is interrupted at the new insertion site. Transposon activity contributes to genome evolution by diversifying the genome. L1, the only active contemporary human transposon contributes to both genetic innovation and disease pathogenesis. 124 cases of disease originating from the L1 activity are known so far which include cancer. Many other roles of L1 activity is being revealed, such as, in developmental biology. L1 inserts into new genomic sites in the copy-paste mechanism where they make an RNA copy of the original L1, this RNA travels to a new genomic site where it is parallelly reverse transcribed into a cDNA and inserted. Interestingly, L1 transposons not only move themselves, but they can also mobilize other human transposon families as well as cellular mRNAs.

In this book chapter, transposon classes, their basic characteristics, mechanisms, and genomic distributions have been presented briefly. The life cycle of an L1 has been described along with how they use various cellular factors as their associates and how their activity is regulated by cellular factors to reduce the mutagenic load. Once disregarded as junk, transposons are now emerging as an attractive arena in genomics and epigenetics!

---

[*] Corresponding Author's E-mail: taniasultana2004@gmail.com.

1. Transposable Elements Shape the Genomes of Living Organisms

Transposable elements (TEs) are repeated DNA sequences in the genome, which have the ability to move from one location of the genome to another. They are also known as various other names, such as, mobile genetic elements (MGEs), mobile DNA, transposons, jumping genes etc. When a TE transposes or moves to a new site in the genome, it can cause various rearrangements at its newly inserted genomic site. Such rearrangements have a number of consequences with crucial impact on species evolution including diversification of genome, expression of new trait (for example, shape of tomato, taste of orange, quality of grapes etc.) as well as disease pathogenesis (section 5). In 1950, Barbara McClintock first coined the notion of mobile genetic elements in the genome. She published the presence of certain 'controlling elements' in maize genome, Ac and Ds. These controlling elements have the ability to move or transpose from one location in the genome to another and can regulate genes nearby their new location (Barbara McClintock 1950). Her fellow scientists were skeptical of her ideas since genes were widely accepted to be stable and fixed on the chromosomes at that time. However, in the next few years, such events of transpositions by mobile genetic elements in the genome were confirmed by other independent studies. At that time TEs were widely considered as the 'junk' of the genome as they were not known to have any beneficial contribution to the genome. They have also been cited in literature as 'selfish DNA' or parasitic DNA as TEs use cellular machinery to replicate or propagate without considering its effect on its host cell. As in 2019, that stigma is over and TEs are known for their diverse role in the genome. To date, transposons have been found in almost all species including humans, with variable occupancy levels, structures and consequences.

2. Transposons Are Diverse in Structure and Mechanism of Mobilization

Transposable elements can be either 'autonomous', possessing all the active genes essential for transposition, or 'non-autonomous', requiring assistance from the machinery of autonomous transposons. Nevertheless, no transposon is strictly autonomous, rather interacts with cellular factors during or at least for part of its life cycle. Transposons can be classified in many ways. The basic classification is based on the nature of transposition mechanism: DNA transposons and retrotransposons.

2.1. DNA Transposons

DNA transposons consist of transposase encoding gene surrounded by two inverted terminal repeats (Figure 1). They transpose by a 'cut and paste' mechanism where transposase enzyme, coded by the DNA transposon itself, excise the transposon from its original location and help it to insert into a new genomic location (Figure 2). They are predominant in bacteria but are also found in fungi, plants, fish and some mammals. Insertion Sequences were the first discovered transposons in bacteria, responsible for horizontal transmission of antibiotic resistance between individuals in a bacterial population. Since the original piece of transposon excise itself out of the genome and re-insert in a new location, DNA transposons generally do not increase their total number in the genome with few exceptions (Curcio and Derbyshire 2003).

Transposases encoded by various DNA transposons vary in their molecular mechanisms but in general, they recognize the short inverted repeat (IR) on both ends of the DNA transposon to excise it out of its original site (Figure 2). Transposases are bound to transposon DNA ends until they reach the target DNA. Transposase cleaves the target DNA strands in staggered fashion resulting in target-site duplication (TSD) of a size typical of 4–8 bp. The Ac/Ds transposition system discovered by Barbara McClintock, is a DNA transposon (Barbara McClintock 1950). DNA transposons occupy 3% of the human genome but none of them present any evidence of recent activity (Lander et al. 2001).

2.2. Retrotransposons

Retrotransposons replicate in the genome by a copy-and-paste mechanism (Figure 2). This means that the actual fragment of a retrotransposon is not altered. Instead it is first transcribed into an intermediate RNA copy, the RNA copy is reverse transcribed into a cDNA, which gets integrated into a new genomic location. Apart from this basic property, retrotransposons can vary by their structure (Figure 1) and by their mechanism of transposition. The two main classes of retrotransposon differ by the presence of long terminal repeat (LTR) at their extremities, and are thus called LTR and non-LTR retrotransposons. Together, retrotransposons and retroviruses will be called as 'retroelements' in this chapter. Retroviruses share a common evolutionary ancestry with LTR-retrotransposons and are assumed to originate from LTR-retrotransposons. One major difference between LTR-retrotransposons and retroviruses is that retroviruses have acquired an envelope gene (env) over the evolutionary period (Figure 1) (H S Malik, Henikoff, and Eickbush 2000).

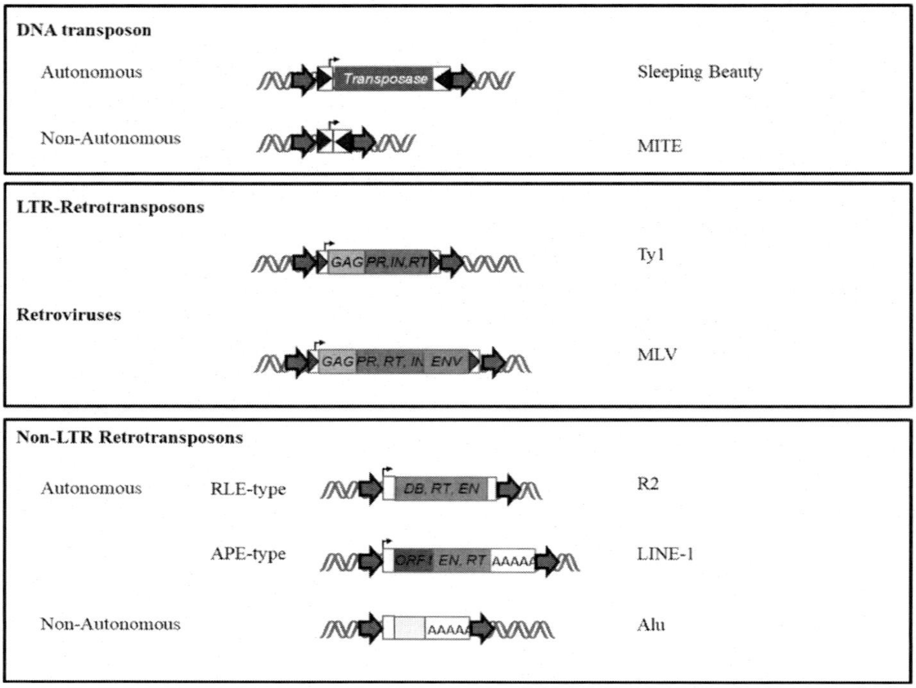

Figure 1. Transposon structures. Prototypes of various classes of transposable elements are presented with examples next to them. Coding sequences are as follows: GAG, gag protein; PR, protease; IN, integrase; RT, reverse transcriptase; ENV, envelope protein; DB, DNA binding domain; EN, endonuclease; ORF1, open reading frame 1. Blue arrows represent target site duplication, boxed triangles represent repeats. Adapted from (Sultana et al. 2017).

2.2.1. Non-LTR Retrotransposons

Non-LTR retrotransposons contain no long-terminal repeat. Evolutionarily they are thought to evolve from the RNA viruses and are the likely ancestor of the LTR-retrotransposons. They can be either autonomous or non-autonomous. Autonomous non-LTR retrotransposons, such as L1, encodes a protein harboring endonuclease (EN) and reverse transcriptase (RT) activities (Figure 1). Some Non-LTR elements also express a protein that can bind RNA and escort it. RNA and proteins encoded by a non-LTR retrotransposon assemble in the cytoplasm to form a ribonucleoprotein particle (RNP), which enters the nucleus, access the genome and integrates in the target DNA by a mechanism called Target-Primed Reverse Transcription (TPRT). The most detailed model on TPRT is derived from studies of the R2 non-LTR element in *Bombyxmori* and *Drosophila melanogaster*. As the name implies, in this process, the nick in the target DNA generated by the EN is used as reverse transcription start site. EN liberates a 3' hydroxyl group in DNA which is used as a primer to start the local reverse transcription of the retrotransposon RNA (Figure 2, 5). Non-LTR retrotransposons vary in their endonuclease domains which belong to two distinct enzymatic classes: restriction enzyme-like EN (RLE) and apurinic/apyrimidinic EN (APE). A number of non-LTR retrotransposons are specific in selecting their target DNA to insert itself (see section 3).

Endonuclease class is one of the factors contributing to this selection by favoring certain DNA sequences over others to nick. For example, L1 APE recognizes and nicks defined consensus sequences at the genomic DNA target (5′-TTTT/A-3′; slash indicates the scissile phosphate).

2.2.2. LTR Retrotransposons

LTR retrotransposons evolved from the non-LTR ones by acquiring two long direct repeats on both terminals. They are particularly abundant in eukaryotes, especially in plants where they are the dominating group of transposons. They contain open reading frames (ORFs) that encode Gag and Pol proteins, and are flanked by direct long terminal repeats (LTR) on each end (Figure 1). The gag gene encodes the structural components of the cytoplasmic particles called 'virus-like particles' (VLPs) where the reverse transcription of an LTR-retrotransposon RNA takes place (Figure 2). The pol gene encodes a polyprotein with multiple protein domains (protease, integrase, reverse transcriptase, and RNase H), which is further processed into individual mature proteins by the enzymatic activity of the protease.

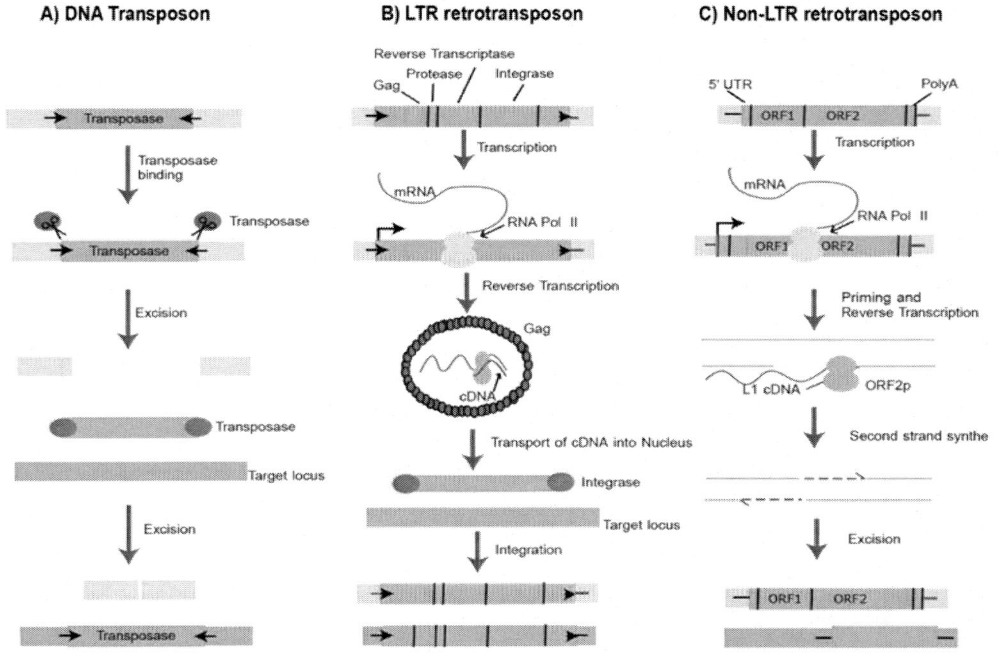

Figure 2. Replication models for the three major classes of transposable elements. TEs are presented as light green bars, terminal inverted repeats in DNA transposon and LTR in LTR-retrotransposon as black arrows, transposase and integrase proteins as green circles, transposon donor sites as light pink bars, novel integration sites as dark pink bars, target site duplications (TSD) as black horizontal lines, RNA polymerase II in yellow color, Gag proteins as dark green circles, reverse transcriptase protein in orange shape and color, RNA transcript of TEs as green waves and their reverse complement cDNAs as blue waves.

LTR-retrotransposons have been extensively studied in *Saccharomyces cerevisiae* (Baker's yeast) and *Drosophila melanogaster* (fruit fly). According to the sequence similarity of reverse transcriptase among the LTR-retrotransposons and the order of the protein domains in the Pol gene, LTR-retrotransposons have been classified in two groups: Ty1/Copia and Ty3/Gypsy. Yeast LTR-retrotransposons, Ty1, Ty2, Ty4 and Ty5 fall in the Ty1/Copia group and Ty3 falls in the Ty3/Gypsy group (reviewed in (Eickbush and Jamburuthugoda 2008)).

3. Genomic Distribution of Transposons

TE integration is a major threat for the maintenance of genome integrity and host survival, especially in compact genomes. On the other hand, death of the host ultimately affects the survival of TEs. Thus, evolutionary strategies have emerged from both sides to allow propagation of TEs while minimizing the genetic damages to the host. Due to the consequences of a TE integration (see section 5), it is intriguing to understand if transposons, when they move to a new location, have any preferred genomic sequence or feature. We have discussed in the previous section that transposons possess distinct mobilization machineries with unique enzymatic properties. In this section, we will discuss how the unique properties of a transposon type play key role in determining the preference for given DNA sequences and/or chromatin structures.

3.1. The Genomic Distribution of a Given TE Results from a Two-Step Process

While some TEs favor integration into specific genomic regions or features, others show a more dispersed pattern of insertions (reviewed in (Sultana et al. 2017)). The genomic distribution of a given TE results from a two-step process: first, site-specific (or not) integration directing the initial allocation of the insertions, and second, selective pressures leading to the loss of harmful transposition events and perpetuation of insertions that benefit the host. This is particularly important in organisms such as *S. cerevisiae*, whose genome contain approximately 70% of protein coding genes, and only 3% of TEs. In these organisms, TEs insert either in heterochromatin, i.e., sub-telomeres or centromeres, or close to/in tDNA or rDNA (Sandmeyer, Patterson, and Bilanchone 2015; Spaller et al. 2016). tDNAs and rDNAs are multicopy genes and thus individually non-essential for the survival of yeast cells. TEs may have two distributions in respect to the time course: the first one is the immediate distribution right after it inserts i.e., the distribution of novel insertions; the second one is what we see in the contemporary genome, the result of host-transposon co-evolution. Since evolution is a continuous process, insertion distribution by a particular class of transposon may vary according to the time past after the insertion event has occurred. Thus it can be said that coevolution of host and transposon impacts the target site preference and the ultimate distribution of transposon in the genome.

3.2. The Importance of Studying *De Novo* Insertions

The distribution of novel versus fixed insertions, or younger versus older insertions of a particular transposon class evidenced the differences in insertion distribution (Brady et al. 2009; Ovchinnikov, Troxel, and Swergold 2001; Barr et al. 2005). For example, comparison between the patterns of novel and the fixed insertion of human endogenous retrovirus (HERV-K) showed that the novel insertions were slightly enriched in transcription units and gene-rich regions. In contrast, the fixed insertions were found preferentially outside the transcription units (Brady et al. 2009). Interestingly, a pattern of intermediate genomic distribution between the novel and fixed HRRV-K insertions was found for the youngest HERV-K elements confirming the changes in genomic distribution of TE over time. Thus to know the initial distribution of a transposon, in other words to know the genomic site preference of a transposon, one must look at the distribution of the newly inserted transposons rather than the fixed or older insertions in the genome.

The importance of studying novel insertions is also illustrated by the distinct distribution of L1 and Alu sequences in the human genome. Fixed endogenous L1 and Alu insertions are enriched in opposite DNA isochores: L1 elements show a bias for AT-rich regions whereas Alu sequences are enriched in GC-rich regions (Lander et al. 2001). Alu elements are non-coding and are mobilized by the L1 retrotransposition machinery, thus L1 and Alu elements may have integration preference for common sites. Consistently, and in contrast to fixed copies, experimentally induced *de novo* Alu insertions are detected in the same AT-rich isochore as L1 elements (Wagstaff et al. 2012). Thus, study of *de novo* insertions may shed light on the preferred integration sites of a transposon class.

3.3. Transposons Vary in Their Target Site Preferences

Advances in sequencing technologies and the availability of annotated genomic features in the reference genomes in the past decade have made it possible to study the integration site selectivity. Transposons are found in four major locations or macrofeatures: 1) in or near gene-rich regions, 2) in telomeric regions, 3) in heterochromatin, 4) dispersed across chromosome.

Many TEs integrate into gene-dense regions, although most of the events occur in sites that prevent disruption of open reading frames (ORFs). For example, the P element, a DNA transposon from *D. melanogaster* avoids disruption of ORFs by integrating 500bp upstream of transcription start sites (Bellen et al. 2011). Yeast LTR-retrotransposons, Ty1, Ty2, Ty3 and Ty4, integrate upstream of RNA polymerase III transcripts, namely tDNA genes, while Tf1 and Tf2 yeast LTR-retrotransposons integrate upstream of RNA polymerase II transcripts. A number of non-LTR retrotransposons integrate specifically in telomeres. For example, TRAS1 and SART1 in silkworm, involved in telomere maintenance, integrate into the 'TTAGG' repeats of the telomeres. Likewise, Het-A, TART, and TAHRE elements in Drosophila are located at the extreme ends of the telomeres (Biessmann and Mason 2003; Pardue and DeBaryshe 2000; Svetlana Rashkova et al. 2002; S Rashkova, Karam, and Pardue 2002). Some transposons target heterochromatin which contains relatively few genes. For example, the yeast Ty5 LTR-retrotransposon integrates preferentially into heterochromatin. Approximately 75% of Ty5integration events occur within the telomeric and sub-telomeric heterochromatin while the rest integrates in easily accessed sites in open chromatin (Baller, Gao, and Voytas 2011). No pattern in genomic distribution has been found yet for a number of transposons, which are thought to have no particular preference. In a recent study on human L1 retrotransposition in cell culture, L1 showed no significant bias to any particular functional genomic elements although replication timing of target DNA may influence choice of L1 integration site (Sultana et al., n.d.).

Transposons that integrate randomly in the genome have lower chance of landing in regions affecting genes. Such transposons can be manipulated to be used as gene delivery vectors for functional genomics study and for clinical gene therapy. DNA transposons from the mariner superfamily, including the Sleeping Beauty, are potential gene delivery vectors under study.

Besides their primary choice of target sites, some TEs also show secondary preferences for alternative chromatin features in response to physiological stimuli, such as stress (J. Dai et al. 2007). Additionally, a microfeature within a preferred macrofeature can participate in integration site selection. For example, yeast Ty5 integrates into heterochromatin, but more specifically in nucleosome free regions and open sites within the heterochromatin (Baller, Gao, and Voytas 2011).

3.4. Selection of Transposition Site May Take Place in Multiple Level

The transcription of transposition machinery in eukaryotes takes place inside the nucleus from where the transposon RNA is transported to the cytoplasm where the transcript will be translated and form the transposition machinery. The transposition machinery travels towards its new destination, a DNA sequence located inside the nucleus. During this whole process, a number of factors in multiple levels define the ultimate target site preference of a number of transposons.

3.4.1. The Nuclear Entry Route May Shape Integration Site Selection

Many retroelements including L1, yeast LTR-retrotransposons and HIV, can integrate in cells that do not divide. Nuclear pore complex (NPC) is the gateway for them to enter the nucleus. Gag and integrase of some retroelement transposition machineries interact with nucleoporins, components of the nuclear pore complex, facilitate the access of machineries inside the nucleus, and notably guide the element to integrate into a particular chromatin context nearby the pore complex. Thus, the organization of chromatin in the vicinity of the NPC influences integration site selection. Among the retroelements, HIV integration complexes localize preferentially to areas of euchromatin that are close to the nuclear envelope (Lelek et al. 2015) and preferentially direct integration into actively transcribed genes near NPCs. By contrast, transcriptionally repressed lamina-associated heterochromatin domains or transcriptionally active regions located in the centre of the nucleus are disfavoured (Benleulmi et al. 2015).

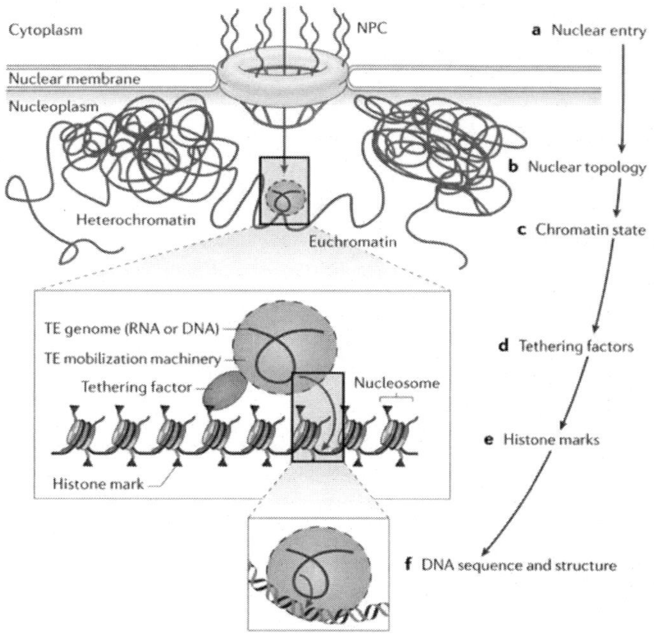

Figure 3. Multiple levels of target site selection. Nuclear pore complex and the surrounding chromatin organization may influence integration site selection (a,b,c). Inside the nucleus, proteins from transposition machineries can interact with chromatin reader host factors to guide transposons towards target DNA (d, e). Once at the target site, transposon integrates if the sequence and the structure of target DNA are favorable (f). Which of the above factors would define a target site depends on a particular transposon type. Figure source (Sultana et al. 2017).

3.4.2. Partner Host Factors Guide Transposons to Target Site

3.4.2.1. Integrase Mediated Targeting

Transposition machineries interact with transposon-specific cellular proteins which escort the machinery to site of transposition. These proteins are supposed to have multiple domains; at least one of them would bind to or nearby the target DNA and another, transposition machinery. Thus the transposition site pattern for a number of transposon classes aligns with the DNA binding pattern of the cellular partner protein. During the integration of LTR retroelements, integrase catalyses the processing of viral DNA ends and their joining to target DNA. Integrases from both fission and baker's yeast as well as from a number of retroviruses interacts with cellular proteins which bind to the DNA at the integration site. These partner proteins are transposon specific which can be evidenced from the various partners retroelements have. Ty1 and Ty3 integrase bind with subunits of RNA polymerase III transcription complex, AC40 and TFIIIB-C respectively (Bridier-Nahmias et al. 2015) while Ty5 integrase interacts with Sir4 (Xie et al. 2001), responsible to maintain heterochromatin. In each case, the host factors bound to the transposition machineries, guide them to target DNA where the host factor bind to the target DNA while the transposition machineries integrate themselves.

3.4.2.2. ORF1p Mediated Targeting

Evidence for the tethering of non-LTR retrotransposons is scarce, but all known cases to date involve an RNA-binding protein called ORF1p. ORF1ps are sometimes called Gag because of their analogy with LTR retroelements. ORF1p of *Drosophila melanogaster* non-LTR retrotransposons, telomere-associated retrotransposon (TART) and telomere-associated and HeT-A-related element (TAHRE), rely on heterochromatic telomeric retrotransposon A (HeT-A) ORF1p to access telomeres (Gijsbers et al. 2010; Silvers et al. 2010; Shun et al. 2007). HeT-A ORF1p is recruited to telomere by Verrocchio (Ver) protein, which is essential for telomere protection (Shun et al. 2007). Similarly, ORF1p from SART1 in *Bombyxmori* tethers SART1 transposon machinery to telomere. Whether integrase-mediated tethering to target can be generalized to most LTR elements and ORF1p-mediated to non-LTR retrotransposons remains unknown. Human L1 ORF1p interacts with a large number of cellular proteins although none of them have been proved to guide L1 transposition machinery to its target site (Pizarro and Cristofari 2016).

3.4.3. Sequence Preference of Transposition Machineries at Target DNA

Most transposition machineries whether it consists of transposase, integrase or endonuclease, exhibit preferences to very short sequences often limited to flexible pyrimidine–purine dinucleotide between the two staggered cuts (Gogol-Döring et al. 2016; Fungtammasan et al. 2012; Ciuffi et al. 2005; Bridier-Nahmias et al. 2015). Since such sequences are commonly found in the genome, local sequence preference may not be the key player in target sequence selection. Bending of the central nucleotide at the target sequence widens the groove of the DNA allowing the catalytic residues of the transposition machinery to access the scissile phosphodiester bond within the active site of the enzyme (Chatterjee et al. 2014; Jacobs et al. 2015; Anzai et al. 2005). Interactions between amino acid residues of transposition machineries and the target DNA nucleotides may be the final factor to determine target site (Jacobs et al. 2015; Feng et al. 1996; J. Yang, Malik, and Eickbush 1999; Christensen and Eickbush 2005).

Sequence specificity of a non-LTR retrotransposon is aided by the specificity of its endonuclease. Two classes of EN domain found in non-LTR elements are: restriction enzyme-like EN (RLE) and apurinic/apyrimidinic EN (APE). APE domains in L1 and R1 directly contribute to the DNA sequence preference of the target site. The DNA-binding surface of APEs has a protruding β-hairpin loop which contacts the DNA minor groove adjacent to the scissile bond and participates in sequence recognition at the cleavage site (Shukla et al. 2013; Rodić et al. 2015; Streva et al. 2015; Chatterjee et al. 2014). In contrast, RLE encoding transposons recognize target DNA through one (or several) independent DNA-binding domain(s) located outside of a non-specific EN domain (Christensen and Eickbush 2005; Thompson and Christensen 2011; Shivram, Cawley, and Christensen 2011).Most of the RLE elements, such as R2 retrotransposons in insects, insert in specific target sites while only a few of APE-encoding elements is site-specific (Hickey et al. 2015).

4. L1 Is the Only Autonomously Active TE Family in Human Genome

Almost half of the human genome is composed of repeat elements. The initial analysis and sequencing of human genome in 2001 revealed unpredicted information on human genome composition (Lander et al. 2001). 45% of the human genome is composed of TEs. The protein coding sequences occupies only 2% of the genome whereas 45% of the human genome is occupied by repeat elements derived from transposon activities. DNA transposons represent 3% of our genome, LTR-retrotransposons 8%, and non-LTR retrotransposons 34%. Among the non-LTR retrotransposons, LINEs and SINEs comprise 21% and 13% respectively (Figure 4). LINE-1 is the only autonomously active TE family in the contemporary human genome. Other retrotransposons, for e.g., SINEs (Alu and SVA) elements are also active and employ the LINE-1 machinery for mobilization. Among the SINE class of transposons, Alu is the most abundant transposon class in human genome. SVA (SINE-VNTR-Alu) elements are compound repeat elements i.e., they are composed of other repeats. SVAs are the youngest active family of mobile elements in humans. LINE-1 has been amplifying in mammals for around 160 million years (Smit et al. 1995). The human specific L1 subfamily, L1HS, emerged only ~4 millions of years ago. L1 sequences accumulate mutations in a neutral rate, thereby older sequences are proportionately more divergent from the active L1 consensus sequence (J. Lee et al. 2007; Boissinot, Chevret, and Furano 2000). L1 elements are active in germ cells and early embryo (Brook Brouha et al. 2002), some somatic tissues such as brain (Erwin et al. 2016; Erwin, Marchetto, and Gage 2014; Upton et al. 2015; Baillie et al. 2011), epithelial tumors (Ewing et al. 2015; Helman et al. 2014; Tubio et al. 2014; Solyom et al. 2012; Shukla et al. 2013; E. Lee et al. 2012; Scott et al. 2016; Doucet-O'Hare et al. 2015; Rodić et al. 2014). L1 mobility in embryo and somatic cells makes an individual somatically mosaic.

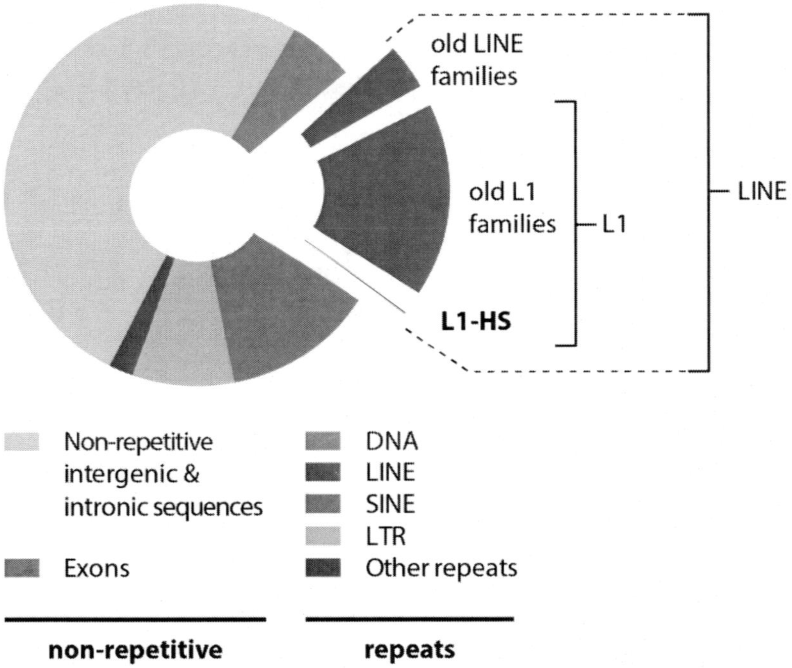

Figure 4. Proportion of transposable elements in the human genome. Half of our genome is occupied by repeat elements. Human specific L1 (L1HS) forms a tiny fraction of genome and solely contribute to the total pool of retrotransposition activity. Figure source: http://eul1db.unice.fr/.

4.1. L1 Replicates by an RNA-Mediated Copy-and-Paste Mechanism, TPRT

L1 is a 6kb DNA sequence. A prototype human L1 has a 5' untranslated region (UTR) with a weak promoter activity for RNA polymerase II, two open reading frames, ORF1 and ORF2, a 3'UTR ending with a weak polyadenylation signal, and a long poly(A) tail of variable length (Figure 5). ORF1 codes for ORF1p, a 241 amino acid protein with nucleic acid binding (Kolosha and Martin 2003) and chaperone activities (S. L. Martin and Bushman 2001). The second L1 ORF, ORF2 codes for ORF2p protein has an N-terminal apurinic/apyrimidic endonuclease (APE) like domain and a reverse transcriptase domain (RT). L1 promoter drives synthesis of numerous copies of the ORF1p per L1 RNA but only one or two copy of ORF2p(Wei et al. 2001). L1 RNA associates with its encoded proteins, several ORF1p homotrimers and at least 2 ORF2p (ORF2p dimer) if L1 follows the same model as R2 (Christensen and Eickbush 2005) to form a ribonucleoprotein particle (L1 RNP) in the cytoplasm (D. a. Kulpa and Moran 2006; D. A. Kulpa 2005) (Figure 6). It is suggested that ORF1p polymerizes at the site of translation, which facilitates their binding to their own RNA to form RNP complex, a phenomenon known as *cis* preference. Most L1 integration takes place via the classical endonuclease dependent target-primed reverse transcription (TPRT). In an alternative pathway, L1 can integrate at pre-existing DNA lesions, and does not require any endonuclease cleavage. This is called the endonuclease-independent (ENi) retrotransposition or non-classical L1 insertion (Morrish et al. 2002; Sen et al. 2007).

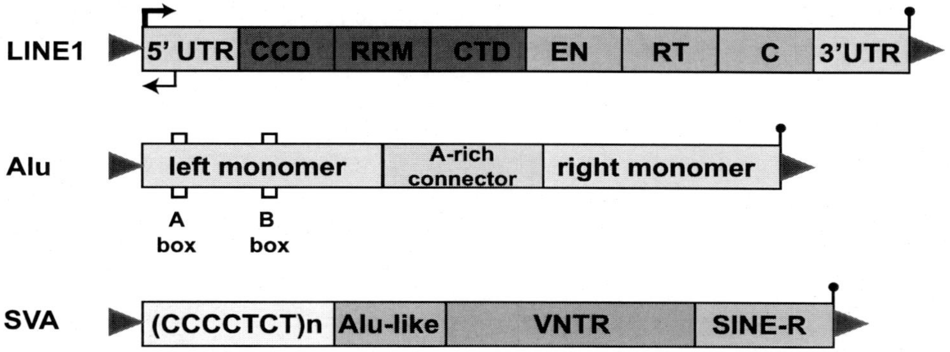

Figure 5. Structures of active human transposable elements. LINE1, Alu and SVA elements are illustrated. L1 ORF1 domains are presented in magenta, L1 ORF2 domains are in light green, L1 untranslated regions are in light grey, target site duplications are in black arrowheads. UTR, untranslated region; CCD, coiled coil domain; RRM, RNA recognition motif; CTD, carboxy-terminal domain; EN, endonuclease domain, RT, reverse transcriptase domain; C, cysteine rich domain; VNTR, variable number of GC-rich tandem repeats; lollipop, polyadenylation signal.

L1 EN domain contains several activities: 3' exonuclease, 3' phosphatase, and RNaseH activities originating from a single active site (Barzilay and Hickson 1995). L1 endonuclease recognizes and nicks defined consensus sequences at the genomic DNA target (5′-TTTT/A-3′; slash indicates the scissile phosphate), which liberates a 5′ phosphate and 3′ hydroxyl group (Feng et al. 1996; G J Cost and Boeke 1998) (figure 6). Variations of this consensus motif are often observed, although a number of pyrimidine (Y) before the scissile bond followed by purines (R) are almost always observed (5'-(Y)n/(R)n-3'). The endonuclease liberated 3' hydroxyl group is used as a primer by the ORF2p RT activity to initiate reverse transcription of the L1 RNA, an event from where the mechanism is named. Reverse transcription starts at the polyA tail of L1 RNA (Doucet, Wilusz, et al. 2015; Monot et al. 2013; Viollet, Doucet, and Cristofari 2016). Maximum efficiency of reverse transcription originates from a snap-Velcro model, where besides the recognition and nicking of an EN site (snap), the polyA tail of L1 RNA anneals with the polyT sequence in the target DNA located downstream of the EN site (Viollet, Monot, and Cristofari 2014).

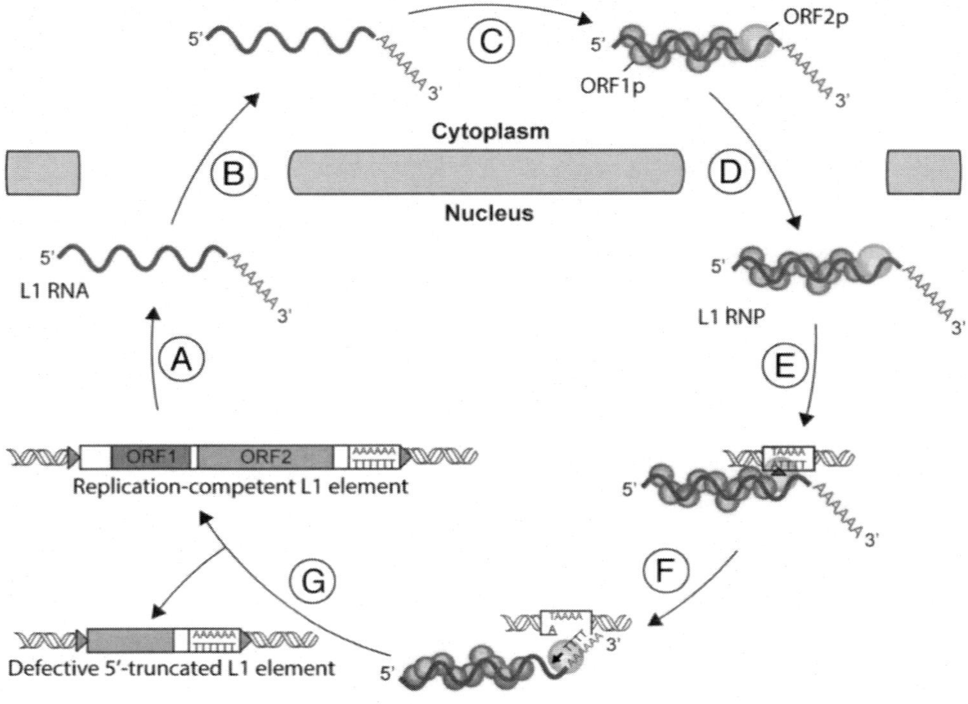

Figure 6. The L1 life cycle. L1 is transcribed to a bicistronic L1 mRNA (A), which is exported to the cytoplasm (B), where ORF1p and ORF2p proteins are translated and bind to the L1 RNA to form L1 ribonucleoprotein particles (RNP) (C). The L1 RNP is imported into the nucleus (D), where L1 ORF2p endonuclease (EN) activity nicks the first target DNA strand (red arrowhead, E) and reverse transcriptase (RT) initiates the reverse transcription of L1 RNA (black arrow, F). The final steps are not completely understood yet (G), L1 reverse transcription is often abortive and results in a 5' truncated L1 progenitor. Progenitor full-length L1 may continue to replicate if the site of integration is open for expression. Figure taken from (Viollet, Monot, and Cristofari 2014).

While the first strand of the inserted L1 DNA thus originates from the cDNA synthesized by L1 RT, the second strand is synthesized in a mechanism yet to be revealed. Cleavage of the second strand of the target DNA occurs following initiation of the first-strand cDNA synthesis (Dustin C. Hancks and Kazazian 2016). The second strand cleavage sites do not show any sequence preferences unlike the first strand (Gregory J Cost et al. 2002). However, the second strand cleavage position may be influenced by the distance from the first strand cleavage position as target site duplications generally range from 4-20 bp (Lander et al. 2001; Dustin C. Hancks and Kazazian 2016; Nicolas Gilbert, Lutz, and Moran 2005). The length of the TSDs at the L1 integration site is equal to the nucleotide number between the first and the second nick. A newly integrated L1 remains active providing that: (i) it is full-length and does not contain any mutation that would hamper its replicative machinery; (ii) it can be transcribed in a timely manner (Philippe et al. 2016); and (iii) its cellular environment is permissive.

4.1.1. Hallmarks of L1-Mediated Integration

A newly integrated L1 displays typical signs of TPRT mediated integration. For example, insertion at a consensus L1 EN recognition sequence, target site duplications on both side of inserted L1 originating from the staggered nick by L1 EN, 5′-truncated L1 as the reverse transcription is often not complete, poly(A) tail of variable length as the adenylation is added later, and absence of introns. The 3'UTR of L1 has a weak transcription termination signal for RNA pol II (AATAAA). This weak signal is often bypassed by RNA pol II during transcription. In such cases, transcription continues until the polymerase finds a downstream termination signal, the resulting L1 RNA will have non-L1 genomic sequence at its 3' terminal. During the reverse transcription of L1 RNA to L1 cDNA by L1RT, this non-L1 sequence will also be reverse transcribed. Integration of such L1 in the DNA, where the newly generated L1 copy in the genome also carries non-L1 sequence are called 3' transduction. Almost one third of L1 retrotransposition events in somatic cells carry 3' transduced sequences (Tubio et al. 2014; J L Goodier, Ostertag, and Kazazian 2000). 3′ transductions contribute to genomic expansion and gene duplications (Moran, DeBerardinis, and Kazazian 1999; Xing et al. 2006).

4.2. L1 Mediated Retrotransposition in *Cis* and *Trans*

Most of 500,000 L1 sequences in the current human genome have lost their activity to retrotranspose as mutations have accumulated in the L1 body(Lander et al. 2001). A set of approximately 80 to 100 elements are full length and capable of expression of transposition machineries and retrotransposition (Beck et al. 2010; B Brouha et al. 2003). This set of active L1 varies from person to person. These retrotransposition-competent L1s are the only source of transposition events in the current human genome. Depending on the analysis method, there is 1 inheritable L1 insertion in every 20 to 200 human newborn(Ewing et al. 2015). L1 encoded proteins preferentially mobilize their own mRNA, a phenomenon known as *cis* preference (Wei et al. 2001). Besides their autonomous activity, L1 proteins can also occasionally act *in trans* to mobilize non-autonomous non-LTR retrotransposons (e.g., human Alu and SVA elements) (Raiz et al.

2012; Dewannieux, Esnault, and Heidmann 2003; D C Hancks et al. 2011) and cellular mRNAs (Wei et al. 2001; Esnault, Maestre, and Heidmann 2000).

4.2.1. L1 Mobilizing Other Non-LTR Retrotransposons in Trans

Alu is the most abundant TE family in humans and hijacks L1 proteins for mobilization. Alu elements are the most abundant retrotransposons by copy number, occupying 11% of the genome with 1 million copies (Lander et al. 2001). Alu elements retrotranspose more frequently compared to other TE in humans, with an estimated retrotransposition rate of one event in every twenty human newborns (Cordaux et al. 2006). Alu elements lack conventional RNA polymerase III termination signal. This allows polymerase III to bypass the signal and the transcript includes a unique flanking genomic sequence until an RNA polymerase III termination signal (a stretch of four to six consecutive thymidine) is encountered (Chu, Liu, and Schmid 1995). Hence, each Alu mRNA is unique and varies in length. Alu elements do not encode any protein and use the L1 retrotransposition machinery to integrate into the target sites, this is why sometimes they are called 'a parasite's parasite'(Weiner 2002). The sequence PolyA at the end of RNA is the site where the reverse transcription is initiated (Doucet, Wilusz, et al. 2015; Kajikawa and Okada 2002). The poly(A) tail of the Alu RNA competes with L1 RNA for the L1 ORF2p reverse transcriptase (Doucet, Wilusz, et al. 2015; Dewannieux, Esnault, and Heidmann 2003; Boeke 1997). ~2700 copies of SVA has been identified in the human genome which represents 0.2% of it (Wang et al. 2005). SVA retrotransposition shows the hallmarks of L1-mediated mobilization(D C Hancks et al. 2012; Raiz et al. 2012). However, some differences have been found between L1 and SVA retrotransposition. For example, transduction of 5' flanking sequences is more frequent for SVA elements (10%) as compared to L1s (Damert et al. 2009; D C Hancks et al. 2009)

4.2.2. L1 Mobilizing Cellular mRNAs in Trans

Besides retrotransposon RNAs, L1 can also mobilize protein coding mRNAs (Esnault, Maestre, and Heidmann 2000; Wei et al. 2001) and small nuclear RNAs, such as U6 (Doucet, Droc, et al. 2015). As the RNAs are reverse transcribed, the integrated copies of the mobilized genes lack intron and promoter, and thereby are called processed pseudogenes. Like Alu and SVA elements, processed pseudogenes exhibit the regular hallmarks of L1-mediated TPRT mechanism. The human reference genome contains ~8,000 to 17,000 processed pseudogenes(Torrents et al. 2003), of which ribosomal protein genes are the most abundant (Zhang, Harrison, and Gerstein 2002). Although most of the processed pseudogenes are non-functional because of the loss of regulatory sequences by 5' truncations, genomic rearrangements, and the absence of promoter, some of them became functional and have provided new cellular function adding diversity to the genome. This has been demonstrated by the integration of a cyclophilin Apesudogene inside the TRIM5 gene in primates within the last 6My. Both of these genes are antiviral restriction factors and give protection against retroviruses through different mechanisms. Remarkably the resulting fusion protein is functional and provided new defense mechanism against exogenous viruses (Sayah et al. 2004).

5. L1-Mediated Genomic Rearrangements Shape Genome Architecture

Approximately 0.3% of all human mutations originate from L1-mediated novel retrotransposition events (Callinan and Batzer 2006). Although frequency of retrotransposition might appear limited, L1 insertions can have much more consequences than point mutations. The high number of non-LTR retrotransposons in the genome as well as their activity has affected genome evolution. L1 Integration results in a variety of rearrangements of the sequence at the target site (Figure 7). The extent of the effect due to DNA rearrangements depends on the genomic features around the integration sites.

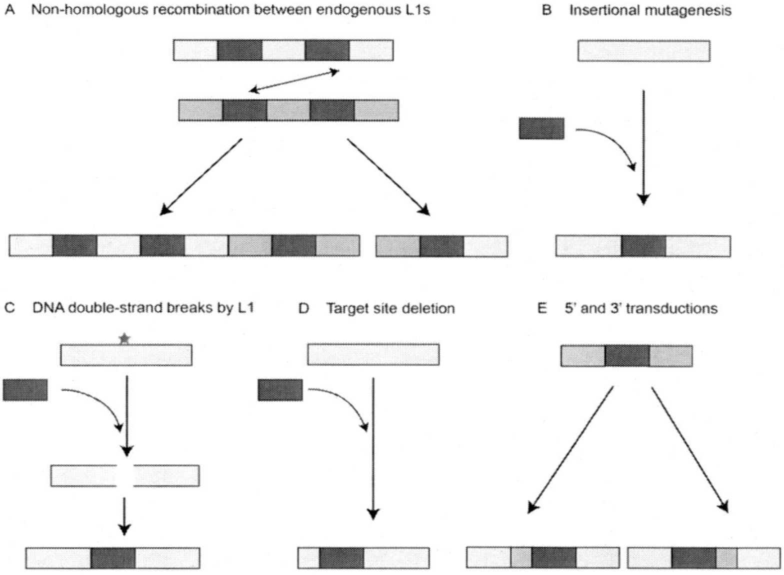

Figure 7. L1 impact on human genome structure. (A) Yellow-magenta and green-magenta boxes show non-allelic homologous retrotransposons. Double arrowhead shows recombination between them. Rearrangements, such as deletions (left arrow) or duplications (right arrow) of intervening sequences may take place at the site of recombination. (B) Retrotransposon (magenta box) insertion at a genic site (yellow) may cause mutagenesis. (C) L1 EN (magenta oval) may create nick at target DNA (broken yellow area). At the same time, existing DSBs may be repaired by insertion of a retrotransposon without requiring any action by EN. (D) Large or small sequence deletion may occur at the site of retrotransposition. (E) 3′ flanking sequence (green box) downstream of L1 or the 5′ flanking sequence (orange box) upstream of L1 may also be copied along with the L1 sequence (known as 3′ or 5′ transduction, respectively). DSB, double-stranded break; EN, endonuclease.

Most rearrangements have no immediate effect, if they are distant from genes and regulatory elements. However, sometimes, such rearrangements may have positive or negative impact on the genome. Transposition causing deleterious rearrangements in the genome will eventually be cleared out while the non-deleterious transposition events will be sustained in the genome. Thus, over an evolutionary period, accumulations of L1-mediated genomic alterations diversify the genome and contribute to genome dynamics.

Apart from their mobility, the mere presence of thousands of L1 and Alu sequences may cause deletions and inversions in the genome by homologous recombination between them. L1 and the other non-LTR retrotransposon sequences contain a number of regulatory sequences and numerous splice sites on their body which can give rise to novel RNAs when mobilized. These RNAs may affect various cellular functions. Also, the endonuclease activity of L1 may cause breaks in the genome even if those breaks would not be used for any retrotransposition events. Such breaks may cause sequence deletion. No matter through which mechanism a genomic rearrangement takes place, a particular retrotransposition event can be pathogenic resulting in a genetic disorder due to the disruption of DNA sequences necessary for cellular functions.

5.1. Non-Homologous Recombination between L1 Copies Causes Deletions and Inversions of Genomic Segments

Regardless the inability of most L1 copies in the genome to replicate, their high density impacts the genome through a variety of rearrangements caused by ectopic recombination between homologous copies (Figure 7).Such recombination events between two L1 elements may result in deletions (Burwinkel and Kilimann 1998) and inversions (J. Lee et al. 2008; K. Han et al. 2008) of intervening genomic sequences. Recombination-mediated deletions are generated via homologous recombination of two retrotransposon sequences in the same orientation on the same chromosome, whereas inversions are generated from crossing over between two retrotransposon sequences inverted relative to each other (Cordaux and Batzer 2009). Since the divergence of human and chimpanzee genomes, recombination events among L1 and Aluhave caused one fifth of the total inversions in the human (J. Lee et al. 2008). In general, recombination-mediated rearrangements are more frequent for Alu elements compared to L1 due to their very high density. More than seventy cases of Alu recombination-mediated deletions responsible for various cancers and genetic disorders have been reported (Callinan and Batzer 2006) while only three such cases are known for L1 (K. Han et al. 2008). Compared to Alu, L1 recombination-mediated deletions are larger and are seen more frequently in regions which are devoid of genes. This suggests that L1 mediated long deleterious deletions are prone to negative selection in human (Song 2007). Over millions of years, such deletions together have eliminated around 1 Mb sequence from the human genome (K. Han et al. 2008).

5.2. L1 Mediated DNA Double-Strand Breaks (DSBs) and Target Site Deletions Destabilize Human Genome

Deletions due to L1 activity can arise from two events. First, deletions due to L1 EN mediated strand breakage; second, transposition mediated deletions at the target sites. L1 mediates genomic instability by EN-mediated DNA breaks across the genome and integration-mediated deletions at integration sites (Figure 7). DSBs are susceptible to recombination and recombination-mediated deletions. Retrotransposition-independent DSBs have been found on L1 body itself since it contains sequence motif recognized by L1. L1 EN expression rise during neural differentiation and induce DSBs at the L1 sequences (Erwin et al. 2016). These breaks are associated with deletion of genomic regions near L1 loci. Association of DSBs has been evidenced in cancerous (Belgnaoui et al. 2006) and aging cells (Erwin, Marchetto, and Gage 2014). The second process leading to deletions comes directly from L1-mediated insertions. Target site deletions of variable sizes originates from the variable position of second strand cleavage and subsequent processing of double strand breaks by a 5'-3' exonuclease activity of unknown origin (N Gilbert, Lutz-Prigge, and JV. 2002). In cell culture-based assay, this phenomenon can lead to deletions of a few base pairs to as long as 71kb (N Gilbert, Lutz-Prigge, and JV. 2002).

5.3. L1 Contributes to Variations of the Human Transcriptome and Proteome

L1-mediated variations of the transcriptome may take place in a number of ways, collectively decided by the composition of L1 body and their flanking genomic sequences. For example, the most immediate phenotypic impact is visible when the transcriptome is affected by insertions in coding or regulatory sequences. L1 insertions in genic regions in antisense orientation can cause gene breakage producing two smaller transcripts: the first one contains the upstream exon and terminates in the major polyadenylation site of L1, the second one includes the exons of the inserted genes which are downstream of the L1 insertion site, transcribed by the L1 antisense promoter (Wheelan et al. 2005). Variations of L1 mobilized sequence composition greatly influence its mutagenic effect. For example, regulatory sequences within the L1 body or in the 3' or 5' L1 transduced sequences can elevate or repress expression of upstream and downstream genes. The L1 antisense promoter in the 5'UTR may drive the transcription of the 5' flanking genomic sequence giving rise to ectopic non-coding RNAs (Criscione et al. 2016). Such alternative transcription initiation by the L1 antisense promoter can alter tissue-specific gene expression (Mätlik, Redik, and Speek 2006), which increases the transcriptional flexibility of several human genes. Non-coding or chimeric transcripts can form double-stranded RNA with complementary mRNA, which can trigger both the protein kinase R degradation pathway and the RNAi pathway (see section 0) (N. Yang and Kazazian 2006).

Another type of regulatory elements in L1 sequences are the canonical and non-canonical internal polyadenylation signals in both forward and reverse direction (Perepelitsa-Belancio and Deininger 2003). These signals minimize accumulation of full length L1mRNA transcripts. Their presence in the body of L1 copies integrated in genes can lead to alternative transcript formation or premature termination (J. S. Han, Szak, and Boeke 2004). Besides the direct influence of L1 sequence, epigenetic changes of L1 elements may also influence the expression of surrounding sequences through changes in their chromatin status. Hypomethylation of the L1 promoter is known to activate alternate transcripts leading to pathological conditions (Wolff et al. 2010).

5.4. L1 Can Mediate Genetic Innovation

New genes are created in various ways over evolutionary period. L1 contribute to the rise of new genes by three known mechanisms: formation of pseudogene, 5' and 3' transduction, and exonization.

5.4.1. Gene Duplication

L1-mediated retrotransposition of cellular mRNAs and small RNAs gives rise to a copy of the encoding. The new gene, which is called a retropseudogene, lacks regulatory sequences required for its expression. It can nevertheless be expressed if it acquires regulatory sequences or from regulatory sequences nearby the site of integration. An example has been described in section 4.2.2. L1 mediated retrotransposition events are responsible for emerging new genes in primates (Babushok et al. 2007; Sayah et al. 2004; Kaessmann, Vinckenbosch, and Long 2009; Marques et al. 2005). At least one novel gene was generated in human every million years over the past ~65 Myr (Marques et al. 2005).

5.4.2. Transduction

The flanks of a progenitor L1 sequence may contain exons or regulatory sequences. During L1 transcription, this flanking non-L1 sequences may also be transcribed due to an upstream promoter or to the weak transcription termination signal of L1 (termed 5' and 3' transductions respectively). When such extended L1 transcripts are mobilized by the retrotransposition machinery, the flanking genic or regulatory sequences can be copied to new genomic locations, thereby giving rise to new gene isoforms by exon or regulatory sequence shuffling (Moran, DeBerardinis, and Kazazian 1999; Tubio et al. 2014) or creating new genes by integration of regulatory sequences. L1-mediated transductions may have contributed to 0.6–1% of total human DNA (Lander et al. 2001; J L Goodier, Ostertag, and Kazazian 2000; Pickeral et al. 2000). A recent analysis of SVA retrotransposons has showed the importance of transductions in genome evolution. Transduction has created three (AMAC1, AMAC1L1 and AMAC1L2) of the four members of the acyl-malonyl condensing enzyme 1 (AMAC1) gene family (Xing et al. 2006). These three retropseudogenes are intronless, which is a mark of L1-mediated retrotransposition. The original AMAC1L3 gene carries two introns.

5.4.3. Exonization

Exonization is the creation of a new exon from intronic sequences. L1 mediated insertions in introns may exonize part of the intron by transcribing it from one of the L1 promoters giving rise to new transcripts (Wheelan et al. 2005). L1 contains numerous functional splice donor and acceptor sites in both forward and reverse orientations though most of them are weak (Belancio, Hedges, and Deininger 2006). Each Alu sequence encompasses 9 GT dinucleotides (splice donor sites) and 14 AG dinucleotides (splice acceptor sites)(Sorek, Ast, and Graur 2002). L1-mediated insertions of functional splice sites in intronic sequences may disrupt normal gene expression or forms alternative mRNA transcripts (Belancio, Hedges, and Deininger 2006). L1-mediated indirect exonization by Alu elements are more frequent than L1-mediated exonization and occurred consistently during primate evolution (Krull, Brosius, and Schmitz 2005).

5.5. Other Roles of L1

Despite their deleterious effect, L1 transcription is de-repressed in embryos and germ cells although the rate of retrotransposition remains low (Kano et al. 2009; Richardson et al. 2017). Maximum L1 retrotransposition rate in somatic tissues has been found in neural progenitor cells generating neural diversity in the CNS (Muotri et al. 2005). Thus, apart from contributing in genome evolution, there may remain other functions of L1 in species development. One of the reasons for nuclear abundance of L1 RNA in early embryos is that L1 RNA plays distinct role in early embryonic development (Percharde et al. 2018). Dux expression is the master activator of 2-cell embryo specific transcriptome. In ESCs, nuclear scaffolds consisted of L1 RNAs and its protein partners (nucleolin and KAP1/TRIM28 protein) repress Dux expression regulating exit from 2-cell state and at the same time/simultaneously induce hypertranscription of rRNAs essential for ESC self-renewal.

Apart from the high expression of L1 in particular tissue or developmental state, L1 expression has been found to reactivate under environmental stress (Terasaki et al. 2013; Stribinskis and Ramos 2006). Numerous studies in organisms including plants, flies, mice, and humans have demonstrated that transposons such as L1 and others are reactivated under conditions of environmental stress, suggesting an inherent mechanism to increase genetic diversity in order to adapt to harsh conditions. In yeast and plants, stress can activate TE transcription. For example, in melon female, stress conditions induce sex reversion by derepression of a TE adjacent to a female specific gene allowing male flower development and seed production (A. Martin et al. 2009). It is plausible that de-repression of L1 in environmental stresses can have both epigenetic and genetic impacts in pluripotent cells. This de-repression can induce new insertion events, impacting the regulation of nearby genes and the fitness of the organism. Study of L1 expression and activity under stressed conditions, such as nutritional deficiency or exposure to certain chemicals, at different developmental stages and cell types, may confirm if L1 could alter epigenome.

5.6. L1-Mediated Genomic Rearrangements Occasionally Result in Disease

Genomic rearrangements caused by L1 may affect the transcriptome and the proteome by a variety of mechanisms (see section 5.3), which occasionally leads to novel genetic diseases. The first report of L1-mediated disease came in 1988 from the Kazazian lab, demonstrating that L1s are still actively replicating in human somatic cells (Haig et al. 1988). So far, 124 L1-mediated insertions have been reported to cause genetic diseases. In most of these diseases, the functional protein was not active due to either insertional mutagenesis in the gene or aberrant splicing of the RNA. Among the disease-causing insertions, 29 are L1 retrotransposition, 77 are L1-mediated Alu retrotransposition, 13 are L1-mediated SVA retrotransposition, and 1 is an L1-mediated retrotransposition of CYBB gene (reviewed in (Dustin C. Hancks and Kazazian 2016)).

Disorders caused by these 124 insertions include cystic fibrosis, Duchenne muscular dystrophy, hemophilia A, hemophilia B, autoimmune diseases, and Neurofibromatosis Type I, Menkes disease, Dent's disease, chronic pancreatitis, severe hereditary nonspherocytic hemolytic anemia, Alstrom syndrome, hereditary cancer, breast-ovary-colon cancer, endometrial carcinoma, leukemia, cholinesterase deficiency, anterior pituitary aplasia, hereditary desmoid disease, chronic hemolytic anaemia, CHARFW syndrome, Walker Warburg syndrome, Aper syndrome, choroideremia, Coffin-Lowry Syndrome, B-thalassemia, rotor syndrome, familial retinoblastoma, chromothripsis, Lynch syndrome, Alport syndrome, Autosomal dominant optic atrophy, Aspartylglucosaminuria, Acute intermittent porphyria, Adrenoleukodystrophy, Autoimmune lymphoproliferative syndrome, Ataxia with oculomotor apraxia

2,Autosomal recessive hypercholesterolemia, Branchio-oto-renal syndrome, Congenital disorders of glycosylation type Ia, Chanarin-Dorfman syndrome, Chronic granulomatous disease, Familial adenomatous polyposis, Fukuyama-type congenital muscular dystrophy, Familial hypocalciuric hypercalcemia and neonatal severe hyperparathyroidism, Glycerol kinase deficiency, Hereditary form of angioedema, Hereditary elliptocytosis and hereditary pyropoikilocytosis, Hyper-immunoglobulin M syndrome, Hereditary non-polyposis colorectal cancer syndrome, Lipoprotein lipase deficiency, Mucolipidosis Type II, Mowat-Wilson syndrome, Pyruvate dehydrogenase complex deficiency, Neutral lipid storage disease with subclinical myopathy, Retinitis pigmentosa, Type 1 antithrombin deficiency, X-linked dystonia-parkinsonism, X-linked agammaglobulinemia, X-linked dilated cardiomyopathy, X-linked retinitis pigmentosa, X-linked severe combined immunodeficiency. Line1 activity is linked to neurological disorders such as Rett syndrome and schizophrenia (Bundo et al. 2014).

Half of all human epithelial cancers have been found to re-express the L1 machinery (Ewing et al. 2015; Helman et al. 2014; Tubio et al. 2014; Solyom et al. 2012; Shukla et al. 2013; E. Lee et al. 2012; Scott et al. 2016; Doucet-O'Hare et al. 2015; Rodić et al. 2014). Besides cancerous and metastatic tissues, the observation of somatic L1 insertions in pre-cancerous lesions and sometimes in the adjacent normal tissue, but not in blood DNA, is consistent with direct involvement of L1 in the early stages of tumorigenesis(Ewing et al. 2015). Somatic L1 retrotranspositions contribute to cancer genome mutagenesis load and can act as drivers of tumorigenesis. Genome-wide sequencing studies have detected extensive somatic insertions in various epithelial carcinomas including colon, pancreas, esophagus, uterus, head and neck, liver, lung, gastrointestinal tract, ovary and prostate (Ewing et al. 2015; Helman et al. 2014; Tubio et al. 2014; Solyom et al. 2012; Shukla et al. 2013; E. Lee et al. 2012; Scott et al. 2016; Doucet-O'Hare et al. 2015; Rodić et al. 2014). While clear driver L1 insertions into –or nearby–genes, which inactivate tumor suppressor genes or activate oncogenes, provide selective advantage and promote tumor growth, others have no defined impact, and might be passenger events. They might also contribute to tumor genome plasticity by shuffling genomic features through transductions of flanking genomic sequences or by pseudogene formation (Tubio et al. 2014; Cooke et al. 2014).

6. Regulators of L1-Mediated Mutagenesis

A good number of cellular L1 interactors have been identified by proteomic studies, such as, affinity chromatography and mass-spectrometry (John L. Goodier, Cheung, and Kazazian 2013; Taylor et al. 2013) although interaction ofL1 with only a few of them have been validated in cell culture. Factors involved in both aiding and limiting L1 retrotransposition have been identified although the mechanism behind their action is not always clear (reviewed in (Pizarro and Cristofari 2016)). Positive regulators of L1 retrotransposition include PABPC1 (a Poly(A) Binding Protein) (L. Dai et al. 2012), PCNA (Taylor et al. 2013), and some of the proline-directed protein kinases who are involved in cell cycle regulation, such as mitogen-activated protein kinases and cyclin-dependent kinases (Cook, Jones, and Furano 2015). Mutation or knock-down of these proteins are associated with either a reduced L1 RNP level (L. Dai et al. 2012) or inhibition of L1 retrotransposition(Taylor et al. 2013; Cook, Jones, and Furano 2015).

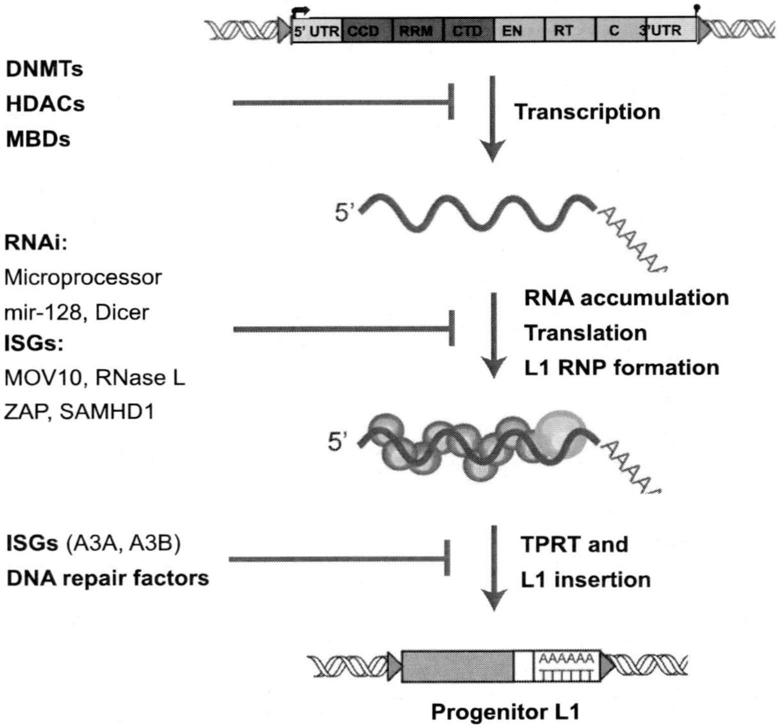

Figure 8. Cellular regulators limit L1 retrotransposition at different level. DNMTs, DNA methyl transferases; HDAC, Histone deacetylases; MBD, methyl-CpG-binding domain proteins, mir, micro-RNA, RNAi, RNA interference; ISG, interferon-stimulated genes; MOV10, Moloney leukemia virus 10; SAMHD1, SAM domain and HD domain 1; ZAP, zinc-finger antiviral protein; RNaseL, ribonuclease L; A3A, APOBEC3A; A3B, APOBEC3B.

Since L1 retrotransposition can cause harmful consequences to the host (see section 5), ranges of defense mechanisms have evolved in the host to protect the genome against the deleterious L1 retrotransposition events. Accumulation of mutations on L1 body can restrict an L1 over an evolutionary period. Also, L1 contains multiple polyadenylation sites which limit the full length transcription of L1 and further mobility (Perepelitsa-Belancio and Deininger 2003). Besides, L1 reverse transcription is often abortive producing 5' truncated progenitors which are incapable to retrotranspose. Apart from these mechanisms, control of L1 retrotransposition takes place at both transcriptional and post-transcriptional levels through the participation of a number of nuclear and cytoplasmic host factors (Figure 8).

6.1. Epigenetic Silencing to Restrict L1 Transcription

L1 transcription is silenced through CpGDNA methylation of L1 promoters and histone modifications (Castro-Diaz et al. 2014; Bestor and Bourc'his 2004). DNA methylation both limits the access of transcription factors to promoters and attracts methyl-CpG-binding proteins (MBDs). MBDs in turn, are bound by the histone deacetylases and other heterochromatin proteins which remodel chromatin state from accessible to inaccessible. Thus heterochromatinization of the surrounding region can limit L1 transcription. De-repression of L1 activity through L1 hypomethylation has been found to be associated with different stages of tumorigenesis(Suter, Martin, and Ward 2004).

6.2. Post-Transcriptional Silencing to Restrict L1 Retrotransposition

Two major mechanisms of post-transcriptional L1 silencing include, 1) sequence specific silencing of L1 mediated by small RNAs, and 2) anti-retroelement responses by interferon-stimulated genes (ISGs).

RNA interference (RNAi) reduces L1 retrotransposition by silencing L1 RNA in cultured somatic or embryonic cells by preventing L1 RNA accumulation (N. Yang and Kazazian 2006). L1 RNA is a direct substrate of the Microprocessor complex (Drosha/DGCR8). This nuclear complex can bind L1 RNA, cleave it to limit L1 retrotransposition(Heras et al. 2013).Lately, a new mechanism has been found where microRNA, miR-128 in complex with the Argonaute (Ago) protein, restricts L1-associated retrotranspositions in cancer cells, cancer-initiating cells and iPS cells by binding the ORF2 region of L1 RNA leading to degradation of L1 RNA (Hamdorf et al. 2015). Another class of small RNA involved in silencing L1 retrotransposition is piRNAs (Piwi-interacting RNAs), they are different from miRNAs in size and sequence. piRNAsare active in the embryonic male germ line where they reduce L1 RNA stability with the help of Piwi proteins (Pezic et al. 2014). piRNAs or rasiRNAs (repeat-associated small interfering RNA) are the sense and antisense L1 transcripts directed by the L1 5′ UTR promoters. The duplex of L1 RNA and piRNA is inhibited by Dicer knockdown.

The second mechanism of L1 post-transcriptional regulation involves interferon response pathways that activate IFN-stimulated genes (ISGs). Many of the ISGs are viral restriction factors and limit L1 retrotransposition as well(John L Goodier et al. 2015). L1 retrotransposition is restricted by a number of interferon-stimulated genes (ISGs), including some of the APOBEC3 (A3) cytidine deaminase family members (A3A, A3B, A3C and A3F), MOV10 (an RNA helicase), BST-2, ISG20, MAVS, MX2, RNase L, SAMHD1 (SAM Domain And HD Domain 1), TREX1 (Three-prime-repair exonuclease 1), and ZAP (zinc-finger antiviral protein) (reviewed in (Pizarro and Cristofari 2016)).

Conclusion

L1, an understudied transposon in the human genome, has been playing key roles in evolution by shaping our genome, regulating development, responding to changing environment, and bestowing individual to individual variation. Due to L1 activity, human genome has had various rearrangements, lose and gain of sequence, and variation in transcriptome and proteome. L1's large sequence size, high abundance in the genome, and very low copy number of ORF2p are the limiting factors in investigating L1. Advances in technologies, such as, development of emulsion PCR (Sultana et al., n.d.), single cell sequencing to scan novel L1 copies (Upton et al. 2015; Erwin et al. 2016) have made it possible to extract unprecedented information on L1. We knew L1 as junk at the beginning or selfish to just sit around there on the genome and replicate to survive. While every biological system highly maintains economy, it would be surprising that half of our genome is there making no significant contribution. Recent findings of the novel roles of L1 have shed lights on how L1 can be beneficial for us. Thus, we may have an intricate symbiotic relationship with L1 rather than a host-parasite one.

References

Anzai, Tomohiro, Mizuko Osanai, Mitsuhiro Hamada, and Haruhiko Fujiwara. 2005. "Functional Roles of 3'-Terminal Structures of Template RNA during in Vivo Retrotransposition of Non-LTR Retrotransposon, R1Bm." *Nucleic Acid Research* 33 (6):1993–2002. https://doi.org/10.1093/nar/gki347.

Babushok, Daria V, Kazuhiko Ohshima, Eric M Ostertag, Xinsheng Chen, Yanfeng Wang, Norihiro Okada, Charles S Abrams, and Haig H Kazazian. 2007. "A Novel Testis Ubiquitin-Binding Protein Gene Arose by Exon Shuffling in Hominoids." *Genome Res.* 17 (8). Cold Spring Harbor Lab:1129–38. https://doi.org/10.1101/gr.6252107.

Baillie, J Kenneth, Mark W Barnett, Kyle R Upton, Daniel J Gerhardt, Todd A Richmond, Fioravante De Sapio, Paul M Brennan, et al. 2011. "Somatic Retrotransposition Alters the Genetic Landscape of the Human Brain." *Nature* 479 (7374):534–37. https://doi.org/10.1038/nature10531.

Baller, Joshua A, Jiquan Gao, and Daniel F Voytas. 2011. "Access to DNA Establishes a Secondary Target Site Bias for the Yeast Retrotransposon Ty5." *Proc. Natl. Acad. Sci. U. S. A.* 108 (51). National Acad Sciences:20351–56. https://doi.org/10.1073/pnas.1103665108.

Barr, Stephen D., Jeremy Leipzig, Paul Shinn, Joe R. Ecker, and Frederic D. Bushman. 2005. "Integration Targeting by Avian Sarcoma-Leukosis Virus and Human Immunodeficiency Virus in the Chicken Genome." *Journal of Virology* 79 (American Society for Microbiology (ASM):12035–44. https://doi.org/10.1128/JVI.79.18.12035-12044.2005.

Beck, Christine R, Pamela Collier, Catriona Macfarlane, Maika Malig, Jeffrey M Kidd, Evan E Eichler, Richard M Badge, and John V Moran. 2010. "LINE-1 Retrotransposition Activity in Human Genomes." *Cell* 141 (7):1159–70.

Belancio, Victoria P, Dale J Hedges, and Prescott Deininger. 2006. "LINE-1 RNA Splicing and Influences on Mammalian Gene Expression." *Nucleic Acids Res.* 34 (5):1512–21. https://doi.org/10.1093/nar/gkl027.

Belgnaoui, S Mehdi, Roger G Gosden, O John Semmes, and Abdelali Haoudi. 2006. "Human LINE-1 Retrotransposon Induces DNA Damage and Apoptosis in Cancer Cells." *Cancer Cell Int.* 6 (1). BioMed Central:13. https://doi.org/10.1186/1475-2867-6-13.

Bellen, H J, R W Levis, Y He, J W Carlson, M Evans-Holm, E Bae, J Kim, et al. 2011. "The Drosophila Gene Disruption Project: Progress Using Transposons With Distinctive Site Specificities." *Genetics* 188 (3):731–43.

Benleulmi, Mohamed Salah, Julien Matysiak, Daniel Rodrigo Henriquez, Cedric Vaillant, Paul Lesbats, Christina Calmels, Monica Naughtin, et al. 2015. "Intasome Architecture and Chromatin Density Modulate Retroviral Integration into Nucleosome." *Retrovirology* 12 (February):13. https://doi.org/10.1186/s12977-015-0145-9.

Bestor, T H, and D Bourc'his. 2004. "Transposon Silencing and Imprint Establishment in Mammalian Germ Cells." *Cold Spring Harb Symp Quant Biol* 69 (0):381–87. https://doi.org/10.1101/sqb.2004.69.381.

Biessmann, H, and J M Mason. 2003. "Telomerase-Independent Mechanisms of Telomere Elongation." *Cellular and Molecular Life Sciences : CMLS* 60 (11):2325–33.

Boeke, Jef D. 1997. "LINEs and Alus {\textemdash} the PolyA Connection." *Nat. Genet.* 16 (1):6–7.

Boissinot, S, P Chevret, and A V Furano. 2000. "L1 (LINE-1) Retrotransposon Evolution and Amplification in Recent Human History." *Mol. Biol. Evol.* 17 (6):915–28. http://eutils.ncbi.nlm.nih.gov/entrez/eutils/elink.fcgi?dbfrom=pubmed&id=10833198&retmode=ref&cmd=prlinks.

Brady, Troy, Young Nam Lee, Keshet Ronen, Nirav Malani, Charles C Berry, Paul D Bieniasz, and Frederic D Bushman. 2009. "Integration Target Site Selection by a Resurrected Human Endogenous Retrovirus." *Genes Dev.* 23 (5):633–42. https://doi.org/10.1101/gad.1762309.

Bridier-Nahmias, Antoine, Aurélie Tchalikian-Cosson, Joshua A. Baller, Rachid Menouni, Hélène Fayol, Amando Flore, Ali Saïb, Michel Werner, Daniel F. Voytas, and Pascale Lesage. 2015. "An RNA Polymerase III Subunit Determines Sites of Retrotransposon Integration." *Science* 348 (6234). American Association for the Advancement of Science:585–88. https://doi.org/10.1126/science.1259114.

Brouha, B, J Schustak, R M Badge, S Lutz-Prigge, A H Farley, J V Moran, and Jr Kazazian H. H. 2003. "Hot L1s Account for the Bulk of Retrotransposition in the Human Population." *Proc Natl Acad Sci U S A* 100 (9). Department of Genetics, University of Pennsylvania School of Medicine, Philadelphia, PA 19104, USA.:5280–85.

Brouha, Brook, Christof Meischl, Eric Ostertag, Martin de Boer, Yue Zhang, Herman Neijens, Dirk Roos, and Haig H Kazazian. 2002. "Evidence Consistent with Human L1 Retrotransposition in Maternal Meiosis I." *American Journal of Human Genetics* 71 (2):327–36. https://doi.org/10.1086/341722.

Bundo, Miki, Manabu Toyoshima, Yohei Okada, Wado Akamatsu, Junko Ueda, Taeko Nemoto-Miyauchi, Fumiko Sunaga, et al. 2014. "Increased L1 Retrotransposition in the Neuronal Genome in Schizophrenia." *Neuron* 81 (2). https://doi.org/10.1016/j.neuron.2013.10.053.

Burwinkel, Barbara, and Manfred W Kilimann. 1998. "Unequal Homologous Recombination between LINE-1 Elements as a Mutational Mechanism in Human Genetic Disease." *J Mol Biol* 277 (3):513–17. https://doi.org/10.1006/jmbi.1998.1641.

Callinan, P A, and M A Batzer. 2006. "Retrotransposable Elements and Human Disease." *Genome Dyn* 1:104–15. https://doi.org/10.1159/000092503.

Castro-Diaz, Nathaly, Gabriela Ecco, Andrea Coluccio, Adamandia Kapopoulou, Benyamin Yazdanpanah, Marc Friedli, Julien Duc, Suk Min Jang, Priscilla Turelli, and Didier Trono. 2014. "Evolutionally Dynamic L1 Regulation in Embryonic Stem Cells." *Genes Dev.* 28 (13). Cold Spring Harbor Lab:1397–1409. https://doi.org/10.1101/gad.241661.114.

Chatterjee, Atreyi Ghatak, Caroline Esnault, Yabin Guo, Stevephen Hung, Philip G McQueen, and Henry L Levin. 2014. "Serial Number Tagging Reveals a Prominent Sequence Preference of Retrotransposon Integration." *Nucleic Acids Res.* 42 (13). Oxford University Press:8449–60. https://doi.org/10.1093/nar/gku534.

Christensen, S M, and T H Eickbush. 2005. "R2 Target-Primed Reverse Transcription: Ordered Cleavage and Polymerization Steps by Protein Subunits Asymmetrically Bound to the Target DNA." *Mol. Cell. Biol.* 25 (15):6617–28.

Chu, W M, W M Liu, and C W Schmid. 1995. "RNA Polymerase III Promoter and Terminator Elements Affect Alu RNA Expression." *Nucleic Acids Res.* 23 (10). Oxford University Press:1750–57. http://www.ncbi.nlm.nih.gov/pubmed/7540287.

Ciuffi, Angela, Manuel Llano, Eric Poeschla, Christian Hoffmann, Jeremy Leipzig, Paul Shinn, Joseph R Ecker, and Frederic Bushman. 2005. "A Role for LEDGF/P75 in Targeting HIV DNA Integration." *Nat Med* 11 (12). Nature Publishing Group:1287–89. https://doi.org/10.1038/nm1329.

Cook, Pamela R, Charles E Jones, and Anthony V Furano. 2015. "Phosphorylation of ORF1p Is Required for L1 Retrotransposition." *Proceedings of the National Academy of Sciences of the United States of America* 112 (14):4298–4303. https://doi.org/10.1073/pnas.1416869112.

Cooke, Susanna L., Adam Shlien, John Marshall, Christodoulos P. Pipinikas, Inigo Martincorena, Jose M C Tubio, Yilong Li, et al. 2014. "Processed Pseudogenes Acquired Somatically during Cancer Development." *Nature Communications* 5:3644. https://doi.org/10.1038/ncomms4644.

Cordaux, Richard, and Mark A Batzer. 2009. "The Impact of Retrotransposons on Human Genome Evolution." *Nat. Rev. Genet.* 10 (10):691–703. https://doi.org/10.1038/nrg2640.

Cordaux, Richard, Dale J Hedges, Scott W Herke, and Mark A Batzer. 2006. "Estimating the Retrotransposition Rate of Human Alu Elements." https://doi.org/10.1016/j.gene.2006.01.019.

Cost, G J, and J D Boeke. 1998. "Targeting of Human Retrotransposon Integration Is Directed by the Specificity of the L1 Endonuclease for Regions of Unusual DNA Structure." *Biochemistry* 37 (51):18081–93. http://www.ncbi.nlm.nih.gov/pubmed/9922177.

Cost, Gregory J, Qinghua Feng, Alain Jacquier, and Jef D Boeke. 2002. "Human L1 Element Target-Primed Reverse Transcription in Vitro." *EMBO J.* 21 (21). European Molecular Biology Organization:5899–5910. https://doi.org/10.1093/emboj/cdf592.

Criscione, Steven W, Nicholas Theodosakis, Goran Micevic, Toby C Cornish, Kathleen H Burns, and Nicola Neretti. 2016. "Genome-Wide Characterization of Human L1 Antisense Promoter-Driven Transcripts." *BMC Genomics* 17 (1). BioMed Central: 740. https://doi.org/10.1186/s12864-016-2800-5.

Curcio, M Joan, and Keith M Derbyshire. 2003. "The Outs and Ins of Transposition: From Mu to Kangaroo." *Nat. Rev. Mol. Cell Biol.* 4 (11):865–77. https://doi.org/10.1038/nrm1241.

Dai, Junbiao, Weiwu Xie, Troy L. Brady, Jiquan Gao, and Daniel F. Voytas. 2007. "Phosphorylation Regulates Integration of the Yeast Ty5 Retrotransposon into Heterochromatin." *Molecular Cell* 27 (2):289–99. https://doi.org/10.1016/j.molcel.2007.06.010.

Dai, Lixin, Martin S Taylor, Kathryn A O'Donnell, and Jef D Boeke. 2012. "Poly(A) Binding Protein C1 Is Essential for Efficient L1 Retrotransposition and Affects L1 RNP Formation." *Molecular and Cellular Biology* 32 (21):4323–36. https://doi.org/10.1128/MCB.06785-11.

Damert, A, J Raiz, A V Horn, J Lower, H Wang, J Xing, M A Batzer, R Lower, and G G Schumann. 2009. "5'-Transducing SVA Retrotransposon Groups Spread Efficiently throughout the Human Genome." *Genome Res.* 19 (11):1992–2008.

Dewannieux, Marie, Cécile Esnault, and Thierry Heidmann. 2003. "LINE-Mediated Retrotransposition of Marked Alu Sequences." *Nature Genetics* 35 (1):41–48. https://doi.org/10.1038/ng1223.

Doucet-O'Hare, Tara T., Nemanja Rodić, Reema Sharma, Isha Darbari, Gabriela Abril, Jungbin a. Choi, Ji Young Ahn, et al. 2015. "LINE-1 Expression and Retrotransposition in Barrett's Esophagus and Esophageal Carcinoma." *Proceedings of the National Academy of Sciences* advanced p:1–7. https://doi.org/10.1073/pnas.1502474112.

Doucet, Aurélien J, Gaëtan Droc, Oliver Siol, Jérôme Audoux, and Nicolas Gilbert. 2015. "U6 SnRNA Pseudogenes: Markers of Retrotransposition Dynamics in Mammals." *Mol. Biol. Evol.* 32 (7):1815–32. https://doi.org/10.1093/molbev/msv062.

Doucet, Aurélien J, Jeremy E Wilusz, Tomoichiro Miyoshi, Ying Liu, and John V Moran. 2015. "A 3' Poly(A) Tract Is Required for LINE-1 Retrotransposition." *Molecular Cell* 60 (5):728–41.

Eickbush, Thomas H, and Varuni K Jamburuthugoda. 2008. "The Diversity of Retrotransposons and the Properties of Their Reverse Transcriptases." *Virus Res.* 134 (1–2):221–34. https://doi.org/10.1016/j.virusres.2007.12.010.

Erwin, Jennifer A, Maria C Marchetto, and Fred H Gage. 2014. "Mobile DNA Elements in the Generation of Diversity and Complexity in the Brain." *Nat. Rev. Neurosci.* 15 (8):497–506. https://doi.org/10.1038/nrn3730.

Erwin, Jennifer A, Apuã C M Paquola, Tatjana Singer, Iryna Gallina, Mark Novotny, Carolina Quayle, Tracy A Bedrosian, et al. 2016. "L1-Associated Genomic Regions Are Deleted in Somatic Cells of the Healthy Human Brain." *Nature Neuroscience*, September.

Esnault, C, J Maestre, and T Heidmann. 2000. "Human LINE Retrotransposons Generate Processed Pseudogenes." *Nature Genetics* 24 (4):363–67. https://doi.org/10.1038/74184.

Ewing, Adam D., Anthony Gacita, Laura D. Wood, Florence Ma, Dongmei Xing, Min Sik Kim, Srikanth S. Manda, et al. 2015. "Widespread Somatic L1 Retrotransposition Occurs Early during Gastrointestinal Cancer Evolution." *Genome Research* 25 (10):1536–45. https://doi.org/10.1101/gr.196238.115.

Feng, Qinghua, John V. Moran, Haig H. Kazazian, and Jef D. Boeke. 1996. "Human L1 Retrotransposon Encodes a Conserved Endonuclease Required for

Retrotransposition." *Cell* 87 (5):905–16. https://doi.org/10.1016/S0092-8674(00)81997-2.

Fungtammasan, Arkarachai, Erin Walsh, Francesca Chiaromonte, Kristin A Eckert, and Kateryna D Makova. 2012. "A Genome-Wide Analysis of Common Fragile Sites: What Features Determine Chromosomal Instability in the Human Genome?" *Genome Res* 22 (6). United States:993–1005. https://doi.org/10.1101/gr.134395.111.

Gijsbers, Rik, Keshet Ronen, Sofie Vets, Nirav Malani, Jan De Rijck, Melissa McNeely, Frederic D Bushman, and Zeger Debyser. 2010. "LEDGF Hybrids Efficiently Retarget Lentiviral Integration into Heterochromatin." *Molecular Therapy : The Journal of the American Society of Gene Therapy* 18 (3). Nature Publishing Group:552–60. https://doi.org/10.1038/mt.2010.36.

Gilbert, N, S Lutz-Prigge, and Moran JV. 2002. "Genomic Deletions Created upon LINE_1 Retrotransposition.Pdf." *Cell* 110 (3):315–25.

Gilbert, Nicolas, Sheila Lutz, and John V Moran. 2005. "Multiple Fates of L1 Retrotransposition Intermediates in Cultured Human Cells." *Mol. Cell. Biol.* 25 (17). American Society for Microbiology:7780–95. https://doi.org/10.1128/MCB.25.17.7780-7795.2005.

Gogol-Döring, Andreas, Ismahen Ammar, Saumyashree Gupta, Mario Bunse, Csaba Miskey, Wei Chen, Wolfgang Uckert, Thomas F Schulz, Zsuzsanna Izsvák, and Zoltán Ivics. 2016. "Genome-Wide Profiling Reveals Remarkable Parallels Between Insertion Site Selection Properties of the MLV Retrovirus and the PiggyBac Transposon in Primary Human CD4+ T Cells." *Molecular Therapy*, 1–15. https://doi.org/10.1038/mt.2016.11.

Goodier, J L, E M Ostertag, and H H Kazazian. 2000. "Transduction of 3'-Flanking Sequences Is Common in L1 Retrotransposition." *Hum. Mol. Genet.* 9 (4):653–57. https://doi.org/10.1093/hmg/9.4.653.

Goodier, John L., Ling E. Cheung, and Haig H. Kazazian. 2013. "Mapping the LINE1 ORF1 Protein Interactome Reveals Associated Inhibitors of Human Retrotransposition." *Nucleic Acids Research* 41 (15). https://doi.org/10.1093/nar/gkt512.

Goodier, John L, Gavin C Pereira, Ling E Cheung, Rebecca J Rose, and Haig H Kazazian. 2015. "The Broad-Spectrum Antiviral Protein ZAP Restricts Human Retrotransposition." Edited by Harmit S. Malik. *PLoS Genetics* 11 (5):e1005252. https://doi.org/10.1371/journal.pgen.1005252.

Haig, H H, Corinne Wong, Hagop Youssoufian, Deborah G Phillips, and Stylianos E Antonarakis. 1988. "Haemophilia A Resulting from de Novo Insertion of L1 Sequences Represents a Novel Mechanism for Mutation in Man." *Nature* 332 (6160). Nature Publishing Group:164–66. https://doi.org/10.1038/332164a0.

Hamdorf, Matthias, Adam Idica, Dimitrios G Zisoulis, Lindsay Gamelin, Charles Martin, Katie J Sanders, and Irene M Pedersen. 2015. "MiR-128 Represses L1

Retrotransposition by Binding Directly to L1 RNA." *Nature Publishing Group* 22 (10):824–31. https://doi.org/10.1038/nsmb.3090.

Han, Jeffrey S, Suzanne T Szak, and Jef D Boeke. 2004. "Transcriptional Disruption by the L1 Retrotransposon and Implications for Mammalian Transcriptomes." *Nature* 429 (6989):268–74. https://doi.org/10.1038/nature02536.

Han, Kyudong, Jungnam Lee, Thomas J Meyer, Paul Remedios, Lindsey Goodwin, and Mark A Batzer. 2008. "L1 Recombination-Associated Deletions Generate Human Genomic Variation." *Proc. Natl. Acad. Sci. U. S. A.* 105 (49). National Acad Sciences:19366–71. https://doi.org/10.1073/pnas.0807866105.

Hancks, D C, A D Ewing, J E Chen, K Tokunaga, and H H Kazazian. 2009. "Exon-Trapping Mediated by the Human Retrotransposon SVA." *Genome Res.* 19 (11):1983–91.

Hancks, D C, J L Goodier, P K Mandal, L E Cheung, and H H Kazazian. 2011. "Retrotransposition of Marked SVA Elements by Human L1s in Cultured Cells." *Hum. Mol. Genet.* 20 (17):3386–3400.

Hancks, D C, P K Mandal, L E Cheung, and H H Kazazian. 2012. "The Minimal Active Human SVA Retrotransposon Requires Only the 5'-Hexamer and Alu-Like Domains." *Mol. Cell. Biol.* 32 (22):4718–26.

Hancks, Dustin C., and Haig H. Kazazian. 2016. "Roles for Retrotransposon Insertions in Human Disease." *Mobile DNA* 7 (1). Mobile DNA:9. https://doi.org/10.1186/s13100-016-0065-9.

Helman, Elena, Michael S Lawrence, Chip Stewart, Carrie Sougnez, Gad Getz, and Matthew Meyerson. 2014. "Somatic Retrotransposition in Human Cancer Revealed by Whole-Genome and Exome Sequencing." *Genome Res.* 24 (7):1053–63. https://doi.org/10.1101/gr.163659.113.

Heras, Sara R, Sara Macias, Mireya Plass, Noemí Fernandez, David Cano, Eduardo Eyras, José L Garcia-Perez, and Javier F Cáceres. 2013. "The Microprocessor Controls the Activity of Mammalian Retrotransposons." *Nature Structural & Molecular Biology* 20 (10):1173–81. https://doi.org/10.1038/nsmb.2658.

Hickey, Anthony, Caroline Esnault, Anasuya Majumdar, Atreyi Ghatak Chatterjee, James R. Iben, Philip G. McQueen, Andrew X. Yang, Takeshi Mizuguchi, Shiv I S Grewal, and Henry L. Levin. 2015. "Single-Nucleotide-Specific Targeting of the Tf1 Retrotransposon Promoted by the DNA-Binding Protein Sap1 of Schizosaccharomyces Pombe." *Genetics* 201 (3):905–24. https://doi.org/10.1534/genetics.115.181602.

Jacobs, Jake Z, Jesus D Rosado-Lugo, Susanne Cranz-Mileva, Keith M Ciccaglione, Vincent Tournier, and Mikel Zaratiegui. 2015. "Arrested Replication Forks Guide Retrotransposon Integration." *Science* 349 (6255). American Association for the Advancement of Science:1549–53. https://doi.org/10.1126/science.aaa3810.

Kaessmann, Henrik, Nicolas Vinckenbosch, and Manyuan Long. 2009. "RNA-Based Gene Duplication: Mechanistic and Evolutionary Insights." *Nat. Rev. Genet.* 10 (1):19–31. https://doi.org/10.1038/nrg2487.

Kajikawa, Masaki, and Norihiro Okada. 2002. "LINEs Mobilize SINEs in the Eel through a Shared 3\textbackslashtextasciiacute Sequence." *Cell* 111 (3):433–44. https://doi.org/10.1016/S0092-8674(02)01041-3.

Kano, Hiroki, Irene Godoy, Christine Courtney, Melissa R Vetter, George L Gerton, Eric M Ostertag, and Haig H Kazazian. 2009. "L1 Retrotransposition Occurs Mainly in Embryogenesis and Creates Somatic Mosaicism." *Genes & Development* 23 (11):1303–12. https://doi.org/10.1101/gad.1803909.

Kolosha, Vladimir O, and Sandra L Martin. 2003. "High-Affinity, Non-Sequence-Specific RNA Binding by the Open Reading Frame 1 (ORF1) Protein from Long Interspersed Nuclear Element 1 (LINE-1)." *J. Biol. Chem.* 278 (10). American Society for Biochemistry and Molecular Biology:8112–17. https://doi.org/10.1074/jbc.M210487200.

Krull, Maren, Jürgen Brosius, and Jürgen Schmitz. 2005. "Alu-SINE Exonization: En Route to Protein-Coding Function." *Mol. Biol. Evol.* 22 (8):1702–11. https://doi.org/10.1093/molbev/msi164.

Kulpa, D A. 2005. "Ribonucleoprotein Particle Formation Is Necessary but Not Sufficient for LINE-1 Retrotransposition." *Hum. Mol. Genet.* 14 (21):3237–48.

Kulpa, Deanna a., and John V Moran. 2006. "Cis-Preferential LINE-1 Reverse Transcriptase Activity in Ribonucleoprotein Particles." *Nat. Struct. Mol. Biol.* 13 (7). Nature Publishing Group:655–60. https://doi.org/10.1038/nsmb1107.

Lander, E S, a Heaford, a Sheridan, L M Linton, B Birren, a Subramanian, a Coulson, et al. 2001. "Initial Sequencing and Analysis of the Human Genome." *Nature* 409 (6822):860–921. https://doi.org/10.1038/35057062.

Lee, Eunjung, Rebecca Iskow, Lixing Yang, Omer Gokcumen, Psalm Haseley, Lovelace J Luquette, Jens G Lohr, et al. 2012. "Landscape of Somatic Retrotransposition in Human Cancers." *Science (New York, N.Y.)* 337 (6097):967–71. https://doi.org/10.1126/science.1222077.

Lee, Jungnam, Richard Cordaux, Kyudong Han, Jianxin Wang, Dale J Hedges, Ping Liang, and Mark A Batzer. 2007. "Different Evolutionary Fates of Recently Integrated Human and Chimpanzee LINE-1 Retrotransposons." *Gene* 390 (1–2):18–27. https://doi.org/10.1016/j.gene.2006.08.029.

Lee, Jungnam, Kyudong Han, Thomas J Meyer, Heui-Soo Kim, and Mark A Batzer. 2008. "Chromosomal Inversions between Human and Chimpanzee Lineages Caused by Retrotransposons." *PLoS ONE* 3 (12):e4047--9. https://doi.org/10.1371/journal.pone.0004047.

Lelek, Mickaël, Nicoletta Casartelli, Danilo Pellin, Ermanno Rizzi, Philippe Souque, Marco Severgnini, Clelia Di Serio, et al. 2015. "Chromatin Organization at the

Nuclear Pore Favours HIV Replication." *Nat. Commun.* 6. Nature Publishing Group:6483. https://doi.org/10.1038/ncomms7483.

Malik, H S, S Henikoff, and T H Eickbush. 2000. "Poised for Contagion: Evolutionary Origins of the Infectious Abilities of Invertebrate Retroviruses." *Genome Research* 10 (9). Cold Spring Harbor Laboratory Press:1307–18. https://doi.org/10.1101/GR.145000.

Marques, Ana Claudia, Isabelle Dupanloup, Nicolas Vinckenbosch, Alexandre Reymond, and Henrik Kaessmann. 2005. "Emergence of Young Human Genes after a Burst of Retroposition in Primates." *PLoS Biol* 3 (11). Public Library of Science:e357. https://doi.org/10.1371/journal.pbio.0030357.

Martin, Antoine, Christelle Troadec, Adnane Boualem, Mazen Rajab, Ronan Fernandez, Halima Morin, Michel Pitrat, Catherine Dogimont, and Abdelhafid Bendahmane. 2009. "A Transposon-Induced Epigenetic Change Leads to Sex Determination in Melon." *Nature* 461 (7267). England:1135–38. https://doi.org/10.1038/nature08498.

Martin, S L, and F D Bushman. 2001. "Nucleic Acid Chaperone Activity of the ORF1 Protein from the Mouse LINE-1 Retrotransposon." *Mol. Cell. Biol.* 21 (2):467–75.

Mätlik, Kert, Kaja Redik, and Mart Speek. 2006. "L1 Antisense Promoter Drives Tissue-Specific Transcription of Human Genes." *J. Biomed. Biotechnol.* 2006 (1). Hindawi Publishing Corporation:71716–53. https://doi.org/10.1155/JBB/2006/71753.

McClintock, B. 1984. "The Significance of Responses of the Genome to Challenge." *Science* 226 (4676). United States:792–801.

McClintock, Barbara. 1950. "The Origin and Behavior of Mutable Loci in Maize." *Proceedings of the National Academy of Sciences* 36 (6). Cambridge University Press:344–55. https://doi.org/10.1073/pnas.36.6.344.

Monot, Clément, Monika Kuciak, Sébastien Viollet, Ashfaq Ali Mir, Caroline Gabus, Jean-Luc Darlix, and Gaël Cristofari. 2013. "The Specificity and Flexibility of L1 Reverse Transcription Priming at Imperfect T-Tracts." *PLoS Genet.* 9 (5):e1003499. https://doi.org/10.1371/journal.pgen.1003499.

Moran, J V, R J DeBerardinis, and H H Kazazian. 1999. "Exon Shuffling by L1 Retrotransposition." *Science* 283 (5407):1530–34. http://eutils.ncbi.nlm.nih.gov/entrez/eutils/elink.fcgi?dbfrom=pubmed&id=10066175&retmode=ref&cmd=prlinks.

Morrish, Tammy A, Nicolas Gilbert, Jeremy S Myers, Bethaney J Vincent, Thomas D Stamato, Guillermo E Taccioli, Mark A Batzer, and John V Moran. 2002. "DNA Repair Mediated by Endonuclease-Independent LINE-1 Retrotransposition." *Nature Genetics* 31 (2):159–65. https://doi.org/10.1038/ng898.

Muotri, Alysson R, Vi T Chu, Maria C N Marchetto, Wei Deng, John V Moran, and Fred H Gage. 2005. "Somatic Mosaicism in Neuronal Precursor Cells Mediated by L1 Retrotransposition." *Nature* 435 (7044):903–10. https://doi.org/10.1038/nature03663.

Ovchinnikov, I, A B Troxel, and G D Swergold. 2001. "Genomic Characterization of Recent Human LINE-1 Insertions: Evidence Supporting Random Insertion." *Genome Res.* 11 (12). Cold Spring Harbor Lab:2050–58. https://doi.org/10.1101/gr.194701.

Pardue, M-L, and P G DeBaryshe. 2000. "Drosophila Telomere Transposons: Genetically Active Elements in Heterochromatin." *Genetica* 109 (1–2):45–52. http://eutils.ncbi.nlm.nih.gov/entrez/eutils/elink.fcgi?dbfrom=pubmed&id=11293794&retmode=ref&cmd=prlinks.

Percharde, Michelle, Chih-Jen Lin, Yafei Yin, Bo Huang, Xiaohua Shen, Miguel Ramalho-Santos, Juan Guan, Gabriel A Peixoto, Aydan Bulut-Karslioglu, and Steffen Biechele. 2018. "A LINE1-Nucleolin Partnership Regulates Early Development and ESC Identity." *Cell* 174:1–15. https://doi.org/10.1016/j.cell.2018.05.043.

Perepelitsa-Belancio, Victoria, and Prescott Deininger. 2003. "RNA Truncation by Premature Polyadenylation Attenuates Human Mobile Element Activity." *Nat. Genet.* 35 (4):363–66. https://doi.org/10.1038/ng1269.

Pezic, Dubravka, Sergei A Manakov, Ravi Sachidanandam, and Alexei A Aravin. 2014. "PiRNA Pathway Targets Active LINE1 Elements to Establish the Repressive H3K9me3 Mark in Germ Cells." *Genes Dev.* 28 (13):1410–28. https://doi.org/10.1101/gad.240895.114.

Philippe, Claude, Dulce B Vargas-Landin, Aurélien J Doucet, Dominic van Essen, Jorge Vera-Otarola, Monika Kuciak, Antoine Corbin, Pilvi Nigumann, and Gaël Cristofari. 2016. "Activation of Individual L1 Retrotransposon Instances Is Restricted to Cell-Type Dependent Permissive Loci." *ELife* 5. eLife Sciences Publications Limited:166. https://doi.org/10.7554/eLife.13926.

Pickeral, O K, W Makałowski, M S Boguski, and J D Boeke. 2000. "Frequent Human Genomic DNA Transduction Driven by LINE-1 Retrotransposition." *Genome Res.* 10 (4). Cold Spring Harbor Laboratory Press:411–15. /pmc/articles/PMC310862/?report=abstract.

Pizarro, Javier G, and Gael Cristofari. 2016. "Post-Transcriptional Control of LINE-1 Retrotransposition by Cellular Host Factors in Somatic Cells." *Frontiers in Cell and Developmental Biology*. http://www.frontiersin.org/Journal/Abstract.aspx?s=1502&name=cellular_biochemistry&ART_DOI=10.3389/fcell.2016.00014.

Raiz, J, A Damert, S Chira, U Held, S Klawitter, M Hamdorf, J Lower, W H Stratling, R Lower, and G G Schumann. 2012. "The Non-Autonomous Retrotransposon SVA Is Trans-Mobilized by the Human LINE-1 Protein Machinery." *Nucleic Acids Res.* 40 (4):1666–83.

Rashkova, S, S E Karam, and M-L L Pardue. 2002. "Element-Specific Localization of Drosophila Retrotransposon Gag Proteins Occurs in Both Nucleus and Cytoplasm."

Proc Natl Acad Sci U S A 99 (6). United States:3621–26. https://doi.org/10.1073/pnas.032071999.

Rashkova, Svetlana, Sarah E Karam, Rebecca Kellum, and Mary-Lou Pardue. 2002. "Gag Proteins of the Two Drosophilatelomeric Retrotransposons Are Targeted to Chromosome Ends." *The Journal of Cell Biology* 159 (3):397–402.

Richardson, Sandra R, Patricia Gerdes, Daniel J Gerhardt, Francisco J Sanchez-Luque, Gabriela-Oana Bodea, Martin Muñoz-Lopez, J Samuel Jesuadian, et al. 2017. "Heritable L1 Retrotransposition in the Mouse Primordial Germline and Early Embryo." *Genome Research* 27 (8):1395–1405. https://doi.org/10.1101/gr.219022.116.

Rodić, Nemanja, Reema Sharma, Rajni Sharma, John Zampella, Lixin Dai, Martin S. Taylor, Ralph H. Hruban, et al. 2014. "Long Interspersed Element-1 Protein Expression Is a Hallmark of Many Human Cancers." *American Journal of Pathology* 184 (5):1280–86. https://doi.org/10.1016/j.ajpath.2014.01.007.

Rodić, Nemanja, Jared P Steranka, Alvin Makohon-Moore, Allison Moyer, Peilin Shen, Reema Sharma, Zachary a Kohutek, et al. 2015. "Retrotransposon Insertions in the Clonal Evolution of Pancreatic Ductal Adenocarcinoma." *Nat. Med.* 21 (9):1060–64. https://doi.org/10.1038/nm.3919.

Sandmeyer, Suzanne, Kurt Patterson, and Virginia Bilanchone. 2015. "Ty3, a Position-Specific Retrotransposon in Budding Yeast." *Microbiol Spectr* 3 (2). United States:MDNA3-0057-2014. https://doi.org/10.1128/microbiolspec.MDNA3-0057-2014.

Sayah, David M, Elena Sokolskaja, Lionel Berthoux, and Jeremy Luban. 2004. "Cyclophilin A Retrotransposition into TRIM5 Explains Owl Monkey Resistance to HIV-1." *Nature* 430 (6999):569–73.

Scott, Emma C, Eugene J Gardner, Ashiq Masood, Nelson T Chuang, Paula M Vertino, and Scott E Devine. 2016. "A Hot L1 Retrotransposon Evades Somatic Repression and Initiates Human Colorectal Cancer." *Genome Res.* https://doi.org/10.1101/gr.201814.115.

Sen, Shurjo K., Charles T. Huang, Kyudong Han, and Mark A. Batzer. 2007. "Endonuclease-Independent Insertion Provides an Alternative Pathway for L1 Retrotransposition in the Human Genome." *Nucleic Acids Research* 35 (11):3741–51. https://doi.org/10.1093/nar/gkm317.

Shivram, Haridha, Dillon Cawley, and Shawn M Christensen. 2011. "Targeting Novel Sites: The N-Terminal DNA Binding Domain of Non-LTR Retrotransposons Is an Adaptable Module That Is Implicated in Changing Site Specificities." *Mob Genet Elements* 1 (3). United States:169–78. https://doi.org/10.4161/mge.1.3.18453.

Shukla, Ruchi, Kyle R. Upton, Martin Muñoz-Lopez, Daniel J. Gerhardt, Malcolm E. Fisher, Thu Nguyen, Paul M. Brennan, et al. 2013. "Endogenous Retrotransposition

Activates Oncogenic Pathways in Hepatocellular Carcinoma." *Cell* 153 (1):101–11. https://doi.org/10.1016/j.cell.2013.02.032.

Shun, Ming-Chieh, Nidhanapati K Raghavendra, Nick Vandegraaff, Janet E Daigle, Siobhan Hughes, Paul Kellam, Peter Cherepanov, and Alan Engelman. 2007. "LEDGF/P75 Functions Downstream from Preintegration Complex Formation to Effect Gene-Specific HIV-1 Integration." *Genes Dev.* 21 (14). Cold Spring Harbor Lab:1767–78. https://doi.org/10.1101/gad.1565107.

Silvers, Robert M, Johanna A Smith, Michael Schowalter, Samuel Litwin, Zhihui Liang, Kyla Geary, and René Daniel. 2010. "Modification of Integration Site Preferences of an HIV-1-Based Vector by Expression of a Novel Synthetic Protein." *Human Gene Therapy* 21 (3):337–49. https://doi.org/10.1089/hum.2009.134.

Smit, Arian F A, Gábor Tóth, Arthur D Riggs, and Jerzy Jurka. 1995. "Ancestral, Mammalian-Wide Subfamilies of LINE-1 Repetitive Sequences." *Journal of Molecular Biology* 246 (3):401–17.

Solyom, Szilvia, Adam D. Ewing, Eric P. Rahrmann, Tara Doucet, Heather H. Nelson, Michael B. Burns, Reuben S. Harris, et al. 2012. "Extensive Somatic L1 Retrotransposition in Colorectal Tumors." *Genome Research* 22 (12):2328–38. https://doi.org/10.1101/gr.145235.112.

Song, Mingzhou. 2007. "Selection against LINE-1 Retrotransposons Results Principally from Their Ability to Mediate Ectopic Recombination." *Gene* 390 (1–2):206–13. https://doi.org/10.1016/j.gene.2006.09.033.

Sorek, Rotem, Gil Ast, and Dan Graur. 2002. "Alu-Containing Exons Are Alternatively Spliced." *Genome Res.* 12 (7). Cold Spring Harbor Lab:1060–67. https://doi.org/10.1101/gr.229302.

Spaller, Thomas, Eva Kling, Gernot Glöckner, Falk Hillmann, and Thomas Winckler. 2016. "Convergent Evolution of TRNA Gene Targeting Preferences in Compact Genomes." *Mob DNA* 7 (1). England:17. https://doi.org/10.1186/s13100-016-0073-9.

Streva, Vincent A, Vallmer E Jordan, Sara Linker, Dale J Hedges, Mark A Batzer, and Prescott L Deininger. 2015. "Sequencing, Identification and Mapping of Primed L1 Elements (SIMPLE) Reveals Significant Variation in Full Length L1 Elements between Individuals." *BMC Genomics* 16. England:220. https://doi.org/10.1186/s12864-015-1374-y.

Stribinskis, Vilius, and Kenneth S Ramos. 2006. "Activation of Human Long Interspersed Nuclear Element 1 Retrotransposition by Benzo(a)Pyrene, an Ubiquitous Environmental Carcinogen." *Cancer Research* 66 (5):2616–20. https://doi.org/10.1158/0008-5472.CAN-05-3478.

Sultana, Tania, Dominic van Essen, Oliver Siol, Marc Bailly-Bechet, Claude Philippe, Amal Zine El Aabidine, Léo Pioger, et al. (in press). "The Landscape of L1 Retrotransposons in the Human Genome Is Shaped by Pre-Insertion Sequence Biases and Post-Insertion Selection."

Sultana, Tania, Alessia Zamborlini, Gael Cristofari, and Pascale Lesage. 2017. "Integration Site Selection by Retroviruses and Transposable Elements in Eukaryotes." *Nature Reviews Genetics* 18 (5):292–308. https://doi.org/10.1038/nrg.2017.7.

Suter, Catherine M, David I Martin, and Robyn L Ward. 2004. "Hypomethylation of L1 Retrotransposons in Colorectal Cancer and Adjacent Normal Tissue." *Int J Colorectal Dis* 19 (2):95–101. https://doi.org/10.1007/s00384-003-0539-3.

Taylor, Martin S., John LaCava, Paolo Mita, Kelly R. Molloy, Cheng Ran Lisa Huang, Donghui Li, Emily M. Adney, et al. 2013. "Affinity Proteomics Reveals Human Host Factors Implicated in Discrete Stages of LINE-1 Retrotransposition." *Cell* 155 (5). Elsevier:1034–48. https://doi.org/10.1016/j.cell.2013.10.021.

Terasaki, Natsuko, John L Goodier, Ling E Cheung, Yue J Wang, Masaki Kajikawa, Haig H Kazazian, and Norihiro Okada. 2013. "In Vitro Screening for Compounds That Enhance Human L1 Mobilization." Edited by Mark A. Batzer. *PloS One* 8 (9):e74629. https://doi.org/10.1371/journal.pone.0074629.

Thompson, Blaine K, and Shawn M Christensen. 2011. "Independently Derived Targeting of 28S RDNA by A- and D-Clade R2 Retrotransposons." *Mobile Genetic Elements* 1 (1). United States:29–37. https://doi.org/10.4161/mge.1.1.16485.

Torrents, David, Mikita Suyama, Evgeny Zdobnov, and Peer Bork. 2003. "A Genome-Wide Survey of Human Pseudogenes." *Genome Res.* 13 (12). Cold Spring Harbor Lab:2559–67. https://doi.org/10.1101/gr.1455503.

Tubio, Jose M C, Yilong Li, Young Seok Ju, Inigo Martincorena, Susanna L Cooke, Marta Tojo, Gunes Gundem, et al. 2014. "Extensive Transduction of Nonrepetitive DNA Mediated by L1 Retrotransposition in Cancer Genomes." *Science* 345 (6196):1251343. https://doi.org/10.1126/science.1251343.

Upton, Kyle R., Daniel J. Gerhardt, J. Samuel Jesuadian, Sandra R. Richardson, Francisco J. S??nchez-Luque, Gabriela O. Bodea, Adam D. Ewing, et al. 2015. "Ubiquitous L1 Mosaicism in Hippocampal Neurons." *Cell* 161 (2):228–39. https://doi.org/10.1016/j.cell.2015.03.026.

Viollet, Sébastien, Aurélien J. Doucet, and Gaël Cristofari. 2016. "Biochemical Approaches to Study LINE-1 Reverse Transcriptase Activity In Vitro." In, 357–76. https://doi.org/10.1007/978-1-4939-3372-3_22.

Viollet, Sébastien, Clément Monot, and Gaël Cristofari. 2014. "L1 Retrotransposition: The Snap-Velcro Model and Its Consequences." *Mobile Genetic Elements* 4 (1):e28907. https://doi.org/10.4161/mge.28907.

Wagstaff, Bradley J, Dale J Hedges, Rebecca S Derbes, Rebeca Campos Sanchez, Francesca Chiaromonte, Kateryna D Makova, and Astrid M Roy-Engel. 2012. "Rescuing Alu: Recovery of New Inserts Shows LINE-1 Preserves Alu Activity through A-Tail Expansion." *PLoS Genet* 8 (8). United States:e1002842. https://doi.org/10.1371/journal.pgen.1002842.

Wang, Hui, Jinchuan Xing, Deepak Grover, Dale J Hedges, Kyudong Han, Jerilyn A Walker, and Mark A Batzer. 2005. "SVA Elements: A Hominid-Specific Retroposon Family." *Journal of Molecular Biology* 354 (4):994–1007.

Wei, W, N Gilbert, S L Ooi, J F Lawler, E M Ostertag, H H Kazazian, J D Boeke, and J V Moran. 2001. "Human L1 Retrotransposition: Cis Preference versus Trans Complementation." *Mol. Cell. Biol.* 21 (4). American Society for Microbiology: 1429–39. https://doi.org/10.1128/MCB.21.4.1429-1439.2001.

Weiner, Alan M. 2002. "SINEs and LINEs: The Art of Biting the Hand That Feeds You." *Curr. Opin. Cell Biol.* 14 (3):343–50. http://eutils.ncbi.nlm.nih.gov/entrez/eutils/elink.fcgi?dbfrom=pubmed&id=12067657&retmode=ref&cmd=prlinks.

Wheelan, Sarah J, Yasunori Aizawa, Jeffrey S Han, and Jef D Boeke. 2005. "Gene-Breaking: A New Paradigm for Human Retrotransposon-Mediated Gene Evolution." *Genome Res.* 15 (8):1073–78. https://doi.org/10.1101/gr.3688905.

Wolff, Erika M, Hyang-Min Byun, Han F Han, Shikhar Sharma, Peter W Nichols, Kimberly D Siegmund, Allen S Yang, Peter A Jones, and Gangning Liang. 2010. "Hypomethylation of a LINE-1 Promoter Activates an Alternate Transcript of the MET Oncogene in Bladders with Cancer." *PLoS Genet.* 6 (4). Public Library of Science:e1000917. https://doi.org/10.1371/journal.pgen.1000917.

Xie, W, X Gai, Y Zhu, D C Zappulla, R Sternglanz, and D F Voytas. 2001. "Targeting of the Yeast Ty5 Retrotransposon to Silent Chromatin Is Mediated by Interactions between Integrase and Sir4p." *Mol. Cell. Biol.* 21 (19):6606–14. https://doi.org/10.1128/MCB.21.19.6606-6614.2001.

Xing, J, H Wang, V P Belancio, R Cordaux, P L Deininger, and M A Batzer. 2006. "Emergence of Primate Genes by Retrotransposon-Mediated Sequence Transduction." *Proc. Natl. Acad. Sci. U. S. A.* 103 (47):17608–13.

Yang, J, H S Malik, and T H Eickbush. 1999. "Identification of the Endonuclease Domain Encoded by R2 and Other Site-Specific, Non-Long Terminal Repeat Retrotransposable Elements." *Proc. Natl. Acad. Sci. U.S.A.* 96 (14). National Acad Sciences:7847–52. https://doi.org/10.1073/pnas.96.14.7847.

Yang, Nuo, and Haig H Kazazian. 2006. "L1 Retrotransposition Is Suppressed by Endogenously Encoded Small Interfering RNAs in Human Cultured Cells." *Nat. Struct. Mol. Biol.* 13 (9):763–71.

Zhang, Zhaolei, Paul Harrison, and Mark Gerstein. 2002. "Identification and Analysis of over 2000 Ribosomal Protein Pseudogenes in the Human Genome." *Genome Res.* 12 (10):1466–82. https://doi.org/10.1101/gr.331902.

In: Trends in Biochemistry and Molecular Biology
Editors: Hossain Uddin Shekhar et al.
ISBN: 978-1-53616-434-3

Chapter 3

ADVANCEMENT OF BIOCHEMISTRY AND MOLECULAR BIOLOGY IN RELATION TO PUBLIC HEALTH

Mohammad D. H. Hawlader**, PhD
Department of Public Health, North South University (NSU), Dhaka, Bangladesh

ABSTRACT

Biochemistry and Molecular Biology has been developed as disciplines in this century. Both the Biochemistry and molecular biology together are making a significant contribution to the aspect of population health. In last decade, the health care and delivery system all over the world became systematic and technologically more advanced. The doorway of the new century is a promising time to review the public health advances over the past 100 years. In the twentieth century, there was a significant improvement in public health methods, practice, and health of the general population. For the assessment and promotion of health and ultimately for the reduction of the burden of morbidity and mortality, there is an emergence of invention of new tools. Some of the contributions of Biochemistry and Bimolecular science are prominently notable for their impact on reduction of mortality and prolonging longevity. Many new and ongoing evolutions which are ultimately giving the guidance to develop a comprehensive health care system in the frame of reference of public health. Both molecular biology and biochemistry have the potentiality to achieve extensive environmental, economic and health benefits. There is a contrast between personal health and public health. The personal health attended as separate from public health. Globally the upcoming challenge is to convince the policy makers that personal health care is not the only option to improve health care and delivery system. Furthermore, allocation of funds, either in research, program operation, monitoring and evaluation, are the prime important to be started urgently and immediately with the novel goal to improve health for the entire population of the world.

* Corresponding Author's E-mail: mohammad.hawlader@northsouth.edu.

So public health incorporation with laboratory techniques will be more effective approach to combate future health challenges.

Keywords: biochemistry, molecular biology, public health, advancement

1. INTRODUCTION

Biochemistry and Molecular Biology has been developed as disciplines in this century. Both the Biochemistry and molecular biology are making a significant contribution to the aspect of population health (Campbell P 1992). In last decade, the health care and delivery system all over the world became systematic and technologically more advanced. The doorway of new century is a promising time to review the public health advances over the past 100 years. In the twentieth century, there was a significant improvement in the method and practices of public health and also the health of the general population. There is also facilitation for effective understanding of the disease process to find out more scopes of prevention through advancement of different methodological approaches. There are stepwise advancement of epidemiological methods, lacking behind the study of the determinants of health. As a result, there is a simplistic formulation of correlating risk factors associated or contributing to different types of diseases, injuries and wellbeing of mankind at the community level, national level and globally (National Academies Press US 2002).

For the assessment and promotion of health and ultimately for the reduction of the burden of morbidity and mortality, there is an emergence of invention of new tools (Sanjiv Kumar 2012). In turns, it will ensure the effective interventions and estimations of the economic effects of different types of investments which will provide the greater efficiency and will be able to provide a comprehensive effective public health programs. To confront unprecedented challenges in public health, including rapidly growing global population, increased number of aged population, and some unprecedented changes in key environmental health indicators, there is an unavoidable need for fragmentation of infrastructure of the public health (Fielding JE 1999).

Some of the contributions from Biochemistry and Molecular Biology sciences are remarkably notable for their impact on reduction of mortality and prolonging longevity. Those are, invention of new vaccines, complete eradication of smallpox, remarkable reductions of communicable disease epidemics, and controlling over several chronic diseases. So many advancement enabled the understanding of basic disease mechanisms and provided tools to measure health in better ways and finding out appropriate intervention (Robert J. A 2010). Progressive advances of invention in public health arena are many, however few of those can be described in brief.

2. Advances in Methods

In last decade, positive progresses in advancing public health has been achieved. Through the remarkable advances of the last few decades in the field of biomedical sciences, it gradually entered from basic biological science through clinical research to comprehensive public health services. These developments have gone through the transformation of both large and small-scale biological and biomedical researches in a dramatic way (National Academies Press, US; 2003). Laboratory and clinical investigations are most essential components of health sector. There are several scientific breakthroughs which have important contributions in public health perspective in a way of how research is conducted. After gaining knowledge from population-based epidemiological studies and clinical trials, integration and correlation of knowledge regarding molecular research is important for the public health researchers. From where they are guided to learn what works and what does not work, what is safe and what is not safe. Now a days, there is an increasing demand for focusing on multidisciplinary research including genetic variation, cell dynamics, and metabolic, nutritional, environmental, and pharmaceutical variables. Also greater emphasis must be given on behavioral social sciences and health services research for more effective treatment of diseases and improvement of quality of life (Nass SJ 2009). New and ongoing evolutions which are ultimately giving the guidance to develop a comprehensive health care system in the frame of reference of public health are:

2.1. CRISPR Genome Editing Has Foreseeable Implications for Plastic and Reconstructive Surgery

The CRISPR (a family of DNA sequence) genome editing technique is affectively used in almost every branch of medicine now a day. It is the new era of "transformative leap" in genetic engineering and treatment (http://home.lww.com/news.entry.html/2018/10/30).

2.2. Scientists Investigate How Broad-Spectrum Antibiotic Therapy Affects Interaction of Gut Bacteria

Billions of beneficial/symbiotic bacteria habitats in human gut. They play active roles in digestion of foods, activating the immune system and helping in production of vitamins. But antibiotic therapies usually destroy these gut bacteria. Side by side, their presence facilitates the spread of pathogens which is called opportunistic infection. These intestinal micro biome, is prone to disruptions. When the gut bacteria is out of balance, then the risk of infection increases and the diabetes become more prevalent, as well as inflammatory and neurological diseases (https://www.news-medical.net/news/20181107).

2.3. Scientists Discover Cell Adhesion Mechanism Used by Mycoplasma Genitalium

Globally, Chlamydia trachomatis bacterial infection is one of the most common causes of sexually transmitted infections (STI). It infects the human cells; by the mechanism of hijacking some parts of the host cell to create defensive layers around the Chlamydia trachomatis (https://www.news-medical.net/news/20181029).

2.4. Genetic Defect Linked to ALS Causes Sugar-Starved Cells to Overproduce Lipids

A variety of diseases like neurodegenerative and mental illness can changes in the pathways of lipid metabolism in the sugar deprived cells. A new study led by a group of researchers from the Bloomberg School of Public Health, Johns Hopkins University shows the findings could lead to new targets to treat these diseases, currently those have no cure or effective treatments (https://www.jhsph.edu/news/news-releases/2018).

2.5. Cryo-Electron Microscopy Helps Study Proteins Responsible for Amyloid-Based Diseases

In the past century, the aim was to understand the contribution of protein with amyloid-based diseases such as, Huntingdon's, Alzheimer's and Parkinson's diseases. Lots of initiative and steps have been taken in the last one year. There is a great revolution by the scientists to use a powerful microscopy technique (https://www.news-medical.net/news/20181030).

2.6. Insights into Tumor Initiation Pave Way for New Skin Cancer Treatments

In case of melanoma; the salient vulnerability is discovered by the researchers. It was found that the melanoma-specific long non-coding RNA SAMMSON in tumor initiation was responsible for melanoma. According to the researchers, it stimulates the protein synthesis in different cellular compartments. This threat begins when there is modification of protein synthesis from normal cells (https://www.news-medical.net/news/20181029).

2.7. Insulin Discovery Could Improve Treatment for Diabetes

Now a days, a group of researchers discovered the therapeutic Insulin through an international collaboration at the Walter and Eliza Hall. This therapeutic insulins are more effective. It is better as resembling the way actually insulin works in human body (Charilaos S 2009).

2.8. Salk Researchers Advance Sonogenetics Technology for Treating Neurological Conditions

Recently the technology of sonogenetics has been emerged. In this technique, cells are controlled by sound. It will open a new era where incase of epilepsy, PTSD (post-traumatic stress disorder), Parkinson disease, the pharmaceutical drugs or invasive surgical treatments can be replaced by sonogenetic technology (Stuart I 2015).

2.9. Understanding of Metal-Free Enzymes Used by Bacteria Could Lead to New Effective Antibiotics

In case of some bacterial pathogens, it is observed that, streptococcus pneumonia has the ability to generate the components mandatory to replicate their DNA without the presence of any required metal ions (Jose M 2016).

2.10. Stem Cell Treatments for Chronic Disease and Anti-Aging

"Alarm signals": The term indicates the process of self-repair. The living organisms have stem cells both centrally and peripherally. These stem cells can be mesmerized to sites of injured tissues. In this manner, these cells ultimately go through the proliferation, migration, and accumulation process at injured or damaged sites gradually. With the sustention of this condition of "alarm," the stem cells become exhausted permanently. That ultimately leads to irreversible disease transformed from their original locations.

Basically, it depends upon the quantity of stem cell and during active regeneration, on its effective availability until a certain time period. The function of these adult stem cells therapy is the re-population of the stem cell pool. Ultimately, this improves the immune system. The presumed consequences of this fact could be the insufficiency of an adequate amount of stem cells for the tissue replacement and regeneration. Which in turn responsible for development of diseases and ageing. An irreversible and premature stem cell exhaustion could be promoted by this process. So, in turn, make the self-repair and survival process of organism impossible. As we grow older, our circulating stem cells become diminished. So the existing cells to our body's circulation can improve health and repopulate our stem cell pool by introducing new stem (Anastasia Y. E 2015).

2.11. Researchers Closer to Gonorrhea Vaccine after Exhaustive Analysis of Proteins

In a study of proteins, November 8, 2018 Oregon State University researchers have reached closer to an invasion of vaccine for gonorrhea. Also proceed towards interpreting why the bacteria that cause the disease are so good at shielding off antimicrobial drugs (https://www.sciencedaily.com/releases/2018/11/181108105945).

2.12. The Food Biochemistry

Focuses on different ideas and schemes to intensify the knowledges and understanding of composition of foods, and particularly give emphasis on food components that have beneficial effects on health. The Food Biochemistry encompasses the implementation of different method including both chemical and biochemical analytical methods of food components (20). It also focuses on their reactions, model systems, and various statistical tools for data analysis efficiently to find out the maximum informative value. And finally apply those knowledges on public health sector.

Recent scientific era has witnessed that there is progressive development in the fields of biotechnology and biochemistry which leads to a tremendous success in the field of medical and public health science. The achievement and advancement created in the domain of biotechnology has resulted in evolution of newer classes of potential drugs. Lots of newly designed subunit vaccines are developed from the contribution of expression of various prophylactically important proteins, polypeptides and cloning. New techniques are constantly being introduced leading to production of phenomenal and curative molecules which help in diagnosis of different diseases, invention of preventive measures and eventually unveiling the ways to cure the diseases (https:// www.

encyclopedia. com/ food/ encyclopedias- almanacs- transcripts- and- maps/ nutritional-biochemistry).

2.13. Vaccination

Recently DNA vaccines have been introduced in the arena of vaccinology, as a new prophylactic strategy enriching the public health sector. The genetic material expresses an antigen which is originated from pathogen. It plays a role to activate host immune system. Plasmid administration, is the main involving strategy (Kishwar H 2013). This plasmid usually conform of a strong viral promoter which drive the transcription and translation of the gene in vivo. The superior one is multicistronic vectors that have the potentiality of expression of more than one immunogen. Sometimes it expresses an immunogen and also an immunostimulatory protein. Definitely this multicistronic vectors are better than those that can express only one protein.

2.14. Treatment of Genetic Disorder

During the past decade, there has been a breakthrough in the field of genome editing and genome regulation. It made a remarkable easier way of involving various technologies. Among them, the CRISPR 9 bacterial system, one of the recently adopted technologies, which has evolved as a simple, RNA-guided phenomenon (Gupta R. M 2014). This is used for an efficient and specific genome editing and regulation in case of both bacteria as well as higher eukaryotes. The technology is emanated as a revolutionary tool in biomedical research and field. It has opened a new glimpse for treatment of genetic disorders. The process involves getting the guidance from RNA. The target sequence is opened by the breaking down of the host genome endonuclease. The defective repair mechanism of the generated double strand breakdown process emanates in insertion or deletion mutations. CRISPR mediated engineering usually expedite the initiation of specific point mutations or insertions in the target DNA. By using Cas9 Multiple genomic loci can be efficiently altered at a time through the introduction of several sgRNAs concurrently. For large-scale chromosomal rearrangements, this approach can be used. This technology is also used to achieve the disease models for genetic disorders. Besides other diseases such as carcinoma, which help us for easy understanding of pathological processes in the aspect of molecular basis (Nejat M 2013).

2.15. Treatment of Cancer

Recently, there is a new era of treatment of various types of cancer by the immune-based therapeutics, which is becoming the leading asset in the treatment of cancer. Antibody mediated checkpoint blockades in cell cycle can be used effectively for the management of carcinoma now a days (https://www.cancer.gov/about-cancer/treatment/types/immunotherapy). Expression of the programmed death-1 (PD-1) receptor is stimulated by cancer cells, used as an approach to downplay adaptive immune response of the host. Antibody mediated blocking of PD-1 can aid in revitalizing the adaptive immune response against cancer (Han Y 2014).

2.16. Antibiotic Resistant

There is imminent upswing of pathogens which are antibiotic-resistant. Side by side, development of new antibiotics has become sluggish even though there is elevated demand for substitute therapeutics (Richard J. F 2014). Now a day's probiotics are providing a promising approach. It provides their efficacy is enhanced by introducing new genetic circuits to deliver drug biomolecules to the specific sites. The term 'designer probiotics' are basically recombinant probiotics, which has the capability to reduce the gap between increasing antibiotic resistance and the inadequacy of new antibiotics. Probiotics are emanating living therapeutic agent. The prototype of probiotics can be used for disease diagnosis and prevention. The best documented effects including intestinal disorders, infectious diarrhoea and diarrhoea associated with antibiotic use, allergic conditions (Maria K 2013).

2.17. Zika Virus

Zika virus (ZIKV) belongs to Flaviviridae family. Nowadays, the emerging pattern of Zika virus is considered as a major public health concern. Globally, the researchers gave emphasis on ZIKV pathogenesis. To inhibit Zika virus, DNA therapy may result in an emerging therapeutic or prophylactic intervention. This knowledge will help to identify fixed targets for drug design. In different animal models, lots of studies have highlighted the role of DNA vaccine, inactivated vaccine, and viral-vector vaccine (John J. S 2017). These are capable of producing adequate antibody responses for the protection against ZIKV challenge.

2.18. Therapeutic Recombinant Proteins

Very recently, the most example of therapeutic usage of recombinant technology is recombinant proteins. Normally, if there are excess accumulation of protein, it results in their misfielding that ultimately causes the development of inclusion bodies. There are several ways by which one can regulate higher recovery of bioactive protein from inclusion bodies.

2.19. Use of Nanotechnology

There is a very important role of nanotechnology in different sectors of medical science. Now a days, some techniques are actually being used in some cases, while some are at different stages of experimentation and some of them are still hypothetical only. There is a revolution of medical science based on the use of nanotechnology. Nanotechnology is a very prospective way for diagnosis and treatment of different diseases of human body. Interesting thing is that, various techniques are assumed in few years back but very recently it is making striking progress towards becoming existence.

2.20. Nanocrystalline Silver

A Nano sized metal based formulation is Nano crystalline silver. This product is already marketed in United States and many other region. It exploits powerful antibacterial effect. The ionic silver has a potent antibacterial activity, especially against Methicillin Resistant Staphylococcus Aureus (MRSA) that may contaminate wound dressing. There is a rapid and early growth of Nano medicine marketing (Valentina M 2012). Whereas some are at the stage of clinical trial, while Nano-enhanced drug delivery products are commercially produced in the meantime, moreover there is a progressive development of advanced nanotech-based medical devices in current year. At present, Nano medicine has a large therapeutic areas in the field of treatment of carcinoma. In this aspect, Nano medicine has made lots of contributions. Worldwide Abraxane, Depocyt, Oncospar, Doxil, and Neulasta are used for the purpose of treatment of carcinoma. Nano-enabled medical products are potentially blotting out other forms of medicines and drug formulations. Starting from control of infection to treatment of CNS diseases, cardiovascular disease, even carcinoma as well, Nano-medical products are widely used. According to public health researchers, social and political scientists, hindrance of new molecular biological approaches can be refined by assessing the community health. This is a dreadful but fundamental challenges to analyze a complete dimension of determinants of health of population.

Both Biomolecular and biochemistry have the potentiality to achieve extensive environmental, economic and health benefits. There is a contrast between personal health and public health. The personal health attended as separate from public health. Besides range, drug treatment or establishing a national home visiting program for new families, it is also essential to expand opportunities for awareness, physical activity, reducing illiteracy, school dropout rates and female empowerment. Globally the upcoming challenge is to convincing the policy makers that personal health care is not the only determinant of health. Further allocation in research, program, or evaluation are the prime important to be started urgently and immediately with the novel goal to improve health for the entire population all over the world.

References

Anastasia Y. E, Tatiana N. K, Zhanna A. A and Yelena V. P. Autologous Stem Cell Therapy: How Aging and Chronic Diseases Affect Stem and Progenitor Cells. *Biores Open Access*. 2015; 4(1): 26–38.

Campbell P. N. Biochemistry and Molecular Biology. *Biochemical Education*. 1992 July; 20(3): 158-65.

Charilaos S and Christopher K. The introduction of successful treatment of diabetes mellitus with insulin. *J R Soc Med*. 2009 Jul 1; 102(7): 298–3.

Fielding JE. Public health in the twentieth century: advances and challenges. *Annu Rev Public Health*. 1999; 20: xiii-xxx.

Gupta R. M, Musunuru K. Expanding the genetic editing tool kit: ZFNs, TALENs, and CRISPR-Cas9. *J Clin Invest*. 2014 Oct; 124(10): 4154-61.

Han Y, Huanbin W, Chushu L, Jing-Y. F, and Jie X. Cancer Cell-Intrinsic PD-1 and Implications in Combinatorial Immunotherapy. *Front Immunol*. 2018; 9: 1774.

https://www.cancer.gov/about-cancer/treatment/types/immunotherapy.

https:// www.encyclopedia.com/ food/ encyclopedias-almanacs-transcripts-and-maps/ nutritional-biochemistry.

https:// www.jhsph.edu/news/news-releases/2018/mutation-associated-with-ALS-causes-sugar-starved-cells-to-overproduce-lipids-study-shows.html.

http://home.lww.com/news.entry.html/2018/10/30/crispr_gene_editing-LZi0.html.

https:// www.news-medical.net/ news/ 20181107/ Scientists-investigate-how-broad-spectrum-antibiotic-therapy-affects-interaction-of-gut-bacteria.aspx.

https:// www.news-medical.net/ news/ 20181029/ Scientists-discover-cell-adhesion-mechanism-used-by-mycoplasma-genitalium.aspx.

https:// www.news-medical.net/ news/ 20181030/ Cryo-electron-microscopy-helps-study-proteins-responsible-for-amyloid-based-diseases.aspx.

https:// www.news-medical.net/news/20181029/Insights-into-tumor-initiation-pave-way-for-new-skin-cancer-treatments.aspx.

https://www.sciencedaily.com/releases/2018/11/181108105945.htm.

John J. S, James A. W, Connie S. S. Advancements in DNA vaccine vectors, non-mechanical delivery methods, and molecular adjuvants to increase immunogenicity. *Hum Vaccin Immunother*. 2017 Dec; 13(12): 2837–48.

Jose M. M and Cesar A. A. Mechanisms of Antibiotic Resistance. *Microbiol Spectr*. 2016 Apr; 4(2): 10.1128/microbiolspec.VMBF-0016-2015.

Kishwar H. Khan. DNA vaccines: roles against diseases. *Germs*. 2013 Mar; 3(1): 26–35.

Maria K, Dimitrios B, Stavroula K, Dimitra D, Konstantina G, Nikoletta S et al. Health Benefits of Probiotics: A Review. *ISRN Nutr*. 2013; 2013: 481651.

Nass SJ, Levit LA, Gostin LO. *The Value, Importance, and Oversight of Health Research.* Beyond the HIPAA Privacy Rule: Enhancing Privacy, Improving Health Through Research. Institute of Medicine (US) Committee on Health Research and the Privacy of Health Information: The HIPAA Privacy Rule. Washington (DC): National Academies Press (US); 2009.

Nejat M and Bahareh R. An Overview of Mutation Detection Methods in Genetic Disorders. *Iran J Pediatr*. 2013 Aug; 23(4): 375–88.

New Opportunities, New Challenges: The Changing Nature of Biomedical Science. National Research Council (US) and Institute of Medicine (US) Committee on the Organizational Structure of the National Institutes of Health. Washington (DC): National Academies Press (US); 2003.

Richard J. F and Yitzhak T. Antibiotics and Bacterial Resistance in the 21st Century. *Perspect Medicin Chem*. 2014; 6: 25–64.

Robert J. A. Improving health outcomes with better patient understanding and education. *Risk Manag Healthc Policy*. 2010; 3: 61–72.

Sanjiv Kumar and GS Preetha. Health Promotion: An Effective Tool for Global Health. *Indian J Community Med*. 2012 Jan-Mar; 37(1): 5–12.

Stuart I, Ada T, Carolyn S, Sadik E, Sreekanth H. C. Sonogenetics is a non-invasive approach to activating neurons in *Caenorhabditis elegans*. *Nat Com* 2015; 6: 8265.

The Future of the Public's Health in the 21st Century. Institute of Medicine (US) Committee on Assuring the Health of the Public in the 21st Century. Washington (DC): National Academies Press (US); 2002.

Valentina M, Alessandro T, Carlo B. P, Jason H. S, Marco A and Ennio T. Nanotechnology in Medicine: From Inception to Market Domination. *J Drug Deliv*. 2012; 2012: 389485.

In: Trends in Biochemistry and Molecular Biology
Editors: Hossain Uddin Shekhar et al.
ISBN: 978-1-53616-434-3

Chapter 4

NUTRACEUTICALS AND FUNCTIONAL FOODS: THE FUTURE DIETARY APPROACH

***Sharmin Jahan**, PhD**
Institute of Food Science and Technology,
Bangladesh Council of Scientific and Industrial Research, Dhaka, Bangladesh

ABSTRACT

Currently, food is not just considered as a means for necessary nutrients to secure proper growth and development, but as a way to the finest wellness. According to consumers opinion it is acceptable that certain foods have health benefits that may reduce the risk of chronic disease or other health issues. There are lots of studies that support the role of diet in prevention of disease and promotion of health that widen the market for functional foods, nutraceuticals, and dietary supplement. People fascinated with self-medication and self-care are increasingly looking for diet based solutions to maintain their health and well-being. All of the phenomenon extrapolate that nutraceuticals and functional foods are one of the most promising, interesting and innovative areas in the food industry. In addition with the progressions in the qualitative and quantitative determining parameters, the demand of these health products has been found to be augmented. Attributable to this, the nutraceutical market has become a million dollar industry at a global level.

Nutraceuticals, dietary supplements and functional foods, or other terms are interchangeable, though these have significant differences that are sometime not always evident. The current chapter aims to get insight the concept of these interchangeable terms. Therefore, a brief introduction, differentiation and the prospect of these health foods are illustrated on the basis of their health benefit and disease management along with their future research direction.

* Corresponding Author's E-mail: shammy79@gmail.com.

Keywords: nutraceuticals, functional foods, dietary supplement, active compound, medicinal food, disease management, health benefit

1. Introduction

Eating behavior and trends in food production and consumption have enormous impact on health, environment and society. The human health and wellness is largely determined by the consumption of nutritious foods. The production of foods with health benefits offers an excellent opportunity to improve the public health; and therefore the phenomenon have received huge attention in recent years by the scientific community, consumers, and food manufacturers as well.

"Food may hold the aptitude to prevent or treatment of diseases" this belief is an old-proclaimed by our ancestors. The father of modern medicine Hippocrates (460–377 BC), proposed the correlation between intake of proper diet for health and their therapeutic conveniences about 3000 years ago. The principal truth depicted in his statement that "Let food be thy medicine and medicine be thy food" is widely applied nowadays (Ahmad F, 2013; Bagchi D, 2006). As a consequence, from the observational findings, it was believed that our nature has so many natural therapies to offer. Interestingly, from the ancient times for the treatment of cancer such natural botanical components were applied as therapeutics on the basis of surveillance. Correspondingly, there are a lot of chemotherapies derived from plant origin such as *Vinca* and *Taxus brevifolia* species to treat cancer and related consequences. Additionally, Ginseng has been another such traditional drug in China which is in use from the time of Liang dynasty and applied even in today's time as a chemotherapeutic agent.

Over the past three decades, an increasing number of dietary supplements have become available in superstores and health food shops, which are also obtainable for purchase in pharmacies. In general the term "nutraceutical" is used to explain these medicinally or nutritionally functional foods. The word "nutraceutical" also recognized as functional foods and nutritional supplements, as well as medical foods, designer foods, and phytochemicals, which comprise everyday food products ranging from "bio" yoghurts to fortified breakfast cereals, in addition to vitamins, herbal remedies and even genetically modified foods and supplements. Many different terms and definitions are applied for nutraceuticals in different countries, which is resulted in confusion. However, the term "nutraceuticals" has huge prospect for growth and development on the basis of health benefits (Ahmad F, 2013; Saika D, 2011). Depending on the nature and the alternative use as the modern medicines, it has shown subsequent results in reducing the need for conventional medicines and has reduced the chances of adverse effects (Chintale AG, 2013).

Nutraceuticals and functional foods have appeared as beneficial health products reached from many food, herbal, and pharmaceutical manufacturing industries. The impact of such medicinal products have been coupled with the treatment of many disorders such as cancer, coronary heart disease, metabolic problems, delayed gastrointestinal emptying, cold and cough, depression, and many more health conditions (Saika D, 2011; Kaur G, 2015).

In these days people are attracted to improve their health by supplementation and by consuming formulated or fortified foods. In addition, the reasons for the growing trend of nutraceuticals are public education, renewable source, cultivation and processing, environmental friendliness and local availability (Keservani RK, 2010). The nutraceutical as well as functional food market have been developed from last few decades due to rising interest of researchers and sophisticated techniques for determination of qualitative and quantitative parameters. It has changed to a million dollar industry at a global level. Global marketing of nutraceuticals was USD 128.4 billion in 2008. Japan holds the largest figures of 70% of marketing share in Asia (Doke RM, 2013; Sapkale AP, 2012). United Kingdom, Germany and France were the first country who considered the diet as a more important factor than exercise or hereditary factors to achieve a sound health. Furthermore, Canada defined nutraceuticals as a product formulated from food stuff however sold in pills, powders, (potions) and other medicinal forms not generally associated with food product.

In this chapter a basic introduction of functional foods, dietary supplements and nutraceuticals are presented followed by their brief narration. Health benefits and applications of nutraceuticals and functional food in disease management are also illustrated in short.

2. Terminology and Definitions

"Functional foods," "dietary supplements" and "nutraceuticals," as well as other terms are found to be overlapping, and differences between them are not always apparent (Hardy G, 2000). The scientists of American Dietetic Association (ADA) consider that all foods serve specific function to sustain life. However, so-called functional foods that include whole foods and fortified, enriched, or enhanced foods have a potentially beneficial effect on health when consumed as part of a varied diet on a regular basis, at effective levels (Position of The American Dietetic Association, 2009*)*. The Dietary Supplement Health and Education Act of 1994 specifically define a dietary supplement as a special category of food, but exclude any product that is "represented for use as a conventional food or as a sole item of a meal or the diet" (Dietary Supplement Health and Education Act of 1994,). While dietary supplements contain nutrients derived from food products and are commonly concentrated in capsule, powder, liquid or pill form (Zeisel

SH. 1999). Furthermore, "Nutraceutical," a portmanteau of the words "nutrient" and "pharmaceutical," was coined by Stephen De Felice in 1989, founder and chairman of the Foundation for Innovation in Medicine, an American organization which encourages medical health research. He defined a nutraceutical as a "food, or parts of a food, that provide medical or health benefits, including the prevention and treatment of disease" (Jack DB, 1995; Brower B, 1998; Mannion M. 1998). In the following section a brief description of these nutritional terminologies are given.

3. Functional Foods

Functional foods are a stimulating current trend in the food and nutrition field (Sloan AE, 2008). However, the term functional food has no legal meaning in the United States. It is currently a marketing, rather than a regulatory, idiom. The rapidly evolving trend of functional foods creates significant new issues and opportunities for public health, especially in providing reliable information.

Table 1. Functional food categories with examples

Functional food category	Selected functional food examples
Conventional foods (whole foods)	Garlic Nuts Tomatoes
Modified foods Fortified	Calcium-fortified orange juice Iodized salt
Enriched	Folate-enriched breads Enhanced Energy bars, snacks, yogurts, teas, bottled water, and other functional foods formulated with bioactive components such as lutein, fish oils, ginkgo biloba, St John's wort, saw palmetto, and/or assorted amino acids
Medical foods	Phenylketonuria (PKU) formulas free of phenylalanine
Foods for special dietary use	Infant foods Hypoallergenic foods such as gluten-free foods, lactose-free foods Weight-loss foods

So what exactly is a functional food? There is no FDA (Food and Drug Administration) or universally accepted definition of this evolving food category (Ross S, 2007). The most accepted definition is given by American Dietetic Association's (ADA), the largest organization of food and nutrition professionals in the United States, classifies all foods as functional at some physiological level because they provide nutrients or other substances that furnish energy, sustain growth, or maintain/repair vital processes (Position of The American Dietetic Association, 2004). However, functional foods move beyond necessity to provide additional health benefits that may reduce disease risk and/or promote optimal health. Functional foods include conventional foods, modified foods (i.e., fortified, enriched, or enhanced), medical foods, and foods for special dietary use. Table 1 Shows the Functional food categories according to ADA.

3.1. Categories of Functional Food

At present due to consumer's interest, the food industries recommend an extraordinary variety of new functional food products (Hsieh YH, 2007). Beside ADA classification there are some other popular definitions of functional food also available. The term functional food is not confined between good foods and bad foods, but rather that the foods are integrated as a healthful or diversified diet. Because of varied diet ADA now further categories the functional food:

3.1.1. Conventional Foods

Conventional foods, which are unmodified whole foods like fruits and vegetables symbolize the simplest example of functional foods. Different kinds of berries, tomatoes, cabbage, or broccoli are considered as functional foods as these are rich in some bioactive components such as lycopene, sulforaphane, lutein, and ellagic acid. Fascinatingly, fruits and vegetables signify as the top functional foods acknowledged by consumer according to the IFIC 2007 "Consumer Attitudes toward Functional Foods/Foods for Health" survey (Consumer attitudes toward functional foods/foods for health - Executive summary, 2009). There is promising evidence that shows the relation of conventional foods to health benefits ranging from diarrhea to cancer. Experimental and epidemiological studies demonstrate that cruciferous vegetables can reduce the risk of several types of cancer (Moriarty RM, 2007). Lycopene-rich vegetables like tomato and its products may reduce the risk of certain cancer such as ovarian, gastric, prostate, and pancreatic cancers (Kavanaugh CJ, 2007). Additionally, citrus fruit may decrease the risk of stomach cancer (Bae J-M, 2008). In case of cardiac health, it is evident that dark chocolate improves endothelial function (Faridi Z, 2008), and hard-shelled nut along with peanuts decline the risk of sudden cardiac arrest (Kris-Etherton PM, 2008). Probiotics a fermented dairy products and cranberry juice improves the irritable bowel syndrome and bacteriuria in maintenance of intestinal health and urinary tract function correspondingly (Quigley EM, 2007 and Nowack R, 2007).

3.1.2. Modified/Fortified/Enriched Foods

Foods that are modified through fortification, enrichment, or enhancement have been also included as functional food. Some common examples of such class functional food include fruit juice fortified with calcium for bone health, breads enriched with folate for proper fetal development. Furthermore, food quality improved by adding bioactive components, like plant stanol or sterol esters added margarines to lower the cholesterol level, or beverages enhanced with energy-promoting factors such as ginseng, guarana, or taurine and marketed to consumers. However, some commercially accessible functional botanical products have been removed from the market due to its fraudulent medical claims and safety concerns such as kava kava, and Echinacea (Jacobson MF, 2000 and Center for Science in the Public Interest, 2009). Captivatingly, genetically engineered food or food modify by biotechnology to increase the nutritional value or health benefit, open a new window for market place in the field of functional foods, e.g., no trans- fat oils, n-3 fatty acid or PUFA enhanced oil (Pew Initiative on Food and Biotechnology, 2009).

3.1.3. Medical Foods

FAO first define the term "medical food" by Orphan Drug Act as "a food which is formulated to be consumed or administered entirely under the supervision of a physician and which is intended for the specific dietary management of a disease or condition for which distinctive nutritional requirements, based on recognized scientific principles, are established by medical evaluation" (Food and Drug Administration. Orphan Drug Act, 2009). Oral supplement such as phenylketonuria formulas free of phenylalanine is an example of medical food. The regulatory status of a medical food is determined by claiming with clinical approval. Likewise, a canned or bottled oral supplement formulation for diabetic, renal, or liver is determined under medical supervision consider as a medical food which is also known as food for special dietary use.

3.1.4. Foods for Special Dietary Use

The Federal Food, Drug, and Cosmetic Act (Section 411(c)(3)) defines "Food for special dietary use" as "a particular use for which a food purports or is represented to be used, including but not limited to the following: 1. Supplying a special dietary need that exists by reason of a physical, physiological, pathological, or other condition; 2. Supplying a vitamin, mineral, or other ingredient for use by humans to supplement the diet by increasing the total dietary intake; 3. Supplying a special dietary need by reason of being a food for use as the sole item of the diet;" (Food and Drug Administration. Orphan Drug Act, 2009). Formulations for infant's hypoallergenic foods like as gluten-free and lactose-free or weight reducing foods are the examples of foods for special dietary use.

4. NUTRACEUTICALS

Fundamentally the meaning of nutraceutical stands for the narrated Hippocratic principle "let food be thy medicine and medicine be thy food," the phrase is believed by both mass people and scientist community (Jones WHS, 1945). In early stage nutraceuticals were considered as the products produced from foods but sold in medicinal forms e.g., capsule, tablet, powder, solution, or suspension, which was not generally allied to the food but had demonstrated physiological benefits and provided protection against particular disease specially the chronic diseases; were also referred to as "natural health products" in Canada (Shahidi, 2004).

Afterward the term "nutraceuticals" has been defined by The European Nutraceutical Association (ENA) as substances that markedly contrast pharmaceuticals, which are "synthetic substances or chemical compounds formulated for specific indications." In 2010 Pandey et.al simply has described nutraceuticals as, "nutritional products that provide health and medical benefits, including the prevention and treatment of disease" (Pandey M, 2010).

Over and above, "Functional food" is a product containing the essential nutrients such as vitamins for survival, whereas according to previous definition "nutraceuticals" are not only complementary to the diet, but also support the prevention or healing of disease or certain physical disorders (Kalra EK, 2003).

More recently, the term "nutraceutical" includes a wide range of products, ranging from purified nutrients (Jacobs DR, 2013) and dietary supplements (Venhuis BJ, 2016), to herbal and botanical products (Nworu C, 2014), or particular processed foods and beverages (Udenigwe CC, 2012) as well.

4.1. Classification of Nutraceuticals

There are number of classifications and categorizations of nutraceuticals, in terms of the food material and nutrient content, effect on the body, and according to chemical constituents or active ingredients. Nutraceuticals can also be broadly categorized into two types on the basis of foods available in the market, and they are "Traditional" and "Non-traditional" nutraceuticals.

4.1.1. Traditional Nutraceuticals

These are basically the food in its natural form with no changes. All Food contains several natural components with its own benefits beyond basic nutrition, for example lycopene in tomatoes, omega-3 fatty acids in walnuts, and salmon, or Mackerel, saponins in beans, peanuts and soy. Benefits of traditional nutraceutical in certain chronic disease control are illustrated in (Table 2). Traditional nutraceuticals can be further sub-grouped on the basis of

a. Chemical elements
   i) Nutrients
   ii) Herbals
   iii) Phytochemicals
b. Pre and Probiotic Microorganisms
c. Nutraceutical Enzymes

4.1.1.1. Chemical Elements

4.1.1.1.1. Nutrients

Vitamins, minerals, amino acids, protein and fatty acids have defined nutritional functions are considered in such class. Fruits, vegetables, wholegrain cereals, animal products such dairy products, fish, meat, poultry, contain a number of physiologically active components and are helpful in curing heart diseases, cancer, stroke, diabetes, osteoporosis, cataracts, intestinal disorder and so on (Block G 1992). Likewise, vitamins and minerals originated in plant, animal or dairy products play vital role in different disease such as anemia, osteoporosis, and are also useful in building strong bones, teeth, muscles, improving nervous system and heart rate (Balsano C, 2009; Cencic A, 2010a). Fatty acids omega-3 PUFAs containing legumes, nuts, flax seed and salmon show potent anti-inflammatory effects, preserve brain function, reduce cholesterol deposition as well (Simopoulos AP, 1991).

4.1.1.1.2. Herbals

Epidemiological evidence data from in vivo, in vitro, and clinical trials indicates that a plant-based diet and herbal product can improve or reduce the risk of chronic diseases. Examples include willow bark (*Salix nigra)*, having active component named salicin act as anti-inflammatory, analgesic, antipyretic, astringent and antiarthritic (Ehrlich SD, 2008). Parsley *(Petroselinum cripsum*) contains flavonoids like apiol and psoralen is diuretic, carminative and anti-pyretic (Farzaei MH, 2013). In addition, peppermint (*Mentha piperita)* contains menthol as an active component and cures cold and flu (Ehrlich SD, 2009).Tannin containing Lavender (*Lavandula angustifolia)* is beneficial in improving hypertension, stress, depression, cold, cough and asthma (Ali B, 2015) and while proanthocyanadin containing cranberries *(Vaccinium erythrocarpum)* are adjuvant in cancer, ulcers and urinary tract infections (Sharma D, 2012).

4.1.1.1.3. Phytochemicals

Health professionals begin to recognize the role of phytochemicals in health enhancement (Howard BV, 1997). It is apparent that the components of plant-based diet, rather than traditional nutrients have the potentiality to reduce the risk of cancer. Steinmetz and Potter have identified around fifteen types of biologically active plant chemicals that are now known as "phytochemicals" (Steinmetz KA, 1991). The main class of phytochemicals includes carotenoids and phenolics in fruits and vegetables, followed by β-glucan, lignans, and inulin in cereal.

Table 2. Application of traditional nutraceutical in chronic disease control

Nutraceutical	Dose/Duration	Effect	References
Allenic carotenoid fucoxanthin (brown seaweeds)	2.4 mg /day	Improves insulin resistance and decreases blood glucose levels through the regulation of cytokine secretions from WAT (white adipose tissues)	Miyashita K, 2011
n-3 PUFAs (polyunsaturated fatty acids)	No data	Prevents several disorders affecting lungs and airways	Fasano E, 2010
ASU (unsaponifiable residues of avocado and soybean oils)	300mg / 3 years	Stimulates synthesis of aggrecan and extracellular matrix component as type II collagen and by reducing the production of catabolic (MMP-3) and pro-inflammatory (IL-8 and IL-6) mediators in OA (osteoarthritis)	Henrotin Y, 2011
CLA (Conjugated linoleic acids)	3 months	Significantly improves AHR (Airway hyper responsiveness) associated with a reduction in leptin/ adiponectin ratio in mild asthma	MacRedmond R, 2010
Siphonaxanthin, a marine carotenoid (green algae)	20μM / within 6h of treatment in HL-60 cells	Induces apoptosis in HL-60 cells by decreasing Bcl-2, and increases activation of caspase-3	Ganesan P, 2010
FPP(Fermented papaya preparation)	6g/day / 6 months	Unregulated TNF-α and thioredoxin (Trx) in liver cirrhosis	Marotta F, 2011
MUFAs (monounsaturated fats)	No data	Lowers CVD (cardiovascular disease) risk and MS (metabolic syndrome)	Kastorini CM, 2011
1,25(OH)2D, or calcitriol	200–600IU/day	Regulates the levels of p21 and p27 and increases expression of BRCA-1 and -2 tumor suppressor genes contributing in the DNA repair mechanism	Bulathsinghala P, 2010
Resveratrol	No data	Chemo sensitizes tumor by modulating drug transporters, cell survival proteins, cell proliferative proteins, and members of the NF-κB and STAT3 signaling pathways	Gupta SC, 2011
Fortified wheat flour	100 to 150 μg/day	Reduces prevalence of NTDs (neural tube defect) at birth and increasing blood folate concentrations	Berry RJ, 2011

Phytochemicals are considered as the most important class of nutraceuticals. These are further classified on the basis of chemical constituents and active ingredients that characterize the phytochemical properties. For example, various fruits, vegetables and egg yolk are the source of carotenoids, boost natural killer immune cells, are anti-carcinogenic, and protect eyes from UV light. Legumes like soybeans and chickpeas, various grains, palm oil contain non-carotenoids, which are anti-carcinogenic and also lower cholesterol level. Flavonoid and polyphenolics compounds are found in different types of berries, vegetables, and legumes are potent antioxidants which are source of phytoestrogens that prevent both prostate and breast cancer, and also control diabetes. Non-flavonoid polyphenolic compounds are found in raisins, berries, black grapes, peanuts, turmeric roots, possesses strong anti-oxidants, anti-inflammatory, and effective anti-clotting effect and play role in reducing cholesterol as well. Phenolic acids, present in tomatoes, blueberries, and bell peppers have the evidence for antioxidant activity and reduce mutagenicity of polycyclic aromatic hydrocarbons. Seeds of *Barbarea verna*, and broccoli comprise isothiocyanates (glucosinolates) and reveal antitumorigenisis activity. Preventing cancer and cardiovascular diseases, reducing tumor frequency, lowering cholesterol, blood pressure, risk of heart disease, and prolonging gastric emptying are some of the significant defensive effects of plant-based foods.

Table 3. Major phytochemicals in foods and their bioactivities

Phytochemicals	Food sources	Bioactivities
Flavonoids Flavones	Celery, parsley	Antioxidant, antiproliferative, anti-hypertensive, anti-carcinogenic, Anti-thrombotic, cell cycle arrest, induction of phase-2 enzymes
Flavanones	Citrus fruits	Inhibition of phase-1 enzymes, inhibition of LDL-oxidation, improvement of vascular tone
Flavonols	Onions, tea, green beans, tomatoes	
Flavan-3-ols	Tea, cocoa, apples, berries, certain beans	
Anthocyanidins	Blueberries, blackcurrants, strawberries	
Isoflavones	Soybeans	
Phenolic acids	Coffee, cereal bran, fruits	Anti-inflammatory
Lignans Linseed,	fruits and vegetables	Estrogenic
Stilbenes	Grapes, peanuts	Antioxidant, cardioprotective, lifespan extension
Phytosterols	Wheat	Cholesterol lowering
Carotenoids	Tomatoes, carrots, bell peppers	Antioxidant, anti-inflammatory, anti-carcinogenic

Adapted from: Gry J, 2007.

4.1.1.2. Pre and Probiotic Microorganisms

Another type of compounds termed as pre and probiotics commonly referred to as "nutraceuticals" which also subsistence a nutritional value. (Burgain J, 2011; Douglas LC, 2008). The term 'Probiotics' mean 'for life' and can be defined as live microorganisms, when consumed in adequate amounts, confer certain health effect of the host (Michail S, 2006). The scientific interest in probiotics was encouraged by Hord, when he observed the action of Metchnikoff to transform the toxic flora of the large intestine into a host-friendly colony of *Bacillus bulgaricus* (Hord NG, 2008). Both beneficially affect the host by "selectively stimulating the growth and activity of one or a limited number of bacteria" in the digestive tract, thus improving the host's health (Gibson GR, 1995). They are known as friendly bacteria that promote healthy digestion and absorption of some nutrients. They act to force out pathogens, such as yeasts, other bacteria and viruses that may otherwise cause disease and develop a mutually advantageous symbiosis with the human gastrointestinal tract (Holzapfel WH, 2001). Moreover, these bacteria show antimicrobial effect via modifying the microflora, competing for nutrients essential for pathogen growth, impeding adhesion of pathogens to the intestinal epithelium, producing an antitoxin effect and reversing certain consequences of infection on the intestinal epithelium, such as secretory changes and neutrophil migration. It is possible to cure lactose intolerance with probiotics like lactobacillus plantarum which produces the β-galactosidase enzyme that can hydrolyze the β glycosidic bond of lactose (Pineiro M, 2007).

The difference between prebiotics and probiotics are in the fact that the former refers to non-digestible fiber. Prebiotics pass undigested and serve as a substrate for the beneficial bacterial growth, accompanied by positive health contributions such as bowel function, energy balance, glycemic control, and immunologic role to sensory perception, and blood pressure regulation (Delzenne NM, 2011; Hord NG, 2008; Rajat S, 2012; Scholz-Ahrens KE, 2016). It is important to note that, the term prebiotics does not confine only to the effects on the gut microbiome, it also includes the other areas of the body, such as skin, because it is observed that moisturizers present in skin can actively play role to the enhancement of the skin microbiome (Schloss, 2014). Up to now this group comprehends diverse ingredients that are not fully understood in regard to their effects, mode of action, and necessary dosages for assessable health benefits (Douglas LC, 2008). Sources of probiotic microorganisms are mentioned in Table 4.

Table 4. Sources of probiotic microorganisms (Holzapfel WH, 2001)

Milk	Yoghurt	Fermented products	Human breast milk	GI tract	vegetables/ grains/fruits
Lactobacillus acidophilus L. lactis	*L.delbrueckii subp bulgaricus*	*L. casei* *L. cellobiosus* *L. curvatus* *L. fermentum* *L. helviticus* *L. farciminis*	*L. reuteri* *L. salivarius*	*L. gasseri* *L. johnsonii*	*L. brevis* *L. plantarum*
	Bifido-bacterium adolescentis	*B. thermophilum* *B. animalis*	*B. infantis* *B. longum* *B. breve* *B. lactis*		
Propioni-bacterium freudenreichii	*Streptococcus thermophilus*	*Enterococcus faecium* *Pediococcus acidilactici*		*Escherichia coli Nissle 1917*	*Leuconstoc mesenteroides*
					S. cerevisiae S. boulardii Mushrooms

4.1.1.3. Nutraceutical Enzymes

To sustain life, enzymes function as catalysts that are necessary for proper metabolic function. Enzyme-based Nutraceuticals become trendier over past few decades due to consumers demand for digestive health products. Digestive health is one of the popular categories for enzyme supplements. As "Bioavailability" becomes a key concern of consumer, enzyme dietary supplements are achieving appreciation to them. Because of the function of Nutraceutical enzyme in digestive health and bioavailability of nutrients from food, is a key concern for athletes and consumers relevant to functional nutrition for growth and development as well as supporting healthy aging. The digestive enzymes are functioning in breakdown of food-derived carbohydrates, proteins and fats into smaller substances, so that our bodies can utilize them. Even though our body produces its own digestive enzymes, which may not enough for complete breakdown of cooked or processed food. Furthermore, during processing or cooking the natural enzymes present in raw food materials are destroyed. Auspiciously, supplements in the form of nutraceutical enzyme can provide additional enzymes and aid to optimize the digestion process which results in the complete release and absorption of food ingredient that can ensure less digestive disorder.

Additionally, it is promising to eradicate the symptoms of disease by enzyme supplements to their diet to the patient suffering from digestive problems, obesity, hypoglycemia, and blood sugar disorders. These enzymes could be derived from microbial, plant and animal sources.

Table 5. List of nutraceutical enzymes from microbes, plants and animals

Microbial Enzymes/source	Plant Enzymes/source	Animal Enzymes/source
Hemicellulase (microorganisms and mushrooms)	Hemicellulase (plant walls)	OxBile (ox)
Catalase	Pectinase (cell wall)	Pancreolipase (pancreatic juice)
Amyloglucosidase (ascomycetes)	α- Galactosidase (beans, cabbage, Brussels sprouts, broccoli, asparagus, other vegetables, and whole grains)	Trypsin (pancreatic juice)
Glucoamylase (*A. niger, Saccharomycopsis fibuligera*)	β-Amylase (higher plants)	Chymotrypsin (all classes of vertebrates)
Cellulase (all living cells)	Bromelain (pineapple)	Pepsin (animals tracheal secretions)
Invertase – Sucrase (yeast)	Biodiastase (soybean)	Lysozyme (saliva, tears, egg white.
Lactase – β-Galactosidase (bacteria)	Glucoamylase (callus and suspension cultures of sugar beets (Beta vulgaris L.) as well as in mature roots)	α –Amylase (saliva)

4.1.2. Non-Traditional Nutraceuticals

Foods that are artificially prepared or produced mainly facilitated with biotechnology and genetic engineering. Foods could be engineered to generate bioactive components containing products for human health and wellness. They are chiefly divided into two classes

i) Fortified nutraceuticals
ii) Recombinant nutraceuticals

Table 6. Product produced by recombinant microorganisms, plants and animals

A. Recombinant microorganisms			
Source	**Enzyme**	**Products**	**References**
Acetobacter xylinum	β-glucuronidase	Kombucha beverage	Malbasa RV, 2011
Escherichia coli K-12	Chymosin	Milk-coagulated products	El-Sohaimy SA, 2010
Fusarium venenatum	Xylanase	Increased bran solubilization	Sibbesen, 2010
Aspergillus oryzae	Esterase–lipase, Aspartic proteinase, Glucose oxidase, Laccase, Lipase, Pectin esterase,	Alcoholic beverages (Sake, koji)	Ghorai S, 2009
Saccharomyces cerevisiae	Stilbene synthase and 4-coumaroyl-CoA	Resveratrol	White E, 2009
Spirulina Pacifica	Indoleamine 2,3-dioxygenase (IDO)	Increased hemoglobin	Liu L, 2016
B. Recombinant plant			
Recombinant	**Deficiency**	**Gene for recombination**	**References**
Gold kiwifruit	Iron	High level of Ascorbic acid, carotenoids lutein and zeaxanthin	Beck K, 2011
Potatoes	Protein	Tuber-specific expression of a seed protein, *AmA1(Amaranth Albumin 1)*	Chakraborty S, 2010
Golden mustard	Vitamin A	Soybean *ferritin* gene	Chow J, 2010
Multivitamin corn	Multivitamin	Vitamins β-carotene corn (*Zea mays*) phytoene synthase (*psy1*) cDNA), ascorbate (rice dehydroascorbate reductase (*dhar*) cDNA), and folate (*E. coli folE* gene encoding GTP cyclohydrolase *(GCH1*)	Naqvi S, 2009
Maize	vitamin A (retinol)	Bacterial genes *crtB* and *crtI*	Aluru M, 2008
Tomato	Folate	Aminodeoxychorismate synthase (*AtADCS*)	de la Garza DRI, 2007
Golden rice	vitamin A (retinol)	Two daffodil genes and one bacterial gene	Burkhardt PK, 1997
Iron rice	Iron deficiency	Soybean *ferritin* gene	Goto F, 1999
C. Recombinant animals			
Fermented soya milk	Calcium deficiency	*Lactobacillus acidophilus* American Type Culture Collection (ATCC) 4962	Cheung AL, 2011
Cattle	human lysozyme	*rHLZ* expression vector *pBC2-HLY-NEOR*	Yang B, 2011
Yogurt	probiotics microorganism	*Bifidobacterium lactis Bb-12* and *Lactobacillus acidophilus LA-5*	Allgeyer LC, 2010
Cows	Lactoferrin deficiency	Recombinant human lactoferrin (rhLf)	Hyvonen P, 2006

4.1.2.1. Fortified Nutraceuticals

Fortified foods are usually enriched with vitamins and minerals at a range up to 100 percent of the RDA (Recommended Daily Allowance) for that nutrient. Fortified foods comprise from agricultural breeding or added nutrients or ingredients, such as calcium fortified orange juice, vitamins or minerals added cereals and folic acid enriched flour. Likewise, milk fortified with cholecalciferol is recommended for vitamin D deficiency (Casey CF, 2010) and prebiotic and probiotic fortified milk with *Bifidobacterium lactis HN019* is suggested in diarrhea, respiratory infections and severe illnesses, in children (Sazawal S, 2010). In 2010 Kumar et.al discovered iron fortification of banana by the expression of "Soybean Ferritin" gene to fight against iron deficiency anemia (Kumar GB, 2010).

4.1.2.2. Recombinant Nutraceuticals

Biotechnology has transformed the potential for plants to be a manufacturing source of pharmaceutical compounds. Alcohol, fermented starch, yogurt, vinegar, cheese, bread, or so on are the energy providing foods could be produced by aiding of biotechnology. In addition, it involves the application of biotechnology and genetic engineering for the production of probiotics and extract bioactive components by enzymatic or fermentation technology. Gold kiwifruit is genetically modified for a high level of ascorbic acid, carotenoids, and lutein and zeaxanthin is a good example of recombinant nutraceuticals (Singh J, 2012; Beck K, 2011).

5. Dietary Supplements

Dietary supplements could be defined by law as products that are anticipated to supplement the diet; but not as drugs and, consequently, are not deliberate to prevent, diagnose, cure, alleviate, or treat diseases. Scientific analysis specifies that the crucial reasons to intake dietary supplements in adults are to improve or maintain sound health, not for prevention or treatment of disease. As a rule, dietary supplements are officially considered as food, but not classified as conventional foods that are consumed in a diet like fruits and vegetables or complete meal. ADA supports the FDA's regulatory definition of dietary supplements as foods, as long as they meet the following criteria outlined in the Dietary Supplement Health and Education Act: a product (other than tobacco) that is intended to supplement the diet, which contains one or more of the following dietary ingredients - a vitamin, a mineral, a herb or other botanical, an amino acid, a dietary substance to supplement the diet by increasing the total daily intake, or a concentrate, metabolite, constituent, extract, or combinations of these ingredients; is ingested in pill, capsule, tablet, or liquid form; is not represented for use as a conventional food or as the sole item of a meal or diet; and is labeled as a "dietary supplement" (Position of The American Dietetic Association: Functional foods, 2009). Unofficially in the United States of America, the term dietary supplements are regard as "nutraceuticals."

Though it is purported that functional foods look like food, while dietary supplements appear drug-like to serve the same purposes, and both can induce drug-like reactions in high concentrations. Dietary supplements are the substances that might be used to append nutrients to the diet or sometimes to lower the health risks, such as arthritis or osteoporosis. Usually, dietary supplements are obtained in the form of capsules, powders, gel tabs, pills, extracts, liquids or suspension. They might enclose vitamins, amino acids, minerals, fiber, herbs or other plant stuff for antioxidant and alkaloids, or enzymes (Zeisel SH, 1999). Occasionally, the ingredients of dietary supplements are added to foods, drinks or beverage. To purchase dietary supplements prescription of doctor is not required.

Eating diversified healthy food is the best way to get required nutrients to maintain proper health. Nevertheless, sometime it is difficult to get enough vitamins and minerals or other nutrient from daily diet and in such case physician may suggest for dietary supplement. Nutrients that might be missing from daily diet could be compensated via dietary supplements.

6. Health Benefits of Nutraceuticals and Functional Foods in Disease Management

6.1. Diabetes

Type 2 diabetes is a metabolic disorder with both short- and long-term undesirable complications. It is characterized by hyperglycemia, impaired insulin secretion or arising insulin resistance, and dysfunction of β-cell (Santaguida PL, 2005; Evans JL, 2002). There has been growing evidence of ineffectiveness of the current medical treatments to manage long-term diabetes complications proves that additional complementary approaches are necessary; utilization of functional foods and bioactive compounds is one of the promising new approaches. Studies on numerous animal models, in vitro, and human trials, have established the beneficial effect of functional foods and nutraceuticals in improving postprandial hyperglycemia and adipose tissue metabolism modulate carbohydrate and lipid metabolism. Functional foods may also improve dyslipidemia, insulin resistance, attenuate oxidative stress and inflammatory processes and subsequently could prevent the development of long-term diabetes complications including cardiovascular disease, neuropathy, nephropathy and retinopathy (Mirmiran P, 2014). Finally, the available data signify that functional foods-based diet could be a novel and comprehensive dietary approach for management of type 2 diabetes. Table 7 shows the bioactive compounds and functional properties of some of favorable functional foods for type 2 diabetes.

Table 7. Bioactive compounds and functional properties of some of favorable functional foods

Possible functional properties in diabetes	Bioactive components and nutraceuticals	Functional foods	Ref.
Improve the features of metabolic syndrome, modulate gut microbiota, regulate satiety and food intake ↑ adiponectin, modulate adipocytokines, induce thermogenesis, lipolysis and β-oxidation ↑ dietary fat excretion ↓ adiposity and body weight ↓ oxidative stress and inflammatory markers, hypo-lipidemic and antithrombotic effects ↑ insulin sensitivity, modulate immune responses in diabetic patients ↑ total antioxidant capacity ↓ lipid peroxidation ↓ HbA1c	Calcium, vitamin B, bioactive proteins such as casein and whey, immunoglobulines, bioactive peptides (α- and β-lactorphines, lactoferrin, lactoferricin, α-lactalbumin, β-lactoglobulin, growth factors), conjugated linoleic acids, lactic acid bacteria and bifidobacteria	Dairy products and probiotics	Al-Salami H,2008; Ataie-Jafari A, 2009; Chang BJ, 2011; Guo C, 2010; Guzel-Seydim ZB, 2011; Lin MY, 1999; Liu S, 2006; Lopitz-Otsoa F, 2006; Marteau PR. 2002; Pfeuffer M, 2007; Shab-Bidar S, 2011; Teruya K, 2002
Improve hypertriglyceridemia and hypertension ↓ cardiovascular disease ↓ insulin resistance and inflammation, improve glycemic management ↓ proteinuria ↓ oxidative stress, inhibit lipogenesis and induce lipolysis, induce PPARα and PPARβ ↓ adiposity and weight management ↑ thermogenesis and energy expenditure, inhibit angiotensin converting enzyme and modulate blood pressure	Bioactive peptides, antioxidant compounds, ω3 fatty acids (docosahexaenoic acid, eicosapentaenoic acid), selenium, taurine	Fish and seafood	Boukortt FO, 2004; Farmer A, 2001; Jacques H, 1995; Lankinen M, 2011; McEwen B, 2010; Pecis M, 1994;
Promote endogenous antioxidant defense system, induce superoxide dismutase and catalase ↓ lipid peroxidation, improve glycemic control ↑ insulin sensitivity ↓ gluconeogenesis ↑ glycogen content ↓ glycation of collagen and fibrosis, protect cardiac muscle, regulate lipid metabolism as well as adipose tissue metabolism, inhibit lipogenic enzymes	Polyphenols, phenolic acids, catechins, epigallocatechin-3-gallat, chlorophyll, carotenoids, pectin, plant sterols	Green tea	Babu PV, 2001; Fiorino P, 2012; Kumar B, 2012; Kim HM, 2013;
↓ satiety ↑ thermogenesis ↓ proliferation and differentiation of adipocytes ↓ pro-inflammatory cytokines ↓monocyte chemotactic protein-1			

Possible functional properties in diabetes	**Bioactive components and nutraceuticals**	**Functional foods**	**Ref.**
↑ Iinsulin sensitivity, improve peripheral uptake of glucose, increase glycolysis and gluconeogenesis, hypoglycemic and hypolipidemic effects, antioxidant and anti-inflammatory properties	Cinnamaldehyde, cinnamic acid, coumarin, catechins, epicatechin, procyanidins B-2	Cinnamon	Anderson RA, 2008; Gruenwald J, 2010; Kirkham S, 2009
Inhibit enzymes involved in inflammation including cyclooxygenase-2, lipoxygenase, and nuclear factor κB, inhibit α-glucosidase and α-amylase activity ↓ postprandial glycemic response ↓ proteinuria, activate PPARγ and regulate carbohydrate and lipid metabolism, prevent diabetic cataract	Curcuminoids, stigmasterol, β-sitosterol, 2-hydroxy methyl anthraquinone, bioactive peptide turmerin	Turmeric	Honda S, 2006; Khajehdehi P, 2011; Lekshmi PC, 2012;
Attenuate oxidative stress, protective effects against oxidative damage ↓ serum creatinine and urea, improve dyslipidemia ↓ atherogenic lipoprotein levels ↓ lipid peroxidation in renal tissue, inhibit α-glucosidase activity ↓ carbohydrate digestion and absorption, protect liver against diabetes-induced oxidative damage	Tannins, flavonoids, anthocyanins, phenolic acid, gallic acid	Sumac	Rayne S, 2007; Shafiei M, 2011; Giancarlo S, 2006; Pourahmad J, 2010

PPAR: Peroxisome proliferator-activated receptor.

6.2. Cardiovascular Disease

Cardiovascular disease (CVD) is a class of diseases that involve the heart or blood vessels. CVD includes coronary artery diseases (CAD) such as angina and myocardial infarction commonly known as a heart attack; leading globally. According to WHO report 17.5 million deaths (31%) were accounted in 2015 which were 12.3 million (25.8%) in 1990 (WHO Cardiovascular diseases (CVDs) 2016). CVDs have taken into account as an endemic disease for economic developed countries, and gone above infectious diseases in mortality rate. Both laboratory and epidemiologic investigation shows the association of reactive oxygen species (ROS) in the pathogenesis of both acute and chronic heart diseases as a consequence of cumulative oxidative stress (Wang CZ, 2007). Specifically, oxidation of low-density lipoproteins (LDL) is responsible for the pathogenesis of cardiovascular heart diseases and atherosclerosis as it initiates the plaque formation process (Wang CZ, 2007). Besides that there are some significant risk factors for CVD includes high blood pressure, high blood cholesterol levels, type-2 diabetes and obesity. Consumption of poor diets is not only the reason of CVD; the increased risk is associated with lifestyle and practice such as alcohol intake, smoking, and physical activities. It is apparent that people who consume healthy food, live an active life, do not smoke and consume excessive alcohol demonstrate lower risk of CVDs (Riccioni G, 2008). It is also scientifically evident that the diets governing to elevate the level of serum total cholesterol, LDL cholesterol, and triacylglycerol, which reduce the level of HDL cholesterol, thereby increases the risk of CVDs.

For prevention of both heart and kidney diseases, regulating the blood pressure plays the critical role. Influencing factors of blood pressure includes imbalances in the rennin angiotensin system, atherosclerosis, and hyperinsulinemia; subsequently increased sodium retention in the body and stimulate atherosclerosis (Lampe JW, 1999). Consequently, a general nutritional plan to minimize hypertension risk includes attaining and maintaining a healthy body weight; consuming a diet rich in calcium, phosphorus, and magnesium; and moderate consumption of sodium and alcoholic beverages (Dwyer J, 1995).

Consuming functional foods or Nutraceuticals in adequate quantities is aid in lowering the risk of CVDs by quite a lot of prospective mechanisms. These include lowering blood lipid levels, decreasing the plaque formation, reducing lipoprotein oxidation, improving arterial compliance, scavenging free radicals, and inhibiting platelet aggregation (Hasler CM, 2000). It is scientifically explicit that consumption of particular plant food can reduce chiefly total cholesterol and LDL, in that way reduce the risk of coronary heart disease (CHD) as well (Truswell AS, 2002). Several examples of health food for cardiovascular disease patient are shown in the following section.

Soybeans are exclusive source of isoflavones like genistein and diadzein which demonstrate frequent biological functions. It is considerate that soybeans play both preventive and therapeutic roles in CVD, cancer, osteoporosis, and the alleviation of menopausal symptoms (Potter SM, 1995). The physiological effect of soy as a cholesterol lowering food is well-established. A 1995 meta-analysis of 38 separate studies (involving 743 subjects) found that the consumption of soy protein resulted in significant reductions in total cholesterol (9.3%), LDL cholesterol (12.9%), and triglycerides (10.5%), with a small but insignificant increase (2.4%) in HDL cholesterol (Anderson JJ, 1995, Anderson JW, 1995). In addition, flax (*Linum usitatissimum*) is an important oil seed crop, is a leading source of omega-3 fatty acid, α–linolenic acid, and phenolic compound lignan, which make them attractive source for the development of nutraceuticals and functional foods with specific health benefits. Especially consumption of flaxseed is well-established to reduce total and LDL cholesterol (Bierenbaum ML, 1993; Cunnane SC, 1993) as well as platelet aggregation (Allman MA, 1995). Wholesome properties of garlic (*Allium sativum*) are plentiful, including antihypertensive, cholesterol-lowering properties, immune stimulation, curing CVDs, anti-infectious properties, cancer chemo-preventive, and free radical scavenging activities, (Srivastava KC, 1995; Borek C, 2006; Singh BB, 2007). Garlic has the potency to fight against lifestyle-related disorders, such as hypercholesterolemia, dyslipidemia, and hypertension that lead to several cardiovascular disorders (Mahmoodi M, 2006; Kojuri J, 2007; Butt MS, 2009). According to clinical trial it is reported that 800-mg garlic/day reduced total cholesterol levels by 12% (Silagy CA, 1994). There is well-built evidence that wine can reduce the risk of CVD. The connection between intake of wine and CVD first came to consideration during 1979 when a strong negative correlation between wine intake and death from ischemic heart disease in both men and women was found (St. Leger AS, 1979). The main ingredient of wine is grapes, contain large amounts of phytochemicals, including resveratrol, which offer considerable health benefits (Pezzuto JM, 2008). Red wine contain 20–50 times higher phenolic content than white wine, in consequence of the incorporation of grape skins into the fermenting grape juice during production. Study illustrated the attribution of the positive benefits of red wine to the ability of phenolic compounds that prevent the oxidation of LDL, a critical event in the process of atherogenesis (Frankel EN, 1993). A number of epidemiological studies have been conducted to investigate the relationship between tea consumption and CVD risk. Hertog et al. attributed that tea consumption that was the major source of flavonoids such as quercetin, kaempferol, myricetin, apigenin, and luteolin was inversely associated with mortality from CHD (Hertog MGL, 1993). Equally, the anti-inflammatory effects of black tea are responsible for preventing CVD development (Steptoe A, 2007).

Even though a vast number of naturally occurring health promoting substances are originated from plant sources, there are several physiologically active components are derived from in animal sources as well which have drawn interest in their prospective role in the finest health. Two most significant functional foods include dairy products and fish. Recent research has exclusively focused on the components of dairy products, particularly fermented dairy products, known as probiotics due to their anti-carcinogenic, hypocholesterolemic, and antagonistic actions against enteric pathogens and other intestinal organisms (Mital BK, 1995). Although fish is the main source of high-quality protein, essential fatty acids, and other nutrients, their consumption is highly allied with certain health benefits as well. Omega-3 (n-3) fatty acids are an essential class of polyunsaturated fatty acids (PUFAs) derived primarily from fish oil plays essential role in protecting against the development of heart diseases. This phenomenon was first brought to light in the 1970s when Bang and Dyerberg reported that Eskimos had low rates of heart disease in spite of consuming a diet which was high in fat (Bang HO, 1972).

6.3. Cancer

The evolvement of cancer is an active and long-term process relating a lot of multifaceted factors with stepwise progression, ultimately leading to an uncontrolled growth of cancer cells all through the body, also termed as metastasis (Cencic A, 2010b; Kalimuthu S, 2013). Epidemiological studies have illustrated the evidence of dietary factors that can alter the development of cancerous cells from normal ones. Moreover, laboratory investigation has confirmed that several bioactive dietary components or natural foodstuffs have the capability to fight against cancer (Liu RH, 2004; Balsano C, 2009). Besides, many undefined food constituents still show nutritional benefits as well as have anti-mutagenic and anti-carcinogenic effects. Recent research offering a strong hold for the acceptability of bioactive components of food as chemo-preventative agents for future therapeutics (Pan MH, 2008). Anti-carcinogenic effects of phytochemicals present in plant foods generally offer a combination of multi effective mechanisms such as amplified enzymes activity for carcinogen detoxification, effect on cell differentiation, inhibition of formation of N-nitrosamine, antioxidant effects, change of colonic environment, conservation of intracellular matrices integrity, alteration of estrogen metabolism, and so on. In addition, Nutraceuticals provide credible aptitude to control the DNA damaging factors in cancer cells and regulate DNA transcription in tumors such as, effect on DNA methylation, maintenance of DNA repair, decrease in cell proliferation and increase in apoptosis of cancer cells (Lampe JW, 1999; Liu RH, 2003; Liu RH, 2004). The roles of some significant functional foods/ active compounds for preventing or combating against cancer are illustrated as follows.

6.3.1. Flaxseed

Flaxseed has been revealed to reduce tumors in the colon and mammary gland in rodents (Thompson LU, 1995) and also in the lungs (Yan L, 1998). Some studies have assessed the effects of flaxseed feeding on risk factor for cancer in human. It is established that the urinary lignan excretion was significantly decreased in postmenopausal breast cancer patients in comparison to the controls eating a normal mixed or a lactovegetarian diet (Adlercreutz H. 1982).

6.3.2. Garlic

Among the many beneficial roles of garlic, inhibition of the growth of cancer is perhaps the most significant one (Matsura N, 2006). Several epidemiological studies have shown that garlic may be effective in reducing human cancer risk (Dorant E, 1993). Anti-tumorigenesis activity has been demonstrated against the gastrointestinal tract cancer, colon cancer, prostate cancer, mammary carcinoma, hepatocellular carcinoma, lung cancer, and sarcoma and squamous cell carcinoma of the skin and esophagus (Kalra N, 2006; Ban JO, 2007; Kim YA, 2007; Zhang ZM, 2007; Butt MS, 2009). However, several epidemiological studies have shown garlic to be protective against carcinogenesis. A 1991 review of 12 case control studies found that eight of these studies showed a negative association, one showed no association, and three studies showed a positive association (Steinmetz KA, 1991).

6.3.3. Soy Isoflavones

Weak estrogenic activity of soy isoflavones offers the potentiality to exert an influence on hormone-dependent cancers such as prostate and breast cancers. The cancer preventive effect, perhaps due to their role in reducing the concentration of unconjugated sex hormones in circulation. Consumption of soy isoflavones has been reported to result in urinary excretion of isoflavone metabolites, whose levels have been inversely linked to the risk of breast cancer. The major soy phytoestrogens are genistein and daidzein which have been used in animal experiments for inhibition of growth of a number of cancer cell lines. It has been shown to inhibit breast and skin cancer and inhibit the formation of aberrant crypts in models of colon cancer. Genistein has been found to be a potent inhibitor of angiogenesis and metastasis, implying pleiotropic effects on inhibition of carcinogenesis and cancer cell growth (Rinaldi A, 2005).

6.3.4. Tea Flavonoids

Flavonoids from both green and black teas have been found to inhibit the mutagenecity of different types of carcinogens. Tea flavonoids have been shown to increase the rate of apoptosis of tumour cells and lead to elimination of neoplastic cell systems (Rinaldi A, 2005).

Most of the research on health benefits of tea has focused on its cancer chemopreventive effects, despite of its indecisive epidemiological studies (Katiyar SK, 1996). Reviewed certain studies on tea consumption and its relationship to cancer found that two-third of the studies show no relationship between tea consumption and cancer risk (Yang CS, 1993). The consumption of five or more cups of green tea per day was associated with decreased recurrence of stage I and II breast cancer in Japanese women (Nakachi K, 1998). Oppositely, several population based studies in Japan and China discovered inverse association of green tea with colon cancer (Khatiwada J, 2006; Xu G, 2010).

6.4. Obesity

Obesity is one of the major global health epidemics of the twenty-first century. In modern world obesity takes the fifth leading risk factor for global deaths (Chuadhary M, 2012). The rising out breaking of obesity is allied with extensive metabolic complications like type 2 diabetes, cardiovascular disease (CVD), hyperlipidemia, and hypertension causes considerable socioeconomic as well as physical burden on the world (Laing P, 2002; Wild S, 2004).

Simply obesity can be defined as a physiological state characterized by the accumulation of excess body fat. Obesity leads to chronic, excessive adipose tissue expansion resulting in an increase in the risk for cardiovascular disease, type 2 diabetes mellitus, and other metabolic abnormalities. First of all it is thought to stem from the low-grade, systemic inflammatory response syndrome that characterizes adipose tissue in obesity. As Obesity is accompanied by adipose tissue hyperplasia and hypertrophy that serves as an important initiator of a chronic low grade systemic inflammatory response (Trayhurn P, 2004). This is characterized by percolation of macrophages and other immune cells with following release of proinflammatory cytokines like interleukin-1 (IL-1), tumor necrosis factor-a (TNF α), plasminogen activator inhibitor-1, monocyte chemoattractant protein-1 (MCP-1), serum amyloid A (SAA), leptin, macrophage inflammatory protein (MIP), retinol binding protein- 4(RBP-4) (Hotamisligil GS, 1993; Sethi JK, 1999; Shoelson SE, 2006).

With a global increase in the prevalence of obesity, nutrition and exercise play a key role in its prevention and treatment. Pharmacological approaches and various surgical procedures at disposal of body fat are not cost effective. Nutritional strategies aimed to reduce positive energy balance by decreasing energy intake, increasing energy expenditure, and suppressing the inflammatory excursions seem to be a very logical and attractive alternative. Natural product like nutraceutical or functional food interventions are currently gained a huge attention and also being investigated on a large-scale basis as potential treatments for obesity and weight management.

Apart from taking care of the imbalance between energy intake and energy output, nutraceuticals should have the potential to improve the oxidative stress condition and inflammation in obesity, thereby become a preventive approach for the onset of obesity complications (Dao HH, 2004). A fiber-rich food is promising in undertaking this threat since fiber contributes less calorie intake. Increasing fiber content in diet is supposed to aid in weight management and helps to reduce obesity complications (Pereira MA, 2001). Phytochemicals and biologically active compounds are beneficial for obesity and weight management for instance curcumin, caspaicin, conjugated linoleic acid, polyunsaturated fatty acids, psyllium fiber with diverse mechanism of action. Curcumin administration in laboratory animal models such as obese and leptin-deficient ob/ob C57BL6/J mice have shown the improvement in diabetes, that indicated by glucose- and insulin-tolerance testing and the percentage glycosylated hemoglobin (Weisberg SP, 2008). Jain *et al.* reported that curcumin supplementation lowered the high glucose-mediated monocyte production of inflammatory cytokines, including TNF-α, IL-6, IL-8, and MCP-1. This same study also showed that blood levels of TNF-α, MCP-1, glucose, and glycosylated hemoglobin were decreased in diabetic rats on a curcumin diet (Jain SK, 2009). Several studies have shown potential benefits of capsaicin a biologically active ingredient found in red chili peppers, for treating obesity and insulin resistance in animal models and clinical studies (Kim CS, 2003; Manjunatha H, 2006) Capsaicin was shown in animal studies to increase the insulin-stimulated uptake of glucose in muscle cells (Park JY, 2004). It has been demonstrated that PUFA markedly decreased the sterol regulatory element-binding protein (SREBP-1) in its mature form and thus down regulated the expression of lipogenic genes such as fatty acid synthase (FAS) and stearoyl-CoA desaturase 1 (SCD1) in the livers of ob/ob mice. Accordingly, the liver triglyceride content and plasma alanine aminotransferase (ALT) levels were decreased. Moreover, both hyperglycemia and hyperinsulinemia in ob/ob mice were improved by PUFA administration, similar to the effect of PPARa activators. Thus, it could be concluded that PUFAs improve obesity-associated symptoms, such as hepatic steatosis and insulin resistance, most probably through both down-regulation of SREBP-1 and activation of PPARa (Sekiya M, 2003). Supplementation of conjugated linolenic acid which found primarily in the seeds of flax, and nut oils, as well as fish, and poultry eggs reduced fat mass of obese individuals (Blankson H, 2000). Laboratory studies have revealed that adding conjugated linolenic acid to a high fat diet fed of rodents effectively prevent the onset of obesity-induced muscle insulin resistance (Lavigne C, 2001). Psyllium fiber extracted from the husks of its seeds, have a lower glycemic index, which has been found to decrease postprandial insulin and glycemic response. Psyllium fiber decreases the rate of glucose absorption. It traps glucose, and slows its absorption. It has been associated with lowering of LDL levels in humans (Kris-Etherton PM, 2002). Resveratrol, a polyphenolic compound found in the skin of grapes and related food products, has been shown to prevent a number of diverse pathologic processes, including CVD, cancer, oxidative stress, and inflammation (Baur JA, 2006). The anti-inflammatory properties of flavonoids another polyphenol subclass, have been extensively studied to establish their potential utility as therapeutic agents in the treatment of inflammatory diseases to control obesity (Garcia-Lafuente A, 2009).

7. Nutraceuticals and Functional Food: Facts and Fiction

Epidemiological studies show a positive correlation between the consumption of plant derived foods and a variety of health benefits. In the last few years, nutraceuticals have appeared in the market. These are available in pharmaceutical forms (pills, powders, capsules, vials, etc.) containing food bioactive compounds specially phytochemicals as active principles. Scientific research proves the biological activity of the most of these phytochemicals, but the health claims credited to the final marketed nutraceutical products have often doubtful scientific basis (Espı´n JC, 2007). A lot of the scientific evidence is derived from in vitro assays and animal testing, while human clinical trials are inadequate and questionable and these may be the reason of such circumstances. In addition, a number of fundamental issues such as bioavailability, metabolism, dose/response and toxicity of these bioactive compounds in the nutraceuticals have been not well established yet. Therefore, consistency between scientific knowledge on the bioavailability and biological activity of these bioactive compounds ('fact'), and the health claims ('fiction') which are not always evident by scientific studies credited to the bioactive containing nutraceuticals, need to be established. For example, the in vitro antioxidant capacity of a bioactive compound often used as a claim, which may irrelevant in terms of in vivo antioxidant effects. As bioavailability, metabolism, and tissue distribution of these bioactive compounds in humans physiology are the key factors that need to be established evidently in association to their biological effects as nutraceuticals.

Most of the nutraceuticals available in the pharmacy have a recommended dose, although it is not apparent what the scientific foundation of this dose recommendation is. Also, it is not identified what is the consequences may be derived from a high intake of bioactive compound such as polyphenols containing supplements. For instance, the anticipated daily intake of dietary anthocyanins may range from several hundred up to a thousand milligram. Additional consumption of one or two pills a day of berry supplements may provide almost up to one more gram of these compounds. That's why consumers should be aware of the risk of ingesting high doses of these supplements, as for most of these natural extracts have the possible toxic effects that have not been properly investigated.

In the case of nutraceuticals and functional foods, however, consumer acceptance is considered to be overwhelmingly positive throughout the US, Canada, Europe and Japan (Siro I, 2008; West, GE, 2002), though Europeans tend to be more critical of new food products and technologies (Lusk JL, 2005). Interestingly, public surveys identify that consumers are eager to purchase and to pay huge amount for these products, particularly when the functional property is added to plant derived foods. Nevertheless, they are less approachable to functional properties incorporated in meat products (West et al., 2002).

Likewise, in recent years it has been striking by key consumer trends that have led to a growing demand for functional foods and neutraceuticals. Increased life expectancy, resulting in an ageing population, as well as the subsequent economic burden placed on national health services, coupled with a rising awareness of the relationship between diet and health and individual pro-activity towards personal health improvement, have led to a more holistic view of medicinal food and food products (Gray J, 2003).

8. Advancement and Future Holds of Nutraceuticals

The increasing preferences of consumers to eat healthy food products and the nutraceuticals showing up to be favorable in preventing as well as curing many diseases impelled scientists and researchers to look for efficient delivery systems. The use of novel drug delivery system to deal with the efficacy issues of the products is drawing more and more attention of the researchers. Drug delivery systems with medicated herbs includes: nanoemulsions, liposomes, phytosomes, microspheres, and transfersomes. Challenges and limitations associated with delivery of nutraceuticals still is a critical issue. The nutraceutical formulations have been considered not as medicines but as diet.

For this reason, unlike the pharmaceutical preparations, nutraceuticals preparations need to fulfill a whole lot more requirements such as the formulation must be of food grade. This limits the options for the researchers and declines the range of innovations in the field of nutraceutical. Thus, choice of appropriate material for the preparation of the formulation becomes a big challenge (Crandell K, 2003). The next challenge is the selection of the delivery system of nutraceutical. Usually, the nutraceuticals compounds comprise the biological products such as proteins, peptides, vitamins, hormones and herbal extracts which have great tendency to degrade readily. The constancy of the formulation of nutraceuticals is a factor which cannot be compromised. Moreover, the active ingredient must be delivered to its target site upon triggering by any external stimulus such as pressure and pH or temperature. Thus, an appropriate delivery system must be selected on the basis of its capability to deliver the food product effectively with its desired effect (Crandell K, 2003). The next challenge is the testing of the products. The *in vitro* examinations were performed to get an insight of the pharmacokinetic mechanism, the rate of release of the core active ingredient at target site. Nevertheless, these experiments have the restrictions as they are unable to provide sufficient data for the active uptake of the product, their metabolic responses, and the biological variability of the nutraceutical product as well. Therefore, complete testing mechanism is necessary to examine the influence of factors such as food on the response of the product (Dev R, 2011). Nutraceuticals is an innovative issue which is not clearly known to the mass people and it still requires attention as well as needs awareness regarding its application and its potentiality.

CONCLUSION

The studies have established the critical role of both nutraceuticals and functional food to prevent and treatment of a range of diseases. It is apparent that health conscious consumers have a preference for the treatment of any illness in a natural way which could be accomplished by nutraceuticals, functional food or commercially available food supplements in regular use. Essentially the consumption of medicinal foods like nutraceuticals or functional foods offer the best option to stay natural and to get better the quality of life. Additionally, functional foods and nutraceuticals provide a potential alternative for the patients who are not agreeable to go through conventional therapy as well. Furthermore, it is also beneficial over other kind of traditional therapies in terms of expense. Both laboratory and epidemiological studies have illustrated the fact that medicinal foods especially nutraceuticals exert special types of biological activities in naturally programmed approach. These characteristics of medicinal foods eventually improve the age related chronic diseases. Currently, the potential of nutraceuticals are gaining huge attention for going through research and development. The marketing graphs for nutraceuticals and food supplements are also growing high all over the world as well. The reported market values demonstrate the public behavior which pays more attention on the exploitation of nutraceuticals which finally is associated to the luminous scope of establishing more nutraceutical industries. On the other hand, there is an enormous possibility for further progression in this lucrative field by introducing delivery carrier and develop awareness about the prospective of such systems. Consequently, the field undoubtedly has the greatest hidden value allied with it, which is just required to be explored to provide easy and better healthy life. In conclusion, the foods whose health benefits are supported by sufficient scientific validation have a great potential to be an increasingly important component of a healthy lifestyle and to be beneficial to the public and the food industry.

REFERENCES

Adlercreutz H, Fotsis T, Heikkinen R, Dwyer JT., Woods M, Goldin BR. and Gorbach SL. Excretion of the lignans enterolactone and enterodiol and of equol in omnivorous and vegetarian postmenopausal women and in women with breast cancer. *The Lancet* 1982; 320:1295–1299.

Ahmad F, Ahmad FA, Azad AA, Alam S, Ashraf AS. Nutraceuticals is the need of hour. *World J Pharm Pharm Sci* 2013; 2:2516-2525.

Ali B, Alial-Wabel N, Shams S, Ahamad A, Khan SA, Anwar F, Essential oils used in aromatherapy: A systemic review, *Asian Pacific Journal of Tropical Biomedicine* 2015; 5 (8) : 601-611.

Allgeyer L C, Miller MJ and Lee SY. Sensory and microbiological quality of yogurt drinks with prebiotics and probiotics. *J. Dairy Sci* 2010; 93: 4471-4479.

Allman MA., Pena MM. and Pang D. Supplementation with flax seed oil versus sunflower seed oil in healthy young men consuming a low fat diet: Effects on platelet composition and function. *Eur. J. Clin. Nutr.*1995; 49:169–178.

Al-Salami H, Butt G, Fawcett JP, Tucker IG, Golocorbin-Kon S, and Mikov M. Probiotic treatment reduces blood glucose levels and increases systemic absorption of gliclazide in diabetic rats. *Eur J Drug Metab Pharmacokinet* 2008; 33: 101-106.

Aluru M, Xu Y, Guo R, Wang Z, Li S, White W, Wang K and Rodermel S. Generation of transgenic maize with enhanced provitamin A content. *J. Exp. Bot* 2008; 59: 3551-3562.

Anderson JJ, Ambrose WW and Garner SC. Orally dosed genistein from soy and prevention of cancellous bone lossin two ovariectomized rat models. Abstract. *J. Nutr.* 1995; 125:799.

Anderson JW, Johnstone BM. and Cook-Newell ME. Metaanalysis of the effects of soy protein in-take on serum lipids. *New Engl. J. Med.* 1995; 333:276–282.

Anderson RA. Chromium and polyphenols from cinnamon improve insulin sensitivity. *Proc Nutr Soc* 2008; 67: 48-53.

Ataie-Jafari A, Larijani B, Alavi Majd H, Tahbaz F. Cholesterol-lowering effect of probiotic yogurt in comparison with ordinary yogurt in mildly to moderately hypercholesterolemic subjects. *Ann Nutr Metab* 2009; 54: 22-27.

Babu PV, Sabitha KE, Srinivasan P, Shyamaladevi CS. Green tea attenuates diabetes induced Maillard-type fluorescence and collagen cross-linking in the heart of streptozotocin diabetic rats. *Pharmacol Res* 2007; 55: 433-440.

Bae J-M, Lee AJ, Guyatt G. Citrus fruit intake and stomach cancer risk: A quantitative systematic review. *Gastric Cancer*. 2008; 11:23-32.

Bagchi D. Nutraceuticals and functional foods regulations in the United States and around the world. *Toxicology* 2006; 221:1-3.

Balsano C and Alisi A. Antioxidant effects of natural bioactive compounds. *Curr. Pharm. Des.* 2009; 15:3036–3073.

Ban JO, Yuk DY, Woo KS, Kim TM, Lee US, Jeong HS, Kim DJ, Chung YB, Hwang BY, Oh KW and Hong JT. Inhibition of cell growth and induction of apoptosis via inactivation of NF-kappaB by a sulphur compound isolated from garlic in human colon cancer cells. *J. Pharmacol. Sci.* 2007; 104:374–383.

Bang HO and Dyerberg J. Plasma lipids and lipoproteins in Greenlandic west-coast Eskimos. *Acta. Med. Scand.* 1972; 192:85–94.

Baur JA, Sinclair DA. Therapeutic potential of resveratrol: The in vivo evidence. *Nat Rev Drug Discov* 2006; 5:493-506.

Beck K, Conlon CA, Kruger R, Coad J and Stonehouse W, Gold kiwifruit consumed with an iron-fortified breakfast cereal meal improves iron status in women with low iron stores: A 16-week randomized controlled trial. *Br. J. Nutr* 2011; 105: 101-109.

Berry RJ, Bailey L, Mulinare J and Bower C. Folic Acid Working Group, Fortification of flour with folic acid. *Food Nutr. Bull* 2011; 31: S22-35.

Bierenbaum ML, Reichstein R and Watkins TR. Reducing atherogenic risk in hyperlipemic humans with flax seed supplementation: A preliminary report. *J. Am. Coll. Nutr.*1993; 12:501–504.

Blankson H, Stakkestad JA, Fagertun H, Thom E, Wadstein J, Gudmundsen O. Conjugated linoleic acid reduces body fat mass in overweight and obese humans. *J Nutr* 2000;130:2943-2948.

Block G, Patterson B and Subar A. Fruit, vegetables, and cancer prevention: A review of the epidemiological evidence. *Nutr. Cancer*. 1992; 18:1–29.

Borek C. Garlic reduces dementia and heart-disease risk. *J. Nutr.* 2006; 136:810S–812S.

Boukortt FO, Girard A, Prost JL, Ait-Yahia D, Bouchenak M, Belleville J. Fish protein improves the total antioxidant status of streptozotocin-induced diabetes in spontaneously hypertensive rat. *Med Sci Monit* 2004; 10: BR397-BR404.

Brower B. Nutraceuticals: Poised for a healthy slice of the market. *Nat Biotechnol* 1998; 16:728-733.

Bulathsinghala P, Syrigos KN and Saif MW, Role of vitamin d in the prevention of pancreatic cancer. *J. Nutr. Metab* 2010; 2010: 721365.

Burgain J, Gaiani C, Linder M, & Scher J. Encapsulation of probiotic living cells: From laboratory scale to industrial applications. *Journal of Food Engineering* 2011; 104: 467-483.

Burkhardt PK, Beyer P, W, Qnn J, Klbti A, Armstrong GA, Schledz M, Lintig JV, Potrykus I. Transgenic rice (Oryza sativa) endosperm expressing daffodil (*Narcissus pseudonarcissus*) phytoene synthase accumulates phytoene, a key intermediate of provitamin A biosynthesis. *The Plant Journal* 1997; 11(5), 1071-1078.

Butt MS, Sultan MT, Butt MS. and Iqbal J. Garlic: Nature's protection against physiological threats. *Crit. Rev. Food Sci. Nutr.* 2009; 49:538–551.

Casey CF, Slawson DC and Neal LR. Vitamin D supplementation in infants, children, and adolescents. *Am. Fam* 2010; 81: 745-748.

Cencic A and Chingwaru W. Antimicrobial agents deriving from indigenous plants. *RPFNA* 2010a. 2:83–92.

Cencic A and Chingwaru W. The role of functional foods, nutraceuticals and food supplements in intestinal health. *Nutrients* 2010b; 2:611–625.

Center for Science in the Public Interest. *Functional food named in the Center for Science in the Public Interest's complaints to the Food and Drug Administration.* Center for Science in the Public Interest Web site. http://www.cspinet.org/reports/funcfoodcomplaint.htm. Accessed January 9, 2009.

Chakraborty S, Chakraborty N, Agrawal L, Ghosh S, Narula K, Shekhar S, Naik P S, Pande PC, Chakrborti SK and Datta A. Next-generation protein-rich potato expressing the seed protein gene AmA1 is a result of proteome rebalancing in transgenic tuber. *Proc. Nat. Acad. Sci.* U S A 2010; 107: 17533-17538.

Chang BJ, Park SU, Jang YS, Ko SH, Joo NM, Kim SI, Kim CH, Chang DK. Effect of functional yogurt NY-YP901 in improving the trait of metabolic syndrome. *Eur J Clin Nutr* 2011; 65: 1250-1255.

Cheung A L, Wilcox G, Walker KZ, Shah NP, Strauss B, Ashton JF, and Stojanovska L. Fermentation of calcium-fortified soya milk does not appear to enhance acute calcium absorption in osteopenic post-menopausal women. *Br. J. Nutr* 2011; 105: 282-286.

Chintale AG, Kadam VS, Sakhare RS, Birajdar GO, Nalwad DN. Role of nutraceuticals in various diseases: A comprehensive review. *Int J Res Pharm Chem* 2013; 3:290-9.

Choudhary M and Grover K. Development of functional food products in relation to obesity. *Funct. Foods Health Dis.*2012; 2:188–197.

Chow J, Klein EY and Laxminarayan R. Cost-effectiveness of "golden mustard" for treating vitamin A deficiency in India. *PLoS One* 2010; 10: 5.

Consumer attitudes toward functional foods/foods for health - Executive summary. International Food Information Council Web site. http://www.ific.org/research/funcfoodsres07.cfm. Accessed January 9, 2009.

Crandell K, Duren SK. Nutraceuticals: What are they and do they Work. *Versailles: Kentucky Equine Research, Inc*. 2003; p- 28.

Cunnane SC, Ganguli S, Menard C, Liede AC, Hamadeh MJ, Chen ZY, Wolever TMS and Jenkins DJA. High-linolenic acid flaxseed (Linum usitatissimum): Some nutritional properties in humans. *Br. J. Nutr.* 1993; 69:443–453.

Dao HH, Frelut ML, Oberlin F, Peres G, Bourgeois P, Navarro J. Effects of a multidisciplinary weight loss intervention on body composition in obese adolescents. *Int J Obes Relat Metab Disord* 2004; 28:290-9.

De la Garza DRI, Gregory JF and Hanson AD. Folate biofortification of tomato fruit. *Proc. Nat. Acad. Sci. U S A* 2007; 104: 4218-4222.

Delzenne NM, Neyrinck AM., B€ackhed F, & Cani PD. (2011). Targeting gut microbiota in obesity: Effects of prebiotics and probiotics. *Nature Reviews Endocrinology* 2011; 7: 639-646.

Dev R, Kumar S, Singh J, Chauhan B. Potential role of nutraceuticals in present scenario: A review. *J Appl Pharm Sci* 2011; 1:26-8.

Dietary Supplement Health and Education Act of 1994, 21 USC §321.

Doke RM, Mahale AM, Karanjkar AR, Amge SA. Recent trends for nutraceuticals in Indian market. *Indo Am J Pharm Res* 2013; 3:4001-4020.

Dorant E, van den Brandt PA., Goldbohm RA., Hermus RJJ, and Sturmans F. Garlic and its significance for the prevention of cancer in humans: A critical review. *Br. J. Cancer* 1993; 67:424–429.

Douglas LC, & Sanders ME. (2008). Probiotics and prebiotics in dietetics practice. *Journal of the American Dietetic Association* 2008; 108: 510-521.

Douglas LC, & Sanders ME. Probiotics and prebiotics in dietetics practice. *Journal of the American Dietetic Association* 2008; 108: 510-521.

Dwyer J. Overview: Dietary approaches for reducing cardiovascular disease risks. *J. Nutr.*1995; 125:656–665.

Ehrlich SD, (Willow bark), private practice specializing in complementary and alternative medicine. *Phoenix, AZ. Review*, VeriMed Healthcare Network, (2008).

Ehrlich SD. Peppermint (Mentha piperita), private practice specializing in complementary and alternative medicine. *Phoenix, AZ. Review*, VeriMed Healthcare Network, (2009).

El-Sohaimy SA, Elsayed E and El-Saadani HMA. Cloning and in vitro-transcription of chymosin gene in *E. coli. The Open Nutraceuticals Journal* 2010; 3: 63-68.

Espín JC, García-Conesa MT, Tomás-Barberán FA. Nutraceuticals: Facts and fiction. *Phytochemistry* 2007; 68: 2986–3008.

Evans JL, Goldfine ID, Maddux BA, Grodsky GM. Oxidative stress and stress-activated signaling pathways: a unifying hypothesis of type 2 diabetes. *Endocr Rev* 2002; 23: 599-622.

Faridi Z, Njike VY, Dutta S, Ali S, Katz DL. Acute dark chocolate and cocoa ingestion and endothelial function: A randomized controlled crossover trial. *Am J Clin Nutr.* 2008; 88:58-63.

Farmer A, Montori V, Dinneen S, Clar C. Fish oil in people with type 2 diabetes mellitus. *Cochrane Database Syst Rev* 2001; (3): CD003205.

Farzaei MH, Abbasabadi Z, Ardekani MRS, Rahimi R, Farzaei F Parsley: a review of ethnopharmacology, phytochemistry and biological activities *J Tradit Chin Med* 2013 December 15; 33(6): 815-826.

Fasano E, Serini S, Piccioni E, Innocenti I and Calviello G. Chemoprevention of lung pathologies by dietary n-3 polyunsaturated fatty acids. *Curr. Med. Chem* 2010; 17: 3358-3376.

Fiorino P, Evangelista FS, Santos F, Motter Magri FM, Delorenzi JC, Ginoza M, Farah V. The effects of green tea consumption on cardiometabolic alterations induced by experimental diabetes. *Exp Diabetes Res* 2012; 2012: 309231.

Food and Drug Admnistration. Orphan Drug Act (as amended). http://www.fda.gov/orphan/oda.htm. Accessed January 9, 2009.

Frankel EN, Kanner J, German JB, Parks E and Kinsella JE. Inhibition of oxidation of human low-density lipoprotein by phenolic substances in red wine. *The Lancet.* 1993; 341:454–457.

Ganesan P, Noda K, Manabe Y, Ohkubo T, Tanaka Y, Maoka T, Sugawara T and Hirata T. Siphonaxanthin, a marine carotenoid from green algae, effectively induces apoptosis in human leukemia (HL-60) cells. *Biochimica et Biophysica Acta* 2010; 1810: 497-503.

Garcia-Lafuente A, Guillamón E, Villares A, Rostagno MA, Martínez JA. Flavonoids as antiinflammatory agents: Implications in cancer and cardiovascular disease. *Inflamm Res* 2009;58:537- 552.

Ghorai S, Banik SP, Verma D, Chowdhury S, Mukherjee S and Khowala S. Fungal biotechnology in food and feed processing, *Food Res. Int,* (2009).

Giancarlo S, Rosa LM, Nadjafi F, Francesco M. Hypoglycaemic activity of two spices extracts: *Rhus coriaria* L. and *Bunium persicum* Boiss. *Nat Prod Res* 2006; 20: 882-886.

Gibson GR, & Roberfroid MB. Dietary modulation of the human colonic microbiota: Introducing the concept of prebiotics. *J Nutr* 1995; 125: 1401-1412.

Goto F, Yoshihara T, Shigemoto N, Toki S & Takaiwa F. Iron fortification of rice seed by the soybean ferritin gene. *Nature Biotechnology* 1999; 17:282-286.

Gray J, Armstrong G, & Farley, H. Opportunities and constraints in the functional food market. *Nutrition & Food Science* 2003; 33: 213-218.

Gruenwald J, Freder J, Armbruester N. Cinnamon and health. *Crit Rev Food Sci Nutr* 2010; 50: 822-834.

Gry J, Black L, Eriksen FD, Pilegaard K, Plumb J, Rhodes M, et al. EuroFIR-BASIS-a combined composition and biological activity database for bioactive compounds in plant-based foods. *Trends in Food Science and Technology* 2007; 18: 434-444.

Guo C, Zhang L. (Cholesterol-lowering effects of probiotics a review). *Weishengwu Xuebao* 2010; 50: 1590-1599.

Gupta SC, Kannappan R, Reuter S, Kim JH and Aggarwal BB. Chemosensitization of tumors by resveratrol. *Ann. NY Acad. Sci* 2011; 1215: 150-160.

Guzel-Seydim ZB, Kok-Tas T, Greene AK, Seydim AC. Review: functional properties of kefir. *Crit Rev Food Sci Nutr* 2011; 51: 261-268.

Hardy G. Nutraceuticals and functional foods: Introduction and meaning. *Nutrition* 2000; 16: 688-689.

Hastler CM, Kundrat S and Wool D. Functional foods and cardiovascular disease. *Curr. Atheroscler. Rep.* 2000; 2:467–475.

Henrotin Y, Lambert C, Couchourel D, Ripoll C, and Chiotelli E. Nutraceuticals: Do they represent a new era in the management of osteoarthritis? – A narrative review from the lessons taken with five products. *Os Car* 2011; 19: 1-21.

Hertog MGL, Feskens EJM, Hollman PCH, Katan MB and Krumhout D. Dietary antioxidant flavonoids and risk of coronary heart disease: The Zutphen Elderly Study. *The Lancet*. 1993; 342:1007–1011.

Holzapfel WH, Haberer P, Geisen R, Bjorkroth J and Schillinger U, Taxonomy and important features of probiotic microorganisms in food and nutrition. *Am. J. Clin. Nutr* 2001; 73: 365S-373S.

Holzapfel WH, Haberer P, Geisen R, Bjorkroth J and Schillinger U. Taxonomy and important features of probiotic microorganisms in food and nutrition. *Am. J. Clin. Nutr* 2001; 73: 365S-373S.

Honda S, Aoki F, Tanaka H, Kishida H, Nishiyama T, Okada S, Matsumoto I, Abe K, Mae T. Effects of ingested turmeric oleoresin on glucose and lipid metabolisms in obese diabetic mice: a DNA microarray study. *J Agric Food Chem* 2006; 54: 9055-9062.

Hord NG. Eukaryotic-microbiota crosstalk: Potential mechanisms for health benefits of prebiotics and probiotics. *Annual Review of Nutrition* 2008; 28: 215-231.

Hotamisligil GS, Shargill NS, Spiegelman BM. Adipose expression of tumor necrosis factor-a: Direct role in obesity-linked insulin resistance. *Science* 1993; 259:87-91.

Howard BV and Kritchevsky D. Phytochemicals and cardiovascular disease – A statement for healthcare professionals from the American Heart Association. *Circulation* 1997; 95:2591–2593.

Hsieh YH, Ofori JA. Innovations in food technology for health. *Asia Pac J Clin Nutr.* 2007; 16(suppl 1):65- 73.

Hyvonen P, Suojala L, Orro T, Haaranen J, Simola O, Rontved C and Pyorala SP. Transgenic cows that produce recombinant human lactoferrin in milk are not protected from experimental Escherichia coli Intramammary Infection. *Infect. Immun* 2006; 74: 6206–6212.

Jack DB. Keep taking the tomatoes - the exciting world of nutraceuticals. *Mol Med Today* 1995; 118-121.

Jacobs DR, & Tapsell LC. Food synergy: The key to a healthy diet. *Proceedings of the Nutrition Society* 2013, 72: 200-206.

Jacobson MF, Silverglade B, Heller IR. Functional foods: Health boon or quackery? *West J Med.* 2000; 172:8-9.

Jacques H, Gascon A, Bergeron N, Lavigne C, Hurley C, Deshaies Y, Moorjani S, Julien P. Role of dietary fish protein in the regulation of plasma lipids. *Can J Cardiol* 1995; 11 Suppl G:63G-71G.

Jain SK, Rains J, Croad J, Larson B, Jones K. Curcumin supplementation lowers TNF-a, IL-6, IL-8, and MCP-1 secretion in high glucose treated cultured monocytes and blood levels of TNF-a, IL- 6,MCP-1, glucose, and glycosylated hemoglobin in diabetic rats. *Antioxid Redox Signal* 2009;11:241-249.

Jones WHS. (1945). *Hippocrates and the corpus Hippocraticum* (G. Cumberlege).

Kalimuthu S and Se-Kwon K. Cell survival and apoptosis signaling as therapeutic target for cancer: Marine bioactive compounds. *Int. J. Mol. Sci.* 2013; 14:2334–2354.

Kalra EK. Nutraceutical-definition and introduction. *Aaps Pharmsci* 2003; 5:27-28.

Kalra N, Arora A and Shukla Y. Involvement of multiple signaling pathways in diallyl sulphide mediated apoptosis in mouse skin tumours. *Asian. Pac. J. Cancer Prev.* 2006; 7:556–562.

Kastorini CM, Milionis HJ, Esposito K, Giugliano D, Goudevenos JA and Panagiotakos DB. The effect of Mediterranean diet on metabolic syndrome and its components: A meta-analysis of 50 studies and 534,906 individuals. *J. Am. Coll. Cardiol* 2011; 57: 1299-1313.

Katiyar SK and Mukhtar H. Tea in chemoprevention of cancer: Epidemiologic and experimental studies (review). *Intl. J. Oncol.* 1996; 8: 221–238.

Kaur G, Mukundan S, Wani V, Kumar MS. Nutraceuticals in the management and prevention of metabolic syndrome. *Austin J Pharmacol Ther* 2015; 3:1-6.

Kavanaugh CJ, Trumbo PR, Ellwood KC. The U.S. Food and Drug Administration's evidence-based review for qualified health claims: Tomatoes, lycopene, and cancer. *J Natl Cancer Inst.* 2007; 99:1074-1085.

Keservani RK, Kesharwani RK, Vyas N, Jain S, Raghuvanshi R, and Sharma A K. Nutraceutical and functional food as future food. *A review. Der. Pharmacia. Lettre* 2010; 2: 106-116.

Khajehdehi P, Pakfetrat M, Javidnia K, Azad F, Malekmakan L, Nasab MH, Dehghanzadeh G. Oral supplementation of turmeric attenuates proteinuria, transforming growth factor-β and interleukin-8 levels in patients with overt type 2 diabetic nephropathy: a randomized, double-blind and placebo-controlled study. *Scand J Urol Nephrol* 2011; 45: 365-370.

Khatiwada J, Verghese M, Walker LT, Shackelford L, Chawan CB and Sunkara R. Combination of green tea, phytic acid, and inositol reduced the incidence of azoxymethane-induced colon tumors in Fisher 344 male rats. *LWT Food Sci. Tech.* 2006; 39:1080–1086.

Kim CS, Kawada T, Kim BS, Han IS, Choe SY, Kurata T, *et al*. Capsaicin exhibits anti-inflammatory property by inhibiting IkB-a degradation in LPS-stimulated peritoneal macrophages. *Cell Signal* 2003;15:299-306.

Kim HM, Kim J. The effects of green tea on obesity and type 2 diabetes. *Diabetes Metab J* 2013; 37: 173-175.

Kim YA, Xiao D, Xiao H, Powolmy AA, Lew KL, Reilly ML, Zeng Y, Wang Z and Singh SV. Mitochondria-mediated apoptosis by diallyl trisulfide in human prostate cancer cells is associated with generation of reactive oxygen species and regulated by Bax/Bak. *Mol. Cancer Ther.* 2007; 6:1599–1609.

Kirkham S, Akilen R, Sharma S, Tsiami A. The potential of cinnamon to reduce blood glucose levels in patients with type 2 diabetes and insulin resistance. *Diabetes Obes Metab* 2009; 11: 1100-1113.

Kojuri J, Vosoughi AR., and Akrami M. Effects of anethum graveolensand garlic on lipid profile in hyperlipidemic patients. *Lipids Health Dis.* 2007; 1:6:5.

Kris-Etherton PM, Hu FB, Ros E, Sabate J. The role of tree nuts and peanuts in the prevention of coronary heart disease: Multiple potential mechanisms. *J Nutr.* 2008; 138(suppl):1746S-1751S.

Kris-Etherton PM, Taylor DS, Smiciklas-Wright H, Mitchell DC, Bekhuis TC, Olson BH, *et al.* High-soluble-fiber foods in conjunction with a telephone-based, personalized behavior change support service result in favorable changes in lipids and lifestyles after 7 weeks. *J Am Diet Assoc* 2002;102:503-10.

Kumar B, Gupta SK, Nag TC, Srivastava S, Saxena R. Green tea prevents hyperglycemia-induced retinal oxidative stress and inflammation in streptozotocin-induced diabetic rats. *Ophthalmic Res* 2012; 47: 103-108.

Kumar GB, Srinivas L and Ganapathi TR, Iron fortification of banana by the expression of soybean ferritin. Biol. Trace Elem. Res, (2010).

Laing P. Childhood obesity: A public health threat. *Paediatr Nurs* 2002;14:14-6.

Lampe JW. Health effects of vegetables and fruit: Assessing mechanisms of action in human experimental studies. *Am. J. Clin. Nutr.* 199; 70:475–490.

Lankinen M, Schwab U, Kolehmainen M, Paananen J, Poutanen K, Mykkänen H, Seppänen-Laakso T, Gylling H, Uusitupa M, Orešič M. Whole grain products, fish and bilberries alter glucose and lipid metabolism in a randomized, controlled trial: the Sysdimet study. *PLoS One* 2011; 6: e22646.

Lavigne C, Tremblay F, Asselin G, Jacques H, Marette A. Prevention of skeletal muscle insulin resistance by dietary cod protein in high fat-fed rats. *Am J Physiol Endocrinol Metab* 2001; 281:E62-71.

Lekshmi PC, Arimboor R, Raghu KG, Menon AN. Turmerin, the antioxidant protein from turmeric (Curcuma longa)exhibits antihyperglycaemic effects. *Nat Prod Res* 2012; 26:1654-1658.

Lin MY, Yen CL. Reactive oxygen species and lipid peroxidation product-scavenging ability of yogurt organisms. *J Dairy Sci* 1999; 82: 1629-1634.

Liu L, Miron A, Klímová B, Wan D, Kuča K. The antioxidant, immunomodulatory, and anti-inflammatory activities of Spirulina: an overview. *Archives of Toxicology* 2016; 90(8): 1817-1840.

Liu RH. Health benefits of fruit and vegetables are from additive and synergistic combinations of phytochemicals. *Am. J. Clin. Nutr.* 2003; 78:517–520.

Liu RH. Potential synergy of phytochemicals in cancer prevention: Mechanism of action. *J. Nutr.* 2004; 134:3479–3485.

Liu S, Choi HK, Ford E, Song Y, Klevak A, Buring JE, Manson JE. A prospective study of dairy intake and the risk of type 2 diabetes in women. *Diabetes Care* 2006; 29: 1579-1584.

Lopitz-Otsoa F, Rementeria A, Elguezabal N, Garaizar J. Kefir: a symbiotic yeasts-bacteria community with alleged healthy capabilities. *Rev Iberoam Micol* 2006; 23: 67-74.

Lusk JL, & Rozan A. Consumer acceptance of biotechnology and the role of second generation technologies in the USA and Europe. *Trends in Biotechnology* 2005, 23: 386-387.

MacRedmond R, Singhera G, Attridge S, Bahzad M, Fava C, Lai Y, Hallstrand TS and Dorscheid DR. Conjugated linoleic acid improves airway hyper-reactivity in overweight mild asthmatics. *Clin. Exp. Allergy* 2010; 40: 1071-1078.

Mahmoodi M, Islami MR, Asadi-Karam GR, Khaksari M, Sahebghadam-Lotfi A, Hajizadeh MR., and Mirzaee MR. Study of the effects of raw garlic consumption on the level of lipids and other blood biochemical factors in hyperlipidemic individuals. *Pak. J. Pharm. Sci.* 2006; 19:295–298.

Malbasa RV, Loncar ES, Vitas JS and Canadanovic-Brunet JM. Influence of starter cultures on the antioxidant activity of kombucha beverage. *Food Chem* 2011; 127: 1727-1731.

Manjunatha H, Srinivasan K. Protective effect of dietary curcumin and capsaicin on induced oxidation of low-density lipoprotein, iron-induced hepatotoxicity and carrageenan-induced inflammation in experimental rats. *FEBS J* 2006; 273:4528-37.

Mannion M. Nutraceutical revolution continues at Foundation for Innovation in Medicine Conference. *Am J Nat Med* 1998; 5:30-33.

Marotta F, Chui DH, Jain S, Polimeni A, Koike K, Zhou L, Lorenzetti A, Shimizu H and Yang H. Effect of a fermented nutraceutical on thioredoxin level and TNF-alpha signaling in cirrhotic patients. *J. Biol. Regul. Homeost. Agents* 2011; 25: 37-45.

Marteau PR. Probiotics in clinical conditions. *Clin Rev Allergy Immunol* 2002; 22: 255-273.

Matsura N, Miryamae Y, Yamane K, Nagao Y, Hamada Y, Kawaguchi N, Katsuki T, Hirata K, Surmi S and Ishikawa H. Aged garlic extract inhibits angiogenesis and proliferation of colorectal carcinoma cells. *J. Nutr.* 2006; 136:842–846.

McEwen B, Morel-Kopp MC, Tofler G, Ward C. Effect of omega-3 fish oil on cardiovascular risk in diabetes. *Diabetes Educ* 2010; 36: 565-584.

Michail S, Sylvester F, Fuchs G and Issenma R. Clinical efficacy of probiotics: Review of the evidence with focus on children, clinical practice guideline. *J Pediatr Gastroenterol Nutr* 2006; 43(4): 550-557.

Mirmiran P, Bahadoran Z, Azizi F. Functional foods-based diet as a novel dietary approach for management of type 2 diabetes and its complications: A Review, *World J Diabetes* 2014 June 15; 5(3): 267-281.

Mital BK and Garg SK. Anticarcinogenic, hypocholesterolemic, and antagonistic activities of Lactobacillus acidophilus. *Crit. Rev. Micro.* 1995;21:175–214.

Miyashita K, Nishikawa S, Beppu F, Tsukui T, Abe M and Hosokawa M. The allenic carotenoid fucoxanthin, a novel marine nutraceutical from brown seaweeds, *J. Sci. Food Agric* 2011; 91: 1166-1174.

Moriarty RM, Naithani R, Surve B. Orgaosulfur compounds in cancer chemoprevention. *Mini Rev Med Chem*. 2007; 7:827-838.

Nakachi K, Suemasu K, Suga K, Takeo T, Imai K and Higashi Y. Influence of drinking green tea on breast cancer malignancy among Japanese. *Jpn. J. Cancer Res.* 1998; 89:254–261.

Naqvi S, Zhu C, Farre G, Ramessar K, Bassie L, Breitenbach JD, Conesa P, Ros G, Sandmann G, Capell T and Christou P. Transgenic multivitamin corn through biofortification of endosperm with three vitamins representing three distinct metabolic pathways. *Proc. Nat. Acad. Sci. U S A* 2009; 106: 7762-7767.

Nowack R. Cranberry juice—a well-characterized folk-remedy against bacterial urinary tract infection. *Wien Med Wochenschr*. 2007; 157:325-330.

Nworu C, Vin-Anuonye T, Okonkwo E, Oyeka C, Okoli U, Onyeto C, et al. Unregulated promotion and sale of herbal remedies: A safety and efficacy evaluation of twelve such commercial products claimed to be beneficial and patronized for a variety of ailments in Nigeria. *Journal of Pharmacovigilance* 2014; S1: 2-9.

Pan MH, Ghai G and Ho CT. Food bioactives, apoptosis, and cancer. *Mol. Nutr. Food Res.* 2008; 52:43–52.

Pandey M, Verma RK., & Saraf SA. Nutraceuticals: New era of medicine and health. *Asian Journal of Pharmaceutical and Clinical Research* 2010; 3: 11-15.

Park JY, Kawada T, Han IS, Kim BS, Goto T, Takahashi N, *et al.* Capsaicin inhibits the production of tumor necrosis factor by LPS-stimulated murine macrophages, RAW 264.7: A PPARγ ligand-like action as a novel mechanism. *FEBS Lett* 2004; 572:266-70.

Pecis M, de Azevedo MJ, Gross JL. Chicken and fish diet reduces glomerular hyperfiltration in IDDM patients. *Diabetes Care* 1994; 17: 665-672.

Pereira MA and Ludwig DS. Dietary fibre and body weight regulation: Observations and mechanisms. *Ped. Clin. N. Am.* 2001; 48:969–980.

Pew Initiative on Food and Biotechnology. *Application of biotechnology for functional foods*. http://pewagbiotech.org/research/functionalfoods/. Accessed January 9, 2009.

Pezzuto JM. Grapes and human health: A perspective. *J. Agric. Food Chem.* 2008; 56:6777–6784.

Pfeuffer M, Schrezenmeir J. Milk and the metabolic syndrome. *Obes Rev* 2007; 8: 109-118.

Pineiro M and Stanton C. Probiotic Bacteria: Legislative framework— Requirements to evidence basis. *J. Nutr 2007*; 137: 850S–853S.

Position of The American Dietetic Association: Functional foods. *J Am Diet Assoc.* 2009; 109:735-746.

Position of The American Dietetic Association: Functional foods. *J Am Diet Assoc*. 2004; 104:814-826.

Potter SM. Overview of possible mechanisms for the hypercholesterolemic effect of soy protein. *J. Nutr.*1995; 125:606–611.

Pourahmad J, Eskandari MR, Shakibaei R, Kamalinejad M.A search for hepatoprotective activity of aqueous extract of Rhus coriaria L. against oxidative stress cytotoxicity. *Food Chem Toxicol* 2010; 48: 854-858.

Quigley EM. Bacteria: A new player in gastrointestinal motility disorders-infections, bacterial overgrowth, and probiotics. *Gastroenterol Clin North Am.* 2007; 6:735-748.

Rajat S, Manisha S, Robin S, & Sunil K. (2012). Nutraceuticals: A review. *International Research Journal of Pharmacy* 2012; 3: 4.

Rayne S, Mazza G. Biological activities of extracts from sumac (Rhus spp.): a review. *Plant Foods Hum Nutr* 2007; 62:165-175.

Riccioni G, Mancini B, Di Ilio E, Bucciarelli T and D'Orazio N (2008). Protective effect of lycopene in cardiovascular disease. *Eur. Rev. Med. Pharmacol. Sci.* 2008; 12:183–190.

Rinaldi A. A nutraceutical a day may keep the doctor away; *EMBO* 2005; 6(8): 708-711.

Ross S. Functional foods: The Food and Drug Administration perspective. *Am J Clin Nutr.* 2000; 71(suppl):1735S-1738S.

Saika D, Deka SC. Cereals: From staple food to nutraceuticals. *Int Food Res J* 2011;18:21-30.

Santaguida PL, Balion C, Hunt D, Morrison K, Gerstein H, Raina P, Booker L, Yazdi H. Diagnosis, prognosis, and treatment of impaired glucose tolerance and impaired fasting glucose. *Evid Rep Technol Assess* (Summ) 2005; 128: 1-11.

Sapkale AP, Thorat MS, Vir PR, Singh MC. Nutraceutical-global status and applications: A review. *Int J Pharm Clin Sci* 2012; 1:1166-1181.

Sazawal S, Dhingra U, Hiremath G, Sarkar A, Dhingra P, Dutta A, Verma P, Menon VP and Black RE. Prebiotic and probiotic fortified milk in prevention of morbidities among children: community-based, randomized, double-blind, controlled trial. *PLoS One* 2010; 5: e12164.

Schloss PD. (2014). Microbiology: An integrated view of the skin microbiome. *Nature* 2014; 514: 44-45.

Scholz-Ahrens KE, Adolphi B, Rochat F, Barclay DV, de Vrese M, Açil Y, et al. (2016). Effects of probiotics, prebiotics, and synbiotics on mineral metabolism in ovariectomized rats impact of bacterial mass, intestinal absorptive area and reduction of bone turn-over. *NFS Journal* 2016; 3: 41-50.

Sekiya M, Yahagi N, Matsuzaka T, Najima Y, Nakakuki M, Nagai R, *et al.* Polyunsaturated fatty acids ameliorate hepatic steatosis in obese mice by SREBP-1 suppression. *Hepatology* 2003; 38:1529- 1539.

Sethi JK, Hotamisligil GS. The role of TNF-a in adipocyte metabolism. *Semin Cell Dev Biol* 1999;10:19-29.

Shab-Bidar S, Neyestani TR, Djazayery A. Efficacy of vitamin D3-fortified-yogurt drink on anthropometric, metabolic, inflammatory and oxidative stress biomarkers according to vitamin D receptor gene polymorphisms in type 2 diabetic patients: a study protocol for a randomized controlled clinical trial. *BMC Endocr Disord* 2011; 11: 12.

Shafiei M, Nobakht M, Moazzam AA. Lipid-lowering effect of Rhus coriaria L. (sumac) fruit extract in hypercholesterolemic rats. *Pharmazie* 2011; 66: 988-992.

Shahidi F. Functional foods: Their role in health promotion and disease prevention. *J. Food Sci.* 2004; 69:146–149.

Sharma D and Tiwari S, Use of cranberry in Urinary Tract Infection of E. coli, *Int. J. of Pharm. & Life Sci. (IJPLS)* 2012 June; 3(6): 1784-1786.

Shoelson SE, Lee J, Goldfine AB. Inflammation and insulin resistance. *J Clin Invest* 2006;116:1793-801.

Sibbesen, Ole, Sorensen, Jens, Frisbaek, Polypeptides with xylanase activity, *WIPO* Patent Application, (2010).

Silagy CA and Neil A. Garlic as a lipid-lowering agent – a metaanalysis. *J. Royal Coll. Physicians Lond.* 1994; 28:39–45.

Simopoulos AP. Omega-3 fatty acids in health and disease and in growth and development. *Am. J. Clin. Nutr.* 1991; 54:438–463.

Singh BB, Vinjamury SP, Der-Martirosian C, Kubik E, Mishra LC., Shepard NP, Singh VJ, Meier M and Madhu SG. Ayurvedic and collateral herbal treatments for hyperlipidemia: A systematic review of randomized controlled trials and quasi-experimental designs. Alter. *Ther. Health Med.* 2007; 13:22–28.

Singh J, Sinha S. Classification, regulatory acts and applications of nutraceuticals for health. *Int J Pharm Biol Sci* 2012; 2:177-87.

Siro I, Kapolna E, Kapolna B, & Lugasi A. Functional food. Product development, marketing and consumer acceptance- review. *Appetite* 2008; 51: 456-467.

Sloan AE. The top 10 functional food trends. *Food Technol.* 2008; 62:24-44.

Srivastava KC, Bordia A and Verma SK. Garlic (Allium sativum) for disease prevention. *S. Afr. J. Sci.* 1995; 91:68–77.

St. Leger AS, Cochrane AL. and Moore F. Factors associated with cardiac mortality in developed countries with particular reference to the consumption of wine. *The Lancet.* 1979; 313:1017–1020.

Steinmetz KA and Potter JD. Vegetables, fruit and cancer II. Mechanisms. *Cancer Causes Control.* 1991; 2:427–442.

Steptoe A, Gibson E, Vuononvirta R, Hamer M, Wardle J, Rycroft J, Martin J and Erusalimsky J. The effects of chronic tea intake on platelet activation and inflammation: A double-blind placebo controlled trial. *Atherosclerosis.* 2007; 193:277–282.

Teruya K, Yamashita M, Tominaga R, Nagira T, Shim SY, Katakura Y, Tokumaru S, Tokumaru K, Barnes D, Shirahata S. Fermented milk, Kefram-Kefir enhances glucose uptake into insulin-responsive muscle cells. *Cytotechnology* 2002; 40:107-116.

Thompson LU. (1995). Flaxseed, lignans, and cancer. In: Flaxseed in Human Nutrition, pp. 219–236. Cunnane, S. and Thompson, L. U., Eds., AOCS Press, Champaign, IL.

Trayhurn P, Wood IS. Adipokines: Inflammation and the pleiotropic role of white adipose tissue. *Br J Nutr* 2004; 92:347-55.

Truswell AS. Cereal grains and coronary heart diseases. *Eur. J. Clin. Nutr.* 2002; 56:1–14.

Udenigwe CC, & Aluko RE. Food protein-derived bioactive peptides: Production, processing, and potential health benefits. *Journal of Food Science* 2012; 77: R11-R24.

Venhuis BJ, van Hunsel F, van de Koppel S, Keizers PHJ, Jeurissen SMF, & De Kaste D. Pharmacologically effective Red Yeast Rice preparations marketed as dietary supplements illustrated by a case report. *Drug testing and analysis* 2016; 8: 315-318.

Wang CZ, Mehendale SR and Yuan CS. Commonly used antioxidant botanicals: Active constituents and their potential role on cardiovascular illness. *Am. J. Chin. Med.* 2007; 35:543–558.

Weisberg SP, Leibel R, Tortoriello DV. Dietary curcumin significantly improves obesity-associated inflammation and diabetes in mouse models of diabesity. *Endocrinology* 2008;149:3549-3558.

West GE, Gendron C, Larue B, & Lambert R. Consumers' valuation of functional properties of foods: Results from a Canada-wide survey. *Canadian Journal of Agricultural Economics/Revue canadienne d'agroeconomie* 2002; 50: 541-558.

White E, Current status of metabolically engineered resveratrol, MMG 445. *Basic Biotech,* 5, 2009.

WHO Cardiovascular diseases (CVDs) 2016 http://www.who.int/mediacentre/factsheets/fs317/en/.

Wild S, Roglic G, Green A, Sicree R, King H. Global prevalence of diabetes: Estimates for the year 2000 and projections for 2030. *Diabetes Care* 2004; 27:1047-1053.

Xu G, Ren G, Xu X, Yuan H, Wang Z, Kang L, Yu W and Tian K. Combination of curcumin and green tea catechins prevents dimethylhydrazine- induced colon carcinogenisis. *Food Chem. Toxicol.* 2010; 48:390–395.

Yan L, Yee JA, Li D, McGuire MH and Thompson LU. Dietary flaxseed supplementation and experimental metastasis of melanoma cells in mice. *Cancer Lett.* 1998; 124:181–186.

Yang B, Wang J, Tang B, Liu Y, Guo C, Yang P, Yu T, Li R, Zhao J, Zhang L, Dai Y and Li N. Characterization of bioactive recombinant human lysozyme expressed in milk of cloned transgenic cattle. *PLoS One* 2011; 6: e17593.

Yang CS and Wang ZY. Tea and cancer. *J. Nat. Cancer Inst.* 1993;85:1038–1049.

Zeisel SH. Regulation of "Nutraceuticals." *Science* 1999; *285*: 185-186.

Zhang ZM, Yang XY, Deng SH, Xu W and Gao HQ. Anti-tumor effects of polybutylcyanoacrylate nanoparticles of diallyl trisulfide on orthotopic transplantation tumor model of hepatocellular carcinoma in BALB/c nude mice. *Chin. Med. J. (Engl).* 2007; 120:1336–1342.

In: Trends in Biochemistry and Molecular Biology
Editors: Hossain Uddin Shekhar et al.
ISBN: 978-1-53616-434-3

Chapter 5

NUTRACEUTICAL ASPECTS OF RICE AND RICE BASED FOOD ITEMS TO MITIGATE MALNUTRITION IN BANGLADESH

Shozib Habibul Bari*[*], *PhD
Grain Quality and Nutrition, Division, Bangladesh Rice Research Institute, Gazipur, Bangladesh

ABSTRACT

Consumption of rice is the most vital source of dietary energy intake in Bangladesh. Although rice consumption among the Bangladeshi population is a decreasing trend, yet more than 60% of daily recommended energy intake comes from rice alone and it varies from rural to urban population. Rice is the only cereal that has diverse applications in our traditional food chain, and it is accessible to all classes of people. Hence, nutraceutical aspects of rice and rice-based food products drew attention for alternative approaches to mitigate malnutrition in Bangladesh. Bangladesh Rice Research Institute (BRRI) plays an essential role in this regard, by characterizing nutraceutical enriched rice varieties, including: antioxidant, low GI, GABA and Zinc enriched rice. Taking these nutraceutical properties into account, BRRI has formulated energy-dense rice biscuit (ED $\geq$5.0) to mitigate malnutrition in Bangladesh, especially among the most vulnerable portion of our population such as street children. BRRI has also formulated rice-based bakery products: such as rice biscuits, cake, bread, etc. to popularize rice powder instead of wheat powder in the bakery industry of Bangladesh. Rice has been considered to be unique among the cereals for its superior nutritional quality, considering its highest biological value (BV), net protein utilization (NPU), true digestibility (TD) and utilizable nitrogen (UN) among all other cereals: such as wheat, corn, barley, millet, and sorghum. Minerals such as Aluminium, Calcium, Chloride, Iron, Magnesium, Manganese, Phosphorous, Potassium, Silicon, Sodium, and Zinc are found in rice. As rice does not have gluten protein, gluten associated diseases, including: Celiac disease, IBS, gluten allergy, etc. are not supposed

[*] Corresponding Author's E-mail: shozib11@gmail.com.

to occur by consuming rice products. Rice products require a lesser amount of rice powder as a carbohydrate source to attain comparatively higher energy; hence it would reduce the amount of rice consumption gradually, which would ultimately neutralize the increasing demand of rice production in Bangladesh.

Keywords: rice nutraceutical properties, low GI rice, Antioxidant rice, GABA rice, Zinc enriched rice.

1. INTRODUCTION

Nutritional aspects of rice mainly deal with the proximate composition of rice and milling process. The effect of post-harvest operations: rice processing, degree of milling and traditional cooking practices also play a vital role in trapping nutrition from rice. First-limiting amino acid, Lysine (g/16gN) and Threonine (g/16gN) are also found in ample amounts in all cereals such as wheat, corn, barley, millet and sorghum (Eggum, 1979). Minerals, especially Aluminium, Calcium, Chloride, Iron, Magnesium, Manganese, Phosphorous, Potassium, Silicon, Sodium, and Zinc are found in rice, and concentration of these minerals are varied on different parts of grain, including: brown rice, milled rice, rice bran, and rice embryo. A considerable portion of the rice ash is accounted for by phosphorus, magnesium, and potassium. But drastic losses of protein, calcium, iron, phosphorus, B-complex vitamins occur while storing, milling, polishing, ringing, and decanting excess cooking water while cooking rice.

Bangladesh is a land of diversified rice cultivation, and rice is treated as one of the most reliable sources of energy from its carbohydrate portion since the prehistoric days in Bangladesh. A total of ninety-two HYVs, including both inbred and hybrids, have been released by Bangladesh Rice Research Institute (BRRI) till now. At present, total milled rice production is about 34,909 thousand metric tons, which is supposed to be enough to meet the domestic requirement of feeding approximately 160 million people. BRRI rice scientists have the responsibility to achieve the goal of adequate rice production with efficient and improved modern HYVs, addressing both abiotic and biotic stress conditions. Since Bangladesh maintained self-sufficient status through surplus rice production of 1,767 thousand metric tons in 2018,thus special emphasis is required on the improvement rice grain quality and its prospective nutraceutical aspects to represent its essential role in combating with increasing rate of non-communicable diseases in Bangladeshi population, including: heart disease, cancer, diabetics etc. Grain Quality and Nutrition (GQN) Division of BRRI has recently identified some promising nutraceutical enriched HYVs: such as black rice, antioxidant-enriched rice, low glycaemic-index rice, anti-depressive alias GABA enriched pregerminated brown rice and Zn enriched rice etc. The nutraceutical food consumption from cereals might play an essential role in curative as well as preventive measure of non-communicable diseases (NCDs). Having resourceful functional properties, rice is unique in digestibility, hypoallergenic, superior nutritional quality- considering its highest biological value, net protein utilization, true digestibility and utilizable nitrogen among all other cereals.

2. MICRONUTRIENT-ENRICHED RICE VARIETIES IN BANGLADESH

Considering the Bangladeshi context, malnutrition problem is measured to be alarmingly high. According to FAO (2017) and World Development Indicators (WDI, 2016),an estimation of 36%, 14% and 33% children under 5 years of age are suffering from stunting, wasting and underweight respectively in Bangladesh, in addition with 24.4 million (15.1%) people under the prevalence of undernourishment (PoU). In Bangladesh, the prevalence of Zn deficiency is 45% and 57% in preschool-age children and non-pregnant, non-lactating women respectively. Shozib et al. 2017a had screened for a mineral profiling of 68 BRRI released HYVs, and reported that both BRRI dhan42 and BRRI dhan43 have higher amounts of Zn ≥27.0 mg kg-1 at 10% degree of milling for Aus season. BRRI dhan43 has other important minerals such as Fe, Ca, P etc. at the concentration of 17.0 mg/kg, 68.1 mg/kg, and 2.5 mg/kg respectively. In accessing bioavailability, the molar ratio of phytic acid to Zn (PA/Zn); Fe (PA/Fe); Ca(PA/Ca) and P (PA/P) are found lower in BRRI dhan43 among tested HYVs by 3.6, 6.9,1.2 and 25.7 respectively. Therefore, both BRRI dhan43 and BRRI dhan42 could potentially be used as parental sources for the development of advanced breeding lines of micronutrient enriched rice (MER), particularly for Aus season. BRRI has released five Zn enriched HYVs, including BRRI dhan62 (19.5 mg kg^{-1}) and BRRI dhan72 (22.8 mg kg^{-1}) for Aman season, and BRRI dhan64 (24.0 mg kg^{-1}), BRRI dhan74 (24.2 mg kg^{-1}) and BRRI dhan84 (27.5 mg kg^{-1}) for Boro season. Rice is considered a good source of water-soluble vitamins such as thiamin and riboflavin among cereals. Shozib et al. 2018 found that BRRI dhan36 has the highest thiamin content ($mg100g^{-1}$) of 1.15 among all tested 35 HYVs in Bangladesh. Considering all nutritional and physicochemical properties such as apparent amylose content, protein content, elongation ratio, imbibition ratio, cooking time, water-soluble vitamins specially thiamin, mineral contents including Fe, Zn, phytic acid and molar ratio of phytic acid to minerals, BRRI dhan42 and BRRI dhan43 were found appropriate to be termed as nutraceutical enriched HYVs in Bangladesh.

3. LOW GLYCEMIC INDEX (GI) RICE VARIETIES IN BANGLADESH

Rice is a good source of dietary energy and protein. We usually consume white rice or polished rice. The degree of polishing is a very important indicator of whether rice serves us nutritiously or not. Shozib et al. 2018 reported that rice grain size and shape had a significant impact on the degree of milling (DOM), and mineral content tends to decrease significantly because of increasing the polishing time. In the transitioning food environment, white rice is categorized as a polished grain, and is thus implicated in the development of non-communicable diseases (NCDs). Physicochemical properties, such as apparent amylose content (AAC), imbibition ratio and dietary fiber content, post-harvest processing as well as cooking methods are influential factors in determining GI variability (SE, 2015). Epidemiological studies, describing the carbohydrate quality in diets consumed by different populations, have utilized the GI tables to quantify daily dietary GI and GL to elucidate the diet-disease relationship (Murakami, 2006). Most of the carbohydrates, especially the rice that we commonly consume, are complex carbohydrates essentially made of starches belonging to the different scales of amylose categories from waxy (0-2% AAC) to high amylase (>25% AAC) (Montignac method, 2017). Glycemic Index (GI) corresponds to carbohydrates' potential to raise blood sugar levels. Shozib et al. 2017b surveyed a total of 72 BRRI HYVs for low GI rice screening and three HYVs such as BRRI dhan69, BRRI dhan46, and BR16 were found as low GI rice (GI $\leq$ 55), 50 HYVs were found as intermediate GI rice (GI 56-69), and the rest 19 HYVs were found as high GI rice (GI $\geq$ 70) at unparboiled milling condition in vivo experimental rat model. Shozib et al. 2017c reported that BR16 is a low GI rice variety (GI 52.4) with the highest content of both α-amylase inhibitory activity, AAIA (122 μg/g AE), and resistant starch (RS) of 4.68% among all tested HYVs. For a better explanation of the GI value of a particular variety, we have to consider a group of parameters consisting of AAIA, imbibition ratio (IR), and RS content, along with apparent amylose content (AAC). Authors had studied the effects of different rice processing methods, including: parboiling, pressure parboiling, double parboiling, brown rice condition on GI in an in-vivo experiment and reported that GI value reduces towards pressure parboiling, double parboiling, parboiling and are mainly reduced at the brown rice than unparboiled mill rice condition. Due to fiber content, brown rice had shown the lowest GI among all rice processing methods. Thermal and pressuring process helped in lowering GI value from original because of retrogradation changes in starch.

4. Antioxidant-Enriched Rice Varieties in Bangladesh

Rice has antioxidant properties due to the existence of phenolic compounds. Antioxidant trends to reduce the formation of the concentrations of reactive free radicals. Epidemiological studies recommend that rice consuming regions of the world have a low frequency of chronic diseases which might possibly be linked with the rice antioxidants. Rice antioxidant activities include flavonoids, phenolic acids, anthocyanins, proanthocyanidins, tocotrienols, tocopherols, phytic acid, and γ-oryzanol. Black rice varieties appear as the maximum antioxidant activities, followed by purple rice, red rice, and brown rice. In general, the milled rice is preferred over the whole grain for its desirable sensory properties and storage stability. The bran layer consists of the bran, including: pericarp, seed coat, nucellus aleurone and the germ (or embryo). Rice bran is the best source of flavonoids, phenolic acids, anthocyanin, proanthocyanidin, and γ-oryzanol compounds. Depending on the rice color, the vitamin E distribution in rice is ranked in the following decreasing order: bran > whole-grain > endosperm > husk. Approximately 90% of the phytate phosphorus (Phytic acid) is concentrated in the bran, 4-5% in the rice endosperm and 4-5% in the husk (Piebiep Goufo, 2014). Total phenolic content (TPC), total flavonoid content (TFC), total anthocyanin content (TAC), and DPPH radical scavenging activity and ferric ion reducing antioxidant power (FRAP) are few common antioxidant parameters which are reported frequently (Hirawan et al. 2011; Saikia et al. 2012). Alok et al. 2012 reported that, among tested HYVs of rice in Bangladesh, BR5 had the highest total phenolic content (TPC), ferric reducing antioxidant power (FRAP), and total antioxidant capacity (TAC). BR5 is an aromatic short bold type of rice grain, with a higher protein content of 9.1%, and apparent amylose content (AAC) of 26.1%. In another study, 50 local Bangladeshi boro rice germplasms were examined for TAC, FRAP and DPPH to assess their antioxidant properties. Shozib et al. 2019 unpublished data reveal that local boro germplasms such as Baulam, Kalijira, Kasra, Basanti and Bashful had a higher value of TAC (μM AAE/100g) 645.79, 627.47, 618.31, 613.73, and 609.15 respectively. Sadatupa showed the highest DPPH (30.60 IC50 mg/mL) and Lafai had shown the highest FRAP (370.49 μM AAE/100g) among all tested local boro germplasms. In 2015, Shozib et al. 2015 demonstrated how dietary administration of antioxidant-enriched rice improves the antioxidant status in blood of the experimental long-evan rat. Hundreds of local pigmented rice germplasms are currently characterizing for antioxidant properties at GQN, Division, BRRI Bangladesh.

5. GABA-Enriched Rice Variety in Bangladesh

GABA (Gamma-aminobutyric acid) enriched rice is made by germinating brown rice to sprout its 0.5-1.0 mm length cereal germ, fully activating the rice's innate nutrition and making it fluffier, tastier and easier to digest. Brown rice (BR) can be soaked in water at 30°C for 3 hours, and sprouting for 21 hours has been found to be optimum for getting the highest GABA content at pregerminated brown rice (Swati, 2011). GABA is reported to be found in significantly higher concentration at PGBR condition than brown rice. PGBR has physiological effects on anti-hypertension and anti-hyperlipidaemia, and plays a vital role in reducing the risk of non-communicable diseases including diabetes, cancer, and Alzheimer's. Siddiquee et al. 2017 had reported that BRRI dhan31 could generate a high level of GABA at PGBR condition among tested local and HYVs in Bangladesh.

BRRI dhan31 is a medium bold type grain, and its protein content is around 8.9% along with higher AAC of 26.5%, elongation ratio (ER) of 1.4, and imbibition ratio (IR) of 4.0. PGBR could be acceptable to consumers as well as the food industry as a promising foodstuff that contains more nutritional proteins, amino acid and bio-functional components than ordinary rice products. PGBR can be a dietary food for health improvement (Muhammad et al. 2017). Diversified use of GABA enriched PGBR can potentially be utilized in rice-based bakery industries in Bangladesh to achieve food safety and nutrition as well.

6. Black Rice Varieties in Bangladesh

Black rice, consisting of black pericarp or black kernel, is especially rich in anthocyanin pigments, phytochemicals, proteins, and nutrient compositions and contain their health benefits. Black (purple) rice is becoming popular not only for the development of functional foods, but also due to the genetic variability of cultivars which causes diversity in pigmentation, nutrition value, and phytochemical properties. In Bangladesh, some local black rice varieties are found in hilly areas of Sylhet and Bandarban districts. The bran hull of black rice is the outermost layer of the rice grain which contains one of the highest levels of the antioxidant anthocyanin found in any known food. Several studies revealed that the anthocyanin is the dominant compound in black rice bran, suggesting that cyanidin and peonidin are additional potential antioxidant compounds for food additive materials (Park at el, 2008, Zhang at el, 2010, Kaneda et al. 2006, Ichikawa at el, 2004, Kim at el, 2014). Anthocyanin compounds in black rice include cyanidin 3-Oglucoside, peonidin 3-O-glucoside, malvidin 3-O-glucoside, pelargonidin 3-O-glucoside, and delphinidin 3-O-glucoside (Kim at el, 2014). Park at el, 2008 reported that the most abundant anthocyanin in black rice bran is cyanidin-3-O-glucoside (~ 95%) followed by peonidin-3-O-glucoside concentrations of less than 5%. Jun et al, 2015 reported that the phenolic compound abundant in black rice bran extract was hydroxycinnamic acids (especially, ferulic acid (4-hydroxy-3-methoxy-cinnamic acid). The concentrations of hydroxycinnamic acids have also been reported to correlate with the antioxidant activity of black rice bran extract. Anthocyanin, rich in plant food, may have therapeutic uses in the treatment of cardiovascular diseases by enhancing cellular cholesterol removal via apolipoproteins. Many studies have used in-vitro cancer cell models, including cell lines derived from breast, colon, cervix, leukaemia, liver, and stomach, in order to evaluate the anti-cancer properties of compounds found in black rice (Iriti et al, 2013, Hudson et al, 2000, Chen et al,2005, Leardkamolkarn et al, 2011, Banjerdpongchai et al, 2013). Black rice contains important antioxidant Vitamin E, which is useful in maintaining eye, skin, and immune health, in addition to other important functions. Black rice decreases dangerous atherosclerotic plaque formation in the arteries, which is very important to keep the arteries clear and prevent heart attacks and stroke. Consuming black rice can help to detox the body and cleanse the liver of harmful toxic build-up. The fiber in black rice helps to prevent constipation, bloating, and other unwanted digestive symptoms. Black rice can also help prevent or cure cases of diarrhea, since fiber adds bulk to stool. Black rice naturally contains no gluten, so consuming black rice seems to test negative for Celiac Disease, including bloating, constipation, diarrhea, nutrient deficiencies, and an increased risk for developing the leaky gut syndrome. Black rice contains the entire bran of the grain where the fiber is stored, and fiber is able to help glucose (sugar) from the grain to be absorbed by the body over a longer period of time. From different districts of Bangladesh, BRRI has collected a few black rice varieties, such as: BK5, BK6, BK7, BK8, BK9, BK10, BK11, aromatic black rice (Sylhet), gabura and muktahar. It is reported that the GI of Black rice (BK11) is found low in both brown rice (GI 48) and milled rice (GI 51). It has a high protein content of 9.5% and 8.8% at brown and milled rice condition respectively. BK11 contains Zn content of 57.29 $mgkg^{-1}$(ppm) in bran followed by brown rice (23.95 ppm), cooked brown rice (22.96 ppm), polished milled rice (20.68 ppm) and rice gruel (15.72 ppm). It has gamma oryzanol of 60-65 mgg^{-1}, which is an excellent antioxidant and makes the rice bran oil special. The magnesium content of BK11 was found to have the highest value (585.0 ppm) among all tested varieties (Shozib, 2016). Fatty acid profiling of BK11 showed a healthier ratio (3.20:1) between unsaturated fatty acid (76.22%) and saturated fatty acid (23.78%). Since the yield of local black rice varieties is very poor and location-specific, thus potential black rice varieties are being selected for developing black pericarp modern inbred HYVs in Bangladesh through both conventional and marker-assisted molecular breeding.

7. Rice Bran Oil Scenario in Bangladesh

In Bangladesh, the consumption of edible oil is 26.57g $capita^{-1}day^{-1}$ or 9.70kg $capita^{-1}annum^{-1}$. Therefore, the total requirement of edible oil becomes approximately 15.09 Lakh million tonnes for a population of 160 million. Since oil seeds such as mustard seed, sesame seed, and groundnut seed provide about 2.19 Lakh MT of edible oil, the country still needs to import 13.44 Lakh million tonnes of crude edible oil for obtaining 12.90 million tonnes of refined edible oil to meet the present local demand. The Perspective Plan of Bangladesh 2010-21 aimed at increasing the production of domestic oilseeds for providing the population with 40 g $capita^{-1}day^{-1}$ or 14.6 kg $capita^{-1}annum^{-1}$ of edible oil in 2021(Ali, 2015). This is the predicted scenario of rice bran oil (RBO) in Bangladesh. In a study, Shozib et al. 2017 extracted RBO from the most antioxidant-enriched rice variety BR5 and further compared dietary intake of RBO with other edible oils such as soybean oil (SBO), mustard oil (MTO) and butter oil (BTO) to evaluate the health effects on an in-vivo experimental long-even rat model. Dietary intake of RBO was potentially found to be the best among other tested edible oils such as mustard oil, soybean oil and butter oil in terms of least increment of blood cholesterol and triglyceride in an in-vivo experimental rat model. Hence, good quality of edible RBO has an immense prospect in Bangladesh, considering both health and economic benefits (Muhammad, 2017). Rice bran oil has a huge prospective as an edible oil in Bangladesh. Along with rice production, rice bran, rice bran oil, and de-oiled rice bran have value-added rice products which might bring economic benefit for all corners related to rice and rice-based industries in Bangladesh. In Bangladesh, several types of rice bran are available, such as: bran from the auto rice mill, bran from semi-auto rice mills, and bran from Engleberg rice bran. RBO industries usually use parboiled rice bran instead of unparboiled. Habibul et al, 2018 reported that unparboiled rice bran is not suitable in RBO mills as oil content% remains lower than parboiled rice bran. Since there is no central collection system of rice bran in Bangladesh, hence RBO millers tend to collect rice bran through the suppliers from auto and semi-auto rice mills in different corners of the country. It is a tricky trade and it takes more than 7 days. It is an expected practice to send the fresh rice bran to the oil extraction mill soon after bran production from rice mill. Otherwise, the quality of rice bran would deteriorate due to the hydrolysis activity of the lipase enzyme which is retained in bran. Lipase enzyme produces 10 - 20% FFA in a day, and this increases up to 80% in a month (Biswas, et al, 2018). It is recommended that in crude and edible oil, the FFA level should not exceed 8% and 3% respectively. Haque et al, 2018 also reported that untreated rice bran, after 28 days of harvesting, had grained the highest FFA% and lowest oil %, compared to freshly harvested rice bran. Heat-treated rice bran, after 28 days of harvesting, had retained similar FFA% and oil % compared to freshly harvested rice bran, and performed better than chemically treated bran. Oil and FFA content of parboiled milling processed rice bran showed significantly greater value than unparboiled milling processed rice bran at all three treatments such as untreated, heat-treated and chemically treated in terms of several oil chemistry parameters, such as: Acid Value (as KOH), mg/g, KOH), mg/g, Free Fatty Acid as oleic (FFA) %, Peroxide Value (PV) meq O2/kg, and Iodine Value (IV) (Hanus method) and oil content % except Saponification Value (SV) (as KOH) mg/g, Refractive Index at 40ºC and Relative Density at 300C which did not shown any difference. Heat treatment at 130 - 1350C for 2 hours was found suitable for stabilizing rice bran from increasing FFA% and lowering oil content % for at least 28 days, and it is expected that lipase activity might possibly be inhibited or at least slowed down by heat treatment. Stabilization of rice bran will allow RBO industries in Bangladesh to phaseout with post-harvest loss minimization of rice bran for quality bran oil production in Bangladesh in the imminent future.

8. Mitigation of Malnutrition with Rice-Based Products in Bangladesh

UNDP in 2001 estimated that the number of street children in Bangladesh is 445226 (of which 75% are in Dhaka city); including 53% boys and 47% girls. The average daily income of street children is approx. USD $0.55. Another statistic estimated that there are over 600,000 street children living in Bangladesh, 75% of these children live in the capital city Dhaka. Official study by the Appropriate Resources for Improving Street children Environment (ARISE- 2002) revealed that some 500000 children are living on street in the country's main cities, and they warned that the number of street children is set to raise as the urban population grows by 9% a year (UK Essays, 2013). However, according to the Bangladesh Institute of Development Studies (BIDS) report in December 2004, the total number of street children in Bangladesh is 6,74,178. In 2015, according to BIDS, the number became 1.5 million, and it will reach to 1.56 million in 2024 (Mohammed, 2016). Rural-urban migration is one of the major reasons for the exposure of the street children to the vulnerabilities. Considering the increasing trend of street children and diversity of the vulnerabilities, it is very much necessary to take proper and effective steps to protect the children from all types of violations to properly ensure their basic necessities, good growth and development. Since rice is an important source of energy and it is hypoallergenic and easily digested, providing protein with higher nutritional quality and has versatile functional nutraceutical properties (Saikia, 2011), so rice-based low cost balanced, nutritious and safe diet formulation will be able to address malnutrition-related problems of the vulnerable section of our population, especially urban street children. Among cereals, rice does not have gluten protein. Since gluten protein of wheat and maize cereals produce hypersensitive reactions to humans and we do not have multiple opinions other than consuming rice only, thus rice-based food items such as rice cookies, rice ball, rice cake, rice noodles etc. can be formulated at a lower cost but higher nutritious value. We have to formulate low cost and nutritionally balanced rice-based food items from nutritionally enriched selected brown and pre-germinated brown rice varieties. Hakim et. al., 2016 in a study on health and nutritional condition of street children of Dhaka City revealed that about 65% of street children are underweight and 22.5% of children eat only two times in a day (Hakim et. al., 2016). Masud et al., 2011 conducted a population-based survey on lives and livelihoods on the streets of Dhaka city, and his data revealed that housing, food, and lack of jobs are the three most common problems for which street people are commonly sought assistance. Although there are several reports, case studies, surveys available on street children of Bangladesh (Neera et al., 2013) but very little information is available regarding food formulation, especially for their nutritional requirement to address malnutrition. BRRI scientists aimed to strategically intervene at this point to make a concrete step to eradicate malnutrition with formulated rice-based food items. Our population target is children at of 4 to10 years of age, and their expected weight ranges from 16.81 to 28.21 kg at Bangladesh population standard including both sexes (DDP, 2013). Energy requirements for the Bangladeshi population were calculated using the FAO recommendations and methodology (FAO, 2004). Occupation has a very significant impact on daily energy expenditure, and thus on per capita energy requirement. This is because individuals engaged in a particular occupation have to remain engaged in a specific activity for one-third of the total daily available time, and the type of occupation determines the mean physical activity level (PAL) of a person. Energy requirements of boys and girls (from 4 years to 10 yrs of age) range from 1362-1890 and 1244-1777 kcal/day according to FAO for Bangladeshi population (DDP, 2013). Rice-based formulated diets for street children with moderate to acute malnutrition must have some important characteristics, including: high content of micronutrients, especially growth (type II) nutrients, high energy density, adequate high protein and fat content, low content of anti-nutrients, low risk of contamination, acceptable taste and texture, cultural acceptability, ease of preparation, affordability and availability etc.. We formulated our rice-based products with several ingredients, including: rice, jackfruit seed, nut, egg, carrot, pumpkin, rice bran oil (RBO), salt, sugar, and food-grade preservatives. Since we are approaching rice-based food items, so we explored micronutrient (Zn, Fe, Ca, Mn and Mg) enriched and high protein content rice (>10% protein) varieties in our research activities. We tried to use pre-germinated brown rice and pre-germinated brown rice where we get higher protein. We have to focus on the energy density of formulated food. If the energy density is too low, the food becomes too bulky, and the children will not be able to eat adequate amounts. We had a target to formulate food which will supplement at least 30-40% of total energy requirement by one meal solution, and the formulated food should be high energy density (ED≥5.0) food. Energy density is the ratio of energy per weight of the food. The energy density is most important for children with wasting, as they have an increased energy need for catch-up growth. The most important factor influencing energy density is the fat content, as the energy density of fat (9 kcal/g), according to the Atwater factors, is more than double than that of protein and carbohydrate (4 kcal/g). Another important factor is the water content. For example, a biscuit will typically have an energy density of 4 kcal/g, whereas the energy density is much lower in gruels. Brown et al. have described in detail how different levels of energy density can influence energy intake and how energy intake is also affected by the number of meals given. In a study on children suffering from Peruvian malnutrition, the energy density increased from 1.0 kcal/g to 1.5 kcal/g, showing a significant impact on the energy intake per kilogram of body weight, which increased by approximately 20% to 25% (Brown, et al.1995). In a review of complementary feeding, nine studies comparing energy intake in malnourished children receiving diets with different energy densities were identified (WHO, 1998). In six of the studies, energy intake was considerably higher when the children were given an energy-dense diet. In most of the studies, the level of the energy density in the low-energy-density diet was about 0.5kcal/g or lower. However, a study of 5 to 18 months old malnourished children in Bangladesh compared a diet with an energy density of 0.92 kcal/g with a diet of 1.47 kcal/g, and also found an increase (about 50%) in energy intake in the group on the energy-dense diet (Brown, et al.1995). Increasing energy densities to above 1.0 kcal/g also resulted in increased energy intake among malnourished children.

9. Wheat-Based HEB and Rice-Based EDRB

World Food Program (WFP) of the United Nations has a wheat-based high energy biscuit (HEB) project operating all around the world. HEB biscuits provide 450 kilocalories of energy and at least 10 grams of protein per 100 grams. High energy biscuits (HEB) are wheat biscuits containing high-protein cereals and vegetable fat. We have formulated a similar type of product, such as energy-dense rice biscuit (EDRB), which provides 500-515 kilocalories of energy and at least 10 grams of protein per 100 grams. We have formulated our rice-based products with several ingredients, including: rice flour (nutraceutical enriched rice as mentioned above), sugar, milk powder, egg, vanilla essence, a lubricating agent such as sagu powder, butter, yeast powder, Rice Bran Oil (RBO). EDRB consists of 57% carbohydrate from rice flour, 27% fat, 10% protein, 1.2% dietary fiber, and 0.8% ash and 3.4 % moisture. It produces 515 kcal energy from consuming 100g serving, which means ED 5.15. According to BSTI (Bangladesh Standards and Testing Institution (BSTI), we have analyzed all required parameters for biscuits, and found no heavy metal such as As, Cr Pb, Cd at the detectable limit (<0.1ppm). The metabolizable energy values of all formulated food ingredients will be given in kilocalories (kcal). The energy values have to be calculated based on protein, fat, available carbohydrates and fiber values and by applying the energy conversion factors such as kcal/g 4; 9; 4 and 2 kcal respectively.

10. Baseline Survey on Street Children in Dhaka City

A semi-structured questionnaire was developed to collect data through a face-to-face interview with the respondents. The questionnaire was pre-tested and revised on the basis of feedback received from field-testing. The quantitative questionnaire mainly covered the information of study respondents including socioeconomic and demographic characteristics, current living conditions and livelihood activities, morbidity and health-seeking behaviour, dietary intake pattern education, and drug use and anthropometrical measurements such as body weight and height, BMI, waist and hip circumferences ratio, mid-arm circumference etc. The target population was children at the age between 4 to10 years and their expected weight ranges from 16.81 to 28.21 kg at Bangladesh population standard including both sexes (Desirable Dietary Pattern for Bangladesh, 2013). Formulated food was prepared based on the calorie gap between recommended calorie intake and actual calorie intake for the age of 4 to 10 years for both sexes. A total of 224 interviewees were questioned during this baseline survey. Among them, 136 are male (60.71%) and 88 were female (39.28%) street children with an age range of 4-10 years old. Primary data on daily dietary intake in kcal for this study group helps us to predict the scenario of the current status of their dietary intake per day and prospective effect of our proposed rice-based food items on their health to mitigate malnutrition. Muhammad et al, 2019 found the RDI gap of male and female were found 33-46% and 40-57% respectively, irrespective to age ranges from 4-10 year old. Since we have formulated rice-based energy-dense biscuits having ED of ≥5.0, so we can predict energy-dense rice biscuit (EDRB) can play a vital role in this regards to minimize the RDI gap by 81 to 100% for male and 74 to 100% for female.

Conclusion

Rice is the most valued cereal crop to combat non-communicable diseases, (NCDS) such as diabetes, hypertension and cancer, with its resourceful nutraceutical properties rooted in it. We tried to demonstrate that the formulation of nutraceutical enriched rice-based food items have the potential to be popular among Bangladeshi people. This might play a fundamental role in steadily decreasing overall rice consumption and preferably help to sustain food security in Bangladesh in a way by properly utilizing the rice to attain the RDI intact. Nutraceutical rice-based food formulations for specific NCDs have the potential for success in the treatment of non-communicable diseases as well as malnutrition management in Bangladesh.

References

[1] Ali S.K., Ubaidullah M.R, Hasan M.M, Hossain M.L. (2015). Prospect of Rice Bran Oil Industries in Bangladesh. *Bangladesh Journal of Tariff and Trade*, (3) 14-33.

[2] Appropriate Resources for Improving Street children Environment (ARISE- 2002) *Annual report 2002.*

[3] Banjerdpongchai R., Wudtiwai, B., and Sringarm, K., (2013). The cytotoxic effect of purple rice extracts on human cancer cells related to their active compounds. *Proceeding of 13th ASEAN Food Conference*, Singapore, 1-10.

[4] Biswas JK. Rice Bran Oil: Prospects and Probabilities in Bangladesh. *The Daily Sun* (2018).

[5] Brown, K.H, Sanchezgrinan, M., Perez, F., Peerson, J.M., Ganoza, L., Stern, J.S., (1995). Effects of dietary energy density and feeding frequency on total daily energy intake of recovering malnourished children. *Am J Clin Nutr.* 62:13–8. 9.

[6] Chen, P.N., Chu, S.C., Chiol, C.L., Yang, S.F., Hsieh, Y.S. (2005). Cyanidin 3-glucoside and Peonidin 3-glucoside inhibit tumor cell growth and induce apoptosis in vitro and suppress tumor growth in vivo. *Nutrition and Cancer.* 53(2): 232-243.

[7] *Desirable Dietary Pattern (DDP) for Bangladesh.* 2013.

[8] Dutta A. K., Partha, S. G., Subrata, B., Sukh, M., Muhammad, A. S., Yearul, K., (2012). Antioxidant properties of ten high yielding rice varieties of Bangladesh. *Asian Pacific Journal of Tropical Biomedicine* 2(1): S99–S103

[9] Eggum, B.O. (1979) *Proc. Workshop.* Chemical aspects of rice grain quality. IRRI, Philippines. p. 91.

[10] FAO (2004). Human energy requirements. Report of a Joint FAO/WHO/UNU Expert Consultation. *FAO Food and Nutrition Technical Report Series 1*, United Nations University, Rome.

[11] Haque M.A., Sayed M. A., Mahfujul A, Saima J. Habibul B.S., Post-Harvest Loss Minimization of Rice Bran for Quality Bran Oil Production in Bangladesh. *EC Nutrition* 13.11 (2018): 667-675

[12] Hakim M.A., Rahman, A., (2016). Health and Nutritional Condition of Street Children of Dhaka City: An Empirical Study in Bangladesh, *Science Journal of Public Health*, 4(1-1): 6-9.

[13] *High Energy Biscuits* (PDF). Archived from the original (PDF) on March 27, 2009. Retrieved August 22, 2008b.

[14] Hirawan R., Jones W. and Beta T., (2011). Comparative evaluation of the antioxidant potential of infant cereals produced from purple wheat and red rice grains and LC-MS analysis of their anthocyanins. *J. Agric. Food Chem.* 59:12330–12341.

[15] Hudson E.A., Dinh P.A., Kokubun T., Simmonds M.S.J., and Gescher A. (2000). Characterization of potentially chemopreventive phenols in extracts of brown rice that inhibit the growth of human breast and colon cancer cells. *Cancer Epidemiology, Biomarkers & Prevention.* 9: 1163–1170.

[16] Ichikawa H, Ichiyanagi T, Xu B, Yoshii Y, Nakajima M, and Konishi T, (2004). Antioxidant activity of anthocyanin extract from purple-black rice. *Journal of Medicinal Food*, 4(4):211-218.

[17] Iriti M., and Varoni E.M., (2013). Chemopreventive potential of flavonoids in oral squamous cell carcinoma in human studies. *Nutrients*. 5: 2564-2576.

[18] Jenkins D.J., Wolever T.M., Taylor R.H., (1981) Glycemic index of foods: a physiological basis for carbohydrate exchange. *Am J Clin Nutr*. (34) 362–366.

[19] Jun, H.I., Shin, J.W., Song, G.S., and Kim, Y.S., (2015) Isolation and identification of phenolic antioxidants in black rice bran. *Journal of Food Science*, 80(2): 262-268.

[20] Kabir S.M., Salam M., Chowdhury A., Rahman N., Iftekharuddaula K., Rahman M., Rashid M., Dipti, S., Islam A., Latif M., Islam A., Hossain M., Nessa B., Ansari T., Ali M. & Biswas, J. (2016). Rice Vision for Bangladesh: 2050 and Beyond. *Bangladesh Rice Journal*, 19(2), 1-18.

[21] Kaneda I., Kubo F., and Sakurai J., (2006), Isolation and Identification of Phenolic Antioxidants in Black Rice Bran. *J Health Sci*, 52: 495– 511.

[22] Kim G.R., Jung E.S., Lee S., Lim S.H., Ha S.H., and Lee C.H. (2014) Combined mass spectrometry-based metabolite profiling of different pigmented rice (Oryza sativa L.) seeds and correlation with antioxidant activities. *Molecules*, 19: 15673-15686.

[23] Leardkamolkarn V., Thongthep W., Suttiarporn P., Kongkachuichai R., Wongpornchai S. and Wanavijitr A. Chemopreventive properties of the bran extracted from a newly-developed Thai rice: The Riceberry. *Food Chemistry*, 2011; 125: 978–985.

[24] Marta S. M., Tomasz R., Justyna D.I., Anna B., Jarosław W., Agnieszka W.K., Paweł Z., Angela B., Marcin M. and Heiner, B.(2016). The link between Food Energy Density and Body Weight Changes in Obese Adults. *Nutrients*, 8, 229

[25] Masud S.A., Shamim H., Antora M. K., Qazi S. I., Muhammad K., (2011). Lives and Livelihoods on the Streets of Dhaka City: Findings from a Population-based Exploratory Survey. *BRAC Working Paper No. 19.*

[26] Mohammed N.A.R. & Mahfuja S., (2016). Street children in Bangladesh: A life of uncertainty, Editorial, *The Independent*, 6 February 2016.

[27] Montignac, *Méthod Montignac 2004-2017*, web link: http://www.montignac.com/ en/ thefactors- that- modify- glycemic- indexes/ nutrimont6, rue du Parc, 74100 Annemasse, France.

[28] Muhammad A.S., Shozib H.B., Jahan S. (2017b). Prospects of Bangladeshi Antioxidant Enriched HYV Rice and Its Bran Oil. *The 4th International Conference on Rice Bran Oil 2017 (ICRBO 2017) Rice Bran Oil Application: Pharma-Cosmetics, Nutraceuticals, and Foods.*

[29] Muhammad A. S., Shozib H.B. Muhammad M.I. (2019). *Annual report on Fortification and standardization of nutritional level in selected human foods and efficacy test of polyphenolic compounds in livestock* (ID099, BRRI component). PIU-BARC, NATP-2, Farmgate, Dhaka.

[30] Murakami K., Sasaki S., Takahashi Y., Okubo H., Hosoi Y., Horiguchi, H., Oguma, E., and Kayama F. (2006) Dietary glycemic index and load in relation to metabolic risk factors in Japanese female farmers with traditional dietary habits. *Am J Clin Nutr.* 83:1161–1169.

[31] Neera F. S., Amit K. B., Muhammad K. S., Abir H., (2014). The Social Life of Street Children in the Khulna City of Bangladesh: A Socio-Psychological Analysis. *Asian Journal of Social Sciences & Humanities* Vol. 3(1) February.

[32] Park S. Y., Kim S.J., Chang H.I., (2008). Isolation of anthocyanin from black rice (Heugjinjubyeo) and screening of its antioxidant activities. *Kor J Microbiol Biotechnol*, 36 (1): 55–60.

[33] Piebiep G. and Henrique T., (2014). Rice antioxidants: phenolic acids, flavonoids, anthocyanins, proanthocyanidins, tocopherols, tocotrienols, c-oryzanol, and phytic acid. *Food Science & Nutrition*, 75-104.

[34] Rahman M.M., Islam M.A., Mahalanabis D., Biswas E., Majid N., Wahed M.A., (1994). Intake from an energy-dense porridge liquefied by amylase of germinated wheat-a controlled trial in severely malnourished children during convalescence from diarrhea. *Eur J Clin Nutr* 48:46–53.

[35] Saikia D. and Deka S.C, Cereals: from staple food to nutraceuticals. *International Food Research Journal* 18: 21-30 (2011).

[36] Saikia S., Dutta H., Saikia D., Mahanta C. L., (2012). Quality characterization and estimation of phytochemicals content and antioxidant capacity of aromatic pigmented and non-pigmented rice varieties. *Food Res. Int.* 46:334– 340.

[37] SE C.H., Khor B.H., and Karupaiah T., (2015). Prospects in the development of quality rice for human nutrition. *Malays. Appl. Biol.* 44(2): 01-33.

[38] Shozib H. B., Jahan S. Suman C. D., Samsul A., Rifat B. A., Mahedi H., Richard, M. and Muhammad A. S., (2017a), Mineral Profiling of HYV Rice in Bangladesh. *Vitam Miner* 6:3

[39] Shozib H. B., Shourab B., Jahan, S., Farzana H., Muhammad S. A., Suman C. D., Shamsul A., Muhamamd M. A., Muhammad A. S., (2017 b) In vivo screening for low glycemic index (GI) rice varieties in Bangladesh and evaluate the effect of differently processed rice and rice products on GI. *Biojournal of Science and Technology*, July Vol 5, Article no: m170001.

[40] Shozib H.B., Jahan S., Muhammad Z. S., Samsul A., Suman C. D., Rifat B. A., Mahedi H., Muhammad A. S., (2018). Nutritional Properties of Some BRRI HYV Rice in Bangladesh. *Vitam Miner*, 7:1

[41] Shozib H.B. and Siddiquee M.A. (2016), Black rice research initiatives in BRRI, *Bangladesh Rice Research Institute (BRRI) Newsletter.*

[42] Shozib H.B., Hossain M.M., Jahan S., Alam M.S., Das S.C., Alam S., Amin R.B., Hasan M.M., Malo R., Islam M.R., Shekhar H.U., Siddiquee M.A., (2017c). Study

of biochemical and cooking quality traits of major rice varieties of Bangladesh. *Malays. Appl. Biol*. 46(4): 55–62.

[43] Shozib H.B., Jahan S., Shourab B., Farzana H., Darmin C., Mahmud H., Mohammad O. F., Muhammad S. R., Muhammad A. S., (2015), Dietary administration of rice in improving the antioxidant status in Long-Evans Rat. *Bio Journal of Science and Technology*, 2:1-7.

[44] Siddiquee M.A., Jahan S., Kabir Y., Shozib H.B., (2017a) BRRI dhan31 generate the elevated level of bioactive component, γ-aminobutyric acid (GABA) at pre-germinated brown rice condition. *International Journal of Scientific Research (IJSR)*, 2017, 6 (7) 511-513

[45] Swati B.P., Muhhanmad K.K., (2011). Germinated brown rice as a value-added rice product: A review. *J Food Sci Technol*. 48(6),661–667.

[46] UK Essays. *Migration and Street Children in Bangladesh Sociology Essay*. November 2013. https://www.ukessays.com/essays/sociology/migration-and-street-children-in-bangladesh-sociology-essay.php?cref=1.

[47] WHO, Nutrition, *Population nutrient intake goals for preventing diet-related chronic diseases*. http://www.who.int/nutrition/topics/5_population_nutrient/en/index4.html

[48] World Bank. 2016. *World development indicators 2016* (English). World Development Indicators. Washington, D.C.: World Bank Group. http://documents.worldbank.org/ curated/ en/ 805371467990952829/ World-development-indicators-2016.

[49] World Health Organization. *Complementary feeding of young children in developing countries: a review of current scientific knowledge*. Geneva: WHO, 1998.

[50] Zhang M.W., Zhang R.F., Zhang F.X., Liu R.H., (2010) Phenolic profiles and antioxidant activity of black rice bran of different commercially available varieties. *J Agric Food Chem*, 58: 7580–7587.

In: Trends in Biochemistry and Molecular Biology
Editors: Hossain Uddin Shekhar et al.
ISBN: 978-1-53616-434-3

Chapter 6

PLANT-DERIVED DRUGS: A NEW PARADIGM OF BIOCHEMISTRY

Md. Atiar Rahman** and Md. Rakibul Hassan Bulbul
Department of Biochemistry and Molecular Biology,
University of Chittagong, Chittagong, Bangladesh

ABSTRACT

Plant-derived drugs are the integrated therapeutic approach to use plants or plant-products as dreadful resource of disease management. The vigorous adverse effects of synthetic drugs, multidrug resistance (MDR) and extremely drug resistance (XDR) to emerging and reemerging microbial strains and aggressively higher costs of synthetic drugs evolved this new avenue as alternative source of treatment in the form of herbal, ayurvedic, Unani, complementary and alternative medicine and some others. The inevitable starting for plant-derived new drug discovery craves the identification of the accurate candidate plant by applying wisdom on traditional medicine, documentation procedure, tribal non-documented information, and exhaustive literature review. All these explore the occurrence of plant secondary metabolites, the pivotal factors for plant-derived drugs function, which are currently being strived to be focused for their possibility of altering the production of bioactive compounds via tissue culture and relevant techniques. Wide spectrum of therapeutic application of plant-derived drugs leads to alter the production of bioactive compounds through an attractive alternative passage of biotechnological production system. Pre-clinical research on plant-derived drugs has made particularly rewarding progress in the important fields of various complicated and life-threatening diseases such as cancer, (e.g., taxoids and camptothecins) malaria (e.g., artemisinin compounds) and HIV. Plant-derived drugs have recently achieved a very big size of market share reflecting the exponential demand of this drug over the world. In association with the purified plant-derived drugs, there is an enormous market for crude herbal medicines, while both have the same

* Corresponding Author's E-mail: atiar@cu.ac.bd.

pharmacoeconomic hurdle against new synthetic pharmaceuticals. Advent of molecular drug designing provided a new hope of higher success for novel approach of integrated drug discovery from plant-derived sources. The integration necessitates the coordination of involved biochemical pathways, cutting edge biochemical tools and techniques to bring a paradigm shift in the evolution of plant sample extraction process from sequential to parallel. This chapter, therefore, discusses the use of plant derived drugs at a glance, its safety and toxicity, perceptual ambiguity on categories of plant-derived drugs, plant-derived drugs as molecular therapeutics, extension of plant-based research, propagation and expansion of bio-market, plant-derived drugs for future genomic hallmark and existing setbacks to introduce plant-derived drugs.

Keywords: plant-derived drugs, molecular therapeutics, bio-market, genomic hallmark

1. INTRODUCTION

Throughout the ages of world's population, humans have relied on plants for the most mere source of life saving drugs (Cragg and Newman, 2005). Plants have made the basis of sophisticated traditional medicine systems that have existed for thousands of years and continued to serve mankind with new remedies. Despite the erroneous role of few of the therapeutic properties attributed by the plants, plant-based therapy is grounded on the empirical observation of hundreds and probably thousands of years of use. Plant derived drugs came into use in modern medicine through the uses of plant material as indigenous cure in traditional or folklore systems of medicine. It is evident that currently about twenty five percent of pharmaceutical prescriptions in the United States is recommended with at least one plant-derived ingredient. In the last century, roughly 121 pharmaceutical products were formulated based on the traditional knowledge from various sources. Currently 80% of the world population is somehow linked with plant-derived medicine for the first line of primary health care for human remedy because it has no or less side effects. Several regulatory models for plant-derived medicines are currently available including prescription drugs, over-the-counter substances (OTC), traditional medicines and dietary supplements. About 500 plants with medicinal use are mentioned in ancient texts and around 800 plants have been used in indigenous systems of medicine. Indian subcontinent is a wide repository of medicinal plants which are used in traditional medical treatments (Chopra 2000), that also forms a rich source of knowledge.

The earliest documentation about the usage of herbal remedies comes from China and dates back to 2800 BC. The ancient record is evidencing their use by Indian, Chinese, Egyptian, Greek, Roman and Syrian dates back to about 5000 years (Table 1).

The various indigenous systems such as Ayurveda, Allopathy, Siddha and Unani use several plant species to treat different ailments (Rabe and van Staden 1997). More than 80,000 plants species are in use throughout the world. In India, almost 20,000 medicinal plant species have been recorded, but more than 500 traditional communities use around 800 plant species to cure different diseases (Kamboj 2000). Traditional medicinal plants are a therapeutic resource used by the population of the African continent specifically for health care, which may also serve as starting materials for drugs (Sofowora 1993). The plant-derived drugs is currently far-reached to the whole globe through its multidimensional ways of preparation, application and uses.

Table 1. Plant-derived natural products approved for therapeutic use in the last thirty years (1984–2014)

Chemical structure and generic name	Plant name (citation)	Trade name (year of introduction)	Uses and indication (mechanism of action)
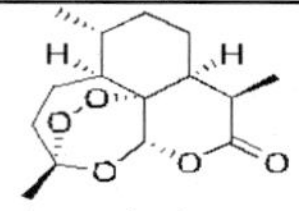 Artemisnin	*Artemisia annua* L.	Artemisin (1987)	Malaria treatment (radical formation)
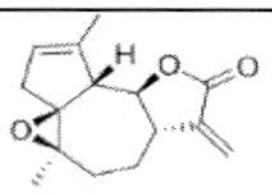 Arglabin	*Artemisia glabella* Kar. et Kir. replaced by *Artemisia obtusiloba* var. glabra Ledeb.	Arglabin (1999)	Cancer chemotherapy (farnesyl transferase inhibition)
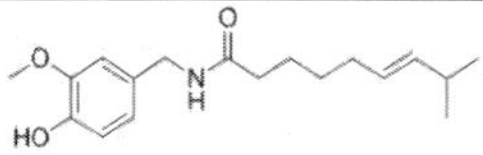 Capsaicin	*Capsicum annum* L., or C. minimum Mill.	Qutenza (2010)	Post herpetic neuralgia (TRPV1 activator)
Colchicine 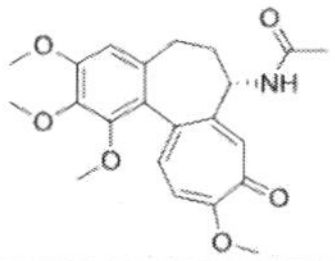	*Colchicum spp.*	Colcrys (2009)	Gout (tubulin binding)
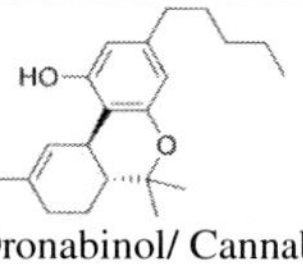 Dronabinol/ Cannabidol Dronabinol	*Cannabis sativa L.*	Sativexb (2005)	Chronic neuropathic pain (CB1 and CB2 receptor activation
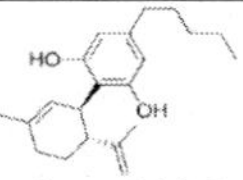 Cannabidol	*Cannabis sativa* L.	Sativex[b] (2005)	
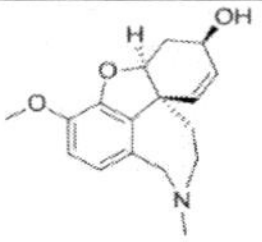 Galanthamine	*Galanthus caucasicus* (Baker) Grossh	Razadyne (2001)	Dementia linked with Alzheimer's disease ((ligand of human nicotinic acetylcholine receptors (nAChRs)
Ingenol mebutate 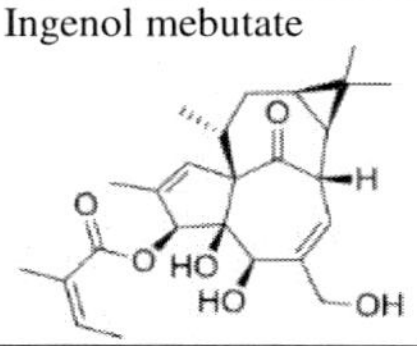	*Euphorbia peplus* L.	Picato (2012)	Actinic or solar keratosis (inducer of cell death)

Table 1. (Continued)

Chemical structure and generic name	Plant name (citation)	Trade name (year of introduction)	Uses and indication (mechanism of action)
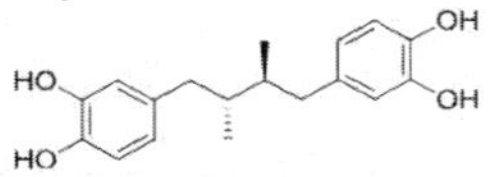 Masoprocol	*Larrea tridentata* (Sessé & Moc. ex DC.) Coville	Actinex (1992)	Chemotherapeutics (lipoxygenase inhibitor)
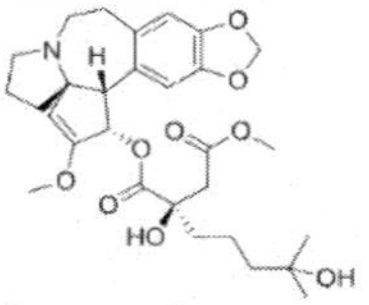 Omacetaxine mepesuccinate (Homoharringtonine)	*Cephalotaxus harringtonia* (Knight ex Forbes) K. Koch	Synribo (2012)	protein translation inhibitor in Oncology
Paclitaxel 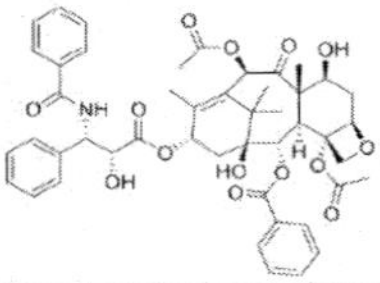	*Taxus brevifolia* Nutt.	Taxol (1993), Abraxane[c] (2005), Nanoxel[c] (2007)	Cancer chemotherapy (mitotic inhibitor)
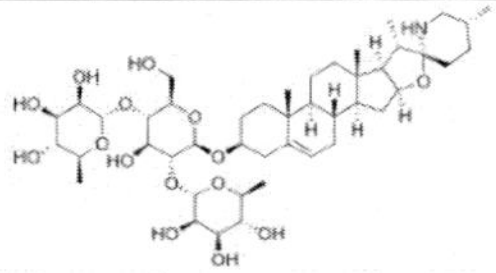 Solamargine	Solanum spp.	Curaderm[d] (1989)	Cancer chemotherapy (apoptosis triggering)

Notes:

a Resources: (1, 58). b Mixture of two compounds.

c Paclitaxel nanoparticles. d Containing both solamargine as well as solasodine glycosides.

According to the World Health Organization (WHO 1977), a medicinal plant is defined as any plant which has one or more organ to contain substances that can be used as therapeutics or which are precursors for the synthesis of useful drugs (WHO 1977). This definition distinguishes those plants whose therapeutic properties and constituents have been established scientifically and plants that are regarded as medicinal but which have not yet been subjected to thorough investigation. Closely conceptualized term "herbal drug" determines the part/parts of a plant used to prepare medicines (for example: barks, flowers, leaves, seeds, roots, stems, etc.) (Anon 2007). Furthermore, WHO (2001) defines medicinal plant as herbal preparations produced by subjecting plant materials to extraction, fractionation, purification, concentration or other physical or biological processes which may be produced for immediate consumption or as a basis for herbal products (WHO 2001). (Figure 1).

It is reported by Iwu et al. (1999) that the primary benefits of using plant-derived medicines are: they are relatively safer than synthetic drugs, full of profound therapeutic benefits and more affordable (Iwu, Duncan and Okunji 1999). The use of medicinal plants in developing countries as a normative basis to maintain good health has been widely observed (UNESCO 1996). Furthermore, the increasing reliance on the use of medicinal plants in the industrialized societies has been increased and traced to the extraction and development of several drugs including chemotherapeutics (UNESCO 1998). Moreover, in these societies, plant-derived remedies have been more popular in the treatment of minor ailments and also on account of increasing costs of personal health maintenance. Survey conducted by the WHO Roll back malaria program in1998, showed that more than 60% of the children with high fever were treated at home with herbal medicines in Ghana, Mali, Nigeria and Zambia (WHO 2004). Hugo and Russell (2003) asserted that as a result of the importance of plant-derived drugs in saving the lives of people, the World Health Organization devoted 27 centers, out of 915 collaborating centers worldwide, for traditional medicine (WHO 2001, Hugo and Russell 2003).

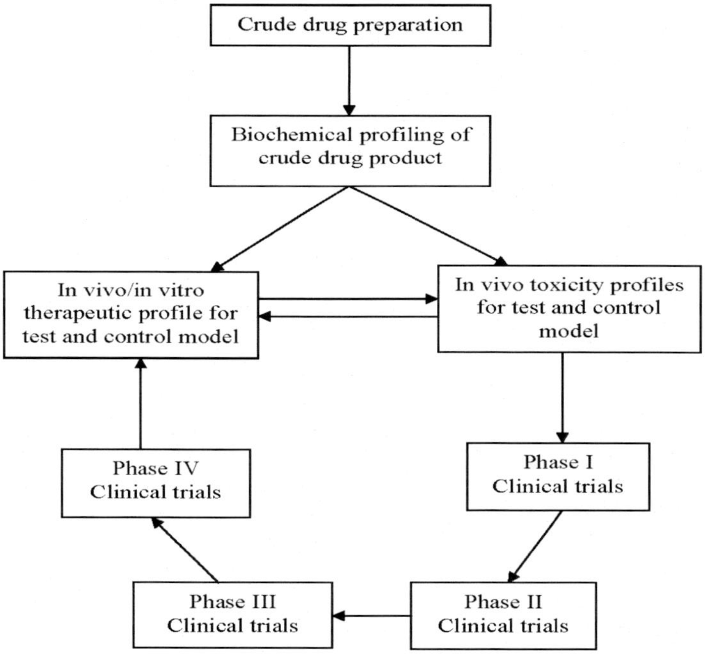

Figure 1. Illustration of the key steps in the development of a drug from a putative drug candidate extract (Sasidharan et al. 2011).

The plant kingdom has served as not only an inexhaustible source of useful drugs but also as foods, additives, flavoring agents, lubricants, coloring agents and gums from time immemorial (Parikh et al. 2006). Plants extracted bioactive compounds currently using as fine chemicals, cosmetics and perfumes, pigments, insecticides, dyes, and food additives (Rao and Ravishankar 2002). The therapeutic power of herbs has been recognized since creation of the universe and herbal medicine is one of the oldest practiced professions by mankind (Kambizi and Afolayan 2001). The medicinal actions of plants are unique to particular plant species or group of plants, consistent with the concept that the combination of secondary metabolites in a particular plant is taxonomically discrete

(Parikh et al. 2006). All these medicinal plants have been used in many forms over the years to treat, manage or control man's ailments. Therefore, any effort to further explore the medicinal or natural products from botanical flora towards improving health care delivery deserves attention (Prescott, Harley and Klein 2002).

Despite the abundance, utility and prospects of plant-derived drugs, the modernization of plant-based drugs, the articulation in their preparation and comprehensive therapeutic approaches have not been evolved till the recent pasts. Fortunately, in recent time, traditional medicine system has become a hot theme for global importance and embolden scientists as well as industrialists to ponder the possibilities into cell cultures as an alternative supply for the production of plant phytochemicals. Biotechnological application into the plant cell cultures provided new way for the commercial purpose of even rare plants. Due to requisition of secondary phytochemicals products as a natural drug, scientist exploring new research methodology to produce large amount of secondary product expression *in vitro*. *In vitro* cultures provide an excellent path for mysterious investigation of metabolic and biochemical pathways and produces only metabolite of interest in specialized plant tissues or glands, for example vanillin and taxol production *in vitro*. Essentially, the discovery of new drugs, incorporation of medicinal chemistry for the characteristics of druggability, chemical synthesis of natural products, preclinical and clinical trials for hundreds of natural product-derived drugs, development of plant-derived drugs for life-threatening diseases, exploration of the biochemical mechanism for all these drugs, categorization of the herbal drugs based on therapeutic classes made them considered as the very indispensable part of modern Biochemistry and allied subjects. This chapter discusses the prospects, probability, multidimensional uses, current demand and distribution, and after all the mechanism of plant-derived drugs for their molecular uses in terms of Biochemical approach.

2. Materials and Methods

All the materials of this book chapter have been perceptualized through a longtime research on medicinal plants, literatures of relevant fields by searching the wider scientific databases including Google scholar, Pubmed, Science direct, Medline and for multidimensional aspects of plant-derived drugs adherent with the modes of Biochemistry and allied biosciences. Some articles were found through tracking citations from other publications or by directly accessing the journals' web-site. They were considered based on the geographical region of their origin. Research activities of the Laboratory of Alternative Medicine and Natural Product Research, University of Chittagong has been incorporated in a paragraph where the major works of that laboratory especially the antidiabetic action of plant-products has been presented graphically. Supplementary information was gathered by using some other keywords combination such as medicinal plant, medicinal uses of plant extract, traditional uses of plants, biochemical mechanism of plant-derived drugs a total of more than 100 research articles that reports on the in vivo and not *in vitro* activity of plant derived products have been recovered and presented in this chapter.

2.1. Plant-Derived Drugs: Integral Part of Biochemistry

Modern drug design and development is complex process that needs accumulation of knowledge from multidiscipline such as chemical, biological and clinical disciplines. Concepts and methodology of Biochemistry and their incorporation in drug discovery process in terms of pathobiochemistry and pharmacological biochemistry effectively elucidate the molecular events of a disease process, evaluate as well as validate the action of therapeutic agents on disease pathology (Stefanovich 1980). Early detection of drug candidates and their molecular targets, validation of drug-receptor relationships, incorporation of biochemical and biophysical techniques for drug delivery system, troubleshooting and implementation throughout the screening and optimization stages of both traditional small molecules and biopharmaceutical products make the biochemistry as an integral part in drug discovery.

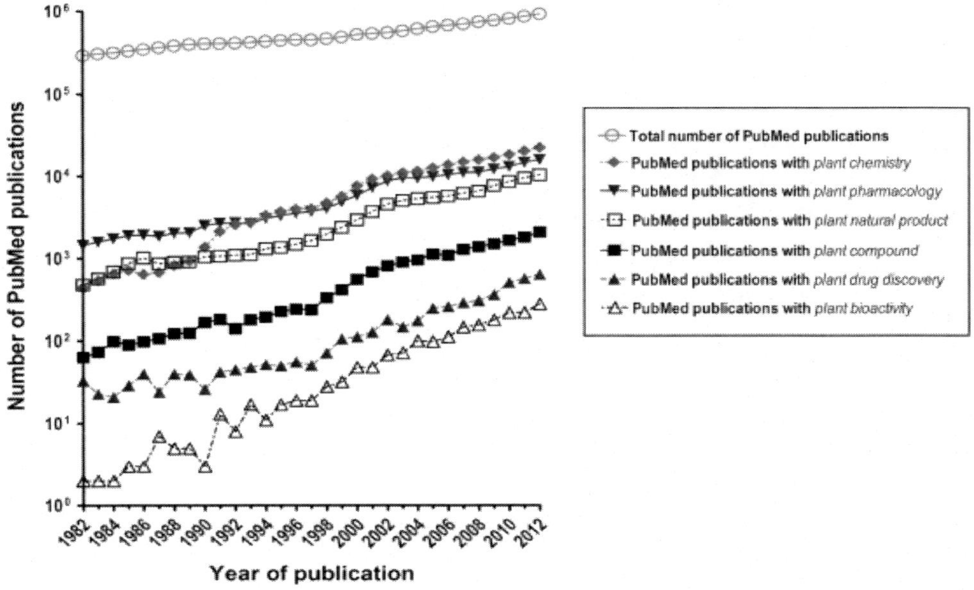

Figure 2. Analysis of PubMed publication trend, demonstrating increased scientific interest in plant-derived natural product pharmacology, chemistry, and drug discovery (Atanasov et al. 2015).

Plant kingdom represents an extensive diversity of pharmacologically active secondary products that can serve as drug precursors and prototypes as well as pharmacological probes (Salim, Chin and Kinghorn 2008). Since natural products are mostly the outcome of plant metabolic system, they are evolutionarily fit to exert biological actions they possess and thus can essentially be considered as drug candidates (Appendino, Fontana and Pollastro 2010). Recent biochemical and biophysical technology advancements speed up the investigation of bioactive compounds from plants and other organisms for extraction, analysis, purification, and structural identification of compound as well as bioactivity screening (Figure 2).

The plant kingdom produces a diversity of bioactive compounds with different chemical scaffolds and their biological activity assaying is rate limiting step in drug discovery process from plants. But this issue can successfully be acknowledged with improved and automated high-throughput techniques, advanced data handling systems and robotics by which thousands of samples that allow rapid screening of plant extracts in the same manner as libraries of pure compounds in just over a week. Different extraction techniques like supercritical-fluid extraction, solid-phase micro-extraction, pressurized-liquid extraction, solid-phase extraction, microwave-assisted extraction, and surfactant-mediated techniques are developed upon analysis of biochemical and biophysical properties of the bioactive compounds from medicinal plants to improve in extraction efficiency, selectivity, and/kinetics of extraction before chromatographic analysis and with automated technology, compound extraction and bio analysis become robust (Huie 2002).

Moreover, data from HPLC-coupled spectroscopy in combination of other detection techniques like diode-array detector (DAD), circular dichroism (CD), mass spectrometry (MS), and nuclear magnetic resonance (NMR), FTIR, etc. have been used in the study of compound characterization, purification, chemical and physical structural identification as well as bioactivity screening from unknown mixtures of plant extracts (Bringmann et al. 1999, Niessen, Lin and Bondoux 2002, Wolfender, Ndjoko and Hostettmann 2003, Eberhardt et al. 2007, Hazra et al. 2007). Introduction of combinatorial biochemistry in plant cell cultures with advance genomic tools allows to synthesize target specific metabolites or to synthesize entire novel compounds that may lead the future drug discovery process (Oksman et al. 2011).

2.2. Drug Discovery from Natural Resources: Advantages and Disadvantages in Context of Biochemistry

Use of plant sources as starting point in the drug development program is strongly associated with few specific advantages: mostly, the choice of a candidate species for investigations can be done on the basis of ethnomedicinal history. This approach is based on a presumption that the bioactive compounds isolated from these plants are very likely to be safer than those produced from plant species with no ethnomedicinal uses. After a certain time-period, one can attempt upon the synthesis of active molecule and lessen pressure on the resource. Drug development from *Digitalis purpurea*, *Rauwolfia serpentine*, etc. in the past fall under this particular category of approach. Sometimes, such approaches lead to the development of novel molecules isolated from the source due to inherent limitations of the original molecule. For instance, podophyllin derived from *Podophyllum hexandrum* was faced with dose-limiting toxicities. Such limitations could be overcome to a great extent by semi-synthesis of etoposide, which continues to be used as anticancer therapy today. Similar case was observed with camptothecin (originally isolated from *Camptotheca* sp. and subsequently from *Mappia* sp.), which led to development of novel anticancer molecules like topotecan and irinotecan. Natural bioresources as starting point has a bilateral promise of delivering the original isolate as a candidate or a semisynthetic molecule development to overcome any inherent limitations of original molecule. On the other hand, drug development from natural resources is also relevant with certain disadvantages. Drug discovery and eventual commercialization sometimes would pressurize the resource and might lead to undesirable environmental concerns. While synthesis of active molecule could be an inevitable option, not every molecule is amenable for complete synthesis. Therefore, certain degree of dependence on the lead resource would continue. For instance, anticancer molecules like docetaxel, etoposide, paclitaxel, irinotecan and topotecan continue to depend upon highly vulnerable plant resources for collecting the starting material since a complete synthesis is not possible. On the contrary, it is expected that some 25,000 plant species would be ceased to exist by the end of this century (Mahidol et al. 1998). Over a period of time, the intellectual property rights protection related to the natural products is going haywire. By and large, the leads are dependent on some linkage to traditional usage. With larger number of countries becoming the parties to the Convention on Biological Diversity (CBD), the process of accessing the basic lead resource as well as the benefit sharing during the commercial phase are found to be highly complex in many countries. These processes tend to impede the pace of discovery process at various stages irrespective of the concerns leading to such processes.

2.3. Biochemical and Phytopharmacological Research on Plant-Derived Drugs

2.3.1. Screening of Drug Candidate and Trends of Plant-Derived Drugs in Use

High throughput screening (HTS) was picked up by major pharmaceutical industries in the 1990s keeping in mind that rapid compound screening from large synthetic compound libraries would produce data for therapeutic candidates but failed to meet the intended expectations. Some important notable causes behind this scenario includes limiting chemical diversity of synthetic compound libraries, similar generation strategies and a negative correlation of cell-free *in vitro* potency and favorable ADMET (absorption, distribution, metabolism, excretion, toxicity) restricted the outcome of new drug candidates to the market (Scannell et al. 2012, Gleeson et al. 2011). For instance, US Food and Drug Administration (FDA) approved 45 new drugs in 1990 but only approved 21 in 2010 (David, Wolfender and Dias 2015, Kingston 2011).

The plant kingdom offers a repository of searching new drug entities as huge number of plant species jointly present a diversity of bioactive compounds full of different chemical scaffolds. Recent scientific and methodological advances in the respective research fields in terms of better understanding of diseases and their underlying mechanisms, advances in screening methods and analytical technique, increasing number of available targets for testing, and improved possibilities for optimization of natural leads using synthetic modification strategies set the renewal of scientific interest in plant-derived natural product-based drug. Recent PubMed publication trends give a quick glimpse that plant-derived drugs is a growing interest in more scientific studies (Figure 3) (Henrich and Beutler 2013, Koehn and Carter 2005, Li and Vederas 2009, Paterson and Anderson 2005).

In the USA, the National Center for Complementary and Alternative Medicine at the National Institutes of Health spent approximately US$ 33 million in fiscal year 2005 on herbal medicines; the National Cancer Institute in 2004 committed nearly US$ 89 million to studying a range of traditional therapies (White, 2006). While this scale of investment pales in comparison to the total research and development expenses of the pharmaceutical industry, it nevertheless reflects genuine public, industry and governmental interest in this area. While public-health entities may be concerned with defining the risks and benefits of herbal medicines already in use, entrepreneurs and corporations hope herbal medicines may yield immediate returns from herbal medicine sales, or yield clues to promising chemical compounds for future pharmaceutical development. They test individual herbs, or their components, analyzed in state-of-the-art high-throughput screening systems, hoping to isolate therapeutic phytochemicals or biologically active functional components. In 2006, Novartis reported that it would invest over US$ 100 million to investigate traditional medicine in Shanghai alone (Zamiska 2006).

2.3.2. Molecular Biology the Pioneer Approach for Modern Use of Plant-Derived Drugs

Genomics knowledge has been initiated in term of "Herb Genome Programme" for the genome sequencing of different medicinal plants and post genomic functional analysis of various secondary metabolite biosynthetic pathways (Chen, Xu and Guo 2011). Some of the commonly used medicinal herbs such as *Ganoderma lucidum*, *Salvia miltiorrihiza* and *Catharanthus roseus* produces active pharmaceutical components, including triterpenes, diterpene quinone and indole alkaloids and the genome sequences of these species are now available (Chen et al. 2015, Giddings et al. 2011, Kellner et al. 2015). With the help of draft genome sequence and bacterial artificial chromosome (BAC) sequencing, biosynthesis of the monoterpene indole alkaloids vinblastine and vincristine has been revealed (Kellner et al. 2015). The chloroplast genome sequences facilitates population, phylogenetic and genetic engineering studies of this medicinal plant. The purpose of genomics in plant metabolism is not only to deciphering the underlying genetic configuration to synthesize bioactive compounds but also to use medicinal plant metabolic pathway enzymes for other works to carry out the new drug discovery. For example, through the use of recombinant technology, combining enzyme engineering, genes from microorganisms and plant genetic engineering, search for new classes of compounds and their potentials as new drugs were carried out (Chakraborty 2018). Techniques in proteomics have been used to analyze the synthesis of bioactive compounds that confer to medicinal plants in their health promoting properties (Kim et al. 2016).

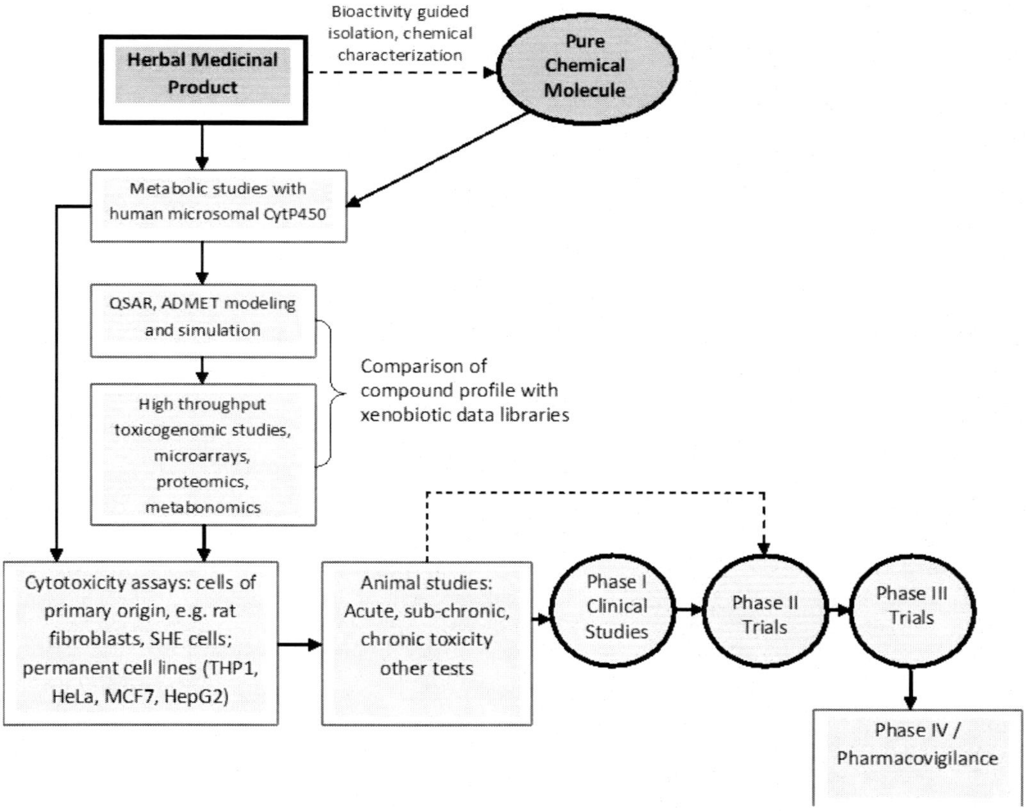

Figure 3. Schematic processes involved in evaluating and establishing the toxicity of medicinal herbs. The broken arrow indicates that for some herbal medicines, phase 1 clinical trials may not always be necessary (Ifeoma and Oluwakanyinsola 2013).

Proteomics offers the advantage of collating a comprehensive collection of physiological information both from medicinal plants and animals that consume the plants. The identification of key enzymes has helped elucidate the multistep, complicated biosynthetic pathways of plant bioactive compounds. External factors that affect the pathways and the physiological changes caused by the factors are described using the proteomic analyses of plants. Responses of the animals treated with plant-derived natural products were shown to include various biological processes. These studies are two sides of the same coin in research on medicinal plants. The biological reaction against the treatment was elucidated by iTRAQ-based proteomic analysis (Chen et al. 2015).

The identification of target pathways by proteomic profiling interprets the various applications in heterogeneous disorders and will help improve the treatment regimen. The therapeutic mechanism underlying *Shufeng Jiedu* comprises eight medicinal herbs, and AKT1 kinase is considered as a pivotal factor in acute lung injury. The hypothesis from proteomics data was proved by gain and loss-of-function analyses of macrophages, displaying that AKT1-mediated biological processes, including apoptosis, oxidative stress, and inflammatory responses were associated with acute lung injury (Tao et al. 2017). For the detection of adverse side effects, biomarkers were searched for geniposide overdose-induced hepatic injuries in animal model. Using the intensity-based absolute quantification (iBAQ) method, glycogen phosphorylase and glycine N-methyltransferase were found to induce hepatotoxicity during the initial stages (Wei et al. 2014).

2.4. Safety and Toxicity of Plant Derived Drugs in Context of Biochemistry

2.4.1. Evolving of Phytotoxicants in Plant-Based Drugs

The aim and objective of toxicity testing is the detection of toxic plant extracts or compounds derived thereof in the early (pre-clinical) and late (clinical) stages of drug discovery and development from plant sources (Figure 4).

Plants have an advantage in finding future drug source based on their long-term use by humans. Any bioactive compounds obtained from plants that has been used in traditional medicine, are expected to produce low human toxicity. A 3000 year history of therapeutic uses of plant-based resources is now globally accepted and considered as safe to use, effective in pharmacologic action, less production of undesirable side effects, minimal toxicity and readily available with low cost. Moreover, structural and chemical diversity of plant products paved the way to explore drug like molecules (Philomena 2011).

During the ancient time, medicinal plants were only applied on an empirical basis without mechanistic knowledge of their active constituents and pharmacological activities (Sneader 2005). Plants and their derivatives are currently the sources of thousands of drugs worldwide. But it does not mean that they all are safe or free of side-effects. Isolated principles from plants such as digitoxin, morphine, reserpine, vinblastine and vincristine are toxic with high concentration. Moreover, the dose is often the differentiator between a medicine and a poison. A small amount of any of these compounds, properly employed in a clinical setting, can benefit health while too much of any of these compounds can lead to death.

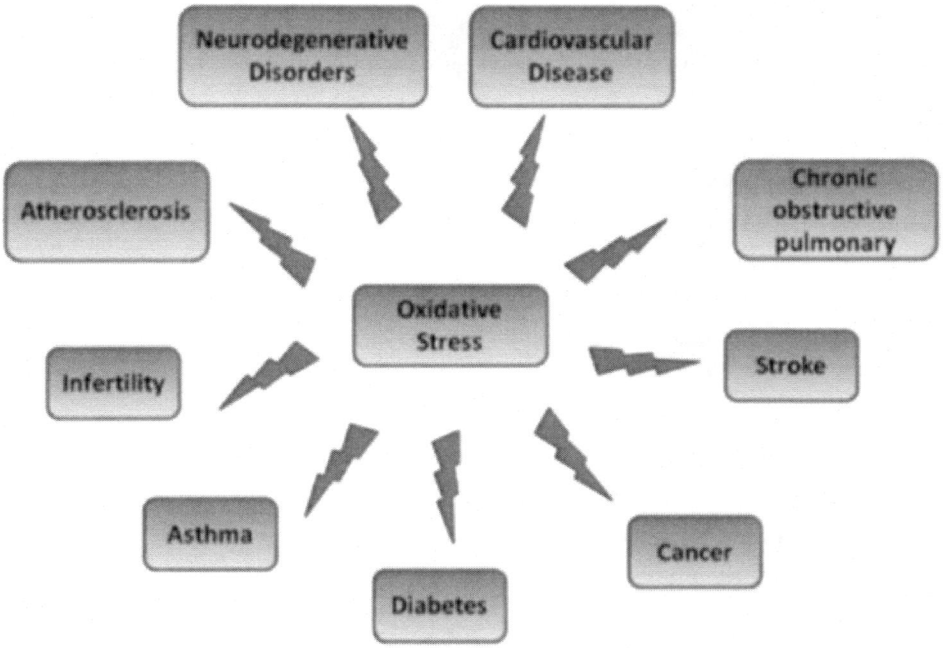

Figure 4. Arrays of diseases caused by oxidative stress in human physiological system (Sharma et al. 2012).

Many of the plants produce toxic secondary metabolites as natural defense from adverse conditions. In some medicinally and toxicologically relevant plant species like *Atropa belladonna, Digitalis purpurea*, *Hyoscyamus niger*, *Physostigma venenosum*, *Podophyllum peltatum* and *Solanum nigrum*, the toxic substances are not distinguished from therapeutically active ingredients. Plants being the stationary autotrophs, have evolved different means of adaptation to adverse and challenging environments and ensured the co-existence with herbivores and pathogenic microorganisms. Thus, they synthesize an array of metabolites characterized as 'phytoanticipins' or generally 'phytoprotectants' those are stored in specialized cellular compartments and released in response to specific environmental stimulation such as damage due to pathogens, herbivores or nutrient depletion (Kennedy and Wightman 2011). Some of the plant metabolites produced against herbivorous insects also end up being harmful to human, because highly conserved biological similarities are shared between both taxa as seen in most pathways involving carbohydrate, protein, nucleic acid, and lipid metabolism (Wink

2003). Ecologically, a good number of alkaloids serve as feeding deterrents via agonistic or antagonistic activity on neurotransmitter systems (Savelev, Okello and Perry 2004). Similarly, some lipid soluble terpenes have shown inhibitory properties against mammalian cholinesterase, whilst some interact with the GABAergic system in vertebrates (Rattan 2010, Francis et al. 2002). In addition to these, saponins are potent surfactants that can disrupt lipid-rich cellular membranes of human erythrocytes and microorganisms which explains the potent antimicrobial properties of this group of phytochemicals (Arit, Stiborova and Schmeiser 2002). Aristolochic acid, a nitrophenanthrene carboxylic acid in *Aristolochia* species and present in some other botanicals has also been identified as a phytochemical toxicant implicated in the development of nephropathies and carcinogenesis (Dwivedi and Dey 2002). Another implication in the toxicity of certain herbs is the presence of toxic minerals and heavy metals like mercury, arsenic, cadmium, and lead (Amster, Tiwari and Schenker 2007). Lead and mercury can cause serious neurological impairment when an herbal product contaminated with these metals is ingested. (Butler, Robertson and Cooper 2014).

2.5. Events Leading the Insecurity of Plant-Derived Drugs

Global acceptance of plant-based drugs is increasing and many plant-based compounds are currently undergoing different phases of clinical trial for the potential treatment of various diseases where most of these are analogs of known anticancer drugs (Winston and Maimes 2007). Although rapid accessibility of new drug candidates in plant origin into the world drug market, there still remains considerable concerns on their use and safety. Less than 10% of herbal products in the world market are truly standardized to known active components and strict quality control measures are not always diligently adhered to (Calixto 2000). Most of the reported side effects for herbal drugs are extrinsic to the preparation and are related to several manufacturing problems such as plant misidentification, lack of standardization, contamination, failure of good manufacturing practice, substitution and adulteration of plants, incorrect preparations and/or dosage, etc. (Cowan 1999).

In many countries including the US, herbal medicines are not subjected to the same regulatory standards as synthetic drugs in terms of efficacy and safety. This raises concern on their safety and implications for their use as medicines. Toxicity testing can reveal some of the risks that may be associated with use of herbs, therefore avoiding potential harmful effects when used as medicine (Gamaniel 2000).

Plant-based herbal products are also expected to have side effects, which may be of an adverse nature. Some of the reported adverse events are attributable to problems of quality. Major causes of such events are adulteration of herbal products with undeclared other medicines and potent pharmaceutical substances, such as corticosteroids and non-steroidal anti-inflammatory agents. Adverse events may also arise from the mistaken use of the wrong species of medicinal plants, incorrect dosing, errors in the use of herbal medicines both by health-care providers and consumers, interactions with other medicines, and use of products contaminated with potentially hazardous substances, such as toxic metals, pathogenic microorganisms and agrochemical residues (WHO 2004).

According to the WHO guidelines on safety monitoring of herbal medicines in pharmacovigilance systems for the safety of those using herbal medicines, four complementary actions are needed:

- Clear identification of the nature of adverse events
- Management of the risks
- Institution of measure to prevent adverse events
- Good communication of the risks and benefits of herbal medicines.

However, the toxicity study facilitates the identification of toxicants which can be discarded or modified during the process and create an opportunity for extensive evaluation of safer, promising alternatives (Benyhe et al. 1997).

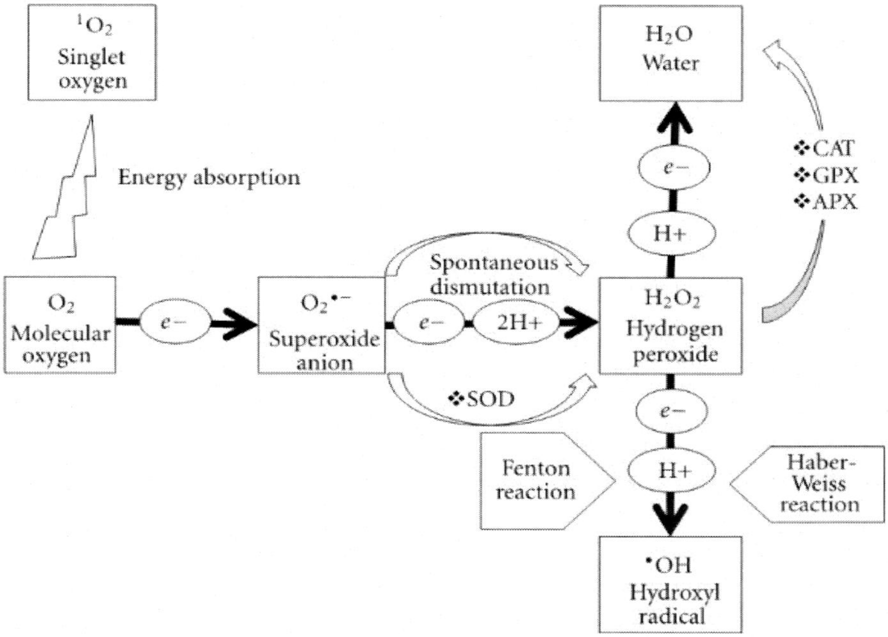

Figure 5. Schematic representation of generation of reactive oxygen species (ROS) in plants. Activation of O2 occurs by two different mechanisms. Stepwise monovalent reduction of O2 leads to formation of O2 •−, H2O2, and •OH, whereas energy transfer to O2 leads to formation of 1O2. O2 •− is easily dismutated to H2O2 either nonenzymatically or by superoxide dismutase (SOD) catalyzed reaction to

H2O2. H2O2 is converted to H2O by catalase (CAT), guaiacol peroxidase (GPX), and ascorbate peroxidase (APX) (Sharma et al. 2012).

2.6. Plant Products Diversity and their Biochemical Mechanism

2.6.1. Optimizing Oxidative Stress Is the Prime Option for Plant Products

Important challenges related with the use of plants as a source for identification of bioactive compounds are related with the accessibility of the starting material. Although many plant-derived natural products have already been isolated and characterized, available compound quantities are often insufficient for testing for a wide range of biological activities. Natural products have provided many effective drugs. These include a wide range of older drugs such as quinine and morphine and newer drugs such as paclitaxel (TaxolTM), camptothecin, etoposide, mevastatin, and artemisinin (Kremsner et al. 1994, Wani et al. 1971, Wall et al. 1966, Keller-Juslén et al. 1971, Klayman 1985, Bucar, Wube and Schmid 2013).

Available plant material often varies on quality and composition and this can hamper the assessment of its therapeutic claims. The chemical composition is not only dependent on species identity and harvest time, but also on soil composition, altitude, actual climate, processing, and storage conditions. Moreover, during extraction, as well as during the isolation processes, transformation and degradation of compounds can occur (Jones and Kinghorn 2012, Corson and Crews 2007). Another aspect determining the chemical composition of the starting plant material is that endophytic organisms, such as fungi and bacteria, might inhabit plants. As a result, natural products present in the collected plant material might be in some occasions metabolites of the endophytic organism, or plant products induced as a result of the interaction with this organism (David, Wolfender and Dias 2015).

Despite a lot of setback on plant product isolation, identification and bioactivity measurement, huge numbers of natural products are in therapeutic uses. Further difficulty is set by the fact that determination of the precise molecular mechanism of action of natural products is a challenging task (Lee and de Beer 2016). However, a detailed knowledge of the interaction of a drug candidate compound with its molecular target is very advantageous for the drug development process, because it allows property optimization by medicinal chemistry approaches, and on some occasions a more appropriate clinical trial design.

Among all other reasons, advanced scientific evidences showed that imbalanced oxidative stress due to the ROS and RNS is the leading cause of several different human diseases including cardiovascular diseases, cancer, neurodegenerative diseases, diabetes, etc. (Figure 5).

Individual attempt to prevent or treat these diseases around these days is less important than the complete therapeutic management to mitigate oxidative stress which is possible to be either attenuated or removed through the activation of biological tools. The generation of molecular oxygen in the form of reactive oxygen species (ROS) is a natural part of aerobic life that is responsible for the manifestation of cellular functions ranging from signal transduction pathways, defense against invading microorganisms and gene expression to the promotion of growth or death (Apostolova, Blas-Garcia and Esplugues 2011). Redox signaling is of essential importance due to the abundance of oxygen in the Earth's atmosphere. Nevertheless, an excessive amount of ROS is highly toxic to cells. Oxidative stress affects the major cellular components: proteins, lipids and DNA (Figure 6).

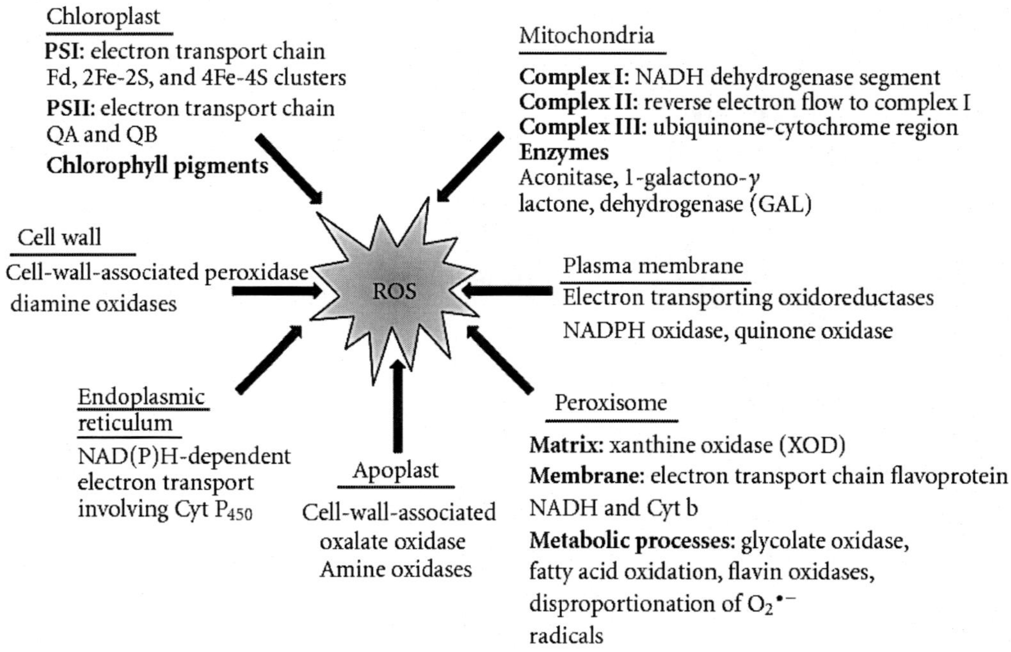

Figure 6. Sites of production of reactive oxygen species (ROS) in plants. ROS are produced at several locations in the cell-like chloroplast, mitochondria, plasma membrane, peroxisomes, apoplast, endoplasmic reticulum, and cell wall (Chong and Pagano 1997).

The importance of oxidative stress is commonly emphasized in the pathogenesis of various degenerative diseases, such as diabetes, cancer, cardiovascular disorders or neurodegenerative diseases (Yi et al. 2014).

2.7. Biochemical Markers and Plant-Derived Products

Liver, the pivotal organ for metabolic biotransformation, is thought to be the most vulnerable organ for oxidative stress. In recent years, a great number of plants have been attempted to eliminate hepatic damage induced by ethanol in animal models, and the bioactive compounds that are responsible for relieving oxidative stress are usually indistinctly ascribed to polyphenols and flavonoids compounds. Major target of the plant-based products are to minimize both the specific and nonspecific liver function markers. For instance, betulinic acid is a pentacyclic lupane-type triterpene, and has a wide range of bioactivities (Zhang et al. 2015) has reported that pre-treatment of betulinic acid could significantly reduce the serum levels of ALT, AST, total cholesterol, and triacylglycerides in the mice treated with alcohol. Hepatic levels of GSH, SOD, GSH-Px, and CAT were remarkably increased, while MDA contents and microvesicular steatosis in the liver were decreased by betulinic acid (Rao and Gan 2014). We have recently observed that oral administration of *Leea macrophylla* root extract potentially minimizes ALT, AST, and LDH along with increasing the mRNA expression of SOD and CAT while no change has been observed for GSH-Px in a four week intervention study to protect carbon tetrachloride induced hepatic damage (Figure 7).

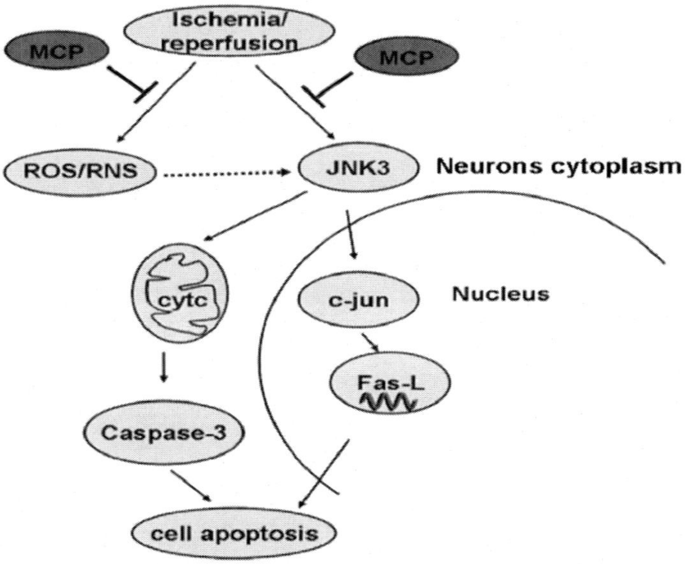

Figure 7. Scheme summarizing the proposed JNK signaling mechanisms underlying neuroprotection of Momordica charantia polysaccharides (MCP) against ischemia/reperfusion brain injury (Gong et al. 2015).

2.8. Neuroprotective Mechanism

The World Health Organization has concerned the increasing prevalence of neurodegenerative diseases accounts for over 11% of the overall burden which sometimes turns to life threatening and negatively affects quality of life (Apostolova, Blas-Garcia and Esplugues 2011). Concerning the high incidence and sophisticated outcome in neurologic disorders, plant-derived drugs have been paid much attention due to their biosafety and efficiency. Plants and herbal extracts have been utilized to treat various central nervous system (CNS) dysfunctions (Beppe et al. 2014, Seoposengwe, Tonder and Steenkamp 2013, Nelson et al. 2013). One of the effective compounds from well-known neurobiologically active plants is polysaccharide (Walker, Liu and Xu 2013) although very little attention has been given to explore the protection mechanism of naturally derived plant polysaccharides. The function of polysaccharides in plants is typically either related to structure or storage.

Promotion of neural cell proliferation, neural plasticity and cell survival are greatly influenced by mitogen- activated protein kinases (MAPKs) signaling (Wong et al. 2015). The polysaccharide from culinary and medicinal mushroom, *Hericium erinaceus* (Bull.: Fr.) Pers. restores the sensory function through the upregulation of Akt and p38 MAPK expression in dorsal root ganglia after peripheral nerve injury (Bie et al. 2015). LBP modulates autophagy and MAPK pathway and improves microglial cell injury (Seow et al. 2015). In BV-2 cells, LBP pretreatment alleviates electric stimulation-induced cellular necrosis and apoptosis, stabilized the expression and cellular localization of light chain 3B (LC3B) after electric stimulation (ES). Moreover, LBP exerts antiapoptotic and antioxidant effects on ES-induced BV-2 cell which are the result of autophagy (Seow et al. 2015). Crude polysaccharides of *L. rhinocerotis* sclerotium regulate the MEK/ERK1/2 signaling pathway in PC12 cells and prompted neuritogenic activity but did not invigorate the generation of NGF (Fan et al. 2012). According to a research, ginseng pectin (GP) reduced up to 26% H_2O_2 mediated oxidative damage on primary cortical neuron cells and human glioblastoma U87 cells and thus, maintained cell integrity and decreased nuclei condensation. Western blot analysis interpreted that pre-treatment with ginseng pectin (GP) also accelerated the phosphorylation of both the extracellular signal-regulated kinases 1 and 2 (ERK1/2) and Akt in cortical neuron cells (Li et al. 2013). Furthermore, LBP can suppress the oxidative stress and the JNK/c-jun pathway, and by briefly increasing production of insulin-like growth factor-1 (IGF-1) and deferred secondary degeneration of retinal ganglion cells (Gong et al. 2015). Attenuation of neuronal death by *M. charantia* polysaccharides (MCP) could involve thrombin in primary hippocampal neurons, and inhibition of JNK3/c-Jun/Fas-L and JNK3/cytochrome C/caspases-3 signaling cascades in ischemic brains *in vivo* (Figure 8) (Duan et al. 2015). MCP could scavenge ROS in intra-cerebral hemorrhage damage in the similar fashion (Deng et al. 2015).

A polysaccharide from the plant *Campanumoea javanica* (CJP-1) having molecular weight of 2900 Da with -d-Fruf (2 → 1)-linkage promotes neurite outgrowth and maturation in neuronal cells. Result from the *in vitro* study suggested that CJP-1 expanded the neurite length, refine the dendritic architectural networks and the density of dendritic spines of hippocampal neurons, as well as the expression of PSD-95 significantly (Xiang et al. 2015). Neurotrophic effects of polysaccharides through the complementation of molecular signals associated to cell survival is vital to suppress the cell death machinery and rectify apoptotic signaling in the nervous system. According to Xiang et al. (Wang et al. 2016), neuronal cells treated with *Rhizoma Dioscoreae* (RDPS) polysaccharides increased the cell viability, and potentially reduced the hypoxia-induced mitochondrial injury and apoptosis.

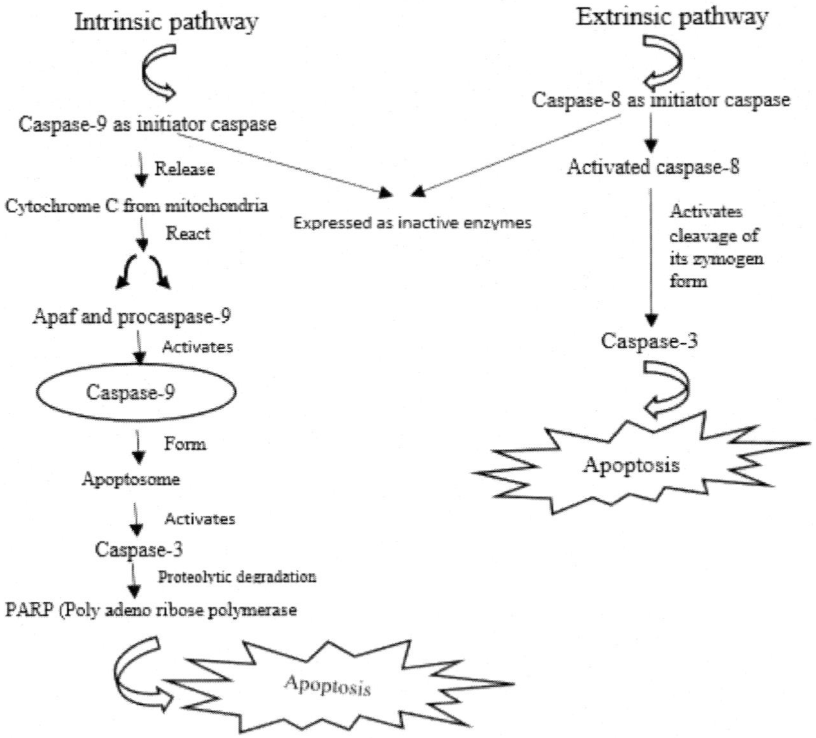

Figure 8. Intrinsic and Extrinsic mechanisms of apoptosis (Ho et al. 2013, Kumar et al. 2007).

Moreover, structural derivatives of polysaccharides with simple and complex glycol-conjugation can cover a wide range of cellular functions, such as antioxidant (Xie et al. 2015), immunomodulation (Jia et al. 2016), anti-tumour (Jin et al. 2013), radioprotection (Eseyin et al. 2014), anti-diabetes (Zeng et al. 2016), hepatoprotection (Shi 2016), anti-virus (Zheng et al. 2010), and anti-fatigue functions (Sharma, Parihar and Parihar 2011). Several mechanisms are involved through which polysaccharides exert CNS protection and were extensively studied previously. This book chapter discusses natural neuroprotective polysaccharides derived from the period of 2010–2016, and mechanisms underlying their multiple neuroprotective activities are classified in the following sections.

2.9. Plant-Derived Drugs as an Anticancer Products

Cancer occurs when clones of mutated cells survive and proliferate infinitely. Generally normal cells multiply when the body requires and die when the body does not need it. The damaged cells dividing inappropriately harm the body and form lumps or a tumor which occurs due to the mutation in cyclin dependent kinases (CDK) which either controls or regulates cell cycle mechanism (Ames, Gold and Willett 1995). Carcinogenesis is due to the presence of toxic chemicals, excessive use of alcohol, exposure to excessive sunlight and environmental toxins, some poisonous mushrooms, inherent genetic problems, ionizing radiation and several viruses (Rajendran and Ramakrishnan 2010). Angiogenesis, invasiveness and metastasis are the regular expression of progressive phase of the malignant disease. Treatment of cancer in developed countries is given by the combination of surgery, chemotherapy and radiation. This sort of treatment destroys the protective mechanism of cancerous cells and its metastasis but their toxic effect kills normal cells along with cancer cells. It decreases the hematological level and immune disorder making the patient liable to infections. Cytotoxicity screening test is the preliminary method for assortment of active plant extracts against cancer. Therapeutic importance of plants is chiefly because of the presence of alkaloids and polyphenols, thus they are much concentrated than other secondary metabolites available in the plants (Akwu 2013). The cytotoxicity and genotoxicity of medicinal plants in terms of herbal extracts, infusions, essential oils and fractions is evaluated using cytogenic assays for their mutagenic and carcinogenic effects (Singh et al. 2012). The search for safety tools for the treatment of malignancy resulted in identifying plants and its compounds with properties like herbal adaptogen, anti-stress and immunomodulators (Qi et al. 2010). However many people still use traditional medicine as an alternative treatment for cancer (Fyhrquist et al. 2014).

A global broad-spectrum thought of medicinal plant users are - the plant derived drugs are always safe because they are from the nature. Plants are generally used in Ayurvedic medicinal practices to promote self-healing, good health and longevity. However, numerous scientific evidences had reported the inclusion of many plants as food or drug in traditional medicine. The below listed plants and its compounds proved to be possessing potent anticancer properties with scientific evidence. Some of its mechanism from reported data has been described in brief. The plants discussed here are least explored in terms of its compounds and mechanism of action, even they might have strong anticancer activity (Table 1).

This book chapter focuses on the study of rarely explored plants and their compounds which have been scientifically proved to be prospective anticancer agents (Table 1). Existing drugs with cytotoxicity adversely affects both the normal and cancerous cells without adequate differentiation. So the strategies for discovering cancer drug mainly reflect on targeting specific proteins implicated in tumor growth and progression. As the use of phytochemicals are promising and escalating rapidly, the appropriate scientific study to extract bioactive chemicals, detail assessment of its role in anticancer treatment and clinical studies might be the appealing issue of upcoming cancer research. This book chapter summarizes proof of benefits by presenting the list of non-toxic and prevailing natural sources of drugs for patients who are in need. Thus, it is worthwhile to overview the traditional and current status on the usage of bioactive compounds in anticancer activity.

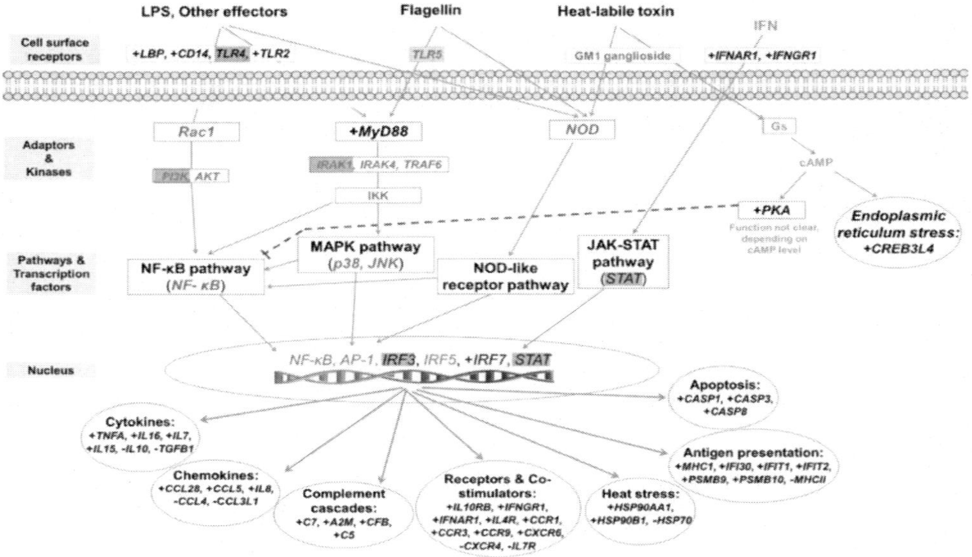

Figure 9. *Escherichia coli* infection induces gut immune responses of pigs. The bolded + sign and bolded − sign represent genes up-regulated and downregulated, respectively. Genes with gray-shaded backgrounds were not affected by *E. coli* at d 5 postinoculation, whereas genes with gray font did not have information from the gene chip (Affymetrix Inc., Santa Clara, CA). The arrow symbol indicates induction or interaction with other molecules; the blocking symbol with dashed line indicates inhibition. Receptor-mediated pathways: nuclear factor kappa-light-chain-enhancer of activated B cells (NF-κB) pathway; mitogen-activated protein kinases (MAPK) pathway; nucleotide-binding oligomerization domain (NOD)-like receptor pathway; Janus kinase-Signal Transducer and Activator of Transcription (JAK-STAT) pathway. A2M = α-2-macroglubin; Akt = serine/threonine protein kinase; AP = activator protein; C = complement component; CASP = caspase, apoptosis-related cysteine peptidase; CCL = C–C motif ligand; CCR = chemokine (C–C motif) receptor; CFB = complement factor B; CREB3L4 = cAMP responsive element binding protein 3-like 4; CXCR = chemokine (C–X–C motif) receptor; Gs = stimulatory guaninie nucleotide-binding protein; HSP70 = heat shock protein 70; HSP90AA1 = heat shock protein 90, α; HSP90B1 = heat shock protein 90, β; IFI30 = interferon, γ-inducible protein 30; IFIT = interferon-induced protein with tetratricopeptide; IFN = interferon; IFNAR = interferon (α and β) receptor 1; IFNGR = interferon γ receptor 1; IKK = IkB kinase; IL4R = interleukin 4 receptor; IL7R = interleukin 7 receptor; IL 10RB = interleukin 10 receptor, β; IRAK = interleukin-1 receptor-associated kinase; IRF = IFN-regulatory factor; LBP =lipopolysaccharide binding protein; LPS = lipopolysaccharide; LT = heat-labile toxin; MAPK = motigen activated kinase-like protein; MHC = major histocompatibility complex; MyD88 = myeloid differentiation factor 88; NOD = nucleotide-binding oligomerization domain; PI3K = phosphoinositide-3-kinase; PKA = cAMP dependent protein kinase; PSMB = proteasome subunit β; Rac1 = ras-related C3 botulinum toxin

substrate 1; STAT = signal transducers and activators of transcription; TGFB1 = transforming growth factor, β 1; TLR = toll-like receptor; TNF = tumor necrosis factor; TRAF = TNF receptor-associated factor (Liu et al. 2014).

Number of plant products isolated from different corners of world have been reported to show anticancer effect (Table 1). The plant *Catharanthus rosea* once used for treating diabetes in Madagascar and found to suppress the bone marrow lately. Careful scientific observation then led to the discovery and the isolation of the anticancer compounds vincristine, vinblastine, and vinorelbine. The anticancer compound taxol was firstly extracted from the Pacific yew, *Taxus brevifolia*. The structural analogs of taxol are paclitaxcel and docetaxel which now are responsible for a new class of broad spectrum anticancer compounds. Plants belong to the Combretaceae family are widely used in traditional medicine in African countries (Moshi and Mbwambo 2005, Mans, da Rocha and Schwartsmann 2000).

However, clinically available anti-cancer drugs of plant origin reside in mainly four major groups. They are vinca alkaloids (e.g., vinblastine and vincristine), epipodophyllotoxin derivatives (e.g., etoposide and teniposide), camptothecin derivatives (e.g., topotecan and irinotecan) and taxanes (e.g., paclitaxel) (Efferth et al. 2007, Wang et al. 2015). Surprisingly, these naturally derived chemical compounds have traditional uses and also known for poisonous herbs (Zhao et al. 2010). They have been produced through sophisticated technologies with a "Fighting fire with fire" framework, a unique concept of Chinese practitioners and this thought has set its legitimate position in cancer treatment (Figure 9).

Drugs that are therapeutically used in various types of cancers coexisted with toxicities or side effects and cannot be entirely denied. Gambogic acid, discharged by various *Garcinia* species including *Garcinia hanburyi* Hook. F. in Southeast Asia, has been considered as strong anti-cancer agent with multiple cellular and molecular targets through various mechanisms. It is additionally a crucial anti-cancer drug candidate for its potent cytotoxic effect against varied malignant tumors (Li et al. 2009). GA has potential growth inhibition effects and arrest cell cycle at G0/G1 in chronic myeloid leukemia cells by interfering steroid receptor coactivator-3 (SRC-3) (Kupchan et al. 1972). Combretastatins, another class of antimitotic drugs that additionally restrain angiogenesis and are at present being produced for the treatment of holding a pledge for new anticancer medications. Triptolide from *Tripterygium wilfordii* Hook F. was first documented in 1972 as antileukemic (Liu 2011) and tested for the anti-tumor activity of triptolide in various tumor models, including promyelocytic leukemia, human hepatocellular carcinoma, cervical adenocarcinoma, pancreatic carcinoma, multiple myeloma (MM), cholangiocarcinoma, and oral cancer cells (Kannaiyan, Shanmugam and Sethi 2011). Tripterine, otherwise called celastrol, is chemically different from triptolide however shares a few regular properties. Celastrol has anti-tumor activity against a broad spectrum of tumors, both *in vitro* and *in vivo* (Sethi et al. 2007), which allocates to its several effects on apoptosis, cell cycle arrest, angiogenesis, and metastasis. Celastrol, modulates a number of signaling biomolecules including the NF-aB (Yang et al. 2013), activation of caspase family proteins (Sun et al. 2006), VEGF receptors, heat shock proteins, potassium and calcium channels (Kim et al. 2009) and the immunoglobulin Fc epsilon receptor I (Fc_RI) (Lee et al. 2011). Furthermore, synergistic effect of celastrol has also been reported when integrated with chemotherapeutic agents, including radiation induced suppression of the proliferation of melanoma, ovarian cancer and lung cancer (Kunin and Lawton 1996).

2.10. Bio-Market Demand of Plant-Derived Drugs

The multifaceted uses of plant materials include flavors, foods, insect deterrents, ornamentals, fumigants, spices, and cosmetics (Pieroni et al. 2004, Runner et al. 2001). In general, the therapeutically important plants are sold as commodities in the market, and those that are sold for curative concerns govern the market (Tripathi and Mukherjee 2003). At present, natural products (along with their derivatives and analogs) constitute over half of all drugs in clinical use, in which natural products obtained from higher plants represent *ca.* 25% of the total. According to a report of World Health Organization, more than 80% population in developing countries depends on conventional remedies such as herbs for their everyday health care needs (Farnsworth and Morris 1976), and about 855 traditional medicines incorporate utilizing crude plant extracts. This implies that about 3.5 to 4 billion of the worldwide population depend on plants resources for medications (Bernama 1999).

2.11. Herbal Market in Asia

Almost 95% of plants utilized in traditional prescriptions are derived from forests and other natural sources. The plants gathered from various sources show wide difference in restorative qualities and therefore, they vary in market values. In the ongoing years in India, there has been more prominent development of indigenous drug industry (Singh and Ghouse 1993). Subsequently the interest for the new material (restorative plants) has hugely expanded. As per most recent gauge, there are around eight thousand authorized drug stores of ISM in the nation, occupied with the production of mass medications to meet the necessity of individuals. The yearly necessity of the crude materials of these drug stores was assessed to be a huge number of quintals. This is directly met by cutting down trees in the woods or removing herbs and bushes either on ostensible installment or unapproved. Further, there is prime need to give authentic medications to produce standard drugs, as accentuated by prior worker (Mitra 1980). The yearly interest of the worldwide market is $32 million of medicinal plants from developing nations. The production of herbal medicines in India has been assessed to be rupees 4000 crores in the year 2000. Out of 15,000 - 20,000 curative plants, rural networks utilize 7,000 - 7,500 therapeutic plants. Around 130 purified natural compounds, which are isolated from 100 species of Indian higher plants, are utilized all through the world. India can play a vital role for providing the crude herbs, standardized extracted materials and pure compounds isolated from natural resources (WHO 2012). The market for Indian herbal formulation is expanded up to US$ 181.45 billion in the year 2012 and expected to be expanded up to US$ 255.6 billion by 2014 (Hettigoda 2007). China was the top nation dominating in the fare of herbals with US$1329.72 million pursued by India with a fare figure of US$ 790.56 million for the year 2010. In 1993, the total sale of natural drugs in China was added up to more than US$ 2.5 billion. The deal expanded up to US$ 47.84 billion of every 2010 (Hettigoda 2007).

The natural market in Sri Lanka has been come to US$ 20 billion of every 2007 and will be reached to US$ 5 trillion in the coming future (Okada 1996). In 2010, the herbal deal in Japan expanded up to US$ 47.84 billion. Market of Kampo medicine in 1996 was expanded up to US$ 56 billion (MaSiuMan 2012). Comprising of the total offers of each of the 148 NHI-listed Kampo items, the extent of the Japanese remedy Kampo item advertise was somewhat under US$ 1.21 billion. This figure stands for a mere 1.3% of Japan's complete professionally prescribed drug market, which was roughly US$ 0.1 trillion as of the end of 2012. Traditional Chinese medicines incorporate proprietary Chinese medicines, raw materials and ingredients, as well as herbal extracts. The estimation of China's fares of the drug came to US$ 2.33 billion in the year 2012, an expansion of 36.2 percent from the prior year. In Malaysia, there are 1,000 of manufacturers involved in production of herbal medicines. Malaysian worldwide natural products market for nutraceuticals is at US$ 70 billion and US$ 20 billion for phytomedicines with a normal development rate between 15 to 20 percent every year in the year 2012 (Suherdjoko 2010). Indonesia is the second biggest biodiversity focus after Brazil. All in all, 40% of Indonesian populace uses herbal medications for health care needs. In Indonesia, the improvement of herbal industry has expanded quickly. In the year 1992, there were 468 enlisted ventures that expanded to 807 of every 2000. This expansion in number of assembling units is the impression of an increment in the utilization of natural drugs in Indonesia. One of the top organizations managing production of 'Jammu' drug, specifically 'PT Jammu Air Mancur', detailed a yearly pay of US$ 360 million, for the most part from local deals and US$ 9 million from fare of Jammu prescriptions in 1995. Indonesia sends out US$10 million worth of Jammu traditional herbal medication every year, while the local markets of Jammu remains at $500 million in the year 2010 (Kochhar 1981). Thailand's Public Health Ministry is focusing on a developing business sector for natural drugs, ascending from US$ 9.69 million to US$ 16.15 million and a market value for herbal dietary supplements from

US$ 0.26 billion to US$ 0.32 billion in the year 2013 from their 1400 species out of 10,000 plant species (Kochhar 1981). Vietnam will create itself 70% of the required prescriptions. Herbal and plant derived medications utilizing the indigenous crude materials will represent 30% of the total estimation of manufactured drugs. Numerous different nations of Asia are overwhelmingly endeavoring to mount their herbal development through farming and marketization as the market request of plant based therapeutics are becoming rapidly.

2.12. Plant-Based Drugs Market in USA

Herbal formulations are getting worldwide significance in view of their medical advantages. Their acceptability is going to build on account of the expanded interest of consumers in natural products as they are viewed as more secure and more economical than synthetic drugs by and large (Bhowmik et al. 2009). Turnover of herbal medicines in developed countries like United State (USA), Europe, Australia, and Canada was around US$ 30 billion in the year 2000, which was increased around 5-15% than the same for the last decade. Botanical sale in USA was US$ 14 billion in the year 2009 and by the year 2050 it had been predicted to be US$ 5 trillion (Eisenberg et al. 1998). In 1990-1997, it was assessed utilization of herbal products had been expanded by 380% alone in USA. Similarly, industrial demand for herbal products was also elevated because of emergence of products like health foods, natural cosmetics, and hygiene products. In general, worldwide exchange in therapeutic plants and their items was US$ 60 billion in the year 2000, with yearly growth rate of 7% and was believed to reach US$ 5 trillion within 2050 (Govt. of India 2000). It strengthens the use of herbs for therapeutic or medicinal purpose. Herb based drug industry is one of the quickest developing enterprises on the planet. The worldwide pharmaceutical market was worth US $550 billion out of 2004 and has been expanded to US $ 710.2 billion by 2012 (PRNewswire 2012). Current esteem deals inside the US herbal products market developed by 3% through 2012, achieving an estimation of US$ 4.4 billion of every 2012. Worldwide market for herbal formulations are assessed to be worth US$ 26 to 30 billion as per different evaluations. The utilization of medications in Latin America was worth US$ 16.5 billion in the beginning of 1990's, somewhat under 5% of the world utilization. The per capita utilization of medications is roughly US$ 21. The Latin America nutraceuticals market volume came to US$ 7.8 billion out of 2011 from US$ 6.6 billion out of 2007 with CAGR of 4.2% amid 2007-2011 (Meyer, M. 2013).

2.13. African Herbal Market

In South Africa, the interest for indigenous medications is extensively higher than for western drugs. In Zambia, 70% of the population depends on conventional medicines and their exchange is worth over US$ 43 million for each annum. There are around 78 species of medicinal plants utilized broadly in Zambia. In Ethiopia, 70 to 90% of the populace depends on conventional medicines particularly from therapeutic plants for their essential healthcare needs. As in numerous nations, alternative medications are broadly accessible in South African supermarkets, drug stores and wellbeing shops, as per one industry body, the Health Products Association of Southern Africa, is worth no less than $250-million per year. Herbal drugs account for major part of traditional medicines for which every year around 20,000 tons of plant based products are derived from the 771 plant species utilized by local healers (Meyer, M. 2013).

2.14. European Herbal Market

Most of driving organizations had practical experience in natural medications, are situated in Germany, France, Italy, and Switzerland. Some of them are more than 100 years of age and many are still exclusive. A portion of the European herb based organizations, for example, Dr. Much (Germany) Pharmaton (Switzerland Quest, Canada), Asta Medica (Germany), Klinge (Germany), Woelm Pharma (Germany), Natterman (Germany), have been purchased by global pharmaceutical organizations, for example, American home products, Boehringer Ingelheim, Degussa, Fujisawa, Johnson and Johnson and so forth. A group of global pharmaceutical organizations markets natural prescriptions in a few nations, which are manufactured by smaller organizations like Boehringer Ingelheim (Sweden) and Schering (Belgium and Hungary) but specialized in herbal production. Europe Union nations obtain a market volume of US$ 15-20 billion (Vasisht and Kumar 2002).

2.15. Genomic Hallmark of Plant-Derived Drugs

The use of herb-based medicines globally is now being accepted and gaining popularity (Tapsell et al. 2006, Iriti et al. 2010). It is scientifically proved that the wide range of plant secondary metabolites like polyphenols have excellent antioxidant, anticarcinogenic, antimutagenic, and anti-inflammatory effects. The comprehensive knowledge of the cellular and molecular mechanisms associated with the pharmacological procedures has expanded in current few decades, and this has allowed the disclosure of many promising focuses for the advancement of new bioactive natural compounds to treat different illnesses. Numerous plant derived compounds have been found to be associated with disease related gene regulation and modulation. *Arctium lappa* (AL), *Camellia sinensis* (CS), *Echinacea angustifolia*, *Eleutherococcus senticosus*, *Panax ginseng* (PG), and *Vaccinium myrtillus* (VM) are involved in the modulation of the expression of inflammatory genes such as *TNFα*, *COX2*, *IL1β*, *NFκB1*, *NFκB2*, *NOS2*, *NFE2L2*, and *PPARγ*. Very particularly, mRNA expression of *COX2, IL1β, NFE2L2, NFκB1, NFκB2, NOS2*, and *TNFα* was induced, while *PPARγ* was depressed. The expression of these genes were regulated by plant extracts after pre-stimulation with H_2O_2. *COX2* was downregulated by AL, PG, and VM. All extracts depressed *IL1β* expression, but upregulated *NFE2L2. NFκB1, NFκB2*, and *TNFα* were downregulated by AL, CS, PG, and VM. More recently, the anti-inflammatory and anti-oxidant effects as well as modulation of gut microbiota provided by cranberry phytochemicals have drawn more attention. The cranberry is rich in nutrients and provides many biomolecules that have health-promoting effects. The motivation behind this present investigations was to decide whether cranberry extracts (CEs) contain phytoconstituents having anti-inflammatory properties (Figure 10).

The research showed that human monocytic cell line THP-1 was treated with two CEs (CE and 90MX) and subsequently challenged with Lipopolysaccharides (LPS). Tumor necrosis factor α (TNFα) expression was decreased in the CE-treated cells, indicative of an anti-inflammatory effect (Sohn et al. 2011, Lin et al. 1996, Chan et al. 2011, Lin et al. 2002, Pomari, Stefanon and Colitti 2013).

2.16. Numbers of Immune Genes Modulated with the Treatment with Dietary Plant Extracts

Dietary plant extracts modulate gene expression profiles in ileal mucosa of weaned pigs after an Escherichia coli infection. This study was conducted to characterize the effects of infection with a pathogenic F-18 *Escherichia coli* and 3 different plant extracts: *capsicum oleoresin*, *garlic botanical*, and *turmeric oleoresin* on gene expression of ileal mucosa in weaned pigs. *Escherichia coli* infection induces gut immune responses of pigs. Extracts had both upregulatory and downregulatory effects on different genes. Capsicum oleoresin: C7, CCL5, CCR3, IFIT1, IFIT2, IFNGR1, IRF7, MHCII, and TNFA; Garlic botanical: C7, CASP3, CASP8, CCL3L1, CCL4, CFB, HSP90AA1, IFIT1, IFIT2, IL7, IRF7, LBP, TGFB1, and TNFA; Turmeric oleoresin: C5, C7, CASP3, CASP8, CCL5, CCL28, CCR9, CXCR4, CFB, HSP90AA1, HSP90B1, IFNAR1, IFNGR1, IFIT1, IFIT2, IL8, IL10, IL16, IL7R, IRF7, LBP, MHCII, TGFB1, and TNFA (Liu et al. 2014).

The modulated immune gene expression indicated that *E. coli* infection induced gut immune responses of pigs fed the CON diet (Figure 1). The *E. coli* infection altered the expression level of genes related to LPS activation (increase in LBP and TLR2), and MyD88, cytokines (increase in IL16, IL7, IL15, and TNFA, decrease in IL10 and TGFB1), chemokines (increase in CCL28, CCL5 and IL8; decrease in CCL4 and

CCL3L1), complement cascades (increased C7, A2M, CFB, and C5), receptors and costimulators (increase in IL10RB, IFNGR1, IFNAR1, IL4R, CCR1, CCR3, CCR9, and CXCR6; decrease in IL7R and CXCR4), heat stress (increase in HSP90AA1 and HSP90B1; decrease in HSP70), antigen presentation (increase in MHCI, IFI30, IFIT1, IFIT2, PSMB9, and PSMB10, decrease in MHCII), apoptosis (increase in CASP1, CASP3, and CASP8), and endoplasmic reticulum stress (increased CREB3L4) (Liu et al. 2014).

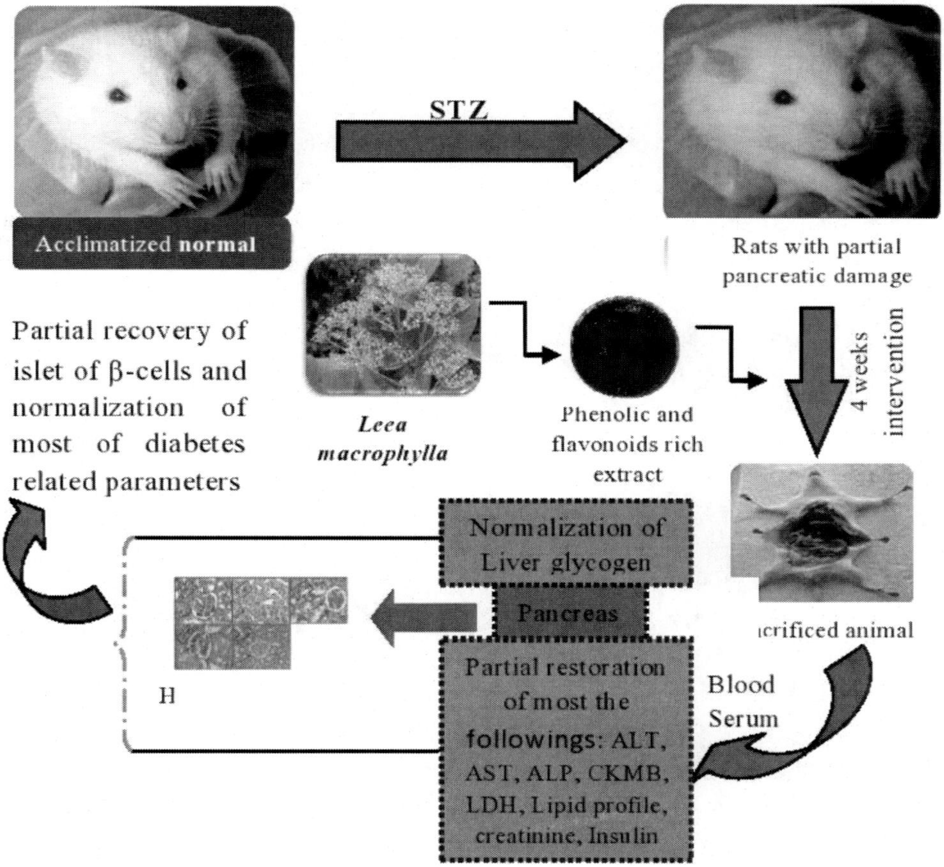

Figure 10. A schematic diagram on how the *Leea macrophylla*, a Bangldeshi plant, is managing the diabetes and diabetes related parameters.

Feeding each of the plant extracts had effects on expression of several genes that were counter to the effects of *E. coli*. Feeding CAP counteracted ($P < 0.05$) effects of *E. coli* on C7, CCL5, CCR3, IFIT1, IFIT2, IFNGR1, IRF7, MHCII, and TNFA. Feeding GAR counteracted ($P < 0.05$) the disease effects on C7, CASP3, CASP8, CCL3L1, CCL4, CFB, HSP90AA1, IFIT1, IFIT2, IL7, IRF7, LBP, TGFB1, and TNFA. Finally, feeding TUR reversed ($P < 0.05$) the effects of E. coli on expression of 24 genes, including C5, C7, CASP3, CASP8, CCL5, CCL28, CCR9, CXCR4, CFB, HSP90AA1, HSP90B1, IFNAR1, IFNGR1, IFIT1, IFIT2, IL8, IL10, IL16, IL7R, IRF7, LBP, MHCII, TGFB1, and TNFA, compared with E. coli-infected control (Figure 10) (Hannon et al. 2017).

2.17. Setbacks to Overcome for Plant Derived Drugs

Plant-derived drugs have been proved to show numerous beneficial effects. Despite their actions for the sake of humanity, they have some limitations. The extraction, processing and preparation of herbal drugs are still not up to the mark to ensure the hygiene. The quality control procedure seems not advanced while the accurate dosage and administration of herbal drugs are quite ambiguous. The molecular approach surrounding the herbal drugs is still at the back. Therefore, research should unfold the most unexplored part of herbal drugs achieving the trusts and belief of the consumer (Rahman and Islam 2014).

Clinical approaches with plant-derived drugs are quite viable and several well-controlled double-blind (placebo-controlled) trials have been completed with specific herbal formulations. A few ongoing surveys published in the Annals of Internal Medicine, the Journal of the American Medical Association (JAMA), the British Medical Journal, the Lancet, and the British Journal of Clinical Pharmacology, among others, affirm the practicality of herbal formulations. However, a few downsides have been directly associated in the development of herbal medicines that needs to be addressed effectively and immediately, for instance:

1. Lack of standardization and quality control of the herbal drugs used in clinical trials;
2. Use of different dosages of herbal medicines;
3. Inadequate randomization in most studies, and patients were not properly selected;
4. Numbers of patients in most trials are insufficient for the attainment of statistical significance;
5. Difficulties in establishing appropriate placebos because of the tastes, aromas, etc;
6. Wide variations in the duration of treatments using herbal medicines.

As an outcome of such challenges, several herbal formulations have been examined sufficiently and well-controlled twofold visually impaired clinical trials to demonstrate their actual safety and viability. In any case, countless clinical trials have been performed just with some herbal medications, including the extracts of *Ginkgo biloba* (utilized for the treatment of CNS and cardiovascular issues) (Brautigam et al. 1998) and *Hypericum perforatum* (St. John's wort, used as a stimulant) (Josey and Tackett 1999). It is essential to underscore that most such clinical examinations have undergone similar critical scientific evaluations from those referenced previously. In spite of the fact that the clinical trials have demonstrated that these plant based drugs are very secure and without resulting serious clinical symptoms, their clinical adequacy still requires well-controlled randomized twofold visually impaired investigations. Plant derived drugs such as *Panax ginseng* (ginseng) used as a tonic, *Tanacetun parthenium* (feverfew) used to treat migraine headache, *Allium sativum* (garlic) used to lower low-density protein cholesterol and some cardiovascular disturbances, *Matricaria chamomilla* (chamomile) recommended as a carminative, anti-inflammatory and antispasmodic, *Silybium marianun* (milk thistle) used for repairing liver function including cirrhosis, *Valeriana officinalis* (valerian) used as a sedative and sleeping aid, Piper methysticum (*Kava kava*) used as an anxiolytic, *Aesculus hippocastanum* (horse chestnut) used for the treatment of chronic venous insufficiency, *Cassia acutfolia* (Senna) and *Rhamnus purshiana* (*Cascara sagrada)* which are used as laxatives, Echinacea purpura Echinacea) used as an anti-inflammatory and immunostimulant, *Arnica montana* (arnica) used to treat post-traumatic and postoperative conditions, and *Serenoa repens* (saw palmetto) used for the treatment of benign prostatic hyperplasia are all herbal drugs with vast global market, and, in spite of the fact that they have been assessed in various clinical trials, more well-controlled and fitting randomized clinical preliminaries are still required so as to demonstrate their genuine adequacy. Notwithstanding, the circumstance will change rapidly on the grounds that the expansion on the global advertise for therapeutic herbs has pulled in a large portion of the biggest local and international pharmaceutical industries, and some of them have lately obtained small manufacturers specialized in producing plant medicines.

3. Our Research

We have been carrying out to investigate the phytopharmacological effects especially the contribution of Bangladeshi medicinal plants/plant products on minimizing the harmful effects of noncommunicable chronic diseases mainly type 2 diabetes and liver diseases. Aside from these, antioxidative possibilities of every one of the twenty plants have been considered both *in vitro* and *in vivo* models. Some of them has chosen for thrombolytic effects while others were evaluated for cytotoxic, membrane stabilizing, anti-inflammatory, analgesic, antinociceptive and antimicrobial effects. The plant materials depending on the extraction conditions are prepared to collect the crude which are then standardized and controlled the quality for further experimental procedures. Among all we studied, few plants especially *Clausina suffriticosa* found to be effective as thrombolytic agent, *Leea indica* and *Leea macrophylla* have shown very strong antioxidative effects both in *in vitro* and *in vivo* models (*Leea macrophylla* leaf extracts have been found to promote the insulin secretion, decrease the ALT and AST levels, lower LDL, total cholesterol and triglycerides in streptozotocin induced Wistar albino rats (Akhter et al. 2015, Rahman et al. 2017). However, the biological antioxidative enzymes catalase, thiolase, superoxide dismutase and glutathione peroxidase are going to be studied whether they are possible to be upregulated by the plant extracts. Some of the plants such as *Litsea glutinosa* was found to show very promising antidiarrheal effects, *Cassia hirsuta* was recorded for antioxidative effects where IC_{50} values were within the cutoff values and cardiac markers are also seen to be optimized by the use of *Tamarindus indicus*. Among all, our research has so far observed that the highest antioxidative plants are the best capable to mitigate the pathological conditions (Figure 11). Therefore, these plants are apprehended to be formulated for future therapeutic drug aligned with the phytopharmacological actions. Bangladeshi indigenous medicinal plant sources have been used to establish those as source of functional food and formulation of therapeutic drugs.

3.1. Plant-Derived Drug as a Molecular Weapon for Treatment

Very beginning of 1882, study of Chinese Medicine (CM) prompted the disclosure of ephedrine from *Ephedra sinica* (Ma-Huang). This compound and its various synthetic analogs are utilized potentially today in anti-asthma treatment. A great deal of researchers are endeavoring to reach into the new era of herbal regimes. Lee (2000) (Lee 2000) announced that he has been planning following new bioactive natural products in his Natural Products Library and their analogs as chemotherapeutic agents:

1. Plant anti-tumour agents–novel plant-derived cytotoxic anti-tumour principals and analogues.
2. Novel anti-tumour etoposide analogues–design and synthesis of novel analogues related to the anti-cancer drug etoposide with better pharmacological profiles.
3. Development of anti-AIDS agents- plant-derived anti-HIV compounds. Two classes of modified natural products from betulinic acid (*Szyigium claviflorum*) and the coumarin suksdorfin (*Lomatium suksdorfii)*
4. Novel anti-malarial agents–anti-malarial analogues related to qinghaosu.
5. Anti-viral hepatitic drugs–design and synthesis of analogues related to DDB, a clinically useful anti-viral hepatitic drug.
6. New non-steroidal anti-inflammatory and anti-arthritis agents.
7. New anti-fungal agents.
8. Novel anti-viral agents–inhibitors of HSV I and II and HCMV.

9. Combinatorial chemistry–focused combinatorial libraries based on bioactive natural products models.
10. Chinese medicines–discovery and development of active principals, fractions, and formulations from traditional Chinese herbal medicines.

In case of anti-cancer therapeutic designing, the Natural Product Laboratory (NPL) has found more than 1000 new cytotoxic anti-tumor principals and their chemical derivatives or manufactured analogs from therapeutic herbs in last three decades. Cytotoxic herbal extracts, different solvent fractions and isolated pure compounds are screened firstly in a primary in-house human tumor cell line (HTCL) panel and after that finished bioactivity-coordinated fractionation and isolation (BDFI), concentrating on a couple of the most dynamic bioassay frameworks. Promising samples are affirmed compelling in forms the assessment of National Cancer Institute (NCI) *in vitro* HTCL and *in vivo* xenograft evaluation (Boyd 1993). As such, BDFI of Chinese sanative herbs, combined with structural alteration and synthesis, have brought about in excess of 100 new cytotoxic anti-tumor compounds and their analogs that are of current demand to NCI and have demonstrated affirmed action in NCI's *in vitro* initial human tumor cell lines bioassay (Figure 1.2). Etoposide is a semisynthetic derivative of the natural product podophyllotoxin, which is extracted and isolated from *Podophyllum peltatum*. Notwithstanding, their disparities include both structural changes at the 4-and 40-position and their corresponding mode of action: podophyllotoxin aims tubulin, while etoposide focuses on the chemical DNA topo II. Etoposide and the related thiophene analog teniposide are broadly used to treat different cancers, including leukemia and lymphomas; in any case, myelosuppression, drug resistance, and poor bioavailability persist as issues related with their utilization (O'Dwyer et al. 1984).

Conclusion

Drugs of plant origin, in addition to their conventional values, also hold substantial public and medical intrigue worldwide as sources of nutraceuticals or novel lead compounds for drug discovery. Herbal therapy insurgency opened new doors as well as served to invigorate research into in the field of direct pertinence to human health insurance. The screening for dynamic phytocompounds will be incredibly facilitated by the blend of different metabolomics approaches with a variety of bioactivity assessments in mammalian frameworks to separate between plant species, tissues, or phytopreparations, and to distinguish novel lead drug candidates for future advancement. In an integral improvement, the utilization of metabolome refined plant extracts with other biochemical components in combination, rather than as isolated single compound (s), may turn out to be extremely valuable as more extensive and comprehensive therapeutic or pharmacological agents for a diversified human medicinal services application. The reconciliation of plant derived drugs, knowing about their biochemical mode of actions, optimization of their dosages and lastly their incorporation into the health system ought to be created in such an approach to bring agreement between the conventional and current human care frameworks with least threats to one another. There might be some who may well dispute at what they see as the unchallenging prospect of working on phytomedicines, since much is as of now known.

REFERENCES

Akhter, S., Rahman, M. A., Aklima, J., Hasan, M. R. and Chowdhury, J. M. (2015). Antioxidative Role of Hatikana (Leea macrophylla Roxb.) Partially Improves the Hepatic Damage Induced by CCl4 in Wistar Albino Rats. *BioMed Research International*, 2015: 356729.

Akwu, N. A. (2013). Comparative study of the cyto-genotoxicity effect of the leaves extracts of *Annona muricata*, *Dacryodes edulis* and *Persea americana* using *Allium cepa* L. assay and cell line culture. *Journal of Cancer Science and Therapy*, 5(10), 241.

Ames, B. N., Gold, L. S. and Willett W. C. (1995). The causes and prevention of cancer. *Proceedings of the National Academy of Sciences*, 92(12): 5258-5265.

Amster, E., Tiwari, A. and Schenker, M. B. (2007). Case report: Potential arsenic toxicosis secondary to herbal kelp supplement. *Environmental Health Perspectives*, 115(4): 606-608.

Anon (2007). *Medicinal, Culinary and Aromatic Plants in the near East-Medic...* http://www.faO.Org/dorep/x5402e16.htm (cited 4 August, 2007).

Apostolova, N., Blas-Garcia, A. and Esplugues, J. V. (2011). Mitochondria sentencing about cellular life and death: a matter of oxidative stress. *Current Pharmaceutical Design*, 17(36): 4047–4060.

Appendino, G., Fontana, G. and Pollastro, F. (2010). 3.08 — Natural products drug discovery. In: Liu, H-W.; Mander, L., editors. *Comprehensive Natural Products II.* Elsevier; Oxford, p. 205-236.

Arit, V. M., Stiborova, M. and Schmeiser, H. H. (2002). Aristolochic acid as a probable human cancer hazard in herbal remedies: a review. *Mutagenesis*, 17(4): 265-277.

Atanasov, A. G., Waltenberger, B., Pferschy-Wenzig, E. M., Linder, T., Wawrosch, C., Uhrin, P., Temml, V., Wang, L., Schwaiger, S., Heiss, E. H., Rollinger, J. M., Schuster, D., Breuss, J.M., Bochkov, V., Mihovilovic, M. D., Kopp, B., Bauer, R., Dirsch, V. M. and Stuppner, H. (2015). Discovery and resupply of pharmacologically active plant-derived natural products: A review. *Biotechnology Advances*, 33(8): 1582–1614.

Benyhe, S., Farkas, J., Tóth, G. and Wollemann, M. (1997). Met5 - enkephalin-Arg6 - Phe7, an endogenous neuropeptide, binds to multiple opioid and nonopioid sites in rat brain. *Journal of Neuroscience Research*, 48(3): 249-258.

Beppe, G. J., Dongmo, A. B., Foyet, H. S., Tsabang, N., Olteanu, Z., Cioanca, O., Hancianu, M., Dimo, T. and Hritcu, L. (2014). Memory-enhancing activities of the aqueous extract of Albizia adianthifolia leaves in the 6-hydroxydopamine-lesionrodent model of Parkinson's disease. *BMC Complementary and Alternative Medicine*, 14:142.

Bernama. (1999). Tapping Multi-billion Ringgit Herbal Industry, *Borneo Post*, 11 Oct. 1999. Available: http://www.borneofocus.com/vaic/Statistics/article3.htm.

Bhowmik, D., Sampath Kumar, K. P., Tripathi, P. and Chiranjib, B. (2009). Traditional Herbal Medicines: An Overview. *Archives of Applied Science Research*, 1(2):165-177.

Bie, M., Lv, Y., Ren, C., Xing, F., Cui, Q., Xiao, J. and So, K. F. (2015). Lycium barbarum polysaccharide improves bipolar pulse current-induced microglia cell injury through modulating autophagy. *Cell Transplantation*, 24(3): 419–428.

Boyd, M. R. (1993). In: Neiderhuber JE, ed. *Current Therapy in Oncology*. Philadelphia, PA: BC Decker, pp.11-22.

Brautigam, M. R. H., Blommaert, F. A., Verleye, G., Castermans, J., Jansen Steur, E. N. H. and Kleijnen, J. (1998). Treatment of age-related memory complaints with Ginkgo biloba extract: a randomized double blind placebo-controlled study. *Phytomedicine*, 5(6): 425–34.

Bringmann, G., Messer, K., Wohlfarth, M., Kraus, J., Dumbuya, K. and Rückert, M. (1999). HPLC-CD on-line coupling in combination with HPLCNMR and HPLC-MS/MS for the determination of the full absolute stereostructure of new metabolites in plant extracts. *Analytical Chemistry*, 71(14): 2678–2686.

Bucar, F., Wube, A. and Schmid, M. (2013). Natural product isolation — how to get from biological material to pure compounds. *Natural Product Reports*, 30(4): 525–545.

Butler, M. S. (2005). Natural products to drugs: natural product derived compounds in clinical trials. *Natural Product Reports*, 22(3): 162–195.

Butler, M. S., Robertson, A. A. and Cooper, M. A. (2014). Natural product and natural product derived drugs in clinical trials. *Natural Product Reports*, 31(11): 1612-1661.

Calixto, J. B. (2000). Efficacy, safety, quality control, marketing and regulatory guidelines for herbal medicines (phytotherapeutic agents). *Brazilian Journal of Medical and Biological Research*, 33(2): 179-189.

Chakraborty, P. (2018). Herbal genomics as tools for dissecting new metabolic pathways of unexplored medicinal plants and drug discovery. *Biochimie Open*, 6: 9-16.

Chan, Y. S., Cheng, L. N., Wu, J. H., Chan, E., Kwan, Y. W., Lee, S. M., Leung, G. P., Yu, P. H. and Chan, S. W. (2011). A review of the pharmacological effects of Arctium lappa (burdock). *Inflammopharmacology*, 19(5): 245–254.

Chen, S., Song, J., Sun, C., Xu, J., Zhu, Y., Verpoorte, R. and Fan, T. P. (2015). Herbal genomics: examining the biology of traditional medicines. *Science*, 347: 527-529.

Chen, S., Xu, L. X. and Guo, Q. L. (2011). An introduction to the medicinal plant genome project. *Frontiers of Medicine*, 5(2): 178-184.

Chong, K. T. and Pagano, P. J. (1997). In vitro combination of PNV-140690, a Human Immunodeficiency Virus type 1 protease inhibitor with Ritonavir against Ritonavir-sensitive and Resistant Clinical Isolates. *Antimicrobial Agents and Chemotherapy*, 41(II), 2367-2377.

Chopra, A. (2000). Ayurvedic medicine and arthritis. *Rheumatic Disease Clinics of North America*, 26(1): 133-144.

Corson, T. W. and Crews, C. M. (2007). Molecular understanding and modern application of traditional medicines: triumphs and trials. *Cell*, 130(5): 769–774.

Cowan, M. M. (1999). Plant Products as Antimicrobial Agents. *Clinical Microbiology Reviews*, 12(4): 564-582.

Cragg, G. M. and Newman, D. J. (2005). International collaboration in drug discovery and development from natural sources. *Pure and Applied Chemistry*, 77 (11): 1923 – 1942.

David, B., Wolfender, J. L. and Dias, D. A. (2015). The pharmaceutical industry and natural products: historical status and new trends. *Phytochemistry Reviews*, 14(2): 299–315.

Deng, J., Yang, J., Wu, L. F., Yu, K. K., Wang, D. P., Yang, X. S. and Chen, F. X. (2015). Structural characterization and neurotrophic activity study of a polysaccharide isolated from Campanumoea javanica. *Journal of Carbohydrate Chemistry*, 34(4): 183–195.

Duan, Z. Z., Zhou, X. L., Li, Y. H., Zhang, F., Li, F. Y. and Qi, S. H. (2015). Protection of Momordica charantia polysaccharide against intra-cerebral hemorrhage-induced. *Journal of Receptor and Signal Transduction Research*, 35(6): 523-529.

Dwivedi, S. K. and Dey, S. (2002). Medicinal herbs: a potential source of toxic metal exposure for man and animals in India. *Archives of Environmental Health*, 57(3): 229-231.

Eberhardt, T. L., Li, X., Shupe, T. F. and Hse, C. Y. (2007). Chinese tallow tree (Sapium Sebiferum) utilization: Characterization of extractives and cell-wall chemistry. *Wood and Fiber Science*, 39(2): 319-324.

Efferth, T., Li, P. C. H., Konkimalla, V. S. B. and Kaina, B. (2007). From traditional Chinese medicine to rational cancer therapy. Trends in Molecular Medicine, 13(8): 353-61.

Eisenberg, D. M., Davis, R. B., Ettner, S. L., Appel, S., Wilkey, S., Van Rompay, M. and Kessler, R. C. (1998). Trends in alternative medicine use in the United States, 1990–1997. Results of a follow-up national survey. *Journal of American Medical Association*, 280(18): 1569-1575.

Eseyin, O. A., Sattar, M. A., Rathore, H. A., Ahmad, A., Afzal, S., Lazhari, M., Ahmad, F. and Akhtar1, S. (2014). Hypoglycemic potential of polysaccharides of the leaf extract of *Telfairia occidentalis. Annual Research & Review Biology*, 4(11): 1813–1826.

Fan, Y., Sun, C., Gao, X. G., Wang, F., Li, X., Kassim, R. M., Tai, G. and Zhou, Y. (2012). Neuroprotective effects of ginseng pectin through the activation of ERK/MAPK and Akt survival signaling pathways. *Molecular Medicine Reports*, 5(5): 1185–1190.

Farnsworth, N. R. and Morris, R. W. (1976). Higher plants-the sleeping giant of drug development. *American Journal of Pharmacy and the Sciences Supporting Public Health*, 148(2): 46-52.

Francis, G., Kerem, Z., Makkar, H. P. S. and Becker, K. (2002). The biological action of saponins in animal systems: a review. *British Journal of Nutrition*, 88(6): 587-605.

Fyhrquist, P., Mwasumbi, L., Haeggstrom, C. A., Vuorela, H., Hiltunen, R. and Vuorela, P. (2004). Antifungal activity of selected species of Terminalia, Pteleopsis and Combretum (Combretaceae) in Tanzania. *Pharmaceutical Biology*, 42(4-5): 308-317.

Gamaniel, K. S. (2000). Toxicity from medicinal plants and their products. *Nigerian Journal of Natural Products and Medicines*, 4(2000): 4-8.

Giddings, L. A., Liscombe, D. K., Hamilton, J. P., Childs, K. L., Della Penna, D., Buell, C. R. and O'Connor, S. E. (2011). A stereoselective hydroxylation step of alkaloid biosynthesis by a unique cytochrome P450 in Catharanthus roseus. *Journal of Biological Chemistry*, 286(19): 16751-16757.

Gleeson, M. P., Hersey, A., Montanari, D. and Overington, J. (2011). Probing the links between in vitro potency, ADMET and physicochemical parameters. *Nature Reviews Drug Discovery*, 10:197–208.

Gong, J., Sun, F., Li, Y., Zhou, X., Duan, Z., Duan, F., Zhao, L., Chen, H., Qi, S. and Shen, J. (2015). Momordica charantia polysaccharides could protect against cerebralischemia/reperfusion injury through inhibiting oxidative stress mediatedc-Jun N-terminal kinase 3 signaling pathway. *Neuropharmacology*, 91:123–134.

Govt. of India. (2000). *Medicinal Plants Introduction*, Indian System of Medicine and Homoeopathy (ISMH), Department of ISMH, Ministry of Health and Family Welfare, Govt. of India. Available at: http://indianmedicine.nic.in/html/plants/mimain.htm.

Hannon, D. B., Thompson, J. T., Khoo, C., Juturu, V., Vanden Heuvel, J. P. (2017). Effects of cranberry extracts on gene expression in THP-1. Cells. Food Science and Nutrition. *Food Science & Nutrition*, 5(1): 148–159.

Hazra, K. M., Roy R. N., Sen S. K. and Laska, S. (2007). Isolation of antibacterial pentahydroxy flavones from the seeds of Mimusops elengi Linn. *African Journal of Biotechnology*, 6(12): 1446-1449.

Henrich, C. J. and Beutler, J. A. (2013). Matching the power of high throughput screening to the chemical diversity of natural products. *Natural Product Reports*, 30(10): 1284–1298.

Hettigoda, A. (2007). *Herbal market will hit $ 5 trillion*. (http://www.sundayobserver.lk).

Ho, J. Y., Hsu, R. J., Wu, C. L., Chang, W. L., Cha, T. L., Yu, D. S. and Yu, C. P. (2013). Ovatodiolide targets β-catenin signaling in suppressing tumorigenesis and overcoming drug resistance in renal cell carcinoma, *Evidence-Based Complementary and Alternative Medicine*, 2013:161628

Hsu, P. and Tien, H. (1974). *Studies on the components of Formosan Solanum species. Part I Alkaloids of Solanum incanum*, Tai-wan Yao Hsueh Tsa Chih 26 p. 102338t.

Hugo, W. B. and Russell, A. D. (2003). *Pharmaceutical Microbiology*; 6th ed. Blackwell Science Publishers p. 91-129.

Huie, C. W. (2002). A review of modern sample-preparation techniques for the extraction and analysis of medicinal plants. *Analytical and Bioanalytical Chemistry*, 373 (1-2): 23-30.

Ifeoma, O. and Oluwakanyinsola, S. (2013). *Screening of Herbal Medicines for Potential Toxicities. New Insights into Toxicity and Drug Testing*, Dr. Sivakumar Gowder (Ed.).

Iriti, M., Vitalini, S., Fico, G. and Faoro, F. (2010). Neuroprotective herbs and foods from different traditional medicines and diets. *Molecules*, 15(5): 3517–3555.

Iwu, M. M., Duncan, A. R. and Okunji, C. O. (1999). New Antimicrobials of plant origin. In Janick, J. (ed) *Perspectives in New crops and New uses*. ASHS Press, Alexandria, V.A. pp. 457-462.

Jia, X., Ma, L., Li, P., Chen, M. and He, C. (2016). Prospects of Poria cocos polysaccharides: isolation process, structural features and bioactivities. *Trends Food Science & Technology*, 54:52–62.

Jin, M., Huang, Q., Zhao, K. and Shang, P. (2013). Biological activities and potential health benefit effects of polysaccharides isolated from Lycium barbarum L *International Journal of Biological Macromolecules*, 54: 16–23.

Jones, W. P. and Kinghorn, A. D. (2012). Extraction of plant secondary metabolites. Methods in Molecular. *Biology*, 864: 341–366.

Josey, E. S. and Tackett, R. L. (1999). St John's Wost: A new alternative for depression. *International Journal of Clinical Pharmacology and Therapeutics*, 37(3): 111-119.

Kambizi, L. and Afolayan, A. J. (2001). An ethnobotanical study of plants used for the treatment of sexually transmitted diseases (njovhera) in Guruve district, Zimbabwe. *Journal of Ethnopharmacology*, 71(1): 5-9.

Kamboj, V. P. (2000). Herbal medicine. *Current Science*, 78(1): 35-39.

Kannaiyan, R., Shanmugam, M. K. and Sethi, G. (2011). Molecular targets of celastrol derived from Thunder of God Vine: potential role in the treatment of inflammatory disorders and cancer. *Cancer Letters*, 303(1): 9–20.

Keller-Juslén, C., Kuhn, M., Von Wartburg, A. and Stähelin, H. (1971). Synthesis and antibiotic activity of glycisidic lignan derivatives related to Podophyllotoxin. *Journal of Medical Chemistry*, 14(10): 936-940.

Kellner, F., Kim, J., Clavijo, B. J., Hamilton, J. P., Childs, K. L., Vaillancourt, B., Cepela, J., Habermann, M., Steuernagel, B., Clissold, L., McLay, K., Buell, C. R. and O'Connor, S. E. (2015). Genome-guided investigation of plant natural product biosynthesis. *The Plant Journal: for cell and molecular biology*, 82(4): 680-692.

Kennedy, D. O. and Wightman, E. L. (2011). Herbal extracts and phytochemicals: plant secondary metabolites and the enhancement of human brain function. *Advances in Nutrition*, 2(1): 32-50.

Kim, S. W., Gupta, R., Lee, S. H., Min, C. W., Agrawal, G. K., Rakwal, R., Kim, J. B., Jo, I. H., Park, S. Y., Kim, J. K., Kim, Y. C., Bang, K. H. and Kim, S. T. (2016). An integrated biochemical, proteomics, and metabolomics approach for supporting medicinal value of Panax ginseng fruits. *Frontiers in Plant Science*, 7: 1-14.

Kim, Y., Kim, K., Lee, H., Han, S., Lee, Y. S., Choe, J., Kim, Y. M., Hahn, J. H., Ro, J. Y. and Jeoung, D. (2009). Celastrol binds to ERK and inhibits Fc epsilon RI signaling to exert an anti-allergic effect. *European Journal of Pharmacology*, 612(1-3): 131–142.

Kingston, D. G. (2011). Modern natural products drug discovery and its relevance to biodiversity conservation. *Journal of Natural Products*, 74(3): 496–511.

Klayman, D. L. (1985). Qinghaosu (Artemisinin): An Antimalarial Drug from China. *Science*, 228(4703): 1049-1055.

Kochhar, S. L. (1981). *Tropical crops: A textbook of economy botany*. Macmillan Pub Ltd. London, 268-71.

Koehn, F. E. and Carter, G. T. (2005). The evolving role of natural products in drug discovery. *Nature Reviews, Drug Discovery*, 4(3): 206–220.

Kremsner, P. G., Winkler, S., Brandts, C., Neifer, S., Bienzle, U. and Graninger, W. (1994). Clindamycin in combination with chloroquine or quinine is an effective therapy for uncomplicated Plasmodium falciparum Malaria in children from Gabon. *Journal Infectious Diseases*, 169(2): 467- 470.

Kumar, R. S., Sunderam, R. S., Sivakumar, T., Sivakumar, P., Sureshkumar, R., Kanagasabi, R., Vijaya, M., Perumal, B. P., Gupta, M., Mazumdar, U. K., Kumar, M. S. and Kumar, K. A. (2007). Effect of Bauhinia racemosa stem bark on N-nitrosodiethylamine-induced hepatocarcinogenesis in rats, *The American Journal of Chinese Medicine*, 35(1), 103-114.

Kunin, W. E. and Lawton, J. H. (1996). Does biodiversity matter? Evaluating the case for conserving species. In: *Biodiversity*, (Editor Gaston, K. J.), Blackwell Science LTD, UK. pp. 283-308.

Kupchan, S. M., Court, W. A., Dailey Jr., R. G., Gilmore, C. J. and Bryan, R. F. (1972). Triptolide and tripdiolide, novel antileukemic diterpenoid triepoxides from Tripterygium wilfordii. *Journal of American Chemical Society*, 94 (20): 7194-7195

Lee, J. H., Choi, K. J., Seo, W. D., Jang, S. Y., Kim, M., Lee, B. W., Kim, J. Y., Kang, S., Park, K. H., Lee, Y. S. and Bae, S. (2011). Enhancement of radiation sensitivity in lung cancer cells by celastrol is mediated by inhibition of Hsp90. *International Journal of Molecular Medicine*, 27(3): 441–446.

Lee, K. H. (2000). Research and future trends in the pharmaceutical development of medicinal herbs from Chinese medicine. *Public Health Nutrition*, 3(4A): 515-522.

Lee, N. and de Beer, T. (2016). The patient with neurological and psychological disorders. *Surgery*, 34(8): 420–424.

Li, H., Liang, Y., Chiu, K., Yuan, Q., Lin, B., Chang, R. C. and So, K. F. (2013). Lycium barbarum (wolfberry) reduces secondary degeneration and oxidative stress, and inhibits JNK pathway in retina after partial optic nerve transaction. *PLoS One*, 8: e68881.

Li, J. W. and Vederas, J. C. (2009). Drug discovery and natural products: end of an era or an endless frontier? *Science*, 325(5937): 161–165.

Li, R., Chen, Y., Zeng, L. L., Shu, W. X., Zhao, F., Wen, L. and Liu, Y. (2009). Gambogic acid induces G0/G1 arrest and apoptosis involving inhibition of SRC-3 and inactivation of Akt pathway in K562 leukemia cells. *Toxicology*, 262(2): 98–105.

Lin, C. C., Lu, J. M., Yang, J. J., Chuang, S. C. and Ujiie, T. (1996). Anti-inflammatory and radical scavenge effects of Arctium lappa. *American Journal of Chinese Medicine*, 24(2): 127–137.

Lin, S. C., Lin, C. H., Lin, C. C., Lin, Y. H., Chen, C. F., Chen, I. C. and Wang, L. Y. (2002). Hepatoprotective effects of Arctium lappa Linne on liver injuries induced by chronic ethanol consumption and potentiated by carbon tetrachloride. *Journal of Biomedical Science*, 9(5): 401–409.

Liu, Q. (2011). Triptolide and its expanding multiple pharmacological functions. *International Immunopharmacology*, 11(3): 377–383.

Liu, Y., Song, M., Che, T. M., Lee, J. J., Bravo, D., Maddox, C. W. and Pettigrew, J. E. (2014). Dietary plant extracts modulate gene expression profiles in ileal mucosa of weaned pigs after an Escherichia coli infection. *Journal of Animal Science*, 92(5): 2050-2062.

Mahidol, C., Ruchirawat, S., Prawat, H., Pisutjaroenpong, S., Engprasert, S., Chumsri, P., Tengchaisri, T., Sirisinha, S. and Pichas, P. (1998). Biodiversity and natural product drug discovery. *Pure and Applied Chemistry*, 70(11): 2065-2072.

Mans, D. R., da Rocha, A. B. and Schwartsmann, G. (2000). Anti-cancer drug discovery and development in Brazil: targeted plant collection as a rational strategy to acquire candidate anti-cancer compounds. *Oncologist*, 5(3): 185–198

MaSiuMan. (2012). *Herbal Market.* November 2012 (http://www.studymode.com).

Meyer, M. (2013). Claim that traditional medicines will be tested is churnalism not journalism. *Africacheck*. (http://www.africacheck.org).

Mitra, R. (1980). Silent pharmacognostical characters of Varuna, important drugs of Ayurveda. *Bulletin of Medicinal and Etheno Botanical Research*, 1: 80-98.

Moshi, M. J. and Mbwambo, Z. H. (2005). Some pharmacological properties of extracts of Terminalia sericea roots. *Journal of Ethnopharmacology*, 97(1): 43-47.

Nelson, E. D., Ramberg, J. E., Best, T. and Sinnott, R. A. (2013). Neurologic effects of exogenous saccharides: a review of controlled human, animal, and in vitro studies. *Nutritional Neuroscience*, 15(4): 149–162.

Niessen, W. M., Lin, J. and Bondoux, G. C. (2002). Developing strategies for isolation of minor impurities with mass spectrometry-directed fractionation. *Journal of Chromatography A*, 970(1-2): 131–140.

O'Dwyer, P. J., Alonso, M. T., Leyland-Jones, B. and Marsoni, S. (1984). Teniposide: a review of 12 years of experience. *Cancer Treatment Reports*, 68(12): 1455-1466.

Okada, F. (1996). Kampo medicine, a source of drugs waiting to be exploited. *The Lancet*, 348 (9019): 5-6.

Oksman, K., Etang, J. A., Mathew, A. P. and Jonoobi, M. (2011). Cellulose nanowhiskers separated from a bio-residue from wood bioethanol production. *Biomass and Bioenergy*, 35(1): 146-152.

Parikh, U. M., Barnas, D. C., Faruki, H. and Mellors, J. W. (2006). Antagonism between the HIV-1 Reverse-Transcriptase Mutation K65R and Thymidine-Analogue Mutations at the Genomic Level. *Journal of Infectious Diseases*, 194(5): 651-660.

Paterson, I. and Anderson, E. A. (2005). Chemistry. The renaissance of natural products as drug candidates. *Science*, 310(5747): 451–453.

Philomena, G. (2011). Concerns regarding the safety and toxicity of medicinal plants - An overview. *Journal of Applied Pharmaceutical Science*, 1(6): 40-44.

Pieroni, A., Quare, C. L., Villanelli, M. L., Mangino, P., Sabbatini, G., Santini, L., Boccetti, T., Profili, M., Ciccioli, T., Rampa, L. G., Antonini, G., Girolamini, C., Cecchi, M. and Tomasi, M. (2004). Ethnopharmacognostic survey on the natural ingredients used in folk cosmetics, cosmeceuticals and remedies for healing sting diseases in the inland Marches, Central-Eastern Italy. *Journal of Ethnopharmacology*, 9(2-3): 331- 344.

Pomari, E., Stefanon, B. and Colitti, M. (2013). Effect of Arctium lappa (burdock) extract on canine dermal fibroblasts. *Veterinary Immunology and Immunopathology*, 156(3-4): 159–166.

Prescott, L. M., Harley, J. P. and Klein, D. A. (2002). *Microbiology* 6th ed. Macgraw Hill Publishers p. 808-823.

PRNewswire. (2012). Nutraceuticals Product Market: Latin America Market Size, Segment and Country Analysis and Forecasts (2007-2017), *Research and Markets*. (http://www.researchandmarkets.com).

Qi, F., Li, A., Inagaki, Y., Gao, J., Li, J., Kokudo, N., Li, X. K. and Tang, W. (2010). Chinese herbal medicines as adjuvant treatment during chemo-or radio-therapy for cancer. *Bioscience Trends*, 4(6): 297-307.

Rabe, T. and van Staden, J. (1997). Antibacterial activity of South African plants used for medicinal purposes. *Journal of Ethnopharmacology*, 56(1): 81-87.

Rahman, M. A. and Islam, M. S. (2014). *Traditional and Folk Herbal Medicine: Prospects, Paradigm and drawbacks of current research on herbal medicine*, V. K Gupta, Daya publishing House New Delhi, Indi, Vol 2.

Rahman, M. A., Chowdhury, J. M. K. H., Aklima, J., Azadi, M. A. (2017). Leea macrophylla Roxb. leaf extract potentially helps normalize islet of β-cells damaged in STZ-induced albino rats. *Food Science & Nutrition*, 6(4): 1-10.

Rajendran, N. K. and Ramakrishnan, J. (2010). Polyphenol analysis and anti-tumor activity of crude extracts from tegmen of Artocarpus heterophyllus. *Medicinal Plants*, 2(1): 63-66.

Rao, P. V. and Gan, S. H. (2014). Cinnamon: a multifaceted medicinal plant. *Evidence-Based Complementary and Alternative Medicine*, 2014:1–12.

Rao, S. R. and Ravishankar, G. A. (2002). Plant cell cultures: chemical factories of secondary metabolites. *Biotechnology Advances*, 20(2): 101–153.

Rattan, R. S. (2010). Mechanisms of action of insecticidal secondary metabolites of plant origin. *Crop Protection*, 29(9):913-920.

Runner, R. T., Majindai, Berhanu, M., Abegaz Bezabih, M., Bonaventure, T., Ngadjui Cornelius, C. W., Wanjala, Ladislaus, K., Bojase, G., Silayo, A., Masesange, I., Samuel, O. and Yeboah, (2001). Recent results from natural product research at the University of Botswana. *Pure and Applied Chemistry*, 73(7): 1197-1208.

Salim, A., Chin, Y. W. and Kinghorn, A. (2008). Drug Discovery from Plants. In: Ramawat K., Merillon J. (eds) Bioactive Molecules and Medicinal Plants. Springer, Berlin, Heidelberg.

Sasidharan, S., Chen, Y., Saravanan, D., Sundram, K. M. and Yoga Latha, L. (2011). Extraction, Isolation and Characterization of Bioactive Compounds from Plants' Extracts. *African Journal of Traditional, Complementary and Alternative Medicine*, 8(1): 1-10.

Savelev, S. U., Okello, E. J. and Perry, E. K. (2004). Butyryl- and acetyl-cholinesterase inhibitory activities in essential oils of Salvia species and their constituents. *Phytotherapy Research*, 18(4): 315-324.

Scannell, J. W., Blanckley, A., Boldon, H. and Warrington, B. (2012). Diagnosing the decline in pharmaceutical R&D efficiency. *Nature Reviews Drug Discovery*, 11: 191-200.

Seoposengwe, K., Tonder, J. J. and Steenkamp, V. (2013). In vitro neuroprotective potential of four medicinal plants against rotenone-induced toxicity in SH-SY5Yneuroblastoma cells. *BMC Complementary and Alternative Medicine*, 13, 353.

Seow, S. L., Eik, L. F., Naidu, M., David, P., David, K. H. and Sabaratnam, V. (2015). Lignosusrhinocerotis (Cooke) Ryvarden mimics the neuritogenic activity of nerve growth factor via MEK/ERK1/2 signaling pathway in PC-12 cells. *Scientific Reports*, 5: 16349.

Sethi, G., Ahn, K. S., Pandey, M. K. and Aggarwal, B. B. (2007). Celastrol, a novel triterpene, potentiates TNF-induced apoptosis and suppresses invasion of tumor cells by inhibiting NF-kappaB-regulated gene products and TAK1-mediated NF-kappaB activation. *Blood*, 109(7): 2727–2735.

Sharma, H., Parihar, L. and Parihar, P. (2011). Review on cancer and anticancerous properties of some medicinal plants. *Journal of Medicinal Plants Research*, 5(10): 1818-1835.

Sharma, P., Jha, A. B., Dubey, R. S. and Pessarakli, M. (2012). Reactive Oxygen Species, Oxidative Damage, and Antioxidative Defense Mechanism in Plants under Stressful Conditions. *Journal of Botany*, 2012:217037.

Shi, L. (2016). Bioactivities, isolation and purification methods of polysaccharides from natural products: a review. *International Journal of Biological Macromolecules*, 92:37–48.

Singh, N., Verma, P., Pandey, B. R. and Bhalla, M. (2012). Therapeutic potential of Ocimum sanctum in prevention and treatment of cancer and exposure to radiation: An overview. *International Journal of Pharmaceutical Sciences and Drug Research*, 4(2): 97-104.

Singh, V. K. and Ghouse, A. K. K. (1993). Plantation of medicinal plants is the need of the day in India. *Glimpses in Plant Research* X: 203-207. (Editors Govil, J. N., Singh, V. K., and Hasmi, S.) Today and Tomorrow's printers, New Delhi.

Sneader, W. (2005). *Drug Discovery: A History*. Wiley.

Sofowora, A. E. (1993). *Medicinal Plants and Traditional Medicines in Africa*. 2nd edition. Spectrum Books, Ibadan, Nigeria. p. 289.

Sohn, E. H., Jang. S. A., Joo, H., Park, S., Kang, S. C., Lee, C. H. and Kim, S. Y. (2011). Anti-allergic and anti-inflammatory effects of butanol extract from Arctium Lappa L. *Clinical and Molecular Allergy*, 9(1): 4.

Stefanovich, V. (1980). The role of biochemistry in drug research. *Current Medical Research and Opinion*, 6(7): 488-499.

Suherdjoko. (2010). 'Jamu' export reaches $10m annually. *The Jakarta Post, Semarang National*, Thu, February 11, 2010 (http://www.thejakartapost.com).

Sun, H., Liu, X., Xiong, Q., Shikano, S. and Li, M. (2006). Chronic inhibition of cardiac Kir2.1 and HERG potassium channels by celastrol with dual effects on both ion conductivity and protein trafficking. *Journal of Biological Chemistry*, 281(9): 5877–5884.

Tao, Z., Meng, X., Han, Y. Q., Xue, M. M., Wu, S., Wu, P., Yuan, Y., Zhu, Q., Zhang, T. J. and Wong, C. C. L. (2017). Therapeutic mechanistic studies of ShuFengJieDu capsule in an acute lung injury animal model using quantitative proteomics technology. *Journal of Proteome Research*, 16(11): 4009–4019.

Tapsell, L. C., Hemphill, I., Cobiac, L., Patch, C. S., Sullivan, D. R., Fenech, M., Roodenrys, S., Keogh, J. B., Clifton, P. M., Williams, P. G., Fazio, V. A. and Inge, K. E. (2006). Health benefits of herbs and spices: the past, the present, the future. *Medical Journal of Australia*, 185(sup 4): S4–S24.

Tripathi, B. K. and Mukherjee, B. (2003). Plant medicines of Indian Origin for wound healing activity: A review. *Lower Extremity Wounds*, 2(1): 25- 39.

UNESCO. (1996). *Culture and Health, Orientation Texts World Decade for cultural Development 1988-1997*. Document CLT/ DEC /PRO –1996, Paris, France. p. 129.

UNESCO. (1998). FIT/ 504-RAF-48. *Terminal Report: Promotion of Ethnobotany and the sustainable use of plant Resources in Africa*. Paris, 1998. p. 60.

Vasisht, K. and Kumar, V. (2002). Trade and production of herbal medicines and natural health products. *Earth, Environmental and marine science and technologies*, Italy. P. 24.

Walker, C. L., Liu, N. K. and Xu, X. M. (2013). PTEN/PI3 K and MAPK signaling in protection and pathology following CNS injuries. *Frontiers in Biology*, 8(4): 421–433.

Wall, M. E., Wani, M. C., Cook, C. E., Palmer, K. H., McPhail, H. T. and Sim, G. A. (1966). Plant antitumor agents. I. The isolation and structure of camptothecin, a novel alkaloidal leukemia and tumor inhibitor from Camptotheca acuminata. *Journal of the American Chemical Society*, 88(16): 3888-3890.

Wang, H., Lau, B. W. M., Wang, N. L., Wang, S. Y., Lu, Q. J., Chang, R. C. C. and So, K. F. (2015). Lycium barbarum polysaccharides promotes in vivo proliferation of adult rat retinal progenitor cells. *Neural Regeneration Research*, 10(12): 1976–1981.

Wang, J., Hu, S., Nie, S., Yu, Q. and Xie, M. (2016). Reviews on mechanisms of in vitro antioxidant activity of polysaccharides. *Oxidative Medicine and Cellular Longevity*, 2016: 1–13.

Wani, M. C., Taylor, H. L., Wall, M. E., Coggon, P. and McPhail, A. T. (1971). Plant antitumor agents, VI: the isolation and structure of taxol, a novel antileukemic and antitumor agent from Taxus brevifolia. *Journal of the American Chemical Society*, 93(9): 2325-2327.

Wei, J., Zhang, F., Zhang, Y., Cao, C., Li, X., Li, D., Liu, X., Yang, H. and Huang, L. (2014). Proteomic investigation of signatures for geniposide-induced hepatotoxicity. *Journal of Proteome Research*, 13(12): 5724–5733.

White, J. (2006). *Public address: Overview of NCI's TCM-related research presented at Traditional Chinese Medicine and Cancer Research: Fostering Collaboration; Advancing the Science, April 10, 2006.* Office of Cancer Complementary and Alternative Medicine (OCCAM).

WHO. (1977). Resolution –Promotion and Development of Training and Research in Traditional Medicine. *WHO document No, 30-49.*

WHO. (2001). Legal Status of Traditional Medicine and Complementary/ Alternative medicine: A worldwide review. WHO Publishing 1.

WHO. (2004). *Guideline on Developing Consumer Information on Proper Use of Traditional, Complementary and Alternative Medicine*. Geneva.

WHO. (2012). *Regional progress in traditional medicine 2011-2010.* (http://www.wpro.who.int).

Wink, M. (2003). Evolution of secondary metabolites from an ecological and molecular phylogenetic perspective. Phytochemistry, 64(1): 3-19.

Winston, D. and Maimes, S. (2007). *Adaptogens: Herbs for strength, stamina and stress relief.* Rochester, Vermont: Healing Arts Press.

Wolfender, J. L., Ndjoko, K. and Hostettmann, K. (2003). Liquid chromatography with ultraviolet absorbance-mass spectrometric detection and with nuclear magnetic resonance spectrometry: a powerful combination for the on-line structural investigation of plant metabolites. *Journal of Chromatography A*, 1000(1-2): 437–455.

Wong, K. H., Kanagasabapathy, G., Bakar, R., Phan, C. W. and Sabaratnam, V. (2015). Restoration of sensory dysfunction following peripheral nerve injury by the polysaccharide from culinary and medicinal mushroom, Hericium erinaceus (Bull.: Fr.) Pers. through its neuroregenerative action, *Food Science and Technology*, 35(4): 712–721.

Xiang, Q., Zhou, W. Y., Hu, W. X., Wen, Z., He, D., Wu, X. M., Wei, H. P., Wang, W. D. and Hu, G. Z. (2015). Neuroprotective effects of Rhizoma Dioscoreae polysaccharides against neuronal apoptosis induced by in vitro hypoxia. *Experimental and Therapeutic Medicine*, 10(6): 2063–2070.

Xie, J. H., Jin, M. L., Morris, G. A., Zha, X. Q., Chen, H. Q., Yi, Y., Li, J. E., Wang, Z. J., Gao, J., Nie, S. P., Shang, P. and Xie, M. Y. (2015). Advances on bioactive polysaccharides from medicinal plants. *Critical Reviews in Food Science and Nutrition*, 56 (sup 1): 60–84.

Yang, Y., Wang, W., Cheng, K. and Zhu, P. (2013). Artemisinic acid: a promising molecule potentially suitable for the semi-synthesis of artemisinin. *RSC Advances*, 2013(3): 7622–7641.

Yi, J., Xia, W., Wu, J., Yuan, L., Wu, J., Tu, D., Fang, J. and Tan, Z. (2014). Betulinic acid prevents alcohol-induced liver damage by improving the antioxidant system in mice. *Journal of Veterinary Science*, 15(1): 141–148.

Zamiska, N. (2006). On the trail of ancient cures. *Wall Street Journal* November 15, B1, B12.

Zeng, B., Su, M., Chen, Q., Chang, Q., Wang, W. and Li, H. (2016). Antioxidant and hepatoprotective activities of polysaccharides from Anoectochilus roxburghii. *Carbohydrate Polymers*, 153: 391–398.

Zhang, P., Qiang, X., Zhang, M., Ma, D., Zhao, Z., Zhou, C., Liu, X., Li, R., Chen, H. and Zhang, Y. (2015). Demethyleneberberine, a natural mitochondria-targeted antioxidant, inhibits mitochondrial dysfunction, oxidative stress, and steatosis in alcoholic liver disease mouse model. *Journal of Pharmacology and Experimental Therapeutics*, 352(1): 139–147.

Zhao, L., Zhen, C., Wu, Z. Q., Hu, R., Zhou, C. L. and Guo, Q. L. (2010). General pharmacological properties, developmental toxicity, and analgesic activity of gambogic acid, a novel natural anticancer agent. *Drug and Chemical Toxicology*, 33(1): 88–96.

Zheng, S. Q., Jiang, F., Gao, H. Y. and Zheng, J. G. (2010). Preliminary observations on the anti-fatigue effects of longan (Dimocarpus longan Lour.) seed polysaccharides. *Phytotherapy Research*, 24(4): 622–624.

In: Trends in Biochemistry and Molecular Biology
Editors: Hossain Uddin Shekhar et al.
ISBN: 978-1-53616-434-3

Chapter 7

SOIL BIOCHEMISTRY

Sirajul Hoque*[*], *PhD
Department of Soil, Water and Environment,
University of Dhaka, Dhaka, Bangladesh

ABSTRACT

Soil consists of inanimate mineral matters, organic matter of different stages of decomposition, water and air and also a vast array of living organisms both micro and macro from all kingdoms of living world. Because of open and complex nature of soil and rapid fluctuations due to changes in environmental factors chemical changes occur rapidly which are responsible for changes in microbial community. As a consequence the products of biochemical reactions also change which influence overall condition of soil. Soils harbor diversified microbial communities which are dependent on the dynamic conditions of soils. A soil may contain aerobic, anaerobic and facultative anaerobic bacteria and also other microbial populations in the same ecosystem based on environmental conditions. The rates of chemical reactions may be slow but can be accelerated depending on the occurrence of biochemical reactions. Soil biochemical reactions, therefore, play dominant roles on the focal point of soil chemistry, the backbone of soil science.

1. INTRODUCTION

Soil is a poly-phasic porous three dimensional entity and a collection of natural bodies consisting of living components such as soil flora and fauna and non-living solid, liquid and gas. The solid is made up of inorganic and organic matter and combined with

[*] Corresponding Author's E-mail: sirajswed@du.ac.bd.

liquid and gases make a heterogeneous system which is located at the top of lithosphere where atmosphere and hydrosphere are highly active. It is a medium for the growth of all terrestrial plants and soil organisms and thus supports the growth of animals and also provides materials to support the growth of aquatic plants and microorganisms and salts to the oceans.

Soils are one of the segments of the environment and their properties are controlled by geological, topographical, climatic, physical, chemical and biological factors and processes. The influence of these factors especially the dominant factors are responsible for the formation of different kinds of soils even in the same region having same parent material. In litho-ecosystem bacteria, algae, fungi, lichens use rock surfaces as a habitat and solubilize minerals through production of carbonic acids, organic acids and chelating agents and thus initiate chemical weathering leading to formation of soil.

Organic matters of soils are derived from plant and animal tissues. When plant tissues are dropped to the soils and microorganisms die a lot of organic compounds of biological origin are added (Table 1). The abundance of organic compounds in plants are in the order of cellulose 45% (20 – 50%) > lignin 20% (10 -30%)> hemicellulose 18% (10 – 30%) > protein 8% (1 – 15%) > sugars and starches 5% (1 – 5%) > fats and waxes 2% (1 – 8%) and polyphenols 2%. Among these groups of substances lignin generally decompose slowly by a specialized group of microorganisms and is considered as recalcitrant component of soil organic matter.

Animal tissues also add several organic compounds. In addition to this soil microbes also contribute to form organic compounds, and some of them are different from those of the compounds of plant and animal origin. All these compounds undergo transformations which are carried out by soil organisms particularly by heterotrophic microorganisms and most of the reactions are biochemical in nature.

Among the environmental segments pedosphere is the best medium for the growth of microorganisms. Some of these organisms are autochthonous which are native to that environment and are autotrophic and others are allochthonous, invaders to that environment and they persist for a short interval of time until the added organic sources are decomposed.

Soil can provide suitable conditions for the growth of most of the recognized nutritional types of microorganisms such as photolithoautotrophs, photoorganoheterotrophs, chemolithoautotrophs, chemoorganoheterotrophs and chemolithoheterotrophs. Among the nutritional groups, chemoorganoheterotrophs dominate in soil organisms.

Soils of the different locations are suitable for the growth of obligate aerobes, obligate anaerobes, facultative anaerobes and microaerophilic organisms and Paul and Clark (1989) mentioned facultative microaerophilic bacteria. Among the bacteria facultative anaerobes are the predominant organisms that can survive under the changing conditions of oxygen content.

Table 1. Compounds added through addition of fresh tissues of plant and microorganisms

Carbohydrates Monosaccharide (Simple sugars) Disaccharides Polysaccharides	 Trioses, pentoses, hexoses, etc. Maltose, lactose and sucrose Starches, hemicelluloses, cellulose, callose and inulin
Proteins Crude protein	170 amino acids are known, 26 are constituents of protein and among them 20 are common
Nucleic acids: DNA and RNA	Purine bases adenine and guanine and pyrimidine bases thymine and cytosine and uracil; pentose sugars ribose and deoxyribose; and phosphate group
Lignin	Coniferyl alcohol (4-hydroxy-3-methoxy phenyl propane), sinapyl alcohol (3,5-dimethoxy-4-hydroxy phenyl propane) and coumaryl alcohol (4-hydoxy phenyl propane)
Sugar acids	D-glucuronic acid, D-muramic acid and D-galacturonic acid
Amino sugars	Glucosamine, Galactosamine and Mannosamine
Peptidoglycan and murein	*N*-acetylglucosamine, *N*-acetylmuramic acid, meso-diaminopimelic acid
Pectins, gums and mucilage	Consist of sugars (arabinose, galactose, xylose and rhamnose) and sugar acids (glucuronic acid and galacturonic acids)
Chitin	Condensation product of acetylglucosamine
Lipids Waxes	Stearic acid and Oleic acid Esters of fatty acids with long chain alcohols
Growth regulators	Indole Acetic Acid, Gibberellins, Cytokinins, Ethylene, Abscisic Acid
Others	Teichoic acid, Chlorophyll, Urea etc.

Soils are also suitable for diverse group of organisms based on their temperature requirement. Psychrophiles are present in low temperature temperate region, thermophiles are dominant where temperature is above 45°C but not common in most of the agricultural soils. Among the soil microorganisms mesophiles are dominant even under less than 20°C cold tolerant mesophiles are very much present. Some organisms can grow under extreme conditions of low pH and are known as acidophiles and other groups as alkaliphiles but for most of the microorganisms the favorable pH range is in between 5.5 and 7.5. Halophiles and halotolerant organisms are found in the respective environment. These happen because the factors controlling the growth of different organisms either in favorable conditions or adaptation to different environmental conditions through developing different mechanisms. The density, composition of microbial populations and their biochemical potentials, however, can be frequently altered to a great extent due to non-biological factors, such as moisture, temperature, organic matter content, pH, inorganic nutrients; presence of growth inhibitory substances, etc. and all the factors are controlled by some other indirect factors like agricultural operations, season and depth of soil. Thus soils act as media for the growth of diversified groups of microorganisms without stressed and with certain stressed conditions. The top soil harbors most of the microorganisms because of the availability of light, oxygen and organic matter, are essential for the growth of photosynthetic, obligate aerobic and heterotrophic organisms.

Soils are the habitats of a large array of macro- and micro-fauna which are herbivores, detritivores, predators, and parasites and thus make an ecosystem of diversified organisms in intimate associations with soil micro-flora (Table 2). Rhizosphere, the area around and on the surface of roots plays a very significant role to make the ecosystem viable. All these organisms transform the components of soil through biochemical reactions directly for their survival and chemical reactions happen indirectly to modify their habitat.

Table 2. List of common soil micro-flora

Bacteria	**Actinomycetes**	**Fungi**	**Algae**
Pseudomonas	*Streptomyces*	*Aspergillus*	*Chlamydomonas*
Rhizobium	*Nocardia*	*Penicillium*	*Chlorella*
Agrobacterium	*Micromonospora*	*Furasium*	*Pinnularia*
Chromobacterium	*Streptosporangium*	*Trichoderma*	*Navicula*
Achromobacter	*Micropolyspora*	*Mucor*	*Heterothrix*
Flavobacterium	*Actinoplanes*	*Chaetomium*	*Heterococcus*
Micrococcus	*Microbispora*	*Saccharomyces*	
Corynebacterium	*Thermoactinomyces*	*Candida*	
Arthrobacter		*Rhizopus*	
Bacillus		*Pythium*	
Clostridium			
Nitrosomonas			
Nitrobacter			
Thiobacillus			
Ferrobacillus			
Desulfovibrio			
Methanobacillus			
Nostoc			
Anabaena			

Source: Alexandar,1977; Killham, 1994; Paul and Clark, 1989.

Also to make the ecosystem stable different types of biological interactions occur among the soil organisms such as symbiosis, mutualism, proto-cooperation, commensalism, synergism, neutralism, antagonism, competition, ammensalism, predation and parasitism (Killham1994).

Soil pore spaces are very important when we consider soil as medium for the growth of plants and soil organisms and thus it differs from the other two segments of environment such as atmosphere and hydrosphere. Pores are classified on the basis of their pore diameter as macro-pores (> 0.08mm) and micro-pores (< 0.08mm). Micro-pores are again classified as mesopores, micro-pores, ultramicropores and cryptomicropores.

Macro-pores are in general occupied by air under typical condition, which is known as soil air or gaseous phase, the components of which are similar to the atmospheric air and act as a source of gaseous components to meet the requirement of the organisms. Depending on their sizes organisms can use these pores as their habitat.

Under typical condition (field moisture condition) micro-pores retain water and act as habitat for soil microorganisms. Mesopores and microspores accommodate microorganisms depending on their sizes. Thus soil environments have different characteristics upon which different types of microorganisms get suitable environment for their growth.

Soil water which is known as soil solution or liquid phase of soil, essential for all microorganisms and also for soil macro-organisms, is generally present in micro-pores (< 0.08mm) and contains different inorganic and organic soluble chemical species in ionic and molecular forms (Table 3).

Table 3. Composition of soil solution

Plant essential elements	Forms present in soil solution	Plant non-essential elements	Forms present in soil solution	Organic acids	
Nitrogen	NO_3^-, NH_4^+, NO_2^-	Sodium	Na^+	Formic acid	Aspartic acid
Phosphorus	$H_2PO_4^-$, HPO_4^{2-}, PO_4^{3-}	Aluminium	Al^{3+}, $Al(OH)^{2+}$, $Al(OH)_2^+$	Acetic acid	Glutamic acid
Potassium	K^+	Cadmium	Cd^{2+}	Lactic acid	Arginine
Sulfur	SO_4^{2-}, SO_3^{2-}	Lead	Pb^{2+}	Oxalic acid	Lysine
Calcium	Ca^{2+}	Chromium	CrO_4^{2-}	Malonic acid	Glycine
Magnesium	Mg^{2+}	Bicarbonate	HCO_3^-	Succinic acid	Alanine
Iron	Fe^{2+}, Fe^{3+}	Mercury	Hg^{2+}	Malic acid	
Manganese	Mn^{2+}			Tartaric acid	
Molybdenum	MoO_4^{2-}			Gluconic acid	
Zinc	Zn^{2+}			α-Ketogluconic acid	
Copper	Cu^{2+}			Citric acid	
Boron	H_3BO_3			Salicylic acid	
Chlorine	Cl^-			Gallic acid	

It also contains suspended colloidal particles. The concentration of soil solution may vary and not homogeneous because of its segregation in different discontinuous pore space.

Such complex solution is the focal point of all chemical and biochemical reactions, product translocation and transformation from one form to another with intimate touch of soil air and solid phase of soil.

In soil most of the chemical transformations are carried out with the intimate touch of biochemical reactions. Soil contains primary producers, consumers and decomposer in the same ecosystem. All this components are dependent on biochemical reactions which are enzyme catalyzed. The added organic matter from different sources are transformed by micro organisms for their survival and thus a lot of changes occur in organic matter and the products released also effect the chemical reactions in soil. Some products are also released to the atmosphere and during the conversion reaction some components of the atmosphere are taken up by organisms. Thus soil acts as a source and sink of many components.

The biochemical reactions are enzyme mediated. The sources of these enzymes are micro-organisms. A single bacterium synthesizes about 200 enzymes for survival and reproduction.

A productive soil may contain 10^8 -10^9 counts/g bacteria, 10^5-10^6 fungi, 10^7 -10^8 actinomycetes, and 10^4 -10^5 algae. To support their growth and reproduction, soil plays vital roles and most of the life associated reactions are biochemical reaction.

Microorganisms carry out many important functions as they interact in ecosystems including

1. Formation of organic matter through photosynthetic and chemosynthetic processes.
2. Decomposition of organic matter with the release of inorganic compounds.
3. Serving as a nutrient-rich food source for other chemoheterotrophic microorganisms including protozoa and animals.
4. Modifying substrates and nutrients used in symbolic growth processes and interactions thereby contributing to biogeochemical cycling.
5. Changing the amounts of materials in soluble and gaseous form. These occur either directly by metabolic processes or indirectly by modifying the environment (e.g.. altering the pH).
6. Producing inhibitory compounds that decrease microbial activities or limit the survival and functioning of plants and animals.
7. Contributing to the functioning of plants and animals through positive and negative interactions.

When organic tissues are added to soil, three general reactions take place (Brady 1995)

1. The bulk of the material undergoes enzymatic oxidation with CO_2, H_2O, energy and heat as the major products

   $$\underset{\text{Carbon- and H-containing compounds}}{-(C, 4H)} + 2O_2 \xrightarrow{\text{enzymatic oxidation}} CO_2 + 2H_2O + \text{energy}$$

2. The essential elements, nitrogen, phosphorus, and sulfur are released and/or immobilized by a series of specific reactions relatively unique for each element.
3. Compounds resistant to microbial actions are formed either through modification of compounds in the original plant tissue or by microbial synthesis.

Each kind of reaction has great practical significance. Several biogeochemical transformations are occurring in the globe. Here only the nitrogen transformation in soil is presented.

2. Biogeochemical Transformation of Nitrogen

Atmospheric molecular nitrogen is available in soil atmosphere which is reduced to ammonia by diazotrophic microorganisms.

This reaction is one of the important reactions by which soil gets nitrogen and is carried out by free-living aerobic *Azotobacter* and *Beijerinckia;* facultative microaerophilic *Klebsiella*, *Azospirillum* and *Bacillus*; anaerobic *Clostridium, Dulfovibrio*; *Rhizobium* strains are aerobic and established relationship with legumes and fix molecular nitrogen from soil atmosphere. Such types of symbiotic relationships are also established by cyanobacteria (*Azolla-Anabaena*), *Nostoc* and others. The reductive reaction by which N≡N is converted to NH_3 which is same to all organisms involved in N-fixation and is carried out by nitrogenase enzyme.

$$\underset{\text{Nitrogen}}{N \equiv N} \overset{2H}{\rightarrow} \underset{\text{Diimide}}{(HN = NH)} \overset{2H}{\rightarrow} \underset{\text{Hydrazine}}{H_2N = NH_2} \overset{2H}{\rightarrow} \underset{\text{Ammonia}}{2NH_3}$$

The product of reaction is converted to amino acid according to the following reaction

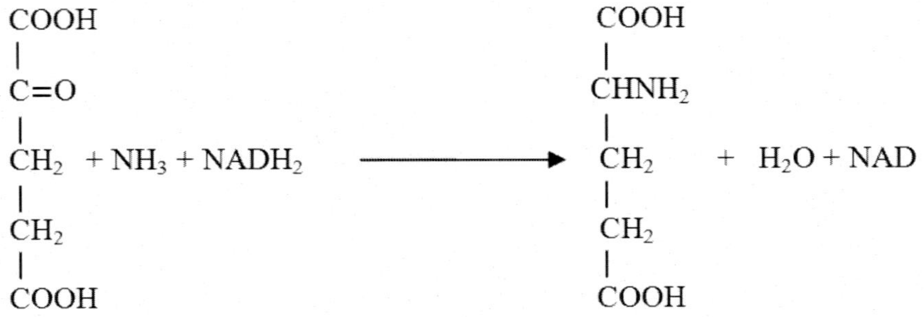

α – Ketoglutaric Acid   Glutamic Acid

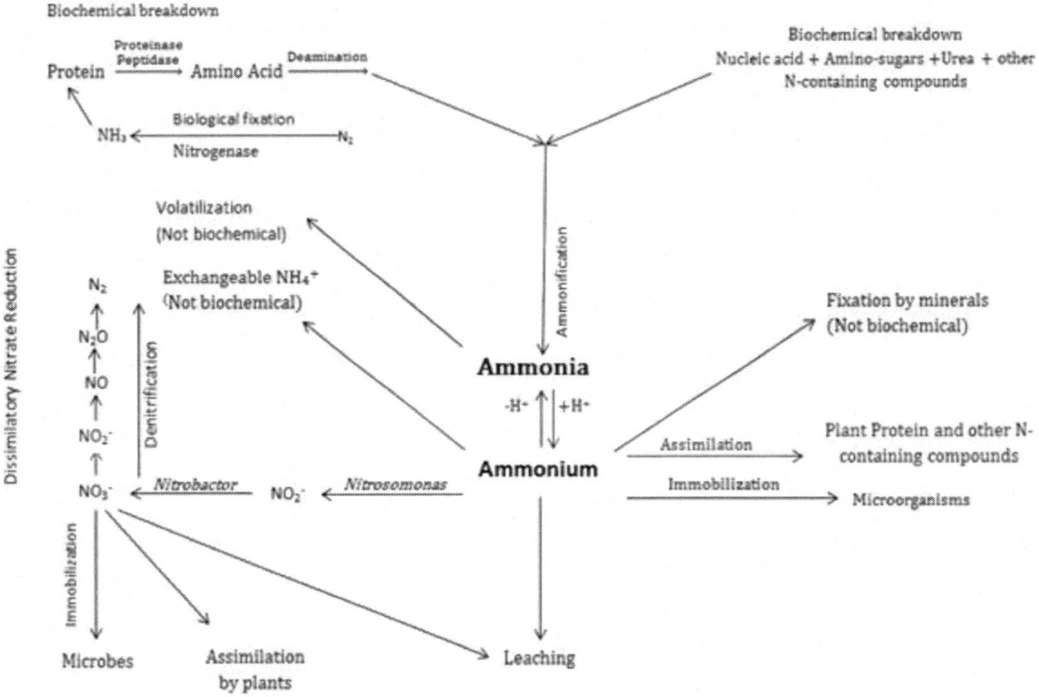

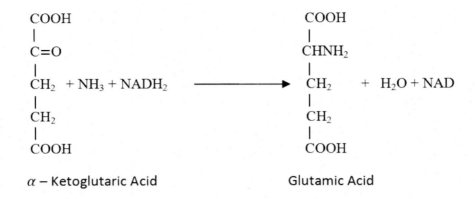

Figure 1. Bio-geochemical transformation of nitrogen in soils (intermediate steps are not shown).

Amino acids are synthesized by transamination reaction in which amino groups are transformed from one kind of a molecule to another.

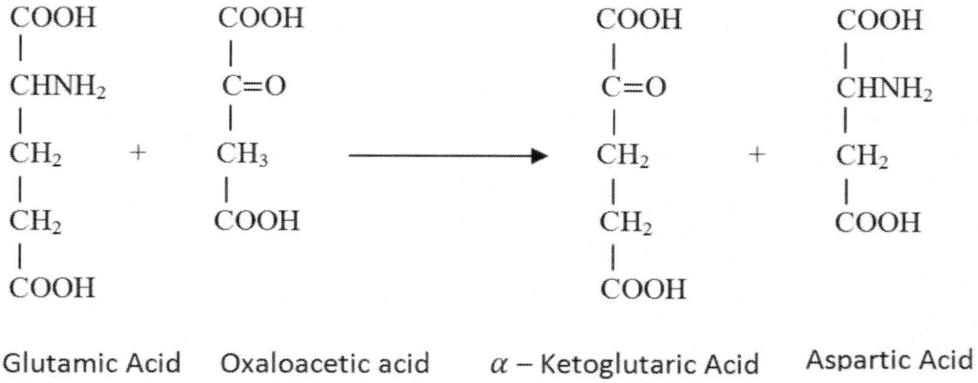

Polymerization of these amino acids yields long-chain protein molecule.

A simple pathway of nitrogen transformation is presented in Figure 1.

REFERENCES

Alexander, M. 1977. *Introduction to soil microbiology*. Wiley Eastern Limited, New Delhi.

Brady, N. C. 1993. *The nature and properties of soils*. Prentice Hall of India Private Limited, New Delhi.

Killham, K. 1994. *Soil ecology*. Cambridge University Press, UK.

Paul, E. A. and Clark, F.E. 1989. *Soil microbiology and biochemistry*. Academic Press Inc. New York.

In: Trends in Biochemistry and Molecular Biology
Editors: Hossain Uddin Shekhar et al.
ISBN: 978-1-53616-434-3

Chapter 8

BETA AND HBE/BETA THALASSEMIA: THE MOST COMMON CONGENITAL HEMOGLOBINOPATHIES IN SOUTH ASIA

Farjana Akther Noor*[1,2]*, Kaiissar Mannoor*[2]*, PhD
and Hossain Uddin Shekhar*[1,*]*, PhD
[1]Department of Biochemistry and Molecular Biology,
University of Dhaka, Dhaka, Bangladesh
[2]Institute for Developing Science and Health Initiatives, Dhaka, Bangladesh

ABSTRACT

Beta (β) thalassemia and hemoglobin E/beta (HbE/β) thalassemia are highly prevalent hemoglobinopathies worldwide and also the most common monogenic autosomal recessive disorders widespread in Mediterranean, African, and Asian countries. β-thalassemia is characterized by a reduced or suppressed expression of the beta (β) globin gene whereas HbE/β-thalassemia acquired by the combination of a β-thalassemia allele inherited from one parent and another allele conatining the structural variant hemoglobin E (HbE) from the other, respectively. β-thalassemia is now regarded as the most common genetic disorder in many South Asian countries including Bangladesh. Like β-thalassemia, HbE is also highly prevalent in Bangladesh. In both the cases, the affected individuals suffer from life-threatening anemia and require regular treatment and management of the complications. Recurrent blood transfusion leads to iron overload which is related to severe complications in vital organs including endocrine complication affecting almost all types of hormone action, myocardiopathy, liver fibrosis and cirrhosis. Pathophysiology of the disease is very complex as it is not a single disorder, rather a group of defects with analogous clinical syndromes. However, knowledge about the prevalence pattern of the disease along with the regional mutation

[*] Corresponding Author's E-mail: shekhardu@hotmail.com.

spectrum could help in finding out the causative genes specific for a particular population which will ultimately help in development of more accurate diagnosis by defusing the confusion that evolves when only clinical descriptions or hematological analysis is considered for detection of the disease. Prevention of the disease is easier and more preferable as it is an incuarble inherited disease. In addition, the available treatments are very expensive and painful as well. Thus, taking preventive measures by population screening, premarital carrier detection following the genetic counseling, and prenatal diagnosis is the only remedy of these deadly diseases. However, it requires a precise knowledge of hemoglobin structure and synthesis, pathophysiology of the disease, as well as molecular charecterization of globin gene among different populations for a successful preventive approach and effective management of the patients as well.

Keywords: thalassemia, hemoglobinopathies, hemoglobin E/beta thalassemia, iron overload

1. Introduction

Hemoglobinopathies are groups of inherited monogenic disorders of the hemoglobin molecule located in red blood cells that affect millions of people worldwide (Modell B et al., 2008; Vichinsky EP., 2005). Approximately 320,000 babies are born every year with a clinically significant hemoglobin disorder (Modell B et al., 2008). Among which, about 80% of the babies are from the developing countries. Highly regarded estimates show that over 360 million people constituting approximately 5.2% of the world population carry genes of hemoglobin variant (Modell B et al., 2008) while the worldwide beta thalassemia carrier frequency is around 1.5% (Colah R et al., 2010).

Homozygous or compound heterozygous states of certain mutant hemoglobin genes may lead to clinical manifestations of hemoglobinopathies which is highly variable, ranging from mild hypochromic anemia to moderately severe hematological disease to lifelong, deadly severe, transfusion-dependent microcytic hypochromic anemia with multi-organ involvement like thalassemia.

The more severe form of thalassemia is referred as thalassemia major, clinical management of which is very expensive including red blood cell transfusion in regular basis until death and iron chelation therapy to remove excess iron introduced and deposited with transfusions. The disease severity is also modified by several genetic factors unlinked to globin gene clusters as well as environmental conditions and the degree of management of the disease. To date, above 400 mutations or defected alleles causing beta thalassemia have been reported (Thein SL, 2004). However, the mutational spectrum is region specific and differs across various geographical areas. Different mutations lead to various degrees of severity and are also associated with different types of thalassemias. Multiple clinical complications are associated with both beta thalassemia and HbE/beta thalassemia including severe anemia, growth retardation, splenomegaly, jaundice, cardiac and liver dysfunctions, expansion of bone-marrow, endocrine disorders, bone deformities and require lifelong blood transfusions at regular intervals to avoid complications.

Without blood transfusion the beta-thalassemia major and HbE/beta-thalassemia patients cannot survive for more than 5 years (Modell B et al., 2008), and the average life expectancy of such patients is about 30 years in spite of taking regular blood transfusions (Mandal PK et al., 2014), particularly in the countries with extremely low resources. In addition to the transfusion-dependent form of β-thalassemia, there are also some milder conditions which might not be detected till adulthood. The β-thalassemias cause the most significant public health problems because of their high incidence rate and severity. Hence the genetically inherited disorders of hemoglobin synthesis are an important catalyst of morbidity and mortality worldwide, putting immense burden not only for the patients but also on their families, their communities and ultimately in the whole country. Although, they are not curable in general, but can be prevented by population screening, genetic counseling and prenatal diagnosis (Balgir, 2002). However, a successful preventive approach for overcoming the problems is not possible without the preliminary knowledge about the structure and synthesis of hemoglobin, pathophysiology of the disease, as well as a spectrum of globin gene mutations among different populations (Caro A 2000). This chapter gives an insight into the pathophysiology of beta thalassemia with a brief discussion on the specialized structure and synthesis of hemoglobin and also depicts the epidemiological aspects and mutation spectrum of thalassemia in South Asia.

2. Synthesis of Hemoglobins: A Highly Regulated Process by Spatiotemporal Expression of Globin Genes

Hemoglobins are the highly specialized protein molecules which act as an oxygen transporter of our body. There are roughly 270 million hemoglobin molecules present in every red blood cell and thus constituting about 27 to 30 picogram in weight per cell (Weatherall DJ, 1989). The matured form of Hb protein present in a healthy adult person has a spheroidal structure that contains four polypeptide chains of two different globin genes and all the subunits arranged like a thick-walled shell with a central cavity (Figure 1).

Two pairs of identical sub-units of each hemoglobin belongs to two groups: one is the alpha-globin cluster, comprising the zeta, ζ- and alpha, α-globin chains, and the second one is the beta-globin cluster, which is composed of several genes for different globin chains like epsilon (ε), gamma (γ), beta (β) and delta (δ). These chains are differentially produced, with different combinations, one after another during ontogeny;

therefore, they are diverse at different stages of development (the embryonic, fetal, and adult stages) (Salzano et al., 2001).

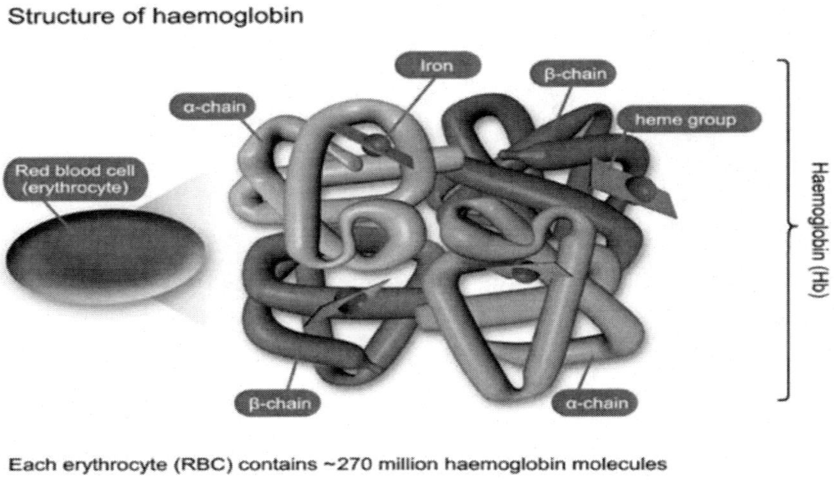

Figure 1. The structure of two alpha and two beta chain contained in the hemoglobin (Hb) teramer. Reprinted from https://thealevelbiologist.co.uk/images/hemoglobin.jpg.

In total, there are four major types of hemoglobins produced during the entire developmental process which requires two switches: the first switching event occurs at 6 weeks of gestation from embryonic to fetal hemoglobin and the second switch is from fetal to adult production observed at birth (Figure 2).

Type-1: "Embryonic" hemoglobin, namely Hb Gower 1 ($\zeta 2\varepsilon 2$) which is gradually replaced by α chains within 8 weeks; the other four embryonic hemoglobins are namely Hb Gower 2 (tetramers of $\alpha 2\varepsilon 2$), Hb Portland 1 (tetramers of $\zeta 2\gamma 2$) and Hb Portland 2 ($\zeta 2\beta 2$ tetramers) which are present only between the third and 10th week of gestation.

Type-2: "Fetal" hemoglobin (HbF), a tetramer of $\alpha 2\gamma 2$ globin chains and the main oxygen carrier molecule during pregnancy.

Type-3: Hb A or "Adult" hemoglobin, a tetramer of two α and two β chains, replaces HbF within six to 12 months after birth; and

Type-4: HbA2 which is a minor adult hemoglobin component consisting two alpha and two delta chain ($\alpha 2\delta 2$).

The whole process of the developmental switching of hemoglobin genes is accomplished within the first 1.5 to 2 years of life and leads to the final pattern in a healthy adult person having a major expression of adult hemoglobin, HbA ($\alpha 2\beta 2$) comprising about 97% of total hemoglobin while only 2-3% is the minor hemoglobin HbA2 ($\alpha 2\delta 2$). Sometimes fetal hemoglobin HbF ($\alpha 2\gamma 2$) is found to be present, but usually less than 1% (Stamatoyannopoulos G et al., 2001).

The exceptionally characteristic structure of globin chains makes them capable of rapid and efficient loading with oxygen in the lung alveoli following the gradual transport of the gas into different tissues throughout the body. The specific genes coding the globin chains are present as two different gene clusters- the alpha gene cluster on chromosomes 16 and the β gene cluster on chromosome 11. In both the gene clusters, flanking upstream region (the 5' side of the gene sequence) and downstream region (the 3' side of the gene cluster) contain several distinguishing nucleotide sequences having "regulatory" function, i.e., they control the gene expression level and also determine which gene is to be turned on and which is to be switched off. However, the globin chains must be synthesized with the accurate structure, and correctly spliced so that the number of α-chains is just equal to that of the β-chains. If there is any imbalance between the two chains, the result is a completely or partially non-functional proteins translated from one or both defected alleles of hemoglobin gene.

Therefore, the synthesis of human globin chains is tightly regulated by multiple fine and intricate mechanisms to assure the accurate and balanced assembly of globin peptide chains at every developmental stage by producing the equal amount of α- globin chains and β-globin chains throughout the ontogeny (Cao, A. and Moi, P. 2000).

All the globin genes have a common pattern of structure which is well-preserved during the course of developmental processes. Their primary transcripts contain three exons (coding region) and two introns (non-coding region). The first exon starts after ~50 bp from the CAP site (the initiation point of the mRNA transcripts) and this upstream region is called the 5' untranslated region (UTR) in which a TATA, CAAT, and duplicated CACCC boxes are present. In addition, there is a long regulatory region, containing several strong enhancers, referred as locus control region (LCR) which occupies around 50 Kb from the beta globin gene.

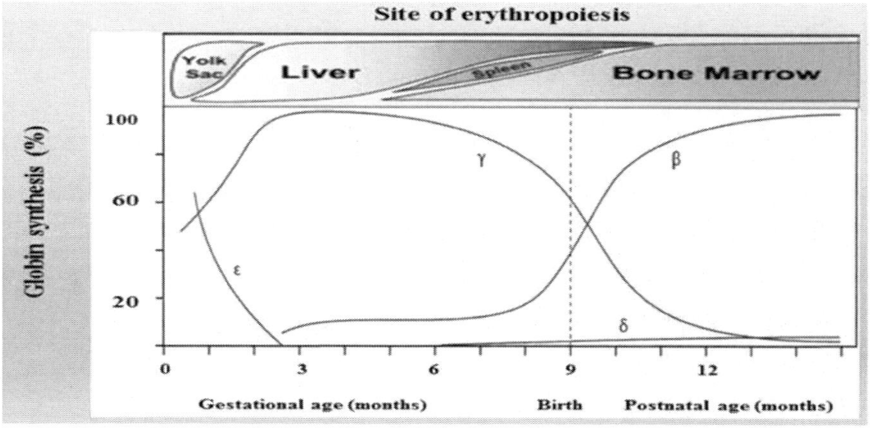

Figure 2. Schematic diagram depicting the normal timing of the developmental hemoglobin switches in humans (Stamatoyannopoulos G, 2001).

Exon 1 encodes amino acids 1 to 31 in the α-globin gene whether in the β-genes, it contains codons encoding 1st –29th amino acid plus the first two bases of codon 30 (Weatherall DJ & Clegg JB, 2001). While exon 2 of the alpha globin gene consists of the codons for amino acids 32–99 and that of the β-globin gene encodes amino acids from 31 to 104 along with the last base of the 30th codon. Exon 2 is involved in binding to the heme protoporphyrin and the interaction between the two chains which leads to α1β2 or α2β1 dimer formation. The rest of the amino acids are encoded by exon 3 that is amino acids 100 to 141 for α chain and 105 to 146 for β chain). In addition, the third exon is ended up with a 100 bp long 3' untranslated region (3' UTR) (Weatherall DJ & Clegg JB 2001).

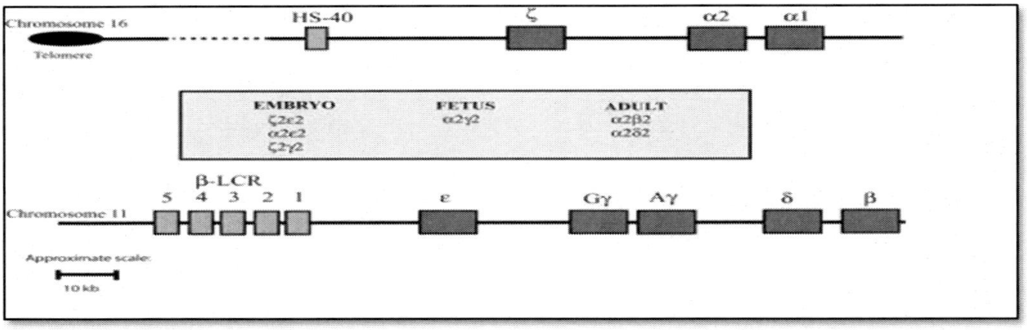

Figure 3. Schematic representation of the human globin gene clusters. The alpha globin cluster (top) is situated near the telomere region of the short arm of chromosome 16 and includes three globin gene which are under the control of an upstream remote regulatory region, HS-40. The beta like globin cluster (bottom) is interstitial and located on the short arm of chromosome 11; expression of the genes in this cluster is under the control of a group of remote regulatory elements/DNase I hypersensitive sites collectively known as the locus control region (LCR). In between are represented the different ontogenetic stages of Hb (Steensma, DP et al., 2005).

The two introns of the alpha globin genes, namely IVS1 and IVS2 are tiny, only 117–149bp long. On the other hand, the second intervening sequence in β-genes, IVS2 is much larger having 850–904 base pairs while the first intron IVS1 is small like the alpha genes i.e., only 122–130bp long. There are some consensus sequences present in the exon-intron junction sites which play the major role during RNA processing leading to exclusion of non-coding intronic parts of the primary transcript and subsequent joining of the exon sequences for production of the mature mRNA. The conserved region at the 5' end of each intron is CAAG or GTAGAGT and includes the invariant GT residues, the splice donor site. At the 3' end of the intervening sequence, the consensus sequence is $(CT)_N$CT AG/G with the acceptor site, an invariant AG dinucleotide preceding the excision site (9). Such arrangements of consensus sequences are strongly sustained as well as the splice donor GT and the acceptor site AG dinucleotides are preserved. Any alteration of the conserved pattern or mutations in these consensus sequences normally leads to unstable or non-functional splice variants of the globin chains which is one of the major causes of thalassemia (Weatherall DJ & Clegg JB. 2001).

The expression of globin genes is influenced by multiple genetic factors and interactions of various trans and cis acting regulatory elements and a "master" controlling region involved in regulating the entire gene clusters encoding the alpha or beta globin chains (Figure 3). Eventually, a very securely regulated procedure for the production of globin chain maintains the exact ratio of α- chains to non-α- chains, at 1.00 (± 0.05). Thalassemia, by impairing this process, distracts this ratio (Amoako YA 2014).

2.1. The Thalassemia: Definition and Phenotypic Classification

The term "Thalassemia" or "Thalassaemia" stands for a group of blood disorders characterized by declined or an absent synthesis of typical globin chains with functional structure. The thalassemias are classified according to the impaired chain whose synthesis or structure is affected due to the corresponding defected gene and named as α-, β-, γ-, δ -, δβ-, or εγδβ-thalassemias accordingly. Most of the thalassemias are inherited as recessive traits. From a clinical point of view, the most relevant types are α- and β-thalassemias, caused from the defects of one of the two types of polypeptide chains (α or β) that are the component of the principal adult human hemoglobin molecule HbA (α2β2).

2.1.1. Beta Thalassemia

It is the most frequent type of thalassemia caused by reduced (β^+) or absent (β^0) synthesis of the β-globin chain of the hemoglobin tetramer ($\alpha_2\beta_2$). In terms of severity, there are three clinical conditions of the syndrome; the beta-thalassemia minor, beta-thalassemia intermedia and beta-thalassemia major (Safizadeh H 2012, Galanello R 2010). The minor is the carrier state which results from heterozygosity for beta-thalassemia and clinically asymptomatic. Thalassemia major is a severe transfusion-dependent anemia caused by two mutant beta globin alleles (β^0). Thalassemia intermedia is characterized by a clinically and genotypically diverse group of thalassemia-like disorders of varying degree of severity ranging from the asymptomatic carrier state to the severe transfusion-dependent type.

Thalassemia major patients are found to have a severe anemia, associated with high RBC numbers and low mean corpuscular volume (MCV) and low mean corpuscular Hb (MCH). Peripheral blood smear shows, microcytosis (smaller in size) and hypochromia, unequal sized RBCs leading to anisocytosis, poikilocytosis that refers to the variation in cell shape (tear-drop and elongated cells) and reticulocytes (nucleated red blood cells). The other two forms can also be defined by specific hematological and clinical features (Brancaleoni V 2016) which is summarized in Table 1.

Table 1. Hematological and clinical feature of beta thalassemia syndromes

	BTT	BTM		BTI
Hb levels	Normal/slightly reduced (<12g/dL)	<7 g/dL		~ 7–10 g/dL
Hemoglobin study	HbF may be increased to 10% HbA2 > 3.5%	β^0/β^0 HbF up to 95% HbA2 >5%	β^0/β^+ HbF 70–90% HbA up to 30%	HbF 10–50% HbA2 > 5%
RBC indices	MCV < 80 fl MCH < 27 pg	MCV < 60 fl MCH < 20 pg		Variable
Transfusion requirements	Not required	TDT		NTDT or rare/ occasional/intermittent transfusions depending on the clinical situations like infections, pregnancy etc.
Clinical presentation	Asymptomatic	Severe microcytic hypochromic anemia, mild to moderate jaundice, and hepatosplenomegaly		

*TDT, transfusion dependent; BTT, beta thalassemia trait; BTM, beta thalassemia major; BTI, beta thalassemia intermedia; NTDT, non-transfusion dependent thalassemia.

2.1.2. HbE/Beta Thalassemia

Hemoglobin E is the common most structural variant with thalassemia causing attributes, particularly when co-inherit with mutant copy of a gene in another allele. The structural variant HbE is formed due to a single nucleotide substitution mutation from G to A at the 79th position of the β-globin genes resulting in the amino acid substitution by lysine for glutamic acid at position 26 of the β-globin chain. This missense mutation stimulates a hidden splice site between codon 24 and codon 25, leading to a different splicing pattern resulting in less production of the variant hemoglobin (HbE) which is the underlying reason of 25-30% HbE of total hemoglobin in individuals with HbE traits or heterozygous HbE allele, instead of the projected 50%. These phenomena suggest that the codon 26 G→A mutation results in not only a qualitative faulty gene from producing a hemoglobin variant, but also a quantitative β-globin gene defects affecting the expression level of HbE protein (Fucharoen S, 2012).

HbE disease is the homozygous status for HbE i.e., when both the beta globin alleles contain the CD 26 (G > A) mutation. Individuals with this genetic makeup are clinically silent with no symptom except insignificant anemia in rare cases. However, the peripheral blood film for the HbE diseased person displays remarkable hematological changes, including reduced size of RBCs with 20-80% of target cells, in CBC analysis low MCV, low MCH and high RDW (Red cell distribution width) is seen while Hb electrophoresis estimates more than 90% of HbE and no HbA. HbF may be present up to 5% (Brancaleoni V, 2016).

HbE is the most observed irregular hemoglobin in South East Asia. The carrier frequency for HbE is as high as 50% in some areas of the region. It is also predominant in parts of the Indian subcontinent, including India, Pakistan, Sri Lanka and Bangladesh.

HbE carriers, i.e., with heterozygous status are clinically healthy and asymptomatic with little manifestation including only nominal changes in RBC indices and a presence of HbE fraction in the hemoglobin electrophoretic study. HbE can also be easily spotted by dichlorophenol indophenol test (DCIP).

Genetic compounds for HbE and β-thalassemia are also common in South East Asia. HbE/beta thalassemia is clinically characterized by marked variability, ranging from non-transfusion dependent asymptomatic anemia to a fatal condition requiring regular blood transfusions from the very early ages until death (Weatherall DJ & Clegg JB 2008, Olivieri NF et al., 2011). These can be categorized into following three different types:

1. Mild form of HbE/β-thalassemia: Around 15% of HbE/beta thalassemia in Southeast Asia are of this category. Such patients always maintain hemoglobin levels at 9-12 g/dl and typically does not develop any significant clinical complications in early ages. Nevertheless, few of the patients can suffer from improper growth, excess iron deposition in the vital organs due to increased iron absorption from food (Zimmermann MB et al., 2008) and other problems similar to those of non-transfusion dependent thalassemia patients.
2. Moderately severe HbE/β-thalassemia: The most of the HbE/β-thalassemia patients have the clinical manifestation of this category. The Hb levels usually stay at 6-7 g/dL and the clinical symptoms are similar to that of beta-thalassemia intermedia whose blood transfusion interval is much longer than the severe groups. Iron overload and the associated long term complications may occur.
3. Severe HbE/β-thalassemia: In the patients with severe form of compound heterozygous HbE/beta thalassemia, hemoglobin level can hit as low as 4-5 g/dL. They show the clinical symptoms similar to major β-thalassemia patients and are treated as transfusion dependent thalassemia patients.

The reasons for such inconsistency in the severity pattern of the HbE/beta thalassemia have only partly been outlined so far. One of the modifying factor is the type of β-thalassemia mutation; for example, β^0 mutation produces no β chains, thus creates the more severe clinical phenotypes when co-inherits with the trans HbE allele. On the other hand, the β+ mutations cause comparatively less clinical severity. Other potential modifiers of the disease severity include coinheritance of α-thalassemia and a distinctive tendency to continue γ-globin expression in adulthood leading to the high HbF percentage than normal (Herbert L et al., 2009; Lettre G et al., 2008 and So CC et al., 2008).

3. PATHOPHYSIOLOGY

The basic pathogenesis of all kinds of thalassemia is a consequence of the synthesis of leftover unsteady globin chains which ultimately precipitates inside the cell and totally distorts the structure of RBCs making them unstable and fragile. However, the precipitation rate and pattern of the unsteady hemoglobin chains on the RBC membrane are variable and thus the effects also fluctuate in different forms of thalassemia, even in patients with the same disease (Shinar E & Rachmilewitz EA, 1990).

When α-globin gene expression is defective, the β-globin chains will be in excess and unused. Eventually the leftover chains lead to α-thalassemia, whereas defected β-globin chain synthesis will result in unassembled α-globin chains in excess leading to β-thalassemia. Beta thalassemia is caused by mutations mainly in the beta subunit gene of the HBB locus but some pathogenic mutations are also reported in the regulatory regions of the locus control region. These mutations are enormously heterogeneous and have been detected to affect gene expression at transcriptional, post-transcriptional level or translational level leading to a variable reduction in β globin output ranging from a little deficit (mild β+ thalassemia alleles) to complete absence (β° thalassemia) with a comparative surplus of α-chains. The direct effects are a clear decline in the hemoglobin production, leading to anemia and a disproportion of the globin chains which drastically affects the red cell precursors by being precipitated inside the cells and creating oxidative stress. Oxidative stress is generally associated with the release of reactive free radicals, which causes severe damage in the lipid bilayer of RBC cell membrane and thus results in their early destruction inside the bone marrow as well as in the extramedullary sites like spleen. This process is referred to as "ineffective erythropoiesis" which is the biochemical hallmark of thalassemia.

Anemia in beta or HbE/beta thalassemia is thus the final outcome of a combination of ineffective erythropoiesis, peripheral breakdown of RBCs and an overall decrease in hemoglobin synthesis. To cope up with this situation, the body triggers erythropoietin production in excessively high level, causing a marked erythroid hyperplasia (increase in the amount of tissues that results from cell proliferation), which, in turn, may cause skeletal deformities, osteoporosis etc. as a consequence of bone marrow expansion. As the spleen is a major degradation site for the nonfunctional and excess RBCs, occurrence of an extra medullary masses is found resulting in enlargement of spleen.

Untreated or poorly transfused thalassemia major patients suffer from severe growth retardation and cardiac enlargement along with cardiac failure due to anemia and the extreme metabolic burden imposed by stimulated erythroid synthetic pathway for compensating the hemoglobin deficit. Ineffective erythropoiesis is also associated with increased absorption of dietary iron in intestine caused by deficiency of the iron regulating hormone, hepcidin which is a 25-amino acid peptide produced in the liver and controls the iron homeostasis. The pathophysiology of β-thalassemia is summarized in Figure 4.

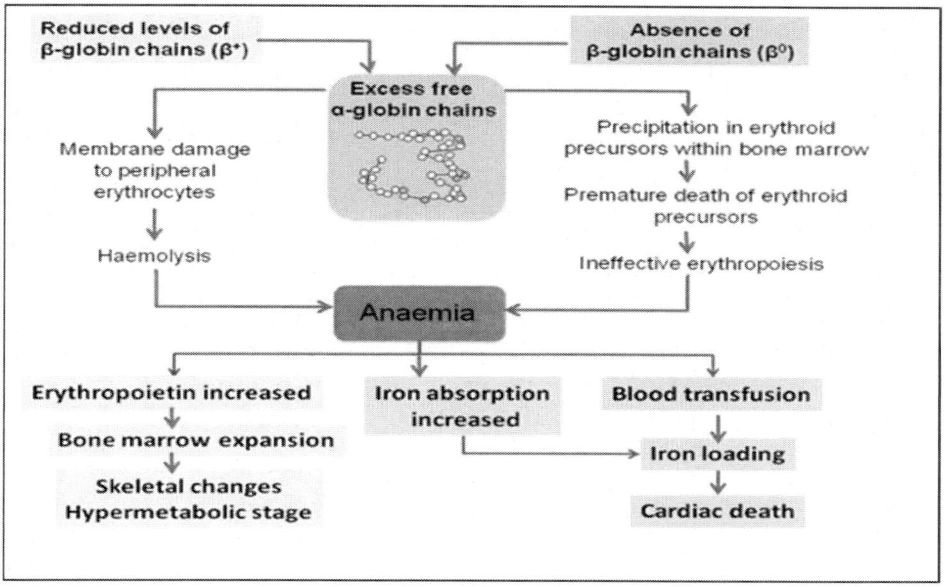

Figure 4. Pathophysiology of beta thalassemia (Cappellini MD, 2014).

4. EPIDEMIOLOGY

4.1. Worldwide Distribution of Beta Thalassemia

Beta thalassemia is the most prevalent non-communicable disease of blood worldwide. The global annual incidence of beta-thalassemia is one in 100,000 (Galanello R and Origa R, 2010). According to the World Health Organization (WHO), around 6.5% of the world populations are currently carriers of different hemoglobinopathies (Modell B et al., 2008). Almost all countries are affected by the fatal disease with more or less frequent carrier status. However, around sixty countries with the highest occurrence of thalassemia in Mediterranean, part of North and West Africa, the Indian subcontinent, Middle East and Southeast Asia together constitute the "Thalassemia Belt." The distribution of the disease is not uniform, even in the countries or regions residing in thalassemia belt.

These variations in frequency of beta thalassemia mainly depend on the ethnic population. The abnormal genes for beta thalassemia are particularly widespread in Italy and Greece. Other regions with the high prevalence are Cyprus with the highest carrier frequency (14%), Greece (5-15%), Sardinia (10.3%), Sicily (10%) and Iran (4-10%) (Flint J et al., 1998). However, at present thalassemia have introduced in almost every country of the world due to increased rate of population immigration and intermarriage between different customs and cultures. For example, in Northern Europe thalassemia was previously absent, but now the country have been reported having patients with thalassemia (Vichinsky EP., 2005). The overall so far estimated β-thalassemia carriers present worldwide is approximately 80–90 million, which is about 1.5% of the total global population from which it can be predicted that at least 60,000 homozygous symptomatic individuals born annually with thalassemia. The incidence of thalassemic baby birth rate has been reported as 1 in 10,000 in Europe (Vichinsky EP. 2005). It is the most common chronic hemolytic anemia in Egypt (85.1%) as well with a projected carrier rate of 9–10.2% from a study on 1000 normal random subjects from different geographic areas of the country (El-Beshlawy A et al., 1999).

Middle East countries are also highly prevalent for beta-thalassemia particularly in Iraq, Lebanon, Egypt and Morocco with moderate to high carrier rate. In Iraq, beta-thalassemia is an evident health problem, specifically the Dohuk region situated in the northern part is mostly affected. According to the latest data published in 2017, the number of thalassemia patients registered in the thalassemia centers of the country until December, 2015 was 11,165 constituting the 66.3% of all registered patients with inherited hemoglobinopathies. The occurrence of thalassemia was 37.1 in 100,000 in 2015 (Kadhim KA, 2017). It is also the most common genetic disorder in Lebanon with a carrier frequency of nearly 2.3% in the general population, even though the rate may be as high as 4.0% to 41.0% in high risk groups (Abi Saad M, 2014).

4.2. Frequency of Beta Thalassemia and HbE Variants in South Asia and Southeast Asia

Thalassemia is very prevalent in Asia, causing a huge public health and socioeconomic burdens. Both beta thalassemia and compound heterozygote HbE/beta thalassemia are the major congenital hemoglobinopathies throughout the region. HbE is the hallmark for South East Asia as well as most common abnormal hemoglobin in many Asian countries. In some regions, the carrier frequency of HbE variant of hemoglobin is as high as around 50%.

The carrier frequency of thalassemia gene is significantly high in Thailand, as a result around 3,000 babies are born with thalassemia per year and the number of existing thalassemia patients is approximately 100,000 in the population (Fucharoen S and Winichagoon P, 1997). Beta thalassemia is also highly widespread in the Indian subcontinent, particularly in many parts of India, Pakistan, Maldives, Bangladesh, Nepal and Sri Lanka. On the other hand, HbE-beta thalassemia accounts for above half of the cases of severe form of thalassemia in Indonesia and Bangladesh and is also very frequently seen in India, Vietnam, Sri Lanka, Cambodia, Laos, Maldives and Malaysia

(Fucharoen, 2000). Table 2 depicts the prevalence of thalassemia in South Asian and South East Asian countries, according to the report of Thalassemia International Federation (TIF) published in 2013.

Table 2. Prevalence of Beta Thalassemia and HbE variants in South Asia and Southeast Asia

Country	Population millions	% HbE carriers	% β-thal carriers	Expected β-thal syndromes Births/year
Bangladesh	161.1	6.1	4.1	5477
Bhutan	0.7	4	na	na
India	1270.0	1	3.9	16200
Maldives	0.4	0.9	18	54
Myanmar	54.6	22	2.2	2398
Nepal	29.4	4.4	4	836
Sri Lanka	21.5	0.5	2.2	64
Brunei Darussalam	0.4		2	1
Indonesia	245.6	6(1-25)	5(3-10)	9619
Thailand	67.1	30	5	6983
Timor Leste	1.2	6	5	65
Cambodia	14.7	30	3	1762
Lao Peop. Dem. Rep.	6.6	18	6	1106
Malaysia	28.7	3.4	4.5	727
Philippines	101.8	0.4	1.2	153
Singapore	4.7	0.64	3	13
Australia	21.6	0.4	0.4	3
Vietnam	90.5	1	2.6	424
Taiwan	22.9	0.027	2	21
Hong Kong	7.05	0.3	3.5	19
Guanxi	49.2	0.42	6.78	902
Guangdong	91.9	0.06	2.54	182
Yunnan	44.5	1.6	3.7	418
Guizhu	39.3	na	1.1	17
Sichuan	87.2	na	2.18	94
Hainan	82.8	na	2.09	132
Macau	0.55	na	3	2
Fujian	36	na	1.32	18

*Thal, Thalassemia; source: //www.ncbi.nlm.nih.gov/books/NBK190463/.

4.3. Prevalence of Thalassemia in Bangladesh

Among the non-communicable diseases, thalassemia is the most common congenital single gene disorder in Bangladesh. As the prevalence pattern of thalassemia in Bangladesh is not well documented, an indirect calculation using the Hardy-Weinberg equation reveals that about 9000 children are born with thalassemia each year considering the fact that Bangladesh has a population of 160 million and a carrier status of 10% based on the last reported study done in 2005 (Khan WA et al. 2005). According to a raw data collected recently from few thalassemia specialized centers in Dhaka dealing with thalassemia patients and other hemoglobinopathies, nearly 200 children are diagnosed with thalassemia every month and around 25 babies are born with thalassemia every day. However, it is a matter of deep concern that thalassemia and other hemoglobinopathies are not recognized as priorities in public health sectors of the country and there are no exact data representing the current prevalence of thalassemia in different regions Bangladesh.

Review of the thalassemia prevalence studies conducted over the last couple of decades in Bangladesh reveals two kinds of data- one is based on the carrier screening approach and another one is from retrospective studies on patient samples obtained from tertiary level hospitals. However, the majority of published and unpublished data suggest that the carrier frequency of β -thalassemia and Hb-E carriers could be as much as 10% or more.

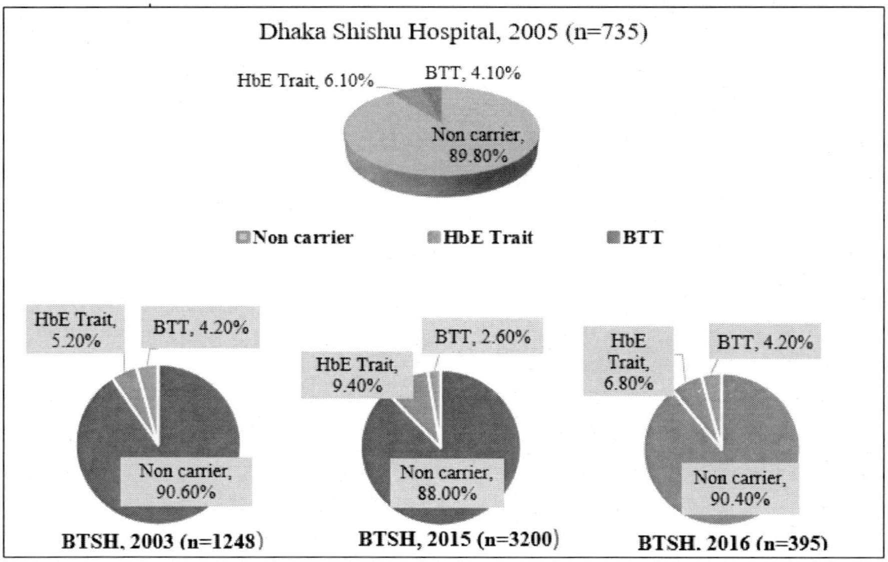

Figure 5. Prevalence of thalassemia carriers in Bangladesh. BTT, Beta thalassemia trait; BTSH, Bangladesh Thalassemia Samity Hospital.

The prevalence of thalassemia carriers in Bangladesh based on population screening data is summarized as pie charts in Figure 5. Among the first category studies, the only published information obtained on the prevalence of thalassemia was a small scale study among (n = 735) school children in Bangladesh covering the whole country which showed the prevalence of the beta-thalassemia trait and the HbE trait were 4.1% and 6.1% respectively. The study also showed the regional variation in prevalence rate in different divisions of the country (Khan WA et al., 2005). Notably, in the hill track, the frequency of beta-thalassemia trait among the tribal population was almost identical, but the prevalence of HbE carriers was considerably higher (41.7%). A study with a small sample size also revealed the same findings on the prevalence rate of HbE which was 39–47% of the tribal population in Bangladesh (Shannon KL, 2015). Bangladesh Thalassemia Samity Hospital (BTSH), a thalassemia specialized center located in Dhaka city conducted three separate studies under different projects from 2003 to 2016 (unpublished, but acknowledged by Thalassemia International Federation, TIF); all of which represent consistently high prevalence of beta thalassemia traits and HbE carriers throughout the country though the situation was not uniform among the eight divisions of Bangladesh. The first study, funded by CIDA (The Canadian International Development Agency) and conducted on 1248 college and high school students in 2003 showed that the prevalence of HbE and beta thalassemia were 5.17% and 4.2% respectively. From the result of a second study conducted in 2015 which was done on larger sample (n = 3200) revealed the total frequency of thalassemia carrier to be as high as 12%. Another study with a small number of samples (n = 395) funded by DGHS (Directorate General of Health Services) of the Ministry of Health & Family Welfare of Bangladesh showed 6.8% and 2.8% occurrence rate of HbE carriers and beta thalassemia traits respectively.

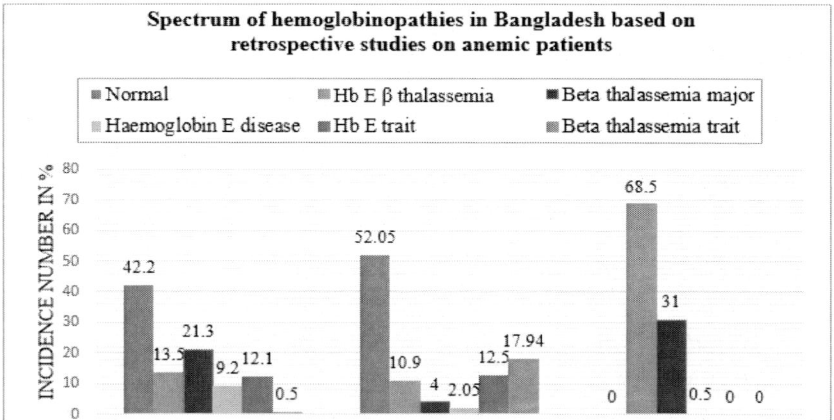

Figure 6. Spectrum of hemoglobinopathies in Bangladesh based on three tertiary level hospital based retrospective studies.

Another date source for hemoglobinopathies spectrum, so far found in Bangladesh is tertiary level hospital based retrospective studies (Figure 6). According to a study of 2011, among 600 anemic patients, β-thalassemia minor (21.3%) was the commonest hemoglobinopathy encountered followed by E-β-thalassemia (13.5%), HbE trait (12.1%), and HbE disease (9.2%) (Shekhar HU et al., 2012). Other two retrospective studies among thalassemic or anemic patients conducted on 2011-2015 (Khan WA et al., 2015) and 2015-2017 (Tahura Sarabon, 2017) study period respectively also revealed the high prevalent cases of homozygotes beta thalassemia (4% and 31% respectively) and compound heterozygotes HbE/beta thalassemia (10.9% and 68.5% respectively). Nevertheless, all of these studies represent too much deviation from each other in the frequency of different hemoglobinopathies and thus, fail to represent the actual prevalence of thalassemia or other hemoglobinopathies in the country.

Table 3. Divisional distribution of carriers of beta thalassemia and HbE variant in Bangladesh reported in the study of 2005 (Khan WA et al., 2005)

Division	Tested	Normal	Beta trait Hb A>3.5	HbE Trait
Dhaka	252	231	8(3.2%)	13(5.2%)
Chittagong	102	94	3(2.9%)	3(2.9%)
Khulna	84	80	2(2.4%)	2(2.4%)
Rajshahi	91	70	5(5.5%)	15(16.5%)
Sylhet	96	85	5(5.2%)	4(4.2%)
Barisal	62	52	5(8.1%)	5(8.1%)
Total	687	612	28(4.1%)	42(6.1%)

All the available published and unpublished data suggest that all the administrative divisions of Bangladesh are significantly prevalent for both β-thalassemia and HbE/β – thalassemia (Khan WA et al., 2005). Although the situation is not uniform throughout the country, it shows almost similar carrier frequency scenario in the different divisions of Bangladesh while the highest frequency of β-thalassemia and HbE carrier was reported in Rajshahi division. Table 3 shows the divisional distribution of carriers of Hb variants in Bangladesh reported in the study of 2005 (Khan WA et al., 2005). However, it was a random screening based data without any statistically calculated sample size for representing the whole country and till date there is no data regarding the current picture of regional distribution of the thalassemia carriers.

5. Molecular Basis of Beta-thalassemia

Beta-thalassemias are enormously diverse at the DNA level and it is one of the earliest diseases to be characterized at the molecular level. At least 400 pathogenic mutations have been so far documented residing in the beta globin gene. Those are mostly point mutations, small insertions or deletions of oligonucleotides leading to frameshift of the genetic code. In a few cases, deletions of moderate or huge length surrounding one or two globin genes cause gross gene deletion or affect the structure and function of the globin chain encoding genes (Chehab F. 2010). A complete updated list of beta-thalassemia mutations is available through the Globin Gene Server Web Site (http://www.globin.cse.psu.edu).

Single nucleotide polymorphisms (SNPs) those are point mutations impairing the synthesis of beta globin chains in terms of either the expression level or the structure, belong to three different classes:

1. Mutations in the regulatory site, in particular the promoter region and untranslated region of the 5'of the mRNA(5' UTR) leading to faulty transcription of the beta-gene;
2. Mutations in consensus sequences, including splice site of intron-exon junction, polyadenylation site, and in the 3' UTR region affecting messenger RNA (mRNA) processing;
3. Nonsense or initiation codon mutations causing abnormal protein synthesis (translation from mRNA).

β^0-thalassemias are characterized as the total lack of beta globin chain which is mainly the consequences of deletional mutations, alteration in start codon or in the splice sites, especially at the splice-site junction in addition to some nonsense and frameshift mutations. However, mutations in the regulatory sites like promoter area, the polyadenylation signal, and the 5' UTR or 3' UTR result in the moderate to severe reduction in beta chain synthesis or abnormalities in the splicing process leading to β^+ -thalassemia.

RNA splicing mutations disturb the splicing procedure at variable amounts depending on the position of mutated nucleotide. Mutations on the consensus sequences residing in the splice site produce less severe form of β-thalassemia (β+ type). Whereas, if the mutation is present in exon or intronic regions, it may cause mild to severe phenotype by activating a cryptic or hidden splicing site which ultimately leads to unusual processing of mRNA processing with the defective spliced product at variable degrees (Cao A and Galanello R, 2010) .

Based on the clinical phenotypes caused by the alterations in the globin gene, all mutations responsible for thalassemia may be categorized as to be severe, mild or silent. The silent mutations result in normal hematological findings. They produce mild imbalance in the ratio of alpha and beta globin chain. Usually the altered bases which are part of the consensus CACCC box, the 5' UTR or the polyA addition site along with some splicing defects are phenotypically silent. The mild mutations display less severe thalassemia-like hematological features and cause moderate imbalance in globin chain synthesis. These mutations are generally present in the proximal CACCC box, TATA box, 5' UTR, or in the first exon, leading to alternative splicing) as well as in the consensus sequences of splicing region, 3'UTR, and the polyadenylation site. Some mutations activating a hidden splicing site in the first exon are linked with a mild or silent phenotype. Here the underlying reason is, the normal splice site is used preferentially over the newly activated splice site. Some example of such abnormal hemoglobin variants are Hb Malay (CD19), HbE (CD 26, highly prevalent in South East Asia), and Hb Knossos (CD 27) (Khan J, 2014; Peixeiro I et al., 2011).

5.1. Spectrum of HBB Gene Mutations in South Asia and Southeast Asia

South Asia consists of eight countries - Bangladesh, India, Pakistan, Bhutan, Maldives, Nepal, Sri Lanka and Afghanistan. In spite of the fact that, a great number of mutations with a huge variation in the β-globin gene of hemoglobin have been described for the various regions all over the world, the β-thalassemia gene carriers from each indigenous population are seen to have only a limited number of mutations including some frequently occurring mutations and some occasional mutations as well. A few number of mutations are found as most frequent in the Indian sub-continent as well as other South Asian countries. For example, above 90% of the beta thalassemia causing mutations in India included only five common mutations. Those are namely- 1.

Table 4. The most common HBB gene mutations in the regional hot-spot of South Asia and Southeast Asia

Countries	Common Mutations
India (Panigrahi I and Marwaha R, 2007; S. Sinha et al., 2009)	IVSI-5 G>C, 619-bp deletion, c.79 G> A, IVSI-1 G>T, c.126_129delCTTT, c.27_28insG, c.47G > A, c.92 G>C, Cap site +1(A>C), c.17_18delCT and c.51delC
Bangladesh (Islam et al., 2018, Chatterjee T et al., 2015; Sultana G et al., 2016; Banu B et al., 2018)	c.79 G> A, IVS1-5 G >C, c.126_129delCTTT, c.92 G>C, c.27_28insG, c.47G > A, c.92 G>C, c.46delT, IVS1-130 G > C, and c.51delC
Pakistan (Ansari SH et al., 2011)	IVS1-5 G >C, c.27_28insG, IVS1-1 G >C, c.126_129delCTTT, c.92 G>C, c.17_18delCT and, c.47G > A
Sri Lanka (Fisher CA 2003)	IVS1-5 G >C, IVS1-1 G >C, c.79 G> A, c.92 G>C, c.27_28insG, c.47G > A, c.51delC, c.126_129delCTTT
Malaysia (George E 2012)	IVS1-5 G >C, IVS1-1 G >C, c.59 A > G, c.126_129delCTTT, c.52A > T, c.27_28insG, c.216_217insA, IVS1-1G >C, c.79G > A
Thailand (Boonyawat B 2014)	c.126_129delCTTT, c.52A > T, c.59A > G, c.27_28insG, IVS1-1 G >C, IVS1-5 G >C, c.108C > A, c.47G > A
Myanmar (Harano T 2002)	IVS1-1 G > T, c.126_129delCTTT, IVS1-5 G >C, c.53A > T, c.135delC, c.108C > A, c.47G > A, c.51delC, c.46delT, c.126delC, IVS1-1 G >C, c.27_28insG

IVSI-5 G > C, 2. IVSI-1 G > T, 3. c.126_129delCTTT or Codon 41/42 (−TCCT), 4. c.27_28insG and 5. 619bp deletion etc. (Thein S et al., 1988 and Varawalla N et al.,

1991) and four of which also predominant in Pakistan and Sri Lanka. Similarly, the most common five mutations accountable for beta-thalassemia incidences in China are c.-78A > G, c.52A > T, c.126_129delCTTT, c.216_217insA and c.316-197C > T in HBB gene (Chan V et al., 1987). The particular region of the gene containing all these common mutations together constitute the mutational hot-spot (c.1 to c.92 of exon 1, c.92 + 1 to c.92 + 130 intron-1 and c.93 – c.217 of exon-2) in the HBB gene for most of the countries of South Asia and Southeast Asia including Bangladesh. Table 4.

5.2. Mutation Spectrum of Beta Thalassemia in Bangladeshi Population

Despite the high prevalence of thalassemia in Bangladesh, the molecular basis that is the responsible gene mutations for beta thalassemia are still quietly unidentified. Only few mutation based studies have been performed so far but all of them recruited small number of samples. However, from all of these studies, it has been found that there are three beta-globin (HBB) gene mutations, namely c.79G > A, c.92 + 5 G >C and c.126_129delCTTT present most commonly in Bangladeshi thalassemia patients constituting ~85% of the pathogenic cases and these mutations are restricted in the identical hot spot region mentioned earlier for the South Asian population (Table 4).

Other mutations residing in the above mentioned hot spot regions, documented in recent studies, are c.46delT, c.47 G > A, c.30G > C, c.92 + 130G > C, c.126delC and c.135delC where the last three mutations are less common (Islam MT et al., 2018, Chatterjee T et al., 2015). In addition to the most common mutations, other two studies also reported some other mutations like c.27_28insG, c.51delC, -90 (C>T) and c.92+1G>T occurring in small extent (Islam MT et al., 2018) including the rare mutations, -29 (G>A) and -92C>G (Sultana G et al., 2016; Banu B et al., 2018 and Ayub MI et al., 2010).

6. Multiple Clinical Complications Associated with the Both Beta Thalassemia and HbE/Beta Thalassemia

Homozygotes for beta-thalassemia or compound heterozygotes like HbE/beta thalassemia may produce significant clinical features representing either deadly severe thalassemia major or moderately severe thalassemia intermedia. Individuals with homozygous thalassemia alleles of fatal phenotypes (thalassemia major) usually develop clinical manifestations drawing medical attention within six months to two years after birth and have a need for regular blood transfusion to survive. On the other hand, thalassemia intermedia patients are diagnosed at comparatively later age and require transfusion occasionally or at longer transfusion intervals. Though the differential diagnosis of thalassemia major and intermedia at presentation is a quite difficult task, it is a serious issue that should be strongly followed so that unnecessary transfusions in thalassemia intermedia could be avoided and early transfusions in case of thalassemia major could be started.

Children with thalassemia major suffer from severe growth retardation and subsequently fail to thrive. They also encounter various critical complications, in particular the difficulties in feeding, irritability, recurrent spells of fever, diarrhea and enlargement of the abdomen caused by splenomegaly along with progressive paleness due to lack of hemoglobin. If the affected infants are left without treatment or with inadequate blood transfusion, they start developing multiple severe clinical conditions other than anemia and growth retardation, namely, pallor, jaundice, splenomegaly, and development of masses from extra medullary hematopoiesis, poor musculature, and skeletal changes due to bone marrow expansion. On the other hand, maintaining an Hb concentration of not less than 9.5 g/dL by a regular and adequate amount of blood transfusion program following proper guideline helps the transfusion dependent major thalassemia patients to thrive with normal growth and development until the age of 10–11 years. However, after the age of 10 to 11 years, affected individuals are at risk of developing severe complications related to post transfusional iron overload, development of allo-immunization and various infections.

Table 5. Complications of vital organs due to iron overload in TDT patients

	Complications
Heart	Dilated myocardiopathy and pericarditis
Liver	Hepatomegaly, chronic hepatitis, fibrosis, and cirrhosis
Endocrine glands	Growth retardation, diabetes mellitus, hypogonadism, hypothyroidism, hypoparathyroidism, pituitary dysfunction and less commonly, adrenal glands
Bone	Orofacial deformities due to bone marrow expansion, Osteoporosis
Spleen	Splenomegaly, hypersplenism, venous thrombosis after splenectomy
Infection	Hepatitis B and C virus and HIV infections are relatively common in old Patients

An unavoidable fate of the patients with thalassemia major after starting regular blood transfusion is the excess deposition of iron in different vital organs of their bodies. The underlying reason behind this consequence is that the human body has no mechanism to excrete left-over iron from the system. A unit of donor blood which is usually processed from 420 mL contains nearly 200 milligram of iron constituting around 0.47 mg/mL of whole donor blood (Zimmermann MB et al., 2008). Iron is very reactive and easily undergo the processes of electron loss and gain resulting in generation of harmful free radicals, which in turn, destroy the lipid membranes of organelles as well as causes DNA damage, thus leading to cell death and the generation of fibrosis. Therefore, iron overload following deposition is highly toxic to many tissues, causing heart failure, liver cirrhosis and carcinoma, growth retardation and multiple endocrine abnormalities (Galanello R et al., 2001; Taher AT et al., 2008 and De Sanctis V et al., 1998). Several studies done in Bangladesh revealed the same degree of clinical situations of almost all kinds exist in both beta and HbE/beta-thalassemia major patients (Ferdaus MJ et al., 2010; Karim MF et al., 2016; Shekhar HU et al., 2005 and Faruk et al., 2015). Table 5 summarizes the adverse effects of post transfusion in severe thalassemic patients.

Liver and heart are the major target organs susceptible to be damaged in transfusion dependent thalassemia patients. There is a high risk of developing hepatocellular carcinoma as a consequence of blood borne infections caused by hepatitis viruses and iron overload (Taher AT et al., 2008; Voskaridou E et al., 2012). Survival period only extends beyond the age of 30 years of severe TDT individuals who have been well transfused and treated with appropriate chelation (Zurlo MG et al., 1989). On the other hand, the heart is seriously affected by post transfusional hemosiderosis and consequently the myocardiopathy which is the most important life-limiting complication of iron overload in beta-thalassemia and reported to cause 71% of deaths from cardiac arrests in beta-thalassemia major patients (Aessopos A et al., 2005).

Conclusion

From the comprehensive review, it can be concluded that beta-thalassemia and HbE/beta thalassemia both are highly prevalent in Bangladesh as well as other countries of South Asia, and create a severe public health burden posing huge clinical complications in the affected individuals. Cross-sectional surveys have revealed that in many regions of the Indian Subcontinent and South Asia, only a few predominant mutations in or around the human beta globin gene are the causative agent for beta-thalassemia in the majority of patients. This conservative mutation pattern advocates that region specific and relatively cost-effective diagnostic methods and disease management policies could be implemented in these areas. Although early diagnosis and supportive treatment measures are available, these are so expensive to afford by the majority of the resource constrained areas. Moreover, some, such as blood transfusion and iron chelation therapies, can have potentially serious complications. Thus, preventative approaches are the most preferable strategy to control the immense burdens on public health caused by these fatal incurable diseases and also needed for saving national resources by limiting new thalassemic cases added annually. Under the circumstances, thalassemia should be placed at the top of the list of genetic disorders that should deserve priority for prevention as the most common single gene disorder. In addition to focusing on controlling major non-communicable diseases (NCDs) like heart attacks and strokes, chronic respiratory diseases, diabetes and cancer, we must have to have specific control strategy for genetic disorders, especially for those with high prevalence in the endemic regions of Bangladesh. To date, there are very few studies on the prevalence of thalassemia in Bangladesh. However, all of these information are discrete, sample sizes are not representative for either the whole country or the regional distribution of the disease prevalence; thus, there is still a lack of definitive data regarding existing thalassemia carrier prevalence in the country. The knowledge of the actual prevalence of β-thalassemia and other hemoglobinopathies in different regions of the country is a prerequisite to launch a national thalassemia prevention program.

Acknowledgments

Bangladesh Thalassemia Samity Hospital, Green Road, Dhaka, Bangladesh.

References

Abi Saad M, Haddad AG, Alam ES, Aoun S, Maatouk P, Ajami N, Khairallah T, Koussa S, Musallam KM, Taher AT. 2014. "Preventing thalassemia in Lebanon: successes and challenges in a developing country." *Hemoglobin.* 38(5):308-11. Epub 2014 Jul 17.

Aessopos A, Farmakis D, Deftereos S, Tsironi M, Tassiopoulos S, Moyssakis I, Karagiorga M. 2005. "Thalassemia heart disease: a comparative evaluation of thalassemia major and thalassemia intermedia." *Chest.* 127: 1523-1530. 10.1378/chest.127.5.1523.

Amoako, YA. & Bedu-Addo, G. 2014. "Hyper-reactive Malarial Splenomegaly (HMS) in a patient with thalassemia syndrome." *Pan African Medical Journal, 19*(310), 19-25.

Ansari SH, Shamsi TS, Ashraf M, Bohray M, Farzana T, Khan MT, et al., 2011. "Molecular epidemiology of β-thalassemia in Pakistan: far reaching implications." *Int J Mol Epidemiol Genet* 2(4):403.

Ayub MI, Moosa MM, Sarwardi G, Khan W, Khan H and Yeasmi S. 2010. "Mutation analysis of HBB gene in selected Bangladeshi β-thalassemic individuals: Presence of rare mutations." *Genet Test Mol Biomarkers* 14:299-302.

Balgir, RS. 2002. "The genetic burden of hemoglobinopathies with special reference to community health in India and the challenges ahead," *Indian Journal of Hematology and Blood Transfusion*, vol. 20, no. 1, pp. 2–7.

Banu B, Khan WA, Selimuzzaman, Sarwardi G, Sadiya S. 2018. "Mutation pattern in beta thalassemia trait population: A basis for prenatal diagnosis." *Bangladesh Medical Research Council Bulletin,* volume 44 65-70.

Boonyawat B, Monsereenusorn C, Traivaree C. 2014. "Molecular analysis of betaglobin gene mutations among Thai beta-thalassemia children: results from a single center study." *Appl Clin Genet* 7:253.

Brancaleoni, V., Di Pierro, E., Motta, I., Cappellini, MD. 2016. "Laboratory diagnosis of thalassemia." *International Journal of Laboratory Hematology*. 32-40.

Cao A. and Galanello R. 2010. "Beta-thalassemia". *Genetics in Medicine* 12 (2); 61–76.

Cao, A. and Moi, P. 2000. "Genetic Modifying Factors in β-Thalassemia." *Clin. Chem. Lab. Med.* 38, 123–132.

Cappellini MD, Cohen A, Porter J, Taher A, Viprakasit V. 2014. *Guidelines for the Management of Transfusion Dependent Thalassemia (TDT)*, 3rd edn. Nicosia, Cyprus: Thalassemia International Federation.

Caro A, Galanello R, Origa R. 2000. "Beta-Thalassemia" *Gene Reviews Initial Posting*: September 28; Last Update: January 24, 2013.

Chan V, Chan T, Chebab F, Todd D. 1987. "Distribution of beta-thalassemia mutations in south China and their association with haplotypes." *Am J Hum Genet.* 41(4):678.

Chatterjee T, Chakravarty A, Chakravarty S, Chowdhury MA, Sultana R. Mutation. 2015. "Spectrum of β-thalassemia and other Hemoglobinopathies in Chittagong, Southeast Bangladesh" *Hemoglobin*. 39(6):389–92.

Chehab, F. 2010. "Molecular Diagnostic Challenges of the Thalassemias." Chapter 36. *Molecular Diagnostics: Techniques and Applications for the Clinical Laboratory* 1ed. By Academic Press. Inc. 441-452.

Colah R, Gorakshakar A, Nadkarni. 2010. "A. Global burden, distribution and prevention of β-thalassemias and hemoglobin E disorders." *Expert Rev Hematol.* 3(1):103–17. Available from: http://www.ncbi.nlm.nih.gov/pubmed/21082937.

De Sanctis V, Tangerini A, Testa MR, Lauriola AL, Gamberini MR, Cavallini AR, Rigolin F. 1998. "Final height and endocrine function in Thalassemia Intermedia." *J Pediatr Endocrinol Metab* 11: 965-971.

El-Beshlawy, A., Kaddah, N., Rageb, L., Hussein, I., Mouktar, G., Moustafa, A., Elraouf, E., Hassaballa, N., Gaafar, T., and El-Sendiony, H. 1999. "Thalassemia Prevalence and Status in Egypt." *Pediatr. Res.* 45, 760–760.

Faruk, Md. Ismail, Arifur Rahman Tanu, Hossain Uddin Shekhar. 2015. "Respiratory burst enzymes, pro-oxidants, and antioxidants status in patients with β-thalassemia major". *North American Journal of Medical Sciences* Volume 7, Issue 6: 253-258. http://www.najms.org on Wednesday, May 11, 2016, IP: 117.103.87.82).

Ferdaus MZ, Mahbub Hasan, AKM, Shekhar HU. 2010. "Analysis of serum lipid profiles, metal ions and thyroid hormones levels abnormalities in b-thalassemic children of Bangladesh." *J Pak Med Assoc* 60(5): 360-365.

Fisher CA, Premawardhena A, De Silva S, Perera G, Rajapaksa S, Olivieri NA, et al., 2003. "The molecular basis for the thalassemias in Sri Lanka." *Br J Haematol.* 121(4):662–71.

Flint, J., Harding, RM, Boyce, AJ, and Clegg, JB. 1998. 1 "The population genetics of the hemoglobinopathies." *Baillieres. Clin. Haematol.* 11, 1–51.

Fucharoen, S., and Winichagoon, P. 1997. "Hemoglobinopathies in Southeast Asia." Hemoglobinopathies in Southeast Asia. *Indian J. Med. Res.* 134, 498–506.

Fucharoen, Suthat MD; Winichagoon, PP. 2000. "Clinical and hematologic aspects of hemoglobin E β-thalassem. *Current Opinion in Hematology. Curr. Opin. Hematol* 7, 106–112.

Galanello R, Piras S, Barella S, Leoni GB, Cipollina MD, Perseu L, Cao A: 2001. "Cholelithiasis and Gilbert's syndrome in homozygous beta-thalassemia." *Br J Haematol.* 2001, 115: 926-928.

Galanello, R., and Origa, R. 2010. "Beta-thalassemia." *Orphanet J. Rare Dis.* 5-11.

George E, Teh L, Rosli R, Lai M, Tan J. 2012. "Beta Thalassemia mutations in Malays: a simplified cost-effective strategy to identify the mutations." *Malays J med. Health Sci* 8(1):1–8.

Harano T, Harano K, Okada S, Shimono KA. 2002. "Wider molecular spectrum of β-thalassemia in Myanmar." *Br J Haemato* 117(4):988–92.

Herbert L. Muncie, JR., MD, and James S. Campbell, MD. 2009. "Alpha and Beta Thalassemia." *Am Fam Physician.* 80(4):339-344.

Islam MT, Kumer SK et al., 2018. "High resolution melting curve analysis targeting the HBB gene mutational hotspot offers a reliable screening approach for all common as well as most of the rare beta-globin gene mutations in Bangladesh" *BMC Genetics* 19:1.

Kadhim KA, Baldawi KH, Lami FH. 2017. "Prevalence, Incidence, Trend, and Complications of Thalassemia in Iraq." *Hemoglobin.* 41(3):164-168. Epub 2017 Aug 24.

Karim MF, Ismail M, Mahbub Hasan AKM, Shekhar HU. 2016. "Hematological and biochemical status of Beta-thalassemia major patients." *International Journal of Hematology, Oncology and Stem Cell Research* 10(1): 08-12.

Khan J. 2014. "Molecular Basis of Thalassemia Intermedia in Pakistan." *JSM Cell Dev Biol.* 2(2): 1010.

Khan WA, Banu B, Amin SK, Selimuzzaman M, Rahman M, Hossain B, et al., 2005. "Prevalence of beta thalassemia trait and Hb E trait in Bangladeshi school children and health burden of thalassemia in our population." *DS HJ.* 21(1):1–7.

Khan WA, Banu B, Sadiya S, Sarwardi G. 2017. "Spectrum of types of thalassemias and hemoglobinopathies study in a tertiary level children hospital in Bangladesh." *Thalassemia Reports* volume 7:6354 18-20.

Lettre, G., Sankaran, VG, BEZERRA, m.A.c., Araújo, AS, Uda, M, Sanna, S, Cao, A., Schlessinger, D., et al., 2008. "DNA polymorphisms at the BCL11A, HBS1L-MYB, and β-globin loci associate with fetal hemoglobin levels and pain crises in sickle cell disease." *Proc Natl Acad Sci USA*, 105(33), 11869-11874.

Mandal PK, Maji SK, Dolai TK. 2014. "Present scenario of hemoglobinopathies in West Bengal, India: An analysis of a large population." *Int J Med Public Heal.* 4(4):496-99.

Modell B, Darlison M. 2008. "Global epidemiology of hemoglobin disorders and derived service indicators." *Bull World Health Organ.*86 (6):480–7.

Nassar AH, Naja M, Cesaretti C, Eprassi B, Cappellini MD, Taher A. 2008 "Pregnancy outcome in patients with beta-thalassemia intermedia at two tertiary care centers, in Beirut and Milan." *Haematologica* 93: 1586-1587.

Olivieri, NF, Pakbaz, Z, Vichinsky, E. 2011. "Hb E/beta-thalassemia: a common &clinically diverse disorder." *The Indian Journal of Medical Research. 134*(4), 522–531.

Panigrahi I, Marwaha R. 2007. "Mutational spectrum of thalassemias in India." *Indian J Hum Genet* 13(1):36.

Peixeiro, I, Silva, AL, Romão, L. 2011. "Control of human β-globin mRNA stability and its impact on beta-thalassemia phenotype." *Haematologica. 96*(6), 905–913.

S. Sinha, Black, ML, Agarwal, S, Colah, R, Das, R, Ryan, K, Bellgard, M, Bittles, AH. 2009. "Profiling β-thalassemia mutations in India at state and regional levels: implications for genetic education, screening and counselling programmes." *Hugo J.* Dec; 3(1-4): 51–62. Published online 2010 Feb 10. doi: 10.1007/s11568-010-9132-3. PMCID: PMC2882644.

Safizadeh, H., Farahmandinia, Z., Nejad, SS., Pourdamghan, N., and Araste, M. 2012. "Quality of life in patients with thalassemia major and intermedia inkerman-iran (I.R.)". *Mediterr. J. Hematol. Infect.* Dis. 4, e2012058.

Salzano, FM & Maria, CB. 2001. "Normal genetic variation at the protein, glycoconjugate and DNA levels." *The Evolution and Genetics of Latin American Populations.* 1st ed. Cambridge: Cambridge University Press. pp. 255-300. *Cambridge Books Online.* Web. 03 March 2015. http://dx.doi.org/10.1017/CBO9780511666100.009.

Shannon KL, Ahmed S, Rahman H, Prue CS, Khyang J, Ram M, et al., 2015. "Hemoglobin E and glucose-6-phosphate dehydrogenase deficiency and Plasmodium falciparum malaria in the Chittagong Hill Districts of Bangladesh." *Am J Trop Med Hyg.* 93(2):281–6.

Shekhar HU et al., 2012. "Pattern of β-Thalassemia and Other Hemoglobinopathies: A Cross-Sectional Study in Bangladesh." *Hematology* Volume 2012, Article ID 659191.

Shekhar HU, Kabir Y, Hossain MM, Mesbah-Uddin M, Khatun-E Jannat K, Hossain MS, et al., 2001. "Blood transfusion mediated viral infections in thalassemic children in Bangladesh." *J Med Sci* 7:131-5.

Shinar, E, & Rachmilewitz, EA. 1990. "Oxidative denaturation of red blood cells in thalassemia." *In Seminars in hematology.* 27(1), 70-82.

So, CC, Song, YQ. et al., 2008. "The HBS1L-MYB intergenic region on chromosome 6q23 is a quantitative trait locus controlling fetal hemoglobin level in carriers of β-thalassemia." *J Med Genet*, 45, No. 11, (Nov), pp 745-775.

Stamatoyannopoulos G, Grosveld F. 2001. "Hemoglobin switching". In: Stamatoyannopoulos G, Majerus PW, Perlmutter RM, Varmus H, eds. *The Molecular Basis of Blood Diseases*. Philadelphia: W.B. Saunders. 135–182.

Steensma, DP, Gibbons, RJ, & Higgs, DR. 2005. "Acquired α-thalassemia in association with myelodysplastic syndrome and other hematologic malignancies." *Blood* 105(2), 443-452.

Sultana G, Begum R, Akhter H, Shamim Z, Rahim M, Chubey G. 2016. "The complete Spectrum of beta (β) thalassemia mutations in Bangladeshi population." *Austin Biomarkers Diagn*. 3:1–6.

Suthat Fucharoen and David J. Weatherall. 2012. "The Hemoglobin E Thalassemias." *Cold Spring Harb Perspect Med* 2:a011734.

Taher AT, Otrock ZK, Uthman I, Cappellini MD. 2008. "Thalassemia and hypercoagulability." *Blood Rev.* 22: 283-292.

Tahura, Sarabon. 2017. "Thalassemia and other Hemoglobinopathies in Bangladeshi Children." *Imperial Journal of Interdisciplinary Research (IJIR)* 180-185.

Thein S, Hesketh C, Wallace R, Weatherall D. 1988. "The molecular basis of thalassemia major and thalassemia intermedia in Asian Indians: application to prenatal diagnosis." *Br J Haematol*. 70(2):225–31.

Thein SL. 2004. "Genetic insights into the clinical diversity of beta thalassemia". *Br J Haematol*. 124: 264-274.

Varawalla N, Old J, Sarkar R, Venkatesan R, Weatherall D. 1991. "The spectrum of β-thalassemia mutations on the Indian subcontinent: the basis for prenatal diagnosis." *Br J Haematol* 78(2):242–7.

Vichinsky EP. 2005. "Changing patterns of thalassemia worldwide". *Ann N Y Acad Sci*. 1054(1):18–24.

Voskaridou E, Ladis V, Kattamis A, et al., 2012. A national registry of hemoglobinopathies in Greece: deducted demographics, trends in mortality and affected births." *Ann Hematol* 91(9):1451-8.

Weatherall DJ & Clegg JB. 2001 "The Thalassemia syndromes." 4th ed. Oxford, U.K: Blackwell Science Ltd. CrossRef. Weatherall DJ, Clegg JB and Naughton MA (1965) "Globin synthesis in thalassemia: an *in vitro* study". *Nature* 208: 1061-1065.

Weatherall DJ, Clegg JB, Higgs DR, Wood WG .1989. *The Hemoglobinopathies*. USA: McGraw-Hill.

Weatherall, DJ & Clegg, JB. 2008. *The thalassemia syndromes.* John Wiley & Sons.

Zimmermann MB, Fucharoen S, Winichagoon P, Sirankapracha P, Zeder C, Gowachirapant S, Judprasong K, Tanno T, Miller JL, Hurrell RF. 2008. "Iron metabolism in heterozygotes for hemoglobin E (HbE), alpha-thalassemia 1, or beta-thalassemia and in compound heterozygotes for HbE/beta-thalassemia." *Am J Clin Nutr.* 88(4):1026-31.

Zurlo MG, De Stefano P, Borgna-Pignatti C, et al., 1989." Survival and causes of death in thalassemia major." *Lancet;* 2(8653):27-30.

In: Trends in Biochemistry and Molecular Biology
Editors: Hossain Uddin Shekhar et al.
ISBN: 978-1-53616-434-3

Chapter 9

THE MOLECULAR DETERMINANTS OF PREECLAMPSIA ACCOUNTABLE FOR MULTIFACTORIAL DISORDER IN PREGNANT WOMEN

Md. Alauddin*[1]*, PhD and Yearul Kabir*[2,*]*, PhD
[1]Department of Nutrition and Food Technology,
Jashore University of Science and Technology, Jashore, Bangladesh,
[2]Department of Biochemistry and Molecular Biology,
University of Dhaka, Dhaka, Bangladesh

ABSTRACT

Multifactorial disorder preeclampsia affects 3-10% of pregnant women worldwide. The main manifestations is oxidative stress, hypertension, micoangiopathy, alteration of micro RNA, amino acid, bioactive compound and proteinuria. Preeclampsia is accompanied by the molecular growth restraint within the uterus and premature birth. Moreover opposing confinement consequences who suffered with preeclampsia have an amplified danger of forthcoming health difficulties. Thus this disorder is a foremost reason of maternal and fetal indisposition and their death worldwide. Recent thrilling scientific research have contributed a better molecular understanding of the preclampsia disorder. Epidemiological along with molecular and therapeutic investigation on preeclampsia-based scientific report have provided persuasive suggestion on circulating metabolic factors to preeclampsia disorder. This review paper highlights the role of key molecular factors as well as biomarkers and their mechanism for preeclampsia. This chapter also summarized the recent knowledge of new potential molecular determinants for better understanding of preeclampsia.

[*] Corresponding Author's E-mail: ykabir@yahoo.com.

1. INTRODUCTION

Multifactorial pregnancy complication preeclampsia affects 3–10% of pregnant women. The concern of preeclampsia with molecular determinants is recognized worldwide and its molecular mechanism is being more interested. Molecular determinant of genetic factors is increased in the preeclamptic placenta due to oxidative impairment and abnormal fetal hemoglobin outflow to the maternal circulation. Free radicals production and abnormal free hemoglobin with its metabolites are toxic in numerous ways for preeclampsia (Schroeder et al., 2002, Berg et al., 2009, Redman, 2011). Moreover, angiogenic factors such as soluble fms-like tyrosine kinase 1 (sFlt-1) and soluble Endoglin (sEng) which is responsible for trap of circulating vascular endothelial growth factor (VEGF) as well as placental growth factor (PGF) and transforming growth factor b (TGFb) which are leading scientific manifestations of preeclampsia disorder. The molecular determinant Flt-1 and sFlt-1 are the products of FLT-1 gene which is generated by differential mRNA processing subsequently triggers the clinical condition of preeclampsia. (Romero and Chaiworapongsa, 2013; Venkatesha et al., 2006; Levine et al., 2006b; Maynard et al., 2003; Powe et al., 2011; Roberts et al., 1989). Interestingly, Flt-1, also known as vascular endothelial growth factor receptor 1 (VEGFR-1) is also involved in signal transmission even though this mechanism is not clear (Kendall and Thomas, 1993; Huckle and Roche, 2004; Thomas et al., 2007; Rahimi et al., 2009; Zhao et al., 2010). However, the mechanisms of molecular determinants faulty expression are a major contributor to the preeclampsia disorder (Zhou et al., 2002).

2. ANGIOGENIC AND OXIDATIVE STRESS AS MOLECULAR DETERMINANTS

Not only angiogenic factors but also free radical and free hemoglobin can cause oxidative stress in preeclampsia. The main culprit is endogenous pro-oxidant heme protein α1-microglobulin. They have the aptitude to frustrate free hemoglobin-induced placental and kidney abnormalities. More specifically, this fetal hemoglobin-induced oxidative anxiety shows a key starring role of preeclampsia (Hansson et al., 2015). The imbalance of oxidant and antioxidant is termed as oxidative anxiety that interrupt redox signaling and control of molecular destruction by the oxidative stress molecule such as reactive oxygen species (ROS) (Sies et al., 2017). Not only ROS but also nitrosative stress (NS) causes oxidative anxiety of preeclampsia. The production of ROS and RNS can affect the structural and functional impairment in preeclampsia patient (Pacher et al., 2007). But the antioxidant mechanism of enzymes such as superoxide dismutase (SOD), hemoxygenase (HO-1), catalase (CAT), glutathione peroxidase (GPx), and thioredoxin

(TRX) split superoxide into H_2O and O_2. This reaction is strongly regulated by their import oxidative condition. The effect of oxidative stress leads to the expression of transcription factors such as AP-1, NRF2, FoxO, CREB, HSF1, HIF-1α, TP53 and CREB-1 in preeclampsia patient (Spinosa-Diez et al., 2015; Aouache et al., 2018). Therefore, expression and methylation of placental genes impairment could contribute to the development of preeclampsia disorder. Additionally, oxidative stress persuades molecular destruction such as increase concentration of 8- hydroxy-2-deoxyguanosine (8-OH-dG) as well as decrease concentration of heme oxygenase and paraoxonase (PON-1) in patient with preeclampsia (Azizi et al., 2018; Barber et al., 2001, Al-Kuraishy et al., 2018, Genc et al., 2011, Many et al., 2000). Moreover, DNA destruction can lead cell death in preeclampsia but the vice versa is found with normal pregnancies (Fujimaki et al., 2011, Wiktor et al., 2004, Tadesse et al., 2014, Shaker et al., 2013, Chiarello et al., 2018). The molecular determinants including malondialdehyde (MDA), protein carbonyl, total peroxide and antioxidants including SOD, GPx, CAT, glutathione (GSH), vitamin E and C provides a scientific support for molecular homeostasis that are disrupted in preeclampsia (Taravati et al., 2018).

3. Obesity and Hypertension as Molecular Determinants of Preeclampsia

Recent study showed that the obese female has been increased by 18-77% worldwide. Obesity creates many complications like impulsive miscarriage, gestational diabetes and hypertensive for the period of pregnancy. High body mass index (BMI) can exacerbate preeclampsia complication (Fernández Alba et al., 2018; Yogev and Catalano, 2009; World Health Organization (WHO), 2011; Wang et al., 2008). The prevalence of preeclampsia is getting higher for mothers who bear BMI over 30 kg/m^2 and this scenario is found worldwide (Bodnar et al., 2005; Bodnar et al., 2007; Hauger et al., 2008). The different mechanisms and physiopathology of preeclampsia are widely distributed. The main important mechanism is hypoxic condition that influences maternal transmission and pro-inflammatory cytokine like tumor necrosis factor alpha (TNF-a) for endothelial dysfunction (Reyes et al., 2012; Roberts et al., 2011; Teran et al., 2001). The obesity is linking to the risk of preeclampsia by affecting placental function and perfusion through alterations of lipid profile, insulin impairment and hyperleptinemia but the clear mechanisms are not understood (Hunkapiller et al., 2011). The serum cholesterol and lipid profile alterations can predict the onset of preeclampsia. The LDL can reduce extravillous cytotrophoblast migration and promotes trophoblast apoptosis (Pavan et al., 2004; Dey et al., 2013; Lopez-Jaramillo et al., 2016; Reyes et al., 2012). Moreover, higher lipid profile and free fatty acid increases the molecular abnormalities by the nuclear transcription factor peroxisome proliferator-activated receptor-γ (PPAR- γ) and the trophoblast cells (Fabbrini et al., 2009; Hubel et al., 1996; Holdsworth-Carson et al., 2010). The proinflammatory cytokines are elevated whereas adiponectin decrease in adipose tissue that are related to metabolic disorder (Teran et al., 2001; Gómez-Arbeláez et al., 2013; Lopez-Jaramillo et al., 2014; Lopez-Jaramillo, 2016). Other metabolic marker acetylcholine controls high blood pressure in preeclampsia (Rueda-Clausen et al., 2010). The molecular determinant of leptin and adiponectin also control the inflammatory action in preeclampsia by regulation of relevant gene expression. Thus, the obesity increases leptin production as well as decrease adiponectin in preeclampsia. Nitric oxide is an important molecular determinant to relaxing the smooth vascular muscle. The imbalanced fat-derived metabolic products leads to endothelial dysfunction (Caballero, 2003; Accini et al., 2001; Toda and Okamura, 2013). Obesity can down regulate the Endothelial nitric oxide synthase (eNOS) protein expression and endothelial NO production thus enhance vasoconstriction and drastically decrease relaxation (Jääskeläinen et al., 2018). Overacting leptin induces activation of the NADPH oxidase which impairs endothelium-dependent vasodilatation by controlling nitric oxide (Fortuño et al., 2010). Multiple genetic and environmental factors are also exacerbating preeclampsia diseases (Spradley et al., 2015). However, the pathophysiological mechanisms showed that only ten percent overweight women are prone to develop preeclampsia (Roberts et al., 2011; Lopez-Jaramillo et al., 2005). Hypertensive disorder of pregnancy complication characterized by new onset hypertension, proteinuria, and edema occurring after 20 weeks of gestation. Women with preeclampsia are more likely to develop cardiovascular disease (CVD) and die compared to normal pregnancy. While the relation between history of preeclampsia and elevated CVD risk is well documented but the mechanism(s) underlying this association remain unclear.

4. The Roles of RNA Molecules as Molecular Determinants of Preeclampsia

Even though the precise pathogenesis of preeclampsia is not yet completely assumed but scientific research are going on preeclampsia to illuminate the mechanisms. One of the most important genetic materials MicroRNAs (miRNAs), is a single stranded RNA molecules derived from noncoding portions of the genome. These miRNAs play important role to regulate gene expressions that have significant roles in the pathological processes of preeclampsia. Down regulation of several miRNAs in preeclampsia are gradually exposed in the different cells and circulation (Hromadnikova et al., 2017; Zhao et al., 2014; Zhou et al., 2017). Some micro RNA (MiR-141, miR-23a, miR-136) are highly expressed in the placenta whereas some are down regulated in hypoxic stress

(Barad et al., 2004; Morales et al., 2013). The microRNAs are down regulated (miR-517a, miR-141, and miR-517b) in placenta and up regulated (miR-499a-5p as well as miRNA cluster such as the miR-371-3 cluster and has-miR-373-3p) in cardiovascular diseases and pregnancy induced preeclampsia (Cronqvist et al., 2014; Hromadnikova, et al., 2015; Morales-Prieto et al., 2013). Even though the miRNA expression profiles in the placenta correlate with an important factor but the miRNA profiles in preeclampsia studies are largely inconsistent due to population genetic characteristics, sample variation or transformations in experimental methods. Thus, the expression profiles and function of miRNA in different disorders may vary widely (Xu et al., 2014).

A large number of long noncoding RNAs (lncRNAs), play an important role in human diseases including preeclampsia (Gao et al., 2012; Zhang et al., 2015; Zuo et al., 2016). The mechanical and serviceable alterations in the placenta promote the pathogenesis of preeclampsia. The functional characteristic of the trophoblast cells is deeply related to placental growth (Appel et al., 2014; Jia et al., 2014; Wang et al., 2015) but several evidence suggests that preeclampsia is characterized by an abnormal trophoblastic infiltration and increased apoptosis of trophoblastic cells (Kaufmann et al., 2003; Myatt, 2002). Genes that are very close in space may also be targeted by lncRNAs regulation. Hypoxia-inducible factor 1-α inhibitor, (HIF1AN)) is a well-recognized inhibitor of hypoxia inducible factor (HIF-1) transcription (Fukushima et al., 2008) and is involved in the regulation of trophoblast invasion and differentiation by hypoxia (Light et al., 2013). The down regulation of FIH1 in preeclampsia placenta and FIH1-mediated pathway of inactivation of HIF-1 was found in placental tissue of preeclampsia. The primary role of lncRNAs in disease progression provides a clear understanding of the mechanisms by which these RNAs find their targets and controls the epigenetic trajectories in preeclampsia individual. However, it is unclear how these RNAs are generated and how they are targeted and controlled by others factor (Amaral and Mattick, 2008; Hung and Chang, 2010) but very few are responsible at posttranscriptional level (Wilusz et al., 2009). The critical role of microRNA in preeclampsia remains unclear even though the target micro RNAs are upregulated and negatively connected to miR-548c-5p in placental cells. In addition, micro RNAs inhibit the proliferation and activation of macrophages and subsequently decrease the cytokines. Thus the microRNA may play as an anti-inflammatory factor in preeclampsia (Wang et al., 2018).

The molecular determinant tissue factor pathway inhibitor 2 (TFPI2) play a role for cell proliferation, differentiation and apoptosis in metabolic disorder (Feng et al., 2016; Sun et al., 2016; Zhao et al., 2011). The potential mechanisms of high expression of this molecular factor in severe cases of preeclampsia has been explored recently. The miRNAs can target multiple genes and the complex network regulates the expression of multiple genes in preeclampsia. Furthermore, the several combinations of miRNAs can also serve to modulate the expression of genes in preeclampsia (Esmaili et al., 1986; Fujita et al., 2010; Lakhter et al., 2018). The micro RNA may play important roles in restraining the main function of target gene. Different binding site of mRNAs resulting complex regulatory network and this complex miRNA expression profile in preeclampsia patients explored association of miRNAs (Farran et al., 2018; Al-Haidari et al., 2018; Zhu et al., 2009). The TFPI2 plays a vigorous role in preeclampsia and therapeutic strategy to mitigate the severity of the disorder (Xu et al., 2018). The micro RNA contributes to sustaining the proliferation and invasion of trophoblast cells through inhibiting Nodal/ALK7 signaling in preeclampsia. MicroRNAs are emerging as important regulators in the pathogenesis of preeclampsia and play an important role in regulating cell proliferation and invasion (Shi et al., 2018).

With the advanced miRNA sequencing technology, a lot of placenta-specific miRNAs have been found to regulate pregnancy process. For example, miR-141 and miR-519d-3p could regulate trophoblast cell proliferation, migration, invasion and intercellular communication. MiR-517a, miR-517b, miR-518b, and miR-519a were the four C19MC members observed in complete hydatidiform moles (CHM). Besides, miR-210 expression was upregulated in placental tissues in preeclampsia patients. The microRNAs (miRNAs) play an important role in the transcriptional regulation of genes through binding to the target gene by way of complementary base pairing, then target gene expression is inhibited post-transcriptionally (Noack et al., 2011; Zhao et al., 2009). MiRNAs are involved in the regulation of about one-third of protein-coding genes in human cells. Multiple mRNAs can serve as target genes for the same miRNA. Therefore, miRNAs and target molecules form a complex regulatory network that affects the vital activities of entire cell. miRNAs can act as regulatory molecules to inhibit the translation process of proteins and thus participate in a series of life activities including cell proliferation, differentiation, and apoptosis (Bartel et al., 2004; Ambrros, 2001; Fang et al., 2018). MicroRNA can affects trophoblast migration and invasion in placental tissue that are linked to preeclampsia (Singh et al., 2017; Gunel et al., 2017; Inoue et al., 2017). More than five hundred different miRNAs expression profile in the placental trophoblasts are found in preeclampsia compare to normal pregnancy (Morales-Prieto et al., 2013; Gao et al., 2018; Han et al., 2017; Lykoudi et al., 2018; Yang et al., 2018).

The extraordinary stability of miRNAs makes them appropriate diagnostic biomarkers for some diseases. Aberrant miR-210 expression is present not only in solid tumors and harmed organs but also secreted into the circulation, allowing detection in plasma. Recent meta-analysis showed that the predictive value of miR-210 is a novel preeclampsia biomarker (Lawria et al., 2008; Koushki et al., 2018). Nuclear factor kappa-light-chain-enhancer of activated B cells (NF-κB) is associated with endothelial dysfunction in preeclampsia. This can be happened due to the negative regulation of endothelial nitric oxide synthase that are also considered as crucial factors in the pathogenesis of preeclampsia. Thus the nuclear transcription factor such as NF-κB elicits endothelial dysfunction via post-transcriptional down regulation of eNOS (Kim et al.,

2018). The major molecular determinants that affect the pathophysiological mechanism of preeclampsia are summarized in Table 1.

Table 1. Deferentially expressed major molecular determinants in preeclampsia

Major molecular determinants	**Preeclampsia**	**Reference**
Micro-RNA (miR-544, miR-3942)	Decrease	Lykoudi et al., 2018
Micro-RNA (miR-152, miR-182, 183, miR-210)	Increase	Li et al., 2015
Total antioxidant level (Vit A,E,C)	Decrease	Negi et al., 2014
Malondialdehyde (MDA)	Increase	Atiba et al., 2016
Plasma protein carbonyl levels	Increase	Tsukimori et al., 2008
Total peroxide level	Increase	Hilali et al., 2013
Nitric oxide level	Decrease	Sharma et al., 2014
Superoxide dismutase activity	Decrease	Pimentel et al., 2013
Glutathione peroxidase activity	Decrease	Malinova et al., 2013
Glutathione reductase (GR)	Decrease	Dordevic et al., 2008
Catalase activity	Increase	Yassaee et al., 2018
Uric acid level	Increase	Asgharnia et al., 2017

5. Amino Acid and Bioactive Compound as Molecular Determinants of Preeclampsia

Most amino acids (AAs) and acylcarnitines (AC) is higher in preeclampsia mothers than control (Liu et al., 2018). Another bioactive compound ferulic acid (FA) that protects cell integrity by catalyzing the formation of phenoxy radicals (Methaw et al., 2004; Chen et al., 2009; Zhao et al., 2014; Gong et al., 2016). FA is reported to inhibit of IL-6 expression (Lampiasi et al., 2016) and may improve lipopolysaccharide-induced inflammation through regulating the secretion of TNF-α, IL-6, and IL-1β (Zhang et al., 2018). In addition, AAs and FA are recognized to prevent cell damage and apoptosis which is induced by oxidative stress and inflammation (Niu et al., 2016; Das et al., 2016; Sadar et al., 2016 Baijnath et al., 2014). The FA significantly reduced blood pressure, urine volume, and urinary protein level in rats with preeclampsia. Further, FA also decrease higher expression of circulating TNF-α, IL-6, IL-1β and PGF, it reduced placental TNF-α, NF-κB and p65. In addition, FA can rescue decreased expression of IL-4 and IL-10 expression in the circulation and placenta of preeclampsia (Chen et al., 2018). Moreover, aspirin may affect lipopolysaccharide-induced preeclampsia by the mRNA expressions of IL-6, IL-1β and MCP-1 and other protein expressions including TLR4, MyD88, NF-κBp65 (Sun et al., 2018).

6. Cellular Activation as Molecular Determinants of Preeclampsia

During pregnancy the molecular genetic as well as ecological factors may affect spiral artery remodeling causing placental hypoxia (Pijnenborg et al., 2006; Redman et al., 2005; Cui et al., 2012; Dong et al., 2014; Red-Horse et al., 2004; Zhou et al., 20013). Oxidative stress trigger local and systemic cellular and inflammatory responses.

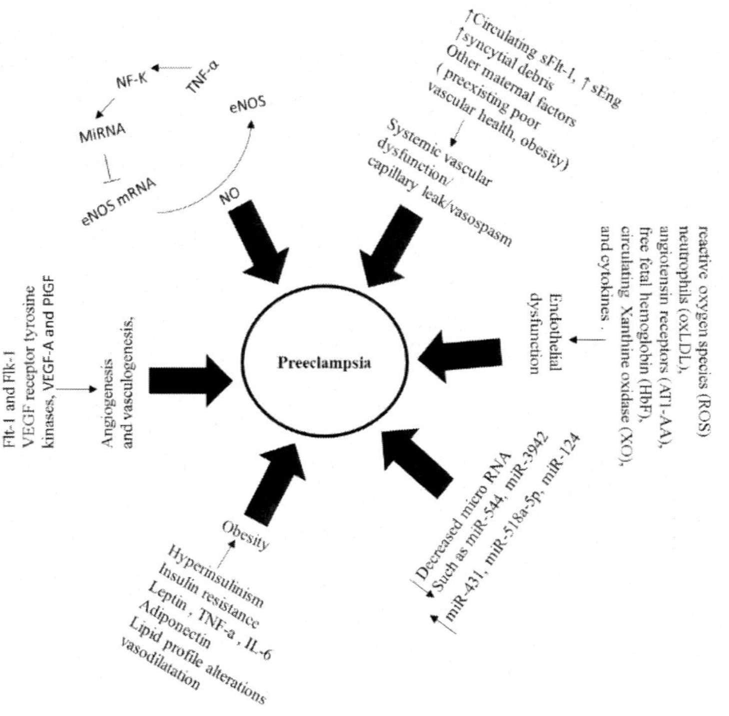

Figure 1. Schematic diagram of major molecular determinants of the pathogenesis of preeclampsia. (Adopted from Lykoudi et al., 2018; Hod et al., 2015; Aouache et al., 2018; Sanchez et al., 2018; Kim et al., 2018; Lv et al., 2018).

The molecular mechanism of high-mobility group box 1 (HMGB1) that promoted endothelial micro-particle (MP) production and blood coagulation in preeclampsia patients. The HMGB1-stimulated endothelial cells activated neutrophils undergo a distinct process of cell death (Hu et al., 2018; Harris et al., 2012; Fuchs et al., 2007). The cell death can lead to arterial and venous thrombosis and the increased neutrophil activation in the maternal circulation found in preeclampsia patients (Caudrillier et al., 2012; Cools-Lartigue et al., 2013; Hakkim et al., 2010; Kessenbrock et al., 2009). The levels of plasma DNA, Myeloperoxidase (MPO) and histones are elevated and associated with enhanced blood coagulation in preeclampsia women. The pro-inflammatory proteins from the hypoxic placenta are responsible for maternal endothelial damage and MP production (Zhou et al., 2002; Jiang et al., 2010; Hu et al., 2018). This review summarizes the recent knowledge of new potential molecular determinants of oxidative stress in preeclampsia, molecular factors of gene expression that are associated with better understanding of preeclampsia. The Figure 1 shows the schematic diagram of major molecular determinants to the preeclampsia.

Conclusion

The pregnancy-specific multifactorial disorder preeclampsia is categorized by oxidative trauma, molecular genetic disproportion with an antiangiogenic factor. Many interrogations remain to be responded, very specifically the upstream and downstream mechanisms of the molecular determinants and deregulation of oxidative stress related gene expression. The major molecular determinants can be summarized in preeclampsia by advanced automated molecular techniques and this information would be useful for pathological decision and therapeutic intensive care in clinical settings for better maternal and neonatal health. The molecular pedagogy of preeclampsia is also concomitant with long-term health jeopardies for both maternal and child health. This review represents summary of molecular determinants of preeclampsia but more scientific experiment is needed to evaluate the exact molecular mechanisms leading to long-term preeclampsia suffering risk for women associated with disease.

References

Accin, J. L., Sotomayor, A., Trujillo, F., Barrera, J. G., Bautista, L. and López-Jaramillo, P., (2001). Colombian study to assess the use of noninvasive determination of endothelium-mediated vasodilatation (CANDEV). Normal values and factors associated. *Endothelium*, *8*(2), 157-166.

Al-Haidari, A., Algaber, A., Madhi, R., Syk, I. and Thorlacius, H., (2018). MiR-155-5p controls colon cancer cell migration via post-transcriptional regulation of human antigen R (HuR). *Cancer letters*, *421*, 145-151.

Al-Kuraishy, H. M., Al-Gareeb, A. I. and Al-Maiahy, T. J., (2018). Concept and connotation of oxidative stress in preeclampsia. *Journal of Laboratory Physicians*, *10*(3), 276.

Amaral, P. P., Clark, M. B., Gascoigne, D. K., Dinger, M. E. and Mattick, J. S., (2010). lncRNAdb: a reference database for long noncoding RNAs. *Nucleic Acids Research*, *39*(suppl_1), D146-D151.

Ambros, V., (2001). microRNAs: tiny regulators with great potential. *Cell*, *107*(7),823-826.

Aouache, R., Biquard, L., Vaiman, D. and Miralles, F., (2018). Oxidative stress in preeclampsia and placental diseases. *International Journal of Molecular Sciences*, *19*(5), 1496.

Appel, S., Ankerne, J., Appel, J., Oberthuer, A., Mallmann, P. and Dötsch, J., (2014). CNN3 regulates trophoblast invasion and is upregulated by hypoxia in BeWo cells. *PloS One*, *9*(7), 103216.

Asgharnia, M., Mirblouk, F., Kazemi, S., Pourmarzi, D., Keivani, M. M. and Heirati, S. F. D., (2017). Maternal serum uric acid level and maternal and neonatal complications in preeclamptic women: A cross-sectional study. *International Journal of Reproductive BioMedicine*, *15*(9), 583.

Atiba, A. S., Abbiyesuku, F. M., Oparinde, D. P., Temitope, A. and Akindele, R. A., (2016). Plasma malondialdehyde (MDA): an indication of liver damage in women with pre-eclamsia. *Ethiopian Journal of Health Sciences*, *26*(5), 479-486.

Azizi, F., Omrani, M. D., Amiri, V., Mirfakhraie, R., Dodangeh, F., Shahmirzadi, S. A. and Gargari, S. S., (2018). Altered methylation and expression patterns of genes regulating placental nitric oxide pathway in patients with severe preeclampsia. *Human Antibodies*, (Preprint), 1-7.

Baijnath, S., Soobryan, N., Mackraj, I., Gathiram, P. and Moodley, J., (2014). The optimization of a chronic nitric oxide synthase (NOS) inhibition model of pre-eclampsia by evaluating physiological changes. *European Journal of Obstetrics & Gynecology and Reproductive Biology*, *182*, 71-75.

Barad, O., Meiri, E., Avniel, A., Aharonov, R., Barzilai, A., Bentwich, I., Einav, U., Gilad, S., Hurban, P., Karov, Y. and Lobenhofer, E. K., (2004). MicroRNA expression detected by oligonucleotide microarrays: system establishment and expression profiling in human tissues. *Genome Research*, *14*(12), 2486-2494.

Barber, A., Robson, S. C., Myatt, L., Bulmer, J. N. and Lyall, F., (2001). Heme oxygenase expression in human placenta and placental bed: reduced expression of placenta endothelial HO-2 in preeclampsia and fetal growth restriction. *The FASEB Journal*, *15*(7), 1158-1168.

Bartel, D. P., (2004). MicroRNAs: genomics, biogenesis, mechanism, and function. *Cell*, *116*(2), 281-297.

Berg, C. J., MacKay, A. P., Qin, C. and Callaghan, W. M., (2009). Overview of maternal morbidity during hospitalization for labor and delivery in the United States: 1993–1997 and 2001–2005. *Obstetrics & Gynecology*, *113*(5), 1075-1081.

Bodnar, L. M., Catov, J. M., Klebanoff, M. A., Ness, R. B. and Roberts, J. M., (2007). Prepregnancy body mass index and the occurrence of severe hypertensive disorders of pregnancy. *Epidemiology*, 234-239.

Bodnar, L. M., Ness, R. B., Markovic, N. and Roberts, J. M., (2005). The risk of preeclampsia rises with increasing prepregnancy body mass index. *Annals of Epidemiology*, *15*(7), 475-482.

Caballero, A. E., (2003). Endothelial dysfunction in obesity and insulin resistance: a road to diabetes and heart disease. *Obesity Research*, *11*(11), 1278-1289.

Caudrillier, A., Kessenbrock, K., Gilliss, B. M., Nguyen, J. X., Marques, M. B., Monestier, M., Toy, P., Werb, Z. and Looney, M. R., (2012). Platelets induce neutrophil extracellular traps in transfusion-related acute lung injury. *The Journal of clinical investigation*, *122*(7), 2661-2671.

Chen, G. P., Ye, Y., Li, L., Yang, Y., Qian, A. B. and Hu, S. J., (2009). Endothelium-independent vasorelaxant effect of sodium ferulate on rat thoracic aorta. *Life Sciences*, *84*(3-4), 81-88.

Chen, Y., Xue, F., Han, C., Yang, H., Han, L., Li, K., Li, J., Xu, Q., Li, Z., Yuan, B. and Yu, L., (2018). Ferulic acid ameliorated placental inflammation and apoptosis in rat with preeclampsia. *Clinical and Experimental Hypertension*, 1-7.

Chiarello, D. I., Abad, C., Rojas, D., Toledo, F., Vázquez, C. M., Mate, A., Sobrevia, L. and Marín, R., (2018). Oxidative stress: Normal pregnancy versus preeclampsia. *Biochimica et Biophysica Acta (BBA)-Molecular Basis of Disease*. Dec, 24.

Cools-Lartigue, J., Spicer, J., McDonald, B., Gowing, S., Chow, S., Giannias, B., Bourdeau, F., Kubes, P. and Ferri, L., (2013). Neutrophil extracellular traps sequester circulating tumor cells and promote metastasis. *The Journal of Clinical Investigation*, *123*(8), 3446-3458.

Cronqvist, T., Saljé, K., Familari, M., Guller, S., Schneider, H., Gardiner, C., Sargent, I. L., Redman, C. W., Mörgelin, M., Åkerström, B. and Gram, M., (2014). Syncytiotrophoblast vesicles show altered micro-RNA and haemoglobin content after ex-vivo perfusion of placentas with haemoglobin to mimic preeclampsia. *PloS One*, *9*(2), 90020.

Cui, Y., Wang, W., Dong, N., Lou, J., Srinivasan, D. K., Cheng, W., Huang, X., Liu, M., Fang, C., Peng, J. and Chen, S., (2012). Role of corin in trophoblast invasion and uterine spiral artery remodelling in pregnancy. *Nature*, *484*(7393), 246.

Das, U., Biswas, S., Sengupta, A., Manna, K., Chakraborty, A. and Dey, S., (2016). Ferulic acid (FA) abrogates ionizing radiation-induced oxidative damage in murine spleen. *International Journal of Radiation Biology*, *92*(12), 806-818.

Dey, M., Arora, D. and Nagarja Narayan, R. K., (2013). Serum cholesterol and ceruloplasmin levels in second trimester can predict development of pre-eclampsia. *North American Journal of Medical Sciences*, *5*(1), 41.

Dong, N., Zhou, T., Zhang, Y., Liu, M., Li, H., Huang, X., Liu, Z., Wu, Y., Fukuda, K., Qin, J. and Wu, Q., (2014). Corin mutations K317E and S472G from preeclamptic patients alert zymogen activation and cell surface targeting. *Journal of Biological Chemistry*, jbc-M114.

Đorđević, N. Z., Babić, G. M., Marković, S. D., Ognjanović, B. I., Štajn, A. Š., Žikić, R. V. and Saičić, Z. S.,(2008). Oxidative stress and changes in antioxidative defense system in erythrocytes of preeclampsia in women. *Reproductive Toxicology*, *25*(2), 213-218.

Esmaili, M. H., Pleuvry, B. J. and Healy, T. E., (1986). Interactions between atracurium and vecuronium on indirectly elicited muscle twitch in vitro. *European Journal of Anaesthesiology*, *3*(6), 469.

Espinosa-Diez, C., Miguel, V., Mennerich, D., Kietzmann, T., Sánchez-Pérez, P., Cadenas, S. and Lamas, S., (2015). Antioxidant responses and cellular adjustments to oxidative stress. *Redox Biology*, *6*, 183-197.

Fabbrini, E., DeHaseth, D., Deivanayagam, S., Mohammed, B. S., Vitola, B. E. and Klein, S., (2009). Alterations in fatty acid kinetics in obese adolescents with increased intrahepatic triglyceride content. *Obesity*, *17*(1), 25-29.

Fang, Y. N., Huang, Z. L., Li, H., Tan, W. B., Zhang, Q. G., Wang, L. and Wu, J. L., (2018). Highly expressed miR-182-5p can promote preeclampsia progression by degrading RND3 and inhibiting HTR-8/SVneo cell invasion. *European Review for Medical and Pharmacological Sciences*, *22*(20), 6583-6590.

Farran, B., Dyson, G., Craig, D., Dombkowski, A., Beebe-Dimmer, J. L., Powell, I. J., Podgorski, I., Heilbrun, L., Bolton, S. and Bock, C. H., (2018). A study of circulating microRNAs identifies a new potential biomarker panel to distinguish aggressive prostate cancer. *Carcinogenesis*, *39*(4), 556-561.

Feng, C., Ho, Y., Sun, C., Xia, G., Ding, Q. and Gu, B., (2016). TFPI-2 expression is decreased in bladder cancer and is related to apoptosis. *J BUON*, *21*, 1518-1523.

Fernández, J. A., Mesa, C. P., Vilar, Á. S., Soto, E. P., González, M. M., Serrano, E. N., Paublete, M. H. and Moreno, L. C., (2018). Overweight and obesity at risk factors for hypertensive states of pregnancy: a retrospective cohort study. *Nutricion hospitalaria*, *35*(4), 874-880.

Fortuño, A., Bidegain, J., Baltanás, A., Moreno, M. U., Montero, L., Landecho, M. F., Beloqui, O., Díez, J. and Zalba, G., (2010). Is leptin involved in phagocytic NADPH oxidase overactivity in obesity? Potential clinical implications. *Journal of Hypertension*, *28*(9), 1944-1950.

Fuchs, T. A., Abed, U., Goosmann, C., Hurwitz, R., Schulze, I., Wahn, V., Weinrauch, Y., Brinkmann, V. and Zychlinsky, A., (2007). Novel cell death program leads to neutrophil extracellular traps. *The Journal of Cell Biology*, *176*(2), 231-241.

Fujimaki, A., Watanabe, K., Mori, T., Kimura, C., Shinohara, K. and Wakatsuki, A., (2011). Placental oxidative DNA damage and its repair in preeclamptic women with fetal growth restriction. *Placenta*, *32*(5), 367-372.

Fujita, Y., Kojima, K., Ohhashi, R., Hamada, N., Nozawa, Y., Kitamoto, A., Sato, A., Kondo, S., Kojima, T., Deguchi, T. and Ito, M., (2010). MiR-148a attenuates paclitaxel-resistance of hormone-refractory, drug-resistant prostate cancer PC3 cells by regulating MSK1 expression. *Journal of Biological Chemistry*, jbc-M109.

Fukushima, K., Murata, M., Hachisuga, M., Tsukimori, K., Seki, H., Takeda, S., Asanoma, K. and Wake, N., (2008). Hypoxia inducible factor 1 alpha regulates matrigel-induced endovascular differentiation under normoxia in a human extravillous trophoblast cell line. *Placenta*, *29*(4), 324-331.

Gao, Y., She, R., Wang, Q., Li, Y. and Zhang, H., (2018). Up-regulation of miR-299 suppressed the invasion and migration of HTR-8/SVneo trophoblast cells partly via targeting HDAC2 in pre-eclampsia. *Biomedicine & Pharmacotherapy*, *97*, 1222-1228.

Gao, Y., She, R., Wang, Q., Li, Y. and Zhang, H., (2018). Up-regulation of miR-299 suppressed the invasion and migration of HTR-8/SVneo trophoblast cells partly via targeting HDAC2 in pre-eclampsia. *Biomedicine & Pharmacotherapy*, *97*, 1222-1228.

Genc, H., Uzun, H., Benian, A., Simsek, G., Gelisgen, R., Madazli, R. and Güralp, O., (2011). Evaluation of oxidative stress markers in first trimester for assessment of preeclampsia risk. *Archives of Gynecology and Obstetrics*, *284*(6), 1367-1373.

Gómez-Arbeláez, D., Lahera, V., Oubiña, P., Valero-Muñoz, M., de las Heras, N., Rodríguez, Y., García, R. G., Camacho, P. A. and López-Jaramillo, P., (2013). Aged garlic extract improves adiponectin levels in subjects with metabolic syndrome: a double-blind, placebo-controlled, randomized, crossover study. *Mediators of Inflammation*, *2013*.

Gong, A. G., Huang, V. Y., Wang, H. Y., Lin, H. Q., Dong, T. T. and Tsim, K. W., (2016). Ferulic acid orchestrates anti-oxidative properties of Danggui Buxue Tang, an ancient herbal decoction: elucidation by chemical knock-out approach. *PloS One*, *11*(11), 0165486.

Gunel, T., Hosseini, M. K., Gumusoglu, E., Kisakesen, H. I., Benian, A. and Aydinli, K., (2017). Expression profiling of maternal plasma and placenta microRNAs in preeclamptic pregnancies by microarray technology. *Placenta*, *52*, 77-85.

Hakkim, A., Fürnrohr, B. G., Amann, K., Laube, B., Abed, U. A., Brinkmann, V., Herrmann, M., Voll, R. E. and Zychlinsky, A., (2010). Impairment of neutrophil extracellular trap degradation is associated with lupus nephritis. *Proceedings of the National Academy of Sciences*, *107*(21), 9813-9818.

Han, L., Zhao, Y., Luo, Q. Q., Liu, X. X., Lu, S. S. and Zou, L., (2017). The significance of miR-145 in the prediction of preeclampsia. *Bratislavske Lekarske Listy*, *118*(9), 523-528.

Hansson, S. R., Nääv, Å. and Erlandsson, L., (2015). Oxidative stress in preeclampsia and the role of free fetal hemoglobin. *Frontiers in Physiology*, *5*, 516.

Harris, H. E., Andersson, U. and Pisetsky, D. S., (2012). HMGB1: a multifunctional alarmin driving autoimmune and inflammatory disease. *Nature Reviews Rheumatology*, *8*(4), 195.

Hauger, M. S., Gibbons, L., Vik, T. and Belizán, J. M., (2008). Prepregnancy weight status and the risk of adverse pregnancy outcome. *Acta Obstetricia et Gynecologica Scandinavica*, *87*(9), 953-959.

Hilali, N., Kocyigit, A., Demir, M., Camuzcuoglu, A., Incebiyik, A., Camuzcuoglu, H., Vural, M. and Taskin, A., (2013). DNA damage and oxidative stress in patients with mild preeclampsia and offspring. *European Journal of Obstetrics & Gynecology and Reproductive Biology*, *170*(2), 377-380.

Holdsworth-Carson, S. J., Lim, R., Mitton, A., Whitehead, C., Rice, G. E., Permezel, M. and Lappas, M., (2010). Peroxisome proliferator-activated receptors are altered in pathologies of the human placenta: gestational diabetes mellitus, intrauterine growth restriction and preeclampsia. *Placenta*, *31*(3), 222-229.

Hromadnikova, I., Kotlabova, K., Hympanova, L. and Krofta, L., (2015). Cardiovascular and cerebrovascular disease associated microRNAs are dysregulated in placental tissues affected with gestational hypertension, preeclampsia and intrauterine growth restriction. *PLoS One*, *10*(9), 0138383.

Hu, Y., Yan, R., Zhang, C., Zhou, Z., Liu, M., Wang, C., Zhang, H., Dong, L., Zhou, T., Wu, Y. and Dong, N., (2018). High-Mobility Group Box 1 From Hypoxic Trophoblasts Promotes Endothelial Microparticle Production and Thrombophilia in Preeclampsia. *Arteriosclerosis, Thrombosis, and Vascular Biology*, ATVBAHA-118.

Hubel, C. A., McLaughlin, M. K., Evans, R. W., Hauth, B. A., Sims, C. J. and Roberts, J. M., (1996). Fasting serum triglycerides, free fatty acids, and malondialdehyde are increased in preeclampsia, are positively correlated, and decrease within 48 hours post-partum. *American Journal of Obstetrics and Gynecology*, *174*(3), 975-982.

Huckle, W. R. and Roche, R. I., (2004). Post-transcriptional control of expression of sFlt-1, an endogenous inhibitor of vascular endothelial growth factor. *Journal of Cellular Biochemistry*, *93*(1), 120-132.

Hung, T. and Chang, H. Y., (2010). Long noncoding RNA in genome regulation: prospects and mechanisms. *RNA Biology*, *7*(5), 582-585.

Hunkapiller, N. M., Gasperowicz, M., Kapidzic, M., Plaks, V., Maltepe, E., Kitajewski, J., Cross, J. C. and Fisher, S. J., (2011). A role for Notch signaling in trophoblast endovascular invasion and in the pathogenesis of pre-eclampsia. *Development*, *138*(14), 2987-2998.

Inoue, K., Hirose, M., Inoue, H., Hatanaka, Y., Honda, A., Hasegawa, A., Mochida, K. and Ogura, A., (2017). The rodent-specific microRNA cluster within the Sfmbt2 gene is imprinted and essential for placental development. *Cell Reports*, *19*(5), 949-956.

Jääskeläinen, T., Heinonen, S., Hämäläinen, E., Pulkki, K., Romppanen, J. and Laivuori, H., (2018). Impact of obesity on angiogenic and inflammatory markers in the Finnish

Genetics of Pre-eclampsia Consortium (FINNPEC) cohort. *International Journal of Obesity*, 1.

Jia, R. Z., Rui, C., Li, J. Y., Cui, X. W. and Wang, X., (2014). CDX1 restricts the invasion of HTR-8/SVneo trophoblast cells by inhibiting MMP-9 expression. *Placenta*, *35*(7), 450-454.

Jiang, R., Teng, Y., Huang, Y., Gu, J. and Li, M., (2010). Protein Kinase C-α Activation Induces NF-κB-Dependent VCAM-1 Expression in Cultured Human Umbilical Vein Endothelial Cells Treated with Sera from Preeclamptic Patients. *Gynecologic and Obstetric Investigation*, *69*(2), 101-108.

Kao, C. K., Morton, J. S., Quon, A. L., Reyes, L. M., Lopez-Jaramillo, P. and Davidge, S. T., (2016). Mechanism of vascular dysfunction due to circulating factors in women with preeclampsia. *Clinical Science*, CS20150678.

Kaufmann, P., Black, S. and Huppertz, B., (2003). Endovascular trophoblast invasion: implications for the pathogenesis of intrauterine growth retardation and preeclampsia. *Biology of Reproduction*, *69*(1), 1-7.

Kendall, R. L. and Thomas, K. A., (1993). Inhibition of vascular endothelial cell growth factor activity by an endogenously encoded soluble receptor. *Proceedings of the National Academy of Sciences*, *90*(22), 10705-10709.

Kessenbrock, K., Krumbholz, M., Schönermarck, U., Back, W., Gross, W. L., Werb, Z., Gröne, H. J., Brinkmann, V. and Jenne, D. E., (2009). Netting neutrophils in autoimmune small-vessel vasculitis. *Nature Medicine*, *15*(6), 623.

Kim, S., Lee, K. S., Choi, S., Kim, J., Lee, D. K., Park, M., Park, W., Kim, T. H., Hwang, J. Y., Won, M. H. and Lee, H., (2018). NF-κB–responsive miRNA-31-5p elicits endothelial dysfunction associated with preeclampsia via down-regulation of endothelial nitric-oxide synthase. *Journal of Biological Chemistry*, *293*(49), 18989-19000.

Koushki, M., Atan, N. A. D., Omidi-Ardali, H. and Tavirani, M. R., (2018). Assessment of Correlation Between miR-210 Expression and Pre-Eclampsia Risk: A Meta-Analysis. *Reports of Biochemistry and Molecular Biology*, *7*(1), 94-101.

Lakhter, A. J., Pratt, R. E., Moore, R. E., Doucette, K. K., Maier, B. F., DiMeglio, L. A. and Sims, E. K., (2018). Beta cell extracellular vesicle miR-21-5p cargo is increased in response to inflammatory cytokines and serves as a biomarker of type 1 diabetes. *Diabetologia*, *61*(5), 1124-1134.

Lampiasi, N. and Montana, G., (2016). The molecular events behind ferulic acid mediated modulation of IL-6 expression in LPS-activated Raw 264.7 cells. *Immunobiology*, *221*(3), 486-493.

Lawrie, C. H., Gal, S., Dunlop, H. M., Pushkaran, B., Liggins, A. P., Pulford, K., Banham, A. H., Pezzella, F., Boultwood, J., Wainscoat, J. S. and Hatton, C. S., (2008). Detection of elevated levels of tumour-associated microRNAs in serum of patients with diffuse large B-cell lymphoma. *British Journal of Haematology*, *141*(5), 672-675.

Levine, R. J., Lam, C., Qian, C., Yu, K. F., Maynard, S. E., Sachs, B. P., Sibai, B. M., Epstein, F. H., Romero, R., Thadhani, R. and Karumanchi, S. A., (2006). Soluble endoglin and other circulating antiangiogenic factors in preeclampsia. *New England Journal of Medicine*, *355*(10), 992-1005.

Li, Q., Long, A., Jiang, L., Cai, L., Xie, L. I., Gu, J. A., Chen, X. and Tan, L., (2015). Quantification of preeclampsia-related microRNAs in maternal serum. *Biomedical Reports*, *3*(6), 792-796.

Light, K. M., Hangasky, J. A., Knapp, M. J. and Solomon, E. I., (2013). Spectroscopic Studies of the Mononuclear Non-Heme FeII Enzyme FIH: Second-Sphere Contributions to Reactivity. *Journal of the American Chemical Society*, *135*(26), 9665-9674.

Liu, G., Deng, W., Cui, W., Xie, Q., Zhao, G., Wu, X., Dai, L., Chen, D. and Yu, B., (2018). Analysis of amino acid and acyl carnitine profiles in maternal and fetal serum from preeclampsia patients. *The Journal of Maternal-Fetal & Neonatal Medicine*, 1-8.

Lopez-Jaramillo, P., (2016). The Role of Adiponectin in Cardiometabolic Diseases: Effects of Nutritional Interventions–3. *The Journal of Nutrition*, *146*(2), 422S-426S.

Lopez-Jaramillo, P., Garcia, R. G. and Lopez, M., (2005). Preventing pregnancy-induced hypertension: are there regional differences for this global problem? *Journal of Hypertension*, *23*(6), 1121-1129.

Lv, Y., Lu, C., Ji, X., Miao, Z., Long, W., Ding, H. and Lv, M., (2019). Roles of microRNAs in preeclampsia. *Journal of Cellular Physiology*, 234(2), 1052-1061.

Lykoudi, A., Kolialexi, A., Lambrou, G. I., Braoudaki, M., Siristatidis, C., Papaioanou, G. K., Tzetis, M., Mavrou, A. and Papantoniou, N., (2018). Dysregulated placental microRNAs in Early and Late onset Preeclampsia. *Placenta*, *61*, 24-32.

Lykoudi, A., Kolialexi, A., Lambrou, G. I., Braoudaki, M., Siristatidis, C., Papaioanou, G. K., Tzetis, M., Mavrou, A. and Papantoniou, N., (2018). Dysregulated placental microRNAs in early and late onset preeclampsia. *Placenta*, *61*, 24-32.

Malinova, M. and Paskaleva, V., (2013). Selenium and glutathione peroxidase in patients with preeclampsia. *Akusherstvo i Ginekologiia*, *52*(5), 3-7.

Many, A., Hubel, C. A., Fisher, S. J., Roberts, J. M. and Zhou, Y., (2000). Invasive cytotrophoblasts manifest evidence of oxidative stress in preeclampsia. *The American Journal of Pathology*, *156*(1), 321-331.

Mathew, S. and Abraham, T. E., (2004). Ferulic acid: an antioxidant found naturally in plant cell walls and feruloyl esterases involved in its release and their applications. *Critical Reviews in Biotechnology*, *24*(2-3), 59-83.

Maynard, S. E., Min, J. Y., Merchan, J., Lim, K. H., Li, J., Mondal, S., Libermann, T. A., Morgan, J. P., Sellke, F. W., Stillman, I. E. and Epstein, F. H., (2003). Excess placental soluble fms-like tyrosine kinase 1 (sFlt1) may contribute to endothelial dysfunction, hypertension, and proteinuria in preeclampsia. *The Journal of Clinical Investigation*, *111*(5), 649-658.

Morales-Prieto, D. M., Ospina-Prieto, S., Chaiwangyen, W., Schoenleben, M. and Markert, U. R., (2013). Pregnancy-associated miRNA-clusters. *Journal of Reproductive Immunology*, *97*(1), 51-61.

Myatt, L., (2002). Role of placenta in preeclampsia. *Endocrine*, *19*(1), 103-111.

Negi, R., Pande, D., Karki, K., Kumar, A., Khanna, R. S. and Khanna, H. D., (2014). Association of oxidative DNA damage, protein oxidation and antioxidant function with oxidative stress induced cellular injury in pre-eclamptic/eclamptic mothers during fetal circulation. *Chemico-Biological Interactions*, *208*, 77-83.

Niu, C., Sheng, Y., Zhu, E., Ji, L. and Wang, Z., (2016). Ferulic acid prevents liver injury induced by Diosbulbin B and its mechanism. *Bioscience Trends*, *10*(5), 386-391.

Noack, F., Ribbat-Idel, J., Thorns, C., Chiriac, A., Axt-Fliedner, R., Diedrich, K. and Feller, A. C., (2011). miRNA expression profiling in formalin-fixed and paraffin-embedded placental tissue samples from pregnancies with severe preeclampsia. *Journal of Perinatal Medicine*, *39*(3), 267-271.

Pacher, P., Beckman, J. S. and Liaudet, L., (2007). Nitric oxide and peroxynitrite in health and disease. *Physiological Reviews*, *87*(1), 315-424.

Pavan, L., Hermouet, A., Tsatsaris, V., Thérond, P., Sawamura, T., Evain-Brion, D. and Fournier, T., (2004). Lipids from oxidized low-density lipoprotein modulate human trophoblast invasion: involvement of nuclear liver X receptors. *Endocrinology*, *145*(10), 4583-4591.

Pijnenborg, R., Vercruysse, L. and Hanssens, M., (2006). The uterine spiral arteries in human pregnancy: facts and controversies. *Placenta*, *27*(9-10), 939-958.

Pimentel, A. M., Pereira, N. R., Costa, C. A., Mann, G. E., Cordeiro, V. S., de Moura, R. S., Brunini, T. M., Mendes-Ribeiro, A. C. and Resende, Â. C., (2013). L-arginine-nitric oxide pathway and oxidative stress in plasma and platelets of patients with pre-eclampsia. *Hypertension Research*, *36*(9), 783.

Powe, C. E., Levine, R. J. and Karumanchi, S. A., (2011). Preeclampsia, a disease of the maternal endothelium: the role of antiangiogenic factors and implications for later cardiovascular disease. *Circulation*, *123*(24), 2856-2869.

Rahimi, N., Golde, T. E. and Meyer, R. D., (2009). Identification of ligand-induced proteolytic cleavage and ectodomain shedding of VEGFR-1/FLT1 in leukemic cancer cells. *Cancer Research*, *69*(6), 2607-2614.

Red-Horse, K., Zhou, Y., Genbacev, O., Prakobphol, A., Foulk, R., McMaster, M. and Fisher, S. J., (2004). Trophoblast differentiation during embryo implantation and formation of the maternal-fetal interface. *The Journal of Clinical Investigation*, *114*(6), 744-754.

Redman, C. W. and Sargent, I. L., (2005). Latest advances in understanding preeclampsia. *Science*, *308*(5728), 1592-1594.

Redman, C. W. G., (2011). Hypertension in pregnancy: the NICE guidelines. *Heart*, *97*(23), 1967-1969.

Reyes, L., Garcia, R., Ruiz, S., Dehghan, M. and López-Jaramillo, P., (2012). Nutritional status among women with pre-eclampsia and healthy pregnant and non-pregnant women in a Latin American country. *Journal of Obstetrics and Gynaecology Research*, *38*(3), 498-504.

Roberts, J. M., Taylor, R. N., Musci, T. J., Rodgers, G. M., Hubel, C. A. and McLaughlin, M. K., (1989). Preeclampsia: an endothelial cell disorder. *American Journal of Obstetrics and Gynecology*, *161*(5), 1200-1204.

Roberts, K. A., Riley, S. C., Reynolds, R. M., Barr, S., Evans, M., Statham, A., Hor, K., Jabbour, H. N., Norman, J. E. and Denison, F. C., (2011). Placental structure and inflammation in pregnancies associated with obesity. *Placenta*, *32*(3), 247-254.

Roberts, K. A., Riley, S. C., Reynolds, R. M., Barr, S., Evans, M., Statham, A., Hor, K., Jabbour, H. N., Norman, J. E. and Denison, F. C., (2011). Placental structure and inflammation in pregnancies associated with obesity. *Placenta*, *32*(3), 247-254.

Romero, R. and Chaiworapongsa, T., (2013). Preeclampsia: a link between trophoblast dysregulation and an antiangiogenic state. *The Journal of Clinical Investigation*, *123*(7), 2775-2777.

Rueda-Clausen, C. F., Lahera, V., Calderón, J., Bolivar, I. C., Castillo, V. R., Gutiérrez, M., Carreño, M., del Pilar Oubiña, M., Cachofeiro, V. and López-Jaramillo, P., (2010). The presence of abdominal obesity is associated with changes in vascular function independently of other cardiovascular risk factors. *International Journal of Cardiology*, *139*(1), 32-41.

Sadar, S. S., Vyawahare, N. S. and Bodhankar, S. L., (2016). Ferulic acid ameliorates TNBS-induced ulcerative colitis through modulation of cytokines, oxidative stress, iNOs, COX-2, and apoptosis in laboratory rats. *EXCLI Journal*, *15*, 482.

Schroeder, B. M., (2002). ACOG practice bulletin on diagnosing and managing preeclampsia and eclampsia. American College of Obstetricians and Gynecologists. *Am Fam Physician*, *66*, 330-331.

Shaker, O. G. and Sadik, N. A. H., (2013). Pathogenesis of preeclampsia: implications of apoptotic markers and oxidative stress. *Human & Experimental Toxicology*, *32*(11), 1170-1178.

Sharma, D., Hussain, S. A., Akhter, N., Singh, A., Trivedi, S. S. and Bhatttacharjee, J., (2014). Endothelial nitric oxide synthase (eNOS) gene Glu298Asp polymorphism and expression in North Indian preeclamptic women. *Pregnancy Hypertension: An International Journal of Women's Cardiovascular Health*, *4*(1), 65-69.

Shi, Z., She, K., Li, H., Yuan, X., Han, X. and Wang, Y., (2019). MicroRNA-454 contributes to sustaining the proliferation and invasion of trophoblast cells through inhibiting Nodal/ALK7 signaling in pre-eclampsia. *Chemico-Biological Interactions*, *298*, 8-14.

Sies, H., Berndt, C. and Jones, D. P., (2017). Oxidative stress. *Annual Review of Biochemistry*, *86*, 715-748.

Singh, K., Williams III, J., Brown, J., Wang, E. T., Lee, B., Gonzalez, T. L., Cui, J., Goodarzi, M. O. and Pisarska, M. D., (2017). Up-regulation of microRNA-202-3p in first trimester placenta of pregnancies destined to develop severe preeclampsia, a pilot study. *Pregnancy Hypertension*, *10*, 7-9.

Spradley, F. T., Palei, A. C. and Granger, J. P., (2015). Increased risk for the development of preeclampsia in obese pregnancies: weighing in on the mechanisms. *American Journal of Physiology-Regulatory, Integrative and Comparative Physiology*, *309*(11), R1326-R1343.

Sun, F. K., Sun, Q., Fan, Y. C., Gao, S., Zhao, J., Li, F., Jia, Y. B., Liu, C., Wang, L. Y., Li, X. Y. and Ji, X. F., (2016). Methylation of tissue factor pathway inhibitor 2 as a prognostic biomarker for hepatocellular carcinoma after hepatectomy. *Journal of Gastroenterology and Hepatology*, *31*(2), 484-492.

Sun, J., Zhang, H., Liu, F., Tang, D. and Lu, X., (2018). Ameliorative effects of aspirin against lipopolysaccharide-induced preeclampsia-like symptoms in rats by inhibiting the pro-inflammatory pathway. *Canadian Journal of Physiology and Pharmacology*, *96*(11), 1084-1091.

Tadesse, S., Kidane, D., Guller, S., Luo, T., Norwitz, N. G., Arcuri, F., Toti, P. and Norwitz, E. R., (2014). In vivo and in vitro evidence for placental DNA damage in preeclampsia. *PLoS One*, *9*(1), 86791.

Taravati, A. and Tohidi, F., (2018). Comprehensive analysis of oxidative stress markers and antioxidants status in preeclampsia. *Taiwanese Journal of Obstetrics and Gynecology*, *57*(6), 779-790.

Teran, E., Escudero, C., Moya, W., Flores, M., Vallance, P. and Lopez-Jaramillo, P., (2001). Elevated C-reactive protein and pro-inflammatory cytokines in Andean women with pre-eclampsia. *International Journal of Gynecology & Obstetrics*, *75*(3), 243-249.

Teran, E., Escudero, C., Moya, W., Flores, M., Vallance, P. and Lopez-Jaramillo, P., (2001). Elevated C-reactive protein and pro-inflammatory cytokines in Andean women with pre-eclampsia. *International Journal of Gynecology & Obstetrics*, *75*(3), 243-249.

Thomas, C. P., Andrews, J. I. and Liu, K. Z., (2007). Intronic polyadenylation signal sequences and alternate splicing generate human soluble Flt1 variants and regulate the abundance of soluble Flt1 in the placenta. *The FASEB Journal*, *21*(14), 3885-3895.

Toda, N. and Okamura, T., (2013). Obesity impairs vasodilatation and blood flow increase mediated by endothelial nitric oxide: an overview. *The Journal of Clinical Pharmacology*, *53*(12), 1228-1239.

Tsukimori, K., Yoshitomi, T., Morokuma, S., Fukushima, K. and Wake, N., (2008). Serum uric acid levels correlate with plasma hydrogen peroxide and protein carbonyl levels in preeclampsia. *American Journal of Hypertension*, *21*(12), 1343-1346.

Venkatesha, S., Toporsian, M., Lam, C., Hanai, J. I., Mammoto, T., Kim, Y. M., Bdolah, Y., Lim, K. H., Yuan, H. T., Libermann, T. A. and Stillman, I. E., (2006). Soluble endoglin contributes to the pathogenesis of preeclampsia. *Nature Medicine*, *12*(6), 642.

Wang, S. S., Huang, Q. T., Zhong, M. and Yin, Q., (2015). AOPPs (advanced oxidation protein products) promote apoptosis in trophoblastic cells through interference with NADPH oxidase signaling: Implications for preeclampsia. *The Journal of Maternal-Fetal & Neonatal Medicine*, *28*(15), 1747-1755.

Wang, Y., Beydoun, M. A., Liang, L., Caballero, B. and Kumanyika, S. K., (2008). Will all Americans become overweight or obese? Estimating the progression and cost of the US obesity epidemic. *Obesity*, *16*(10), 2323-2330.

Wang, Z., Wang, P., Wang, Z., Qin, Z., Xiu, X., Xu, D., Zhang, X. and Wang, Y., (2018). MiRNA-548c-5p downregulates inflammatory response in preeclampsia via targeting PTPRO. *Journal of Cellular Physiology*. *234*(7), 11149-11155.

Wiktor, H., Kankofer, M., Schmerold, I., Dadak, A., Lopucki, M. and Niedermüller, H., (2004). Oxidative DNA damage in placentas from normal and pre-eclamptic pregnancies. *Virchows Archiv*, *445*(1), 74-78.

Wilusz, J. E., Sunwoo, H. and Spector, D. L., (2009). Long noncoding RNAs: functional surprises from the RNA world. *Genes & Development*, *23*(13), 1494-1504.

World Health Organization (WHO) (2011). *Prevalence of obesity and overweight females > 15 years*. https://www.who.int/gho/ncd/risk_factors/ overweight_obesity/ obesity_adults/en/.

Xu, P., Zhao, Y., Liu, M., Wang, Y., Wang, H., Li, Y. X., Zhu, X., Yao, Y., Wang, H., Qiao, J. and Ji, L., (2014). Variations of microRNAs in human placentas and plasma from preeclamptic pregnancy. *Hypertension*, HYPERTENSIONAHA-113.

Xu, Y., Wu, D., Jiang, Z., Zhang, Y., Wang, S., Ma, Z., Hui, B., Wang, J., Qian, W., Ge, Z. and Sun, L., (2018). MiR-616-3p modulates cell proliferation and migration through targeting tissue factor pathway inhibitor 2 in preeclampsia. *Cell Proliferation*, *51*(5), 12490.

Yang, X. and Meng, T., (2019). MicroRNA-431 affects trophoblast migration and invasion by targeting ZEB1 in preeclampsia. *Gene*, *683*, 225-232.

Yassaee, F., Salimi, S., Etemadi, S. and Yaghmaei, M., (2018). Comparison of CAT-21A/T Gene Polymorphism in Women with Preeclampsia and Control Group. *Advanced Biomedical Research*, *7*.

Yogev, Y. and Catalano, P. M., 2009. Pregnancy and obesity. *Obstetrics and Gynecology Clinics*, *36*(2), 285-300.

Zhang, S., Wang, P., Zhao, P., Wang, D., Zhang, Y., Wang, J., Chen, L., Guo, W., Gao, H. and Jiao, Y., (2018). Pretreatment of ferulic acid attenuates inflammation and oxidative stress in a rat model of lipopolysaccharide-induced acute respiratory distress syndrome. *International Journal of Immunopathology and Pharmacology*, *31*, 0394632017750518.

Zhang, Y., Fei, M., Xue, G., Zhou, Q., Jia, Y., Li, L., Xin, H. and Sun, S., (2012). Elevated levels of hypoxia-inducible microRNA-210 in pre-eclampsia: new insights into molecular mechanisms for the disease. *Journal of Cellular and Molecular Medicine*, *16*(2), 249-259.

Zhang, Y., Zou, Y., Wang, W., Zuo, Q., Jiang, Z., Sun, M., De, W. and Sun, L., (2015). Down-regulated long non-coding RNA MEG3 and its effect on promoting apoptosis and suppressing migration of trophoblast cells. *Journal of Cellular Biochemistry*, *116*(4), 542-550.

Zhao, B., Luo, X., Shi, H. and Ma, D., (2011). Tissue factor pathway inhibitor-2 is downregulated by ox-LDL and inhibits ox-LDL induced vascular smooth muscle cells proliferation and migration. *Thrombosis Research*, *128*(2), 179-185.

Zhao, G., Zhou, X., Chen, S., Miao, H., Fan, H., Wang, Z., Hu, Y. and Hou, Y., (2014). Differential expression of microRNAs in decidua-derived mesenchymal stem cells from patients with pre-eclampsia. *Journal of Biomedical Science*, *21*(1), 81.

Zhao, S. and Liu, M. F., (2009). Mechanisms of microRNA-mediated gene regulation. *Science in China Series C: Life Sciences*, *52*(12), 1111-1116.

Zhao, S., Gu, Y., Fan, R., Groome, L. J., Cooper, D. and Wang, Y., (2010). Proteases and sFlt-1 release in the human placenta. *Placenta*, *31*(6), 512-518.

Zhou, C., Zou, Q. Y., Li, H., Wang, R. F., Liu, A. X., Magness, R. R. and Zheng, J., (2017). Preeclampsia downregulates microRNAs in fetal endothelial cells: roles of miR-29a/c-3p in endothelial function. *The Journal of Clinical Endocrinology & Metabolism*, *102*(9), 3470-3479.

Zhou, Y. and Wu, Q., (2013). Role of corin and atrial natriuretic peptide in preeclampsia. *Placenta*, *34*(2), 89-94.

Zhou, Y., McMaster, M., Woo, K., Janatpour, M., Perry, J., Karpanen, T., Alitalo, K., Damsky, C. and Fisher, S. J., (2002). Vascular endothelial growth factor ligands and receptors that regulate human cytotrophoblast survival are dysregulated in severe preeclampsia and hemolysis, elevated liver enzymes, and low platelets syndrome. *The American Journal of Pathology*, *160*(4), 1405-1423.

Zhou, Y., McMaster, M., Woo, K., Janatpour, M., Perry, J., Karpanen, T., Alitalo, K., Damsky, C. and Fisher, S. J., (2002). Vascular endothelial growth factor ligands and receptors that regulate human cytotrophoblast survival are dysregulated in severe preeclampsia and hemolysis, elevated liver enzymes, and low platelets syndrome. *The American Journal of Pathology*, *160*(4), 1405-1423.

Zhu, X. M., Han, T., Sargent, I. L., Yin, G. W. and Yao, Y. Q., (2009). Differential expression profile of microRNAs in human placentas from preeclamptic pregnancies vs normal pregnancies. *American Journal of Obstetrics and Gynecology*, *200*(6), 661-e1.

Zuo, Q., Huang, S., Zou, Y., Xu, Y., Jiang, Z., Zou, S., Xu, H. and Sun, L., (2016). The Lnc RNA SPRY4-IT1 modulates trophoblast cell invasion and migration by affecting the epithelial-mesenchymal transition. *Scientific Reports*, *6*, 37183.

In: Trends in Biochemistry and Molecular Biology
Editors: Hossain Uddin Shekhar et al.
ISBN: 978-1-53616-434-3

Chapter 10

ROLE OF AUTOPHAGY IN CANCER: MECHANISTIC AND THERAPEUTIC UNDERSTANDING FROM THE CELLULAR AND MOLECULAR POINT OF VIEW

Omar Hamza Bin Manjur[1], Akib Mahmud Khan[1], Hamida Nooreen Mahmood[1], Sohidul Islam[2] and Mahmud Hossain[1,*]

[1]Department of Biochemistry and Molecular Biology, University of Dhaka, Dhaka, Bangladesh

[2]Department of Biochemistry and Microbiology, School of Health & Life Sciences, North South University, Bashundhara, Dhaka, Bangladesh

ABSTRACT

For recycling superfluous or damaged cell components, autophagy is a vital cellular process. It happens as part of a cell's daily duties and as a response to traumatic induction, such as starvation. Autophagy shows dual characters in cancer, performing as both a cell survival shunt in an established tumor and as a tumor suppressor in an initial neoplasm. Here, our role is to glimpse autophagy from diverse physiological aspects as well as explaining their knockout consequences in the mouse model. These outcomes can evidently establish autophagy as a therapeutic target for cancer progression and lead to developing several drugs for early phase clinical trials in human patients. Signifying autophagy as cancer's "Achilles' heel" may indulge us new opportunities for obtaining effective cancer therapy.

[*] Corresponding Author's E-mail: mahmudbio1480@du.ac.bd.

Keywords: autophagy, cancer therapy, PI-3K/AKT pathway, apoptosis, macroautophagy, genome damage, oncogene, chloroquine, hydroxychloroquine, autophagosome

1. Introduction

Autophagy is an essential cellular catabolic breakdown response to stress or starvation where cytoplasm, cellular proteins, and organelles are engulfed, digested and recycled to its primary building blocks, and thus carry on cellular metabolism (Levine and Klionsky 2004; Mizushima et al. 2008). Other than maintaining the basic role in protein and organelles turnover, it has numerous physiological and pathophysiological roles (Levine and Klionsky 2004; Meijer and Codogno 2004). Cells often come across environmental stress inducers such as pathogen infection and nutrient starvation, which triggers autophagy to occur, resulting in either cope up to exist in the physiological system or to initiate the cascade of cell death. It is particularly observed in several pathological situations, such as cancer and neurodegenerative diseases (Malicdan et al. 2008). Advances in the conceptual understanding of this process have made progress after the introduction of the term by Christian de Duve in 1963 (Levine and Klionsky 2004). In this decade, however, it was overly highlighted when the Nobel Prize in Physiology was awarded to Yoshinori Ohsumi in 2016 for his effort to explain the mechanisms. Yet, existing findings represent the ambiguous nature of autophagy in cancer. It is reported that autophagy prevents cancer development at an early stage. On the contrary, once cancer is established, autophagic flux is markedly elevated and enhances tumor cell growth (White 2012; Amaravadi, Kimmelman, and White 2016). Thus, rationally, circumstances demand the most important question in this scenario, whether we should try to increase autophagy or plan to reduce it. In this chapter, our particular importance is to take a deeper look in understanding the role of autophagy in cancer development, progression, and therapeutic strategy.

Until now, the major three types of autophagy—macroautophagy, microautophagy, and chaperone-mediated autophagy (CMA) are well known. The term "autophagy" normally points to macroautophagy unless otherwise specified (Mizushima 2007). The apparatus that runs autophagy is evolutionarily conserved, and a number of their components have primarily been characterized in yeast (He and Klionsky 2009). In usual physiological processes, autophagy keeps on at its basal levels, procures nonstop subtraction of damaged and potentially hazardous entities, including organelles or protein portions, and maintains a quality control system (Green, Galluzzi, and Kroemer 2014). However, it's essential to know the underlying processes first to understand the manipulating consequences.

2. Basic Mechanism of Autophagy

Capability to turn over biomolecules and other large cellular organelles are mandatory to the homeostasis of cellular metabolism. Most of the core molecular tools accountable for autophagosome formation have been recognized over the past decade (Ciechanover 2005). Autophagy is maintained by a highly regulated cluster of signaling procedure. It takes place at a minimal level in almost all cells, and is initiated by various molecular signals as well as simultaneous cellular stresses (Towers and Thorburn 2016). In addition, there may be a significant divergence between basal autophagy and stimulus-induced autophagy, but our understanding of these variations is not conclusive yet.

Interestingly, the frequency of autophagic flux can significantly fluctuate in response to several stimuli, like metabolic, nutritional, oxidative, genotoxic, proteotoxic and pathogenic cues (Kroemer, Mariño, and Levine 2010). This process involves autophagy-related evolutionarily conserved genes (*ATG*s) (Mizushima 2007), and characteristically divided into distinct stages. They are briefly described below-

2.1. Autophagy Initiation

Firstly, ATG1/Unc-51-like kinase (ULK) complex formation occurs, which comprises ULK1 (Unc-51 like Autophagy Activating Kinase1) or ULK2, ATG101, ATG13, and FIP200 (in yeast Atg 17) (**Figure 1**).

2.2. Nucleation of the Vesicular Autophagosome

The ULK complex triggers a class III phosphatidylinositol-3-kinase (PI-3K) complex, containing Vps34 (Vacuolar Protein Sorting 34) (in human it's coded by PIK3C3 gene), Vps15, ATG14, and most importantly, ATG6/BECN1 (Simonsen and Tooze 2009). The PI-3K converts PIP2 (phosphatidylinositol(4,5)-bisphosphate) to PIP3 (Phosphatidylinositol (3,4,5)-trisphosphate), followed by elevated PIP3 level recruits WIPI proteins that employs downstream ATG macromolecules (Proikas-Cezanne et al. 2015). Here, both the ULK complex and the PI-3K III complex control the initiation phases of autophagy and induce phagophore formation. **(Figure 1).**

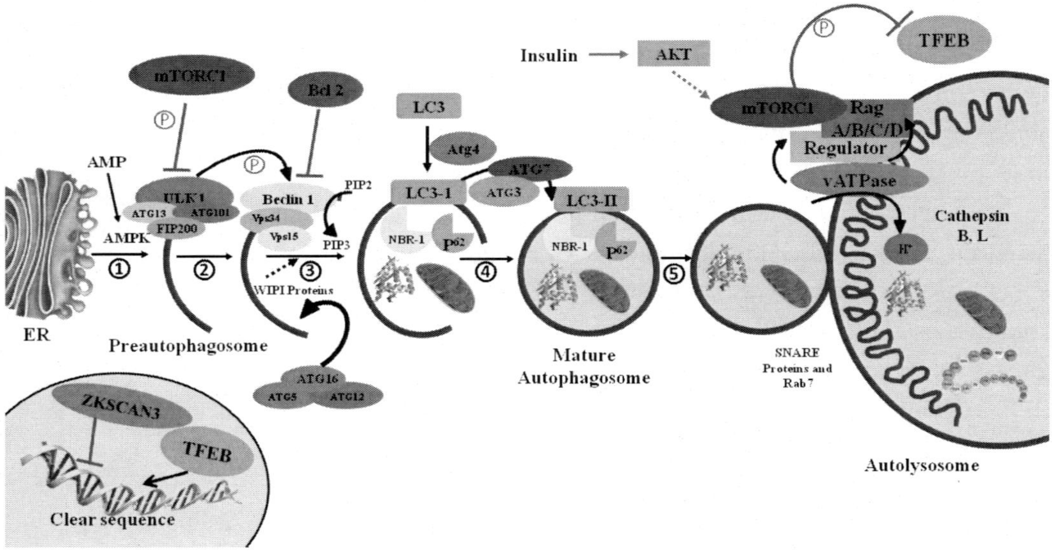

Figure 1. Autophagy mechanism at cellular and molecular level. The process usually starts from Endoplasmic Reticulum (ER) and can be divided into distinct stages. Autophagy initiation is organized by the ULK kinase complex. Sequentially, autophagosome formation can take place from the phagophore and involves Vps34, a PI-3K (class III) that ends up with a complex formation with Beclin1/BECN1. Beclin1 interrelates with factors that modify its binding to Vps34. Autophagosome formation requires the involvement of another two ubiquitin-type protein conjugation systems (ATG12 and LC3). Additionally, the LC3 system is required for autophagosomal transportation and maturation. Mature autophagosomes can easily fuse their outer membranes with lysosomes to maintain cargo catabolism and recycle those components by different cathepsins. In lysosomal membrane, some particular proteins like vesicular ATPase (vATPase) are present and work as a proton pump. It can activate some regulators like Rag proteins by mTOR or TFEB (Transcription Factor EB). mTOR can inhibit TFEB by many mechanisms, one of them is by phosphorylating them. However, if TFEB is not phosphorylated, it's free to bind with the CLEAR sequences in the nucleus. As a transcription factor, TFEB is directly linked to upregulation of autophagy-related genes and lysosomal biogenesis. On the other hand, similar to other transcription factors, ZKSCAN3 plays the opposite role to CLEAR sequence and inhibit autophagy.

2.3. Elongation of the Autophagosomal Bi-Layer

The elongation of phagophore results in the characteristic double-membrane autophagosome development. This step requires two ubiquitin-like conjugation pathways, both primarily catalyzed by ATG7. The first ubiquitin-like system proceeds towards the conjugation of ATG5-ATG12, developing a multimeric complex with another protein, ATG16L. The ATG5-ATG12-ATG16L complex connects with the outer layer of the growing phagophore (Glick, Barth, and Macleod 2010). On the other hand, the second system processes LC3 (also known as MAP1LC3B), encoded by the mammalian homolog of the yeast gene ATG8. In inducing autophagy, LC3B is proteolytically processed by ATG4 to produce LC3B-I (**Figure 1**).

2.4. Closure of the Autophagosome

Resulting LC3B-I is activated by ATG7 homolog, and after that, conjugated to phosphatidylethanolamine (PE) in the membrane portion to create processed LC3B-II. Processed LC3B-II is recruited onto the emerging phagophore and its integration is solely reliant on ATG5-ATG12 (**Figure 1**). The LC3B-II is frequently found on both the inner and outer plane of the autophagosome, which is significantly different from ATG5-ATG12-ATG16L. After the closure of the autophagosome bi-layer, the ATG5-ATG12-ATG16L complex is detached from the vesicle; on the other hand, a segment of LC3B-II remains covalently cogged to the membrane. Hence, it dictates that, LC3B-II can be useful as a biomarker to examine the exact level of autophagy in the cells (He and Klionsky 2009).

2.5. Lysosomal Fusion

After the complete autophagosome formation, LC3B-II is affixed to the outer membrane that is cleaved from phosphatidylethanolamine by ATG4 and returned back to the cell cytosol. The fusion formation between the lysosome and the autophagosome requires LAMP-2 (Lysosome-Associated Membrane Protein 2), the small GTPase Rab7, and different SNARE proteins (**Figure 1**).

2.6. Cargo Catabolism

A number of sequential acid hydrolases participate in the breakdown of the membrane trapped cytoplasmic cargo after the fusion of autophagosome and lysosome. The degraded fragments, particularly amino acids are transferred back to the cell cytosol for further protein synthesis and cellular maintenance.

3. Role of Autophagy in Maintaining Metabolic Features of Cancer

To maintain metabolic homeostasis, autophagy has distinctive features. In the progression of tumor cells, increased metabolic stress is inevitable from oxygen deprivation and nutrient deficiency (Galluzzi et al. 2014). This situation is established by the lack of angiogenesis that leads to the inadequate blood supply in poorly vascularized central tumorigenic region (Jin et al. 2007). Cell-intrinsic stress is solely derived from the high metabolic demand in cell proliferation and inefficient ATP production. Autophagy takes place in hypoxic locations of tumors mostly and maintains a distance from blood vessels, where it helps in tumor cell survival and stress recovery from chemotherapy and radiation (Mathew, Karantza-Wadsworth, and White 2009). On the other hand, peripheral cancer cells carry on to proliferation by maintaining anabolism (Eales, Hollinshead, and Tennant 2016). Transformation of malignancy generally happens with a paradigm shift of cells towards a rapid catabolic uptake of pyruvate (the end product of glycolysis) to maintain an exclusive metabolic pattern. In this case, firstly, glucose consumption is considerably increased to carry on anabolism. Secondly, the mitochondrial respiration rate is elevated to meet up the augmented energy demands. Thirdly, amino acids, mostly serine and glutamine, become vital to adjust with exacerbated metabolic activity (Hanahan and Weinberg 2011; Galluzzi et al. 2014).Finally, p62 degradation by autophagy takes part in a feedback loop to regulate mTORC1 activation in response to adequate nutrient accessibility and tumor progression (Linares et al. 2013).

4. Autophagy Limits Genome Damage

Autophagy-deficient cancer cells can cause genome damage and adaptive tumor generation (Karantza-Wadsworth et al. 2007). This may happen due to protein malfunction, accumulation of toxic protein aggregates, reduction in organelle efficiency, ROS generation, insufficient ATP with low energy homeostasis and lack of proper cellular dynamics required to maintain genome integrity. Upregulation of aneuploidy and DNA damage in autophagy-deficient immortal epithelial cells are highly connected with increased tumorigenesis (Mathew et al. 2007). Autophagy defected abnormal cells are more prone to loss of genome integrity facilitated by cell-cycle checkpoint inhibition (usually through oncogene *RB1* activation and loss of potent tumor suppressor *p53*). Genetic volatility and an increased mutation rate promote the tumor-cell progression, and provides a resistance to chemotherapy and radioactive treatment. Interestingly, autophagy-defective immortal epithelial cells show increased gene amplification rate (Karantza-Wadsworth et al. 2007). From our existing understanding, we know that cancer is a disease with frequent relapsing tendency and when a tumor grows back, it becomes mutated more diversely and emerges with more aggressive features than the original tumor. Though stressful survival with abridged cellular performance is seen by deficient autophagy, superior adaptation through high mutation rate could be a mainstream advantage that enhances tumorigenesis.

5. TARGETING AUTOPHAGY IN CANCER FROM GENETIC STANDPOINT

Initially, autophagy was treated as a cellular defense mechanism that effectively suppresses malignancy and coherently frequent autophagosomal abnormalities are now found to be present in many human tumors. By systemic and tissue-specific knockout of genes involved in autophagy, a significant insight has been gained on the role of autophagy in inducing or inhibiting tumors. Autophagy is a critical process in mammalian adaptation to starvation. Systemic knockouts of most of the *Atg*-related genes in mice model result in embryonic or neonatal lethality, while deficiencies in some of the autophagy genes show no obvious phenotypic abnormality (Kuma, Komatsu, and Mizushima 2017). However, the link between autophagy and cancer has prominently been elucidated through tissue-specific knockout of *Atg*-related genes.

Analysis of *Atg* gene knockout mice has greatly contributed to understanding the physiological roles of autophagy *in vivo*. There are approximately 20 core autophagy-related genes involved in autophagosome formation in mammals. These 'core' Atg genes consist of several functional units; (i) Atg1 kinase and its regulators (Atg1, Atg13, Atg17, Atg29, Atg31), (ii) the phosphatidylinositol 3-kinase (PI-3K) complex (Vps30/Atg6, Atg14), (iii) the Atg12 conjugation system (Atg5, Atg7, Atg10, Atg12, Atg16), (iv) the Atg8 conjugation system (Atg3, Atg4, Atg7, Atg8), (v) Atg9, and (vi) the Atg2-Atg18 complex. So far, 14 of them have been systematically knocked out in mice. Mice deficient in genes functioning upstream of the Atg conjugation system (Atg17-/-, atg13-/-, Atg6-/-, Vps38-/-, Vps34-/-, Atg9a-/-) die *in utero*. Mice deficient in genes involved in the Atg conjugation system (Atg3-/-, Atg5-/-, Atg7-/-, Atg12-/-, Atg16l1-/-) and Ulk1/2 (Atg1) born with normal morphology but die in the following day. Other knockout groups such as- Atg4b-/-, Atg4c-/-, Lc3b-/- (Atg8), Gabarap-/- (Atg8a), Ulk1-/- (Atg1a), and Ulk2-/- (Atg1b) show no noticeable (or weak) phenotypic effects. This is probably because these Atg genes have homologs that function redundantly. Among the viable Atg knockout models, only Atg4c-/- shows increased susceptibility to carcinogen-induced fibrosarcomas. However, substantial knowledge pertaining to the role of ATG genes in cancer has been obtained from tissue-specific deletion of ATG genes (Kuma, Komatsu, and Mizushima 2017).

The mice with systemic mosaic deletion of Atg5 gene in all tissues was created in the Mizushima laboratory which frequently developed benign hepatic adenomas without tumor in any other organ (Takamura et al. 2011). Liver-specific Atg7 ($^{-/-}$) mice also resulted in benign liver tumors in a similar fashion. Another conditional deletion of Atg7 in mice liver developed similar tumor at the Komatsu laboratory (Inami et al. 2011). Notably, tumors from both laboratories were benign liver adenomas and no tendency to develop frank cancer. Mutations in BRAF^{V600E} and KRAS^{G12D} in mice models have been used to develop tissue-specific carcinoma. Lung-specific Atg7 or Atg5 deletions in mice can inhibit BRAF^{V600E}-driven and KRAS^{G12D}-driven lung carcinomas (Strohecker et al.

2013; Rao et al. 2014). Similarly, pancreas restricted knockout of Atg7 or Atg5 also blocks the progression of $KRAS^{G12D}$-driven low-grade pancreatic lesions to high- grade pancreatic carcinoma (Rosenfeldt et al. 2013; A. Yang et al. 2014). Atg7-deficient tumors also activate the Trp53 (Tumor suppressor protein 53; also known as p53) and have less proliferation and more apoptosis, which likely contributes to the significant reduction in tumor burden. In the absence of Trp53, Atg7 deficiency renders the progression of $BRAF^{V600E}$-driven lung tumorigenesis. The results of these experiments suggest that autophagy plays a role in the tumor progression driven by BRAF and KRAS mutations with simultaneous apoptotic deficiency.

Concurrent activation of PI-3K/AKT pathway by acquiring PI-3K alterations, unusual AKT copy number variations, or Phosphatase and Tensin homolog (*PTEN*) loss direct to decreased autophagy in many pathways basically by mTOR activation (Guertin and Sabatini 2007; Díaz-Troya et al. 2008). The PI-3K and mTOR (mammalian target of rapamycin) activation are a very familiar event in most of the human cancers and autophagy is markedly inhibited by mTOR (Kuma et al. 2004). When growth factors and nutrients are abundant, the catabolic demand of autophagy is significantly reduced by mTOR. The activity of the PI-3K pathway and mTOR are turned down in starving stages, which downregulate cell anabolism and proliferation but switch on autophagy (Jin et al. 2007).

Cancer cells gradually develop novel mechanisms for evading chemotherapy-induced apoptosis and autophagy-associated cell death pathways. Key proteins initially considered to be "autophagy-associated proteins" are now to be engaged in either inhibiting or inducing apoptosis. Equally, apoptosis inhibitor proteins can also deregulate autophagy-mediated cell death. Some of the examples are described below-

The BCL-2 (B-cell lymphoma-2) amplification/overexpression is treated as a common alteration in human tumors which potentially inhibit autophagy. Though the lack of proto-oncogenic *BAX* (BCL-2 associated X protein) and *BAK* (Bcl-2 homologous antagonist/killer) gene or the gain of oncogenic *BCL2* or *BCL-XL* (B-cell lymphoma-extra-large) gene function sufficiently concludes why cells systematically unable to apoptosis when severely deprived. It is inadequate to describe how cell viability is retained for weeks during harsh metabolic stress *in vitro* and *in vivo.* It has already been reported that BCL-2 has a significant role to inhibit autophagy, and in some case, through an inhibitory binding to BECN1 (Maiuri et al. 2009; Sinha and Levine 2009). Several studies found that breast cancer cell lines commonly have one allele deletion of BECN1. It is reported that the BECN1 introduction into the MCF7 breast cancer cell line accelerated autophagy and reduced tumorigenicity (Degenhardt et al. 2006). Moreover, BECN1 levels were considerably decreased by 56% breast cancer samples, in comparison with normal epithelial cells from the breast (Degenhardt et al. 2006). Allelic deletion of BECN1 is not only in breast specific, but also found in prostate and ovarian tumors. In the rodent model, *Becn1* (+/–) knockout demonstrated an elevated occurrence of lymphoma, hepatocellular carcinoma, and lung cancer (Mathew et al. 2007; Mathew, Karantza-Wadsworth, and White 2009). The wild-type *BECN1* allele was normally unaffected in both human and mouse tumors, authenticating their haplo-insufficient activity. Heterozygous mice have drastically decreased levels of autophagy compared with wild-type mice; still, they have a considerable autophagic function (Yue et al. 2003). Those heterozygous mice frequently grow a mixture of benign and malignant tumors. Thus, in the context of *BECN1* heterozygosity and considerably reduced autophagy, tumor cells can completely progress (Kimmelman 2011).

The *p53* has a number of mechanisms in showing anticancer activity and it plays a role in genomic stability, apoptosis, and inhibition of angiogenesis. It's one of the most frequently altered genes in human malignancy and has opposing roles in autophagy. Genotoxic stress or nutrient deprivation leads to activation of autophagy by engaging *p53* (Balaburski, Hontz, and Murphy 2010). This can be processed through inhibiting mTOR (Feng et al. 2005) or by activating of DRAM (damage-regulated autophagy modulator) (Crighton et al. 2006). The loss of function in *p53* would then be supposed to take part in decreased autophagy, compatible with the usual role in autophagy as a tumor suppressor gene. Depletion of ULK1, ATG5 or BECN1 in tumor cells tend to gather increased mutant *p53*, whether transgene-driven overexpression of ATG5 or BECN1 results in depletion of mutant *p53* (Rodriguez et al. 2012). In opposing context, verified data at the Kroemer Laboratory also proposed that *p53* can readily suppress autophagy in the basal metabolic state (Tasdemir et al. 2008). Despite in this situation, the suppression is solely reliant on the cytoplasmic pool of *p53*, rather than nuclear one (Tasdemir et al. 2008).

Moreover, profound overexpression of *p62* gene resulting from autophagy inhibition is important in the promotion of tumorigenesis through a wide range of mechanisms, including NF-kB signal inhibition, ROS accumulation, and DNA damage (Mathew, Karantza-Wadsworth, and White 2009). Another mechanism describes that p62 binds to KEAP1 (Kelch-like ECH-associated protein 1), leading to the up-regulation of *NRF2* (also known as NFE2L2), whose activities include the maintenance of an antioxidant defense (Komatsu et al. 2010; Lau et al. 2010). Persistent *NRF2* activation seems to be crucial for the expansion of hepatocellular carcinoma in the context of overexpressing *p62* (Inami et al. 2011). In summary, several reports established that autophagy is engaged in the depletion of oncogenic proteins, including mutant p53 (Rodriguez et al. 2012; Choundhury et al. 2013); p62 (Duran et al. 2008), and BCR-ABL1 (Goussetis et al. 2012).

6. Aberrations of Autophagy Genes Can Lead to Malignant Transformation in Stem Cells

Autophagy acts as a vital part of stem cell maintenance. This is predominantly relevant in hematological malignancies. Deficiency in autophagy causes aberrations in proliferation potential and fragile equilibrium among precursor cells in the bone marrow (Greim et al. 2014). Knockout of Atg7 in rodent hematopoietic stem cells (HSCs) has already been linked to tissue architecture distortion, ultimately consequential expansion with the neoplastic abilities (Mortensen et al. 2011). Similarly, tissue-specific deletion of *FIP200 (Atg17)* gene shrunken the mice fetal HSC exclusive chambers, resulting in drastic anemia and perinatal death (F. Liu et al. 2010). Fascinatingly, *FIP200 (-/-)* HSCs do not show elevated rates of programmed cell death but gained proliferative capacity in the murine model (F. Liu et al. 2010). Postnatal neuronal differentiation abnormalities are also observed in *FIP200* deletion in neuronal stem cells (C. Wang et al. 2013). This effect leads to deregulated redox homeostasis, resulting in activated tumor suppressor *p53*-dependent apoptotic response (C. Wang et al. 2013). Transfection with Atg5 exclusive short-hairpin RNA (shRNA) in human hematopoietic, epidermal and dermal stem cells leads to the reduction in their capability of self-renewal while differentiating into fibroblasts, keratinocytes and neutrophils (Salemi et al. 2012).

7. Ras Mediated Action of Autophagy

Ras is a protein superfamily frequently expressed in all animal cell lines, tissues and organs. It is structurally a small GTPase involved in several signal transduction cascades in cell systems. Ras has a profound role in cell growth, cellular differentiation and tumor progression. Notably, three Ras genes in humans (KRas, HRas, and NRas) are the most prominent oncogenes causing different human cancer (Goodsell 1999). Mutated Ras has been reported in between 20% to 25% of all human cancers and around 90% of cases in pancreatic cancer (Downward 2003). Additionally, several Ras mutated human cancer cell lineages are reported for their dependency on autophagy for tumor growth and survival (S. Yang et al. 2011). Increased baseline autophagy is normally seen in Ras-driven tumor formation (Guo et al. 2011). Sometimes, it's critical in providing necessary metabolic intermediates under the tedious physiological scenario that arises in pancreatic tumor cells (Guo et al. 2011). Previously, it was established that tumorigenesis is lessened in the context of autophagy inhibition. Similarly, mammary epithelial transformation by oncogenic H-Ras and K-Ras exclusively depends on autophagy (Lock et al. 2011; Kim et al. 2011). Furthermore, the genetic constitution and cellular Ras expression level may direct the final biological outcome of autophagy activation. In

HOSE (human ovarian surface epithelial) cells, where elevated H-Ras level leads to a growth seize followed by caspase-induced cell death with features of autophagy (Elgendy et al. 2011). Therefore, the significance of autophagy in driving Ras-induced transformation is evident.

8. Immunological and Microbiological Perspective of Autophagy in Malignant Transformation

Role of autophagy in the different stages of cancer has immunological consequences. Autophagy is important in both innate and adaptive immune response (Ma et al. 2013). These responses are necessary to release an optimal amount of ATP from neoplastic cell death, that recruit antigen presenting cells (APCs) and secrete immuno-stimulatory chemokines (Michaud et al. 2011). In some cases, an inflammatory microenvironment containing a high amount of genotoxic ROS and different mitogenic cytokines is crucial in developing malignant transformation (Coussens, Zitvogel, and Palucka 2013). Both systemic and cancer cell-intrinsic abnormalities in autophagy may confuse the host immune mechanism to selectively recognize and destroy potent malignant cells (Deretic, Saitoh, and Akira 2013). Successful autophagic actions limit inflammation by several pathways- Firstly, autophagy effectively degrades inflammasomes and inhibits secretion of pro-inflammatory Interleukin- 1 beta (IL-1β) and interleukin-18 (IL-18) as well as removes damaged mitochondria, which could secrete inflammasome activators (Nakahira et al. 2011; Zitvogel et al. 2012). Secondly, autophagy is related to the inhibition of some pattern recognition receptor (PRR), such as the RIG-I-like receptors which mediates pro-inflammatory signals (Jounai et al. 2007). Thirdly, it can constrict the relative abundance of BCL10 (B-cell leukemia/lymphoma 10) and inhibit pro-inflammatory NF-kB signaling (Paul et al. 2012). Lastly, it has several mechanisms to inhibit STING (Stimulator of interferon genes) protein, which is involved in pro-inflammatory cue delivery in response to elevated cytosolic nucleic acid concentration (Saitoh et al. 2009).

Autophagy can also suppress mutagenesis by being a part of the first line of immunological defense against microbiological pathogenesis (Deretic, Saitoh, and Akira 2013). Several pathogens with carcinogenic potential effectively activate autophagy for their advantage in the invasion. Those pathogens include, human herpesvirus 8 (in relation to develop Kaposi's sarcoma) (Beral V, Peterman TA, Berkelman RL 1990), hepatitis B virus (related in hepatocellular carcinoma) (Brechot et al. 1980), human papillomavirus type 16 and 18 (exclusive to cervical carcinoma) (Griffin, Cicchini, and Pyeon 2013), and Epstein–Barr virus (may have some role in gastric carcinoma) (Zhang et al. 2014). Bacterial contribution in developing carcinogenesis is not negligible, such as- *Salmonella enterica* (gallbladder carcinoma and colorectal carcinogenesis) (Conway et al. 2013), *Streptococcus bovis* (colorectal carcinoma) (Nakagawa et al. 2004) *Chlamydia pneumoniae* (an etiological determinant of lung cancer) (Yasir et al. 2011), and most prominently, *Helicobacter pylori* (potent role in gastric carcinoma) (Zhang et al. 2014). Xenophagic responses are mandatory for boosting up the clearing process of intracellular pathogens and pathogen- specific immune activity stimulation (Deretic, Saitoh, and Akira 2013; Ma et al. 2013). Therefore, molecular autophagy defects containing epithelial cells, such as mutations in *ATG16L1* and *NOD2* (Nucleotide-binding oligomerization domain-containing protein 2) are relatively more prone to infection by intracellular pathogens (Lassen et al. 2014). A recent study on a cohort of cervical cancer patients suggested that low levels of autophagic markers like BECN1 have been associated with HPV-16 and 18 infections (H. Wang et al. 2014). Thus, autophagy could bring oncosuppressive activity by showing its natural antimicrobial role.

9. Autophagy Mediated Cancer Therapy

Many cancer medications along with ionizing radiation have involved in maintaining the level of autophagy. A diverse group of anti-cancer therapy is used over time, including hormonal agents, anti-metabolites, microtubule-targeted drugs, proteasome inhibitors, kinase inhibitors, death receptor agonists, DNA damaging agents, histone deacetylase inhibitors, and anti-angiogenic agents have all been shown to affect autophagy. Perhaps, no known group of anti-poliferative compounds evidently unresponsive to influence the autophagy.

9.1. Autophagy Inducers

Common cytotoxic medications and isotopic radiation therapy have frequently been reported to enhance autophagy (Amaravadi, Kimmelman, and White 2016). The after-effect of exerting autophagic activity in cancer cells is partially explainable and multifactorial in biochemical and physiological contexts. Increased or durationally sustained autophagy reserves the potential to stimulate tumor cell death.

Among other antineoplastic drugs like imatinib (BCR-ABL tyrosine kinase inhibitor) (Ertmer et al. 2007), cetuximab (the anti–epidermal growth factor receptor) (X. Li and Fan 2010), TRAIL (TNF-related apoptosis inducing ligand) (Han et al. 2008), vorinostat (HDAC inhibitors) (Y.-L. Liu et al. 2010), and OSU-HDAC42 (HDAC inhibitors) (Y.-L. Liu et al. 2010) are reported to induce autophagy **(Figure 2)**. In treating glioma and leukemia via upregulation of mitochondrial stress regulation sensor protein BNIP3 (BCL2/adenovirus E1B 19 kDa protein-interacting protein 3), arsenic trioxide (As_2O_3) was used, and thus, it induces autophagy (Goussetis et al. 2010). Moreover, drugs with a diverse mode of action, including cyclooxygenase inhibitors (Huang and Sinicrope 2010), protease inhibitor (nelfinavir) (Gills, LoPiccolo, and Dennis 2008), and tamoxifen (Selective estrogen receptor modulator) (Bursch et al. 1996) have also been reported to induce autophagy in cancer cells.

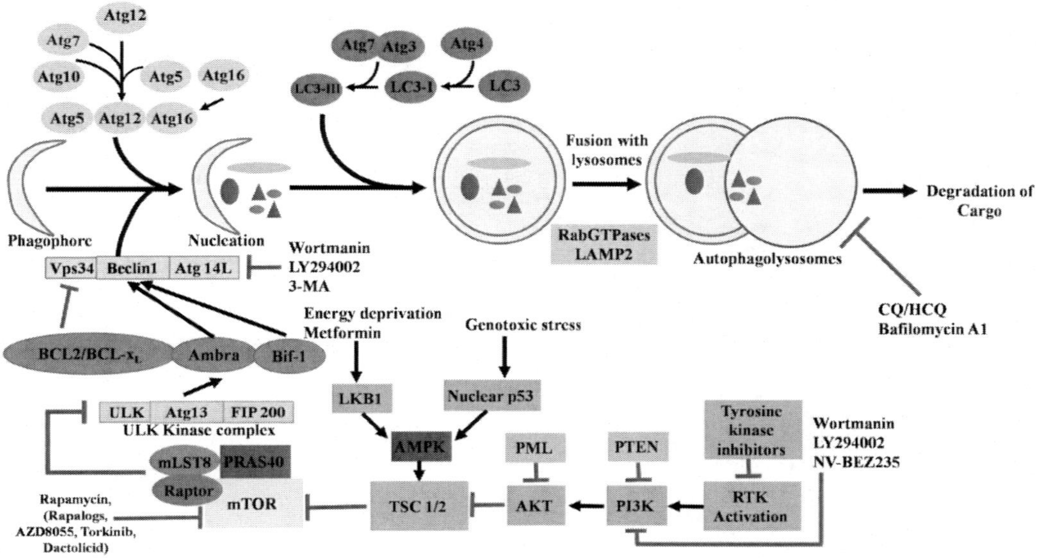

Figure 2. Therapeutic approaches in manipulating autophagy cascade. Autophagy can be upregulated or downregulated depending on needs. Among the autophagy enhancers, Rapamycin allosterically inhibits mTOR cascade, and its artificial analogs exclusively target mTORC1 to elevate autophagy. As their limited success rate, many ATP-competitive inhibitors for both mTORC1 and mTORC2 (Torin1, Torkinib, WYE132 etc.) are developed. Another PI3K-mTOR inhibitor Dactolisib synergized with chloroquine (CQ) to show apoptosis with more rapid action against mTOR. Besides that, more out of the box concept introduces Metformin, a first line antidiabetic type-2 drug, for mTOR signaling inhibition by controlling its upstream mediator, AMPK. Autophagy can also be suppressed by drugs that target early or late stages in Autophagy cascade. Early-stage suppressors include wortmannin, 3-methyladenine and LY294002, which restrict the activity of PI3K class III (VPS34) and inhibit its membrane recruitment for autophagosome elongation. Later stage inhibitors include the antimalarial drugs chloroquine (CQ), hydroxychloroquine (HCQ), monensin, and bafilomycin A1 CQ/HCQ work in inhibiting the autophagolysosome acidification, whose digestive hydrolases works on low pH. Bafilomycin A1 works specifically to prevent vacuolar-ATPase.

The mTOR is central to the cascade mechanism of cell growth and proliferation that has roles in both autophagy and protein translation (Turcotte et al. 2008). Rapamycin (Sirolimus) is a macrolide which allosterically inhibits mTOR and its synthetic analogs such as everolimus, ridaforolimus and temsirolimus exclusively target mTORC1 to upregulate autophagy. Interestingly, rapamycin and rapalogs (rapamycin analogs) have had a limited success rate, exclusively in neuroendocrine carcinomas, renal cell cancers and lymphoma (Meric-Bernstam and Gonzalez-Angulo 2009). They usually do not suppress mTORC2, and are unable to interfere with the S6K-IRS1–mediated negative feedback loop (O'Reilly et al. 2006). These constraints lead to the development of ATP-competitive inhibitors for both mTORC1 and mTORC2 (Torin1, Torkinib, WYE132 etc.). These compounds have shown anticancer activity in pre-clinical settings (Feldman et al. 2009; Chresta et al. 2009), and work as more potent autophagy inducer compared to exclusive mTORC1 inhibitors (Thoreen et al. 2009; Feldman et al. 2009). The activity of dual PI3K-mTOR inhibitor (e.g., PI-103) was shown to induce autophagy in glioma cells by inhibiting their DNA repair mechanism. Additionally, another PI3K-mTOR inhibitor Dactolisib (NVP-BEZ235) synergized with chloroquine (CQ) to show apoptosis in glioma xenograft models in clinical setting (Fan et al. 2010). Metformin, a first line antidiabetic type-2 drug, has been widely reported for mTOR signaling inhibition by its upstream mediator, AMPK, and expresses the cytostatic effect in colon cancer cells (Buzzai et al. 2007). But, in case of prostate cancer cells metformin inhibits 2- deoxyglucose–induced autophagy and reduces *BECN1* expression, and allows cascade towards the cell death (Sahra et al. 2010). Among other stimulators of autophagy include valproic acid (antiepileptic drug) (Fu et al. 2009), SSRI drugs (selective serotonin reuptake inhibitor) such as fluoxetine (Cloonan and Williams 2011), and NERI drugs (norepinephrine reuptake inhibitor) e.g., maprotiline (Cloonan and Williams 2011).

9.2. Autophagy Inhibitors

In preclinical models, autophagy inhibition leads to a better response to DNA alkylating agents in cancer cells (Amaravadi, Kimmelman, and White 2016). It apparently sensitizes apoptosis defective colon cancer and leukemic cell lines to apoptosis (Han et al. 2008). Moreover, inhibition of autophagy by cetuximab, a chimeric antibody against EGFR (Epidermal Growth Factor Receptor) boosted apoptosis in colorectal adeno-carcinoma cells, and vulvar squamous carcinoma cells (X. Li and Fan 2010). Therapeutic autophagy inhibitors can be largely categorized according to their stage of inhibition. Early-stage inhibitors include wortmannin, 3-methyladenine and LY294002, which suppress the activity of PI-3K class III (Vps34) and restrict its membrane recruitment. Later stage inhibitors include the bafilomycin A1, monensin, antimalarial drugs chloroquine (CQ), and hydroxychloroquine (HCQ). Bafilomycin A1 is a macrolide-type antibiotic which specifically inhibits vacuolar-ATPase (Shacka, Klocke, and Roth 2006); whereas, monensin and CQ/HCQ work in preventing the autolysosome acidification, whose digestive hydrolases depends on the low pH. As a part of intracellular transport, microtubules help to move and fuse autophagosomes and lysosomes. Therefore, logically microtubule-disrupting agents broadly taxanes, colchicine, and vincas inhibit autolysosome formation (Dumontet and Jordan 2010). Various anticancer therapies that induce autophagy are listed in **Table 1.**

Table 1. Preclinical and ongoing clinical studies using the autophagy inhibitors Chloroquine (CQ) and Hydroxychloroquine (HCQ) in cancer treatment

Tumor type	Development status	Therapeutic combination	Reference
Breast cancer	Phase II	HCQ only	
Colorectal cancer	*in vitro, in vivo*	CQ + bortezomib	(Cloonan and Williams 2011)
		CQ + vorinostat	(Chresta, Davies et al. 2009)
	Phase II	HCQ + XELOX + bevacizumab	
Chronic lymphocytic leukemia	Phase II	HCQ only	
Chronic myelogenous leukemia	*in vitro*	CQ + vorinostat	(Stein, Lin et al. 2010)
Gastrointestinal stromal tumor	*in vitro, in vivo*	CQ + imatinib	(Ogata, Hino et al. 2006)
Lung cancer	Phase II	HCQ + erlotinib	
Lymphoma	*in vivo*	CQ + cyclophosphamide	(Alexander, Cai et al. 2010)
Multiple myeloma	Phase I/II	HCQ + bortezomib	
Pancreatic cancer	Phase II	HCQ only	
Prostate cancer	*in vitro, in vivo*	CQ + Src kinase inhibitors	(Feldman, Apsel et al. 2009)
Renal cell carcinoma	Phase I	HCQ only	
Vulvar cancer	*in vitro*	CQ + cetuximab	(Shao, Gao et al. 2004)
	Phase II	HCQ + imatinib	

The aptitude of autophagy inhibition to augment chemosensitivity and tumor degeneration has been established in several animal models also. The CQ enhanced with cyclophosphamide leads to autophagy inhibition and cancer cell death in a Myc-induced murine lymphoma model. Similarly, in a colon cancer xenograft murine model, the combination of CQ and vorinostat was reported to considerably reduce tumor size as well as to elevate apoptosis (Carew et al. 2010). Moreover, CQ boosts the therapeutic efficiency of saracatinib (an Src inhibitor) in prostate tumor xenograft rodent model (Wu et al. 2010). Saracatinib plus CQ decreased tumor growth by 64% when the saracatinib-only induced tumor regression is 26% (Wu et al. 2010). Combination of these two drugs suggests that inhibition of autophagy forces cells into apoptosis (Wu et al. 2010). Inhibition of autophagy by nucleoside analog 3-methyladenine upregulate apoptosis in tumor when introduced with 5-fluorouracil in colon cancer xenograft mouse (J. Li et al. 2010). Only CQ and HCQ are evaluated mostly in humans, as they are very common antimalarial and autoimmune drugs. But these drugs can efficiently cross the BBB (blood-brain barrier). Considering side effects, CQ has shown more hazard potential in humans than HCQ (Ruiz-Irastorza et al. 2008). Several phase I and phase II trials are being carried out to measure the different combinatorial performance of HCQ with other cytotoxic drugs in a number of cancer types (**Table 1**). However, the longer half-life of HCQ and micromolar inhibitory concentrations to suppress autophagy can limit its efficacy in human studies and is treated as a major challenge.

Conclusion

Autophagy plays a complex role in cancer and is likely to be dependent on the physiologic and genetic characters of the tumor. We are only in the preliminary phase of discovering the complex relationship between autophagy and cancer, but it seems obvious that autophagy is intensely integrated into the cellular metabolism, concurrent stress response, and cell-death pathways. These responses may usually differ with various cell types and with the type of stress, but effectively reflect the tendency of the mutational occurrences in the cancer cells, mostly by affecting *BECN1* and the universal mTOR pathway. Genetic mutational events and knockout of essential autophagy related genes using mouse models of human cancers have been enlightening and insightful. Current cancer therapeutics indirectly impose metabolic stress, but how these processes are manipulated by the tumor cells for autophagy is not completely known. Therefore, it is vital to distinguish the few tumor cells biochemically with high tumor recovery potential and their connection to tumor dormancy. The complexity of autophagy in cancer can be demonstrated by the recognition of potential action to either promotes or slows down tumorigenesis. Constant activation of autophagy is crucial for the continuous growth of some tumors, helping to both eliminate oxidative stresses and supply key intermediates to carry on cell metabolism. Autophagy can also be stimulated in response to cancer therapeutics where it can perform as a tumor survival mechanism and act to limit drug efficacy. These findings have prompted considerable interest in proposing anti-autophagy therapies as a completely new dimension to cancer management. The regulation and context-dependent biological functions of autophagy should be mastered and simultaneously it is apparent that the modifications in this process will be an attractive area for future cancer therapeutic procedures. Moreover, these topics are especially suitable, with the sustained convergence of a better mechanistic knowledge of how autophagy impose or deny therapeutic responses, both at the intrinsic level of tumor cell and within the host. This convergence will permit us to invent better-targeted approaches to manipulate autophagy and improve clinical outcomes in cancer patients in the future.

References

Amaravadi, Ravi, Alec, C. Kimmelman. & Eileen, White. (2016). "Recent Insights into the Function of Autophagy in Cancer." *Genes & Development*, *30* (17). Cold Spring Harbor Lab, 1913–30.

Balaburski, Gregor M., Robert, D. Hontz. & Maureen, E. Murphy. (2010). "P53 and ARF: Unexpected Players in Autophagy." *Trends in Cell Biology*, *20* (6). Elsevier: 363–69.

Beral, V., Peterman, T. A., Berkelman, R. L. & Jaffe, H. W. (1990). "Kaposi's Sarcoma among Persons with AIDS: A Sexually Transmitted Infection?" *Lancet*, no. 335, 123–28.

Brechot, Christian, Christine, Pourcel, Anne, Louise, Bernadette, Rain. & Pierre, Tiollais. (1980). "Presence of Integrated Hepatitis B Virus DNA Sequences in Cellular DNA of Human Hepatocellular Carcinoma." *Nature*, *286*, (July). Nature Publishing Group: 533. http://dx.doi.org/10.1038/286533a0.

Bursch, Wilfried, Adolf, Ellinger, Harald, Kienzl, Ladislaus, Török, Siyaram, Pandey, Marianna, Sikorska, Roy, Walker. & Rolf, Schulte Hermann. (1996). "Active Cell Death Induced by the Anti-Estrogens Tamoxifen and ICI 164 384 in Human Mammary Carcinoma Cells (MCF-7) in Culture: The Role of Autophagy." *Carcinogenesis*, *17* (8). Oxford University Press: 1595–1607.

Buzzai, Monica., Russell, G. Jones., Ravi, K. Amaravadi., Julian, J. Lum., Ralph, J. DeBerardinis., Fangping, Zhao., Benoit, Viollet. & Craig, B. Thompson. (2007). "Systemic Treatment with the Antidiabetic Drug Metformin Selectively Impairs P53-Deficient Tumor Cell Growth." *Cancer Research*, *67* (14), AACR, 6745–52.

Carew, Jennifer S., Ernest, C. Medina., Juan, A. Esquivel II., Devalingam, Mahalingam., Ronan, Swords., Kevin, Kelly., Hui, Zhang., Peng, Huang., Alain, C. Mita. & Monica, M Mita. (2010). "Autophagy Inhibition Enhances Vorinostat-induced Apoptosis via Ubiquitinated Protein Accumulation." *Journal of Cellular and Molecular Medicine*, *14* (10). Wiley Online Library, 2448–59.

Choundhury, Sujata, Vamsi, Kolukula, Anju, Preet, Chris, Albanese. & Maria, Avantaggiati. (2013). "Dissecting the Pathways That Destabilize Mutant P53: The Proteasome or Autophagy?" *Cell Cycle*, *12* (7). Taylor & Francis, 1022–29.

Chresta, Christine M., Barry, R Davies., Ian, Hickson., Tom, Harding., Sabina, Cosulich., Susan, E Critchlow., John, P Vincent., Rebecca, Ellston., Darren, Jones. & Patrizia, Sini. (2009). "AZD8055 Is a Potent, Selective, and Orally Bioavailable ATP-Competitive Mammalian Target of Rapamycin Kinase Inhibitor with *in Vitro* and *in Vivo* Antitumor Activity." *Cancer Research. AACR*, 8–5472.

Cicchini, Michelle, Rumela, Chakrabarti, Sameera, Kongara, Sandy, Price, Ritu, Nahar, Fred, Lozy, Hua, Zhong, Alexei, Vazquez, Yibin, Kang. & Vassiliki, Karantza. (2014). "Autophagy Regulator BECN1 Suppresses Mammary Tumorigenesis Driven by WNT1 Activation and Following Parity." *Autophagy*, *10* (11). Taylor & Francis, 2036–52.

Ciechanover, Aaron. (2005). "Proteolysis: From the Lysosome to Ubiquitin and the Proteasome." *Nature Reviews Molecular Cell Biology*, *6* (1). Nature Publishing Group, 79.

Cloonan, Suzanne M. & David, Clive Williams. (2011). "The Antidepressants Maprotiline and Fluoxetine Induce Type II Autophagic Cell Death in Drug-resistant Burkitt's Lymphoma." *International Journal of Cancer*, *128* (7). Wiley Online Library, 1712–23.

Conway, Kara L., Petric, Kuballa., Joo–Hye, Song., Khushbu, K. Patel., Adam, B. Castoreno., Omer, H. Yilmaz., Humberto, B. Jijon., Mei, Zhang., Leslie, N. Aldrich. & Eduardo, J. Villablanca. (2013). "Atg16l1 Is Required for Autophagy in Intestinal Epithelial Cells and Protection of Mice from Salmonella Infection." *Gastroenterology*, *145* (6), Elsevier, 1347–57.

Coussens, Lisa M., Laurence, Zitvogel. & Karolina Palucka, A. (2013). "Neutralizing Tumor-Promoting Chronic Inflammation: A Magic Bullet?" *Science*, *339* (6117). American Association for the Advancement of Science, 286–91.Crighton, Diane, Simon, Wilkinson, Jim, O'Prey, Nelofer, Syed., Paul, Smith., Paul, R. Harrison., Milena, Gasco., Ornella, Garrone., Tim, Crook. & Kevin, M. Ryan. (2006). "DRAM, a P53-Induced Modulator of Autophagy, Is Critical for Apoptosis." *Cell*, *126* (1), Elsevier, 121–34.

Degenhardt, Kurt., Robin, Mathew., Brian, Beaudoin., Kevin, Bray., Diana, Anderson., Guanghua, Chen., Chandreyee, Mukherjee., Yufang, Shi., Céline, Gélinas. & Yongjun, Fan. (2006). "Autophagy Promotes Tumor Cell Survival and Restricts Necrosis, Inflammation, and Tumorigenesis." *Cancer Cell*, *10* (1), Elsevier, 51–64.

Deretic, Vojo, Tatsuya, Saitoh. & Shizuo, Akira. (2013). "Autophagy in Infection, Inflammation and Immunity." *Nature Reviews Immunology*, *13* (10), Nature Publishing Group, 722.

Díaz-Troya, Sandra, María, Esther Pérez-Pérez., Francisco, J. Florencio. & José, L. Crespo. (2008). "The Role of TOR in Autophagy Regulation from Yeast to Plants and Mammals." *Autophagy*, *4* (7), Taylor & Francis, 851–65.

Downward, Julian. (2003). "Targeting RAS Signalling Pathways in Cancer Therapy." *Nature Reviews Cancer*, *3* (1), Nature Publishing Group, 11.

Dumontet, Charles. & Mary, Ann Jordan. (2010). "Microtubule-Binding Agents: A Dynamic Field of Cancer Therapeutics." *Nature Reviews Drug Discovery*, *9* (10), Nature Publishing Group, 790.

Duran, Angeles, Juan, F. Linares., Anita, S. Galvez., Kathryn, Wikenheiser, Juana, M Flores, Maria, T Diaz-Meco. & Jorge, Moscat. (2008). "The Signaling Adaptor P62 Is an Important NF-KB Mediator in Tumorigenesis." *Cancer Cell*, *13* (4), Elsevier, 343–54.

Eales, K. L., Hollinshead, K. E. R., & Tennant, D. A. (2016). "Hypoxia and Metabolic Adaptation of Cancer Cells." *Oncogenesis*, *5* (1), Nature Publishing Group, e190.

Elgendy, Mohamed, Clare, Sheridan, Gabriela, Brumatti. & Seamus, J. Martin. (2011). "Oncogenic Ras-Induced Expression of Noxa and Beclin-1 Promotes Autophagic Cell Death and Limits Clonogenic Survival." *Molecular Cell*, *42* (1), Elsevier, 23–35.

Ertmer, A., Huber, V., Gilch, S., Yoshimori, T., Erfle, V., Duyster, J., Elsässer, H. P. & Schätzl, H. M. (2007). "The Anticancer Drug Imatinib Induces Cellular Autophagy." *Leukemia*, *21* (5), Nature Publishing Group, 936.

Fan, Qi-Wen., Christine, Cheng., Chris, Hackett., Morri, Feldman., Benjamin, T. Houseman., Theodore, Nicolaides., Daphne, Haas-Kogan., David James, C., Scott, A. Oakes. & Jayanta, Debnath. (2010). "Akt and Autophagy Cooperate to Promote Survival of Drug-Resistant Glioma." *Sci. Signal.*, *3* (147). American Association for the Advancement of Science: ra81-ra81.

Feldman, Morris E., Beth, Apsel, Aino, Uotila, Robbie, Loewith, Zachary, A. Knight., Davide, Ruggero. & Kevan, M. Shokat. (2009). "Active-Site Inhibitors of MTOR Target Rapamycin-Resistant Outputs of MTORC1 and MTORC2." *PLoS Biology*, *7* (2), Public Library of Science: e1000038.

Feng, Zhaohui, Haiyan, Zhang, Arnold, J. Levine. & Shengkan, Jin. (2005). "The Coordinate Regulation of the P53 and MTOR Pathways in Cells." *Proceedings of the National Academy of Sciences*, *102* (23), National Acad Sciences, 8204–9.

Fu, Jun., Cui-Jie, Shao, Fu-Rong, Chen, Ho-Keung, Ng. & Zhong-Ping, Chen. (2009). "Autophagy Induced by Valproic Acid Is Associated with Oxidative Stress in Glioma Cell Lines." *Neuro-Oncology*, *12* (4). Oxford University Press, 328–40.

Galluzzi, Lorenzo, Federico, Pietrocola, Beth, Levine. & Guido, Kroemer. (2014). "Metabolic Control of Autophagy." *Cell*, *159* (6), Elsevier, 1263–76.

Gills, Joell J., Jaclyn, LoPiccolo. & Phillip, A. Dennis. (2008). "Nelfinavir, a New Anti-Cancer Drug with Pleiotropic Effects and Many Paths to Autophagy." *Autophagy*, *4* (1), Taylor & Francis, 107–9.

Glick, Danielle, Sandra, Barth. & Kay, F. Macleod. (2010). "Autophagy: Cellular and Molecular Mechanisms." *The Journal of Pathology*, *221* (1), Wiley Online Library, 3–12.

Goodsell, David S. (1999). "The Molecular Perspective: The Ras Oncogene." *The Oncologist*, *4* (3), AlphaMed Press, 263–64.

Goussetis, Dennis J., Jessika, K. Altman., Heather, Glaser, Jennifer, L. McNeer., Martin, S. Tallman. & Leonidas, C. Platanias. (2010). "Autophagy Is a Critical Mechanism for the Induction of the Antileukemic Effects of Arsenic Trioxide." *Journal of Biological Chemistry*. ASBMB, jbc-M109.

Goussetis, Dennis J., Elias, Gounaris., Edward, J. Wu., Eliza, Vakana., Bhumika, Sharma., Matthew, Bogyo., Jessica, K. Altman. & Leonidas, C. Platanias. (2012). "Autophagic Degradation of the BCR-ABL Oncoprotein and Generation of Antileukemic Responses by Arsenic Trioxide." *Blood*. Am Soc Hematology, blood-2012.

Green, Douglas R., Lorenzo, Galluzzi. & Guido, Kroemer. (2014). "Metabolic Control of Cell Death." *Science*, *345* (6203). American Association for the Advancement of Science, 1250256.

Greim, Helmut, Debra, A. Kaden., Richard, A. Larson., Christine, M. Palermo., Jerry, M. Rice., David, Ross. & Robert, Snyder. (2014). "The Bone Marrow Niche, Stem Cells, and Leukemia: Impact of Drugs, Chemicals, and the Environment." *Annals of the New York Academy of Sciences*, *1310* (1). Wiley Online Library, 7–31.

Griffin, Laura M., Louis, Cicchini. & Dohun, Pyeon. (2013). "Human Papillomavirus Infection Is Inhibited by Host Autophagy in Primary Human Keratinocytes." *Virology*, *437* (1), Elsevier, 12–19.

Guertin, David A. & David, M. Sabatini. (2007). "Defining the Role of MTOR in Cancer." *Cancer Cell*, *12* (1). Elsevier, 9–22.Guo, Jessie Yanxiang., Hsin-Yi, Chen., Robin, Mathew., Jing, Fan., Anne, M. Strohecker., Gizem, Karsli-Uzunbas., Jurre, J. Kamphorst., Guanghua, Chen., Johanna, M. S. Lemmons. & Vassiliki, Karantza. (2011). "Activated Ras Requires Autophagy to Maintain Oxidative Metabolism and Tumorigenesis." *Genes & Development*. Cold Spring Harbor Lab.

Han, Jie, Wen, Hou, Leslie, A Goldstein, Caisheng, Lu, Donna, B. Stolz., Xiao-Ming, Yin. & Hannah, Rabinowich. (2008). "Involvement of Protective Autophagy in TRAIL Resistance of Apoptosis-Defective Tumor Cells." *Journal of Biological Chemistry*, *283* (28). ASBMB, 19665–77.

Hanahan, Douglas. & Robert, A. Weinberg. (2011). "Hallmarks of Cancer: The next Generation." *Cell*, *144* (5). Elsevier, 646–74.

He, Congcong. & Daniel, J. Klionsky. (2009). "Regulation Mechanisms and Signaling Pathways of Autophagy." *Annual Review of Genetics*, *43*.

Huang, Shengbing. & Frank, Sinicrope. (2010). "Celecoxib-Induced Apoptosis Is Enhanced by ABT-737 and by Inhibition of Autophagy in Human Colorectal Cancer Cells." *Autophagy*, *6* (2). Taylor & Francis, 256–69.

Inami, Yoshihiro., Satoshi, Waguri., Ayako, Sakamoto., Tsuguka, Kouno., Kazuto, Nakada., Okio, Hino., Sumio, Watanabe., Jin, Ando., Manabu, Iwadate. & Masayuki, Yamamoto. (2011). "Persistent Activation of Nrf2 through P62 in Hepatocellular Carcinoma Cells." *The Journal of Cell Biology*, *193* (2). Rockefeller University Press, 275–84.

Jin, Shengkan, Robert, S. DiPaola., Robin, Mathew. & Eileen, White. (2007). "Metabolic Catastrophe as a Means to Cancer Cell Death." *Journal of Cell Science*, *120* (3). The Company of Biologists Ltd, 379–83.

Jounai, Nao., Fumihiko, Takeshita., Kouji, Kobiyama., Asako, Sawano., Atsushi, Miyawaki., Ke-Qin, Xin., Ken, J. Ishii., Taro, Kawai., Shizuo, Akira. & Koichi, Suzuki. (2007). "The Atg5–Atg12 Conjugate Associates with Innate Antiviral Immune Responses." *Proceedings of the National Academy of Sciences*, *104* (35). National Acad Sciences, 14050–55.

Karantza-Wadsworth, Vassiliki, Shyam, Patel, Olga, Kravchuk, Guanghua, Chen, Robin, Mathew, Shengkan, Jin. & Eileen, White. (2007). "Autophagy Mitigates Metabolic

Stress and Genome Damage in Mammary Tumorigenesis." *Genes & Development, 21* (13). Cold Spring Harbor Lab, 1621–35.

Kim, Min-Jung., Soo-Jung, Woo, Chang-Hwan, Yoon, Jae-Seong, Lee, Sungkwan, An., Yung-Hyun, Choi, Sang-Gu, Hwang, Gyesoon, Yoon. & Su-Jae, Lee. (2011). "Involvement of Autophagy in Oncogenic K-Ras-Induced Malignant Cell Transformation." *Journal of Biological Chemistry*. ASBMB, jbc-M110.

Kimmelman, Alec C. (2011). "The Dynamic Nature of Autophagy in Cancer." *Genes & Development, 25* (19). Cold Spring Harbor Lab, 1999–2010.Komatsu, Masaaki, Hirofumi, Kurokawa, Satoshi, Waguri, Keiko, Taguchi, Akira, Kobayashi, Yoshinobu, Ichimura, Yu-Shin, Sou., Izumi, Ueno, Ayako, Sakamoto. & Kit, I. Tong. (2010). "The Selective Autophagy Substrate P62 Activates the Stress Responsive Transcription Factor Nrf2 through Inactivation of Keap1." *Nature Cell Biology, 12* (3). Nature Publishing Group, 213.

Kroemer, Guido, Guillermo, Mariño. & Beth, Levine. (2010). "Autophagy and the Integrated Stress Response." *Molecular Cell, 40* (2). Elsevier, 280–93.

Kuma, Akiko, Masahiko, Hatano, Makoto, Matsui, Akitsugu, Yamamoto, Haruaki, Nakaya, Tamotsu, Yoshimori, Yoshinori, Ohsumi, Takeshi, Tokuhisa. & Noboru, Mizushima. (2004). "The Role of Autophagy during the Early Neonatal Starvation Period." *Nature, 432* (7020). Nature Publishing Group, 1032.

Kuma, Akiko, Masaaki, Komatsu. & Noboru, Mizushima. (2017). "Autophagy-Monitoring and Autophagy-Deficient Mice." *Autophagy, 13* (10). Taylor & Francis. 1619–28.

Lassen, Kara G., Petric, Kuballa., Kara, L. Conway., Khushbu, K. Patel., Christine, E. Becker., Joanna, M. Peloquin., Eduardo, J. Villablanca., Jason, M. Norman., Ta-Chiang, Liu. & Robert, J. Heath. (2014). "Atg16L1 T300A Variant Decreases Selective Autophagy Resulting in Altered Cytokine Signaling and Decreased Antibacterial Defense." *Proceedings of the National Academy of Sciences*. National Acad Sciences, 201407001.

Lau, Alexandria, Xiao-Jun, Wang, Fei, Zhao, Nicole, F. Villeneuve., Tongde, Wu, Tao, Jiang, Zheng, Sun., Eileen, White. & Donna, D. Zhang. (2010). "A Noncanonical Mechanism of Nrf2 Activation by Autophagy Deficiency: Direct Interaction between Keap1 and P62." *Molecular and Cellular Biology, 30* (13). Am Soc Microbiol, 3275–85.

Levine, Beth. & Daniel, J. Klionsky. (2004). "Development by Self-Digestion: Molecular Mechanisms and Biological Functions of Autophagy." *Developmental Cell, 6* (4). Elsevier, 463–77.

Li, Jie, Ni, Hou, Ahmad, Faried, Soichi, Tsutsumi. & Hiroyuki, Kuwano. (2010). "Inhibition of Autophagy Augments 5-Fluorouracil Chemotherapy in Human Colon Cancer *in Vitro* and *in Vivo* Model." *European Journal of Cancer, 46* (10). Elsevier, 1900–1909.

Li, Xinqun. & Zhen, Fan. (2010). "The Epidermal Growth Factor Receptor Antibody Cetuximab Induces Autophagy in Cancer Cells by Downregulating HIF-1α and Bcl-2 and Activating the Beclin 1/HVps34 Complex." *Cancer Research*, *70* (14). AACR, 5942–52.

Linares, Juan F., Angeles, Duran, Tomoko, Yajima, Manolis, Pasparakis, Jorge, Moscat. & Maria, T. Diaz-Meco. (2013). "K63 Polyubiquitination and Activation of MTOR by the P62-TRAF6 Complex in Nutrient-Activated Cells." *Molecular Cell*, *51* (3). Elsevier, 283–96.

Liu, Fei, Jae, Y. Lee., Huijun, Wei, Osamu, Tanabe, James, D. Engel, Sean, J. Morrison. & Jun-Lin, Guan. (2010). "FIP200 Is Required for the Cell-Autonomous Maintenance of Fetal Hematopoietic Stem Cells." *Blood. Am Soc Hematology*, blood-2010.

Liu, Yuan-Ling., Pei-Ming, Yang, Chia-Tung, Shun, Ming-Shiang, Wu, Jing-Ru, Weng. & Ching-Chow, Chen. (2010). "Autophagy Potentiates the Anti-Cancer Effects of the Histone Deacetylase Inhibitors in Hepatocellular Carcinoma." *Autophagy*, *6* (8). Taylor & Francis, 1057–65.

Lock, Rebecca, Srirupa, Roy, Candia, M. Kenific., Judy, S. Su., Eduardo, Salas, Sabrina, and M. Ronen. & Jayanta, Debnath. (2011). "Autophagy Facilitates Glycolysis during Ras-Mediated Oncogenic Transformation." *Molecular Biology of the Cell*, *22* (2). Am Soc Cell Biol, 165–78.

Ma, Yuting, Lorenzo, Galluzzi, Laurence, Zitvogel. & Guido, Kroemer. (2013). "Autophagy and Cellular Immune Responses." *Immunity*, *39* (2). Elsevier, 211–27.

Maiuri, M. C., Tasdemir, E., Criollo, A., Morselli, E., Vicencio, J. M., Carnuccio, R. & Kroemer, G. (2009). "Control of Autophagy by Oncogenes and Tumor Suppressor Genes." *Cell Death and Differentiation*, *16* (1). Nature Publishing Group, 87.

Malicdan, May Christine, Satoru, Noguchi, Ikuya, Nonaka, Paul, Saftig. & Ichizo, Nishino. (2008). "Lysosomal Myopathies: An Excessive Build-up in Autophagosomes Is Too Much to Handle." *Neuromuscular Disorders*, *18* (7). Elsevier, 521–29.

Mathew, Robin, Vassiliki, Karantza-Wadsworth. & Eileen, White. (2009). "Assessing Metabolic Stress and Autophagy Status in Epithelial Tumors." *Methods in Enzymology*, *453*. Elsevier, 53–81.

Mathew, Robin, Sameera, Kongara, Brian, Beaudoin, Cristina, M Karp, Kevin, Bray, Kurt, Degenhardt, Guanghua, Chen, Shengkan, Jin. & Eileen, White. (2007). "Autophagy Suppresses Tumor Progression by Limiting Chromosomal Instability." *Genes & Development*, *21* (11). Cold Spring Harbor Lab, 0.

Meijer, Alfred J. & Patrice, Codogno. (2004). "Regulation and Role of Autophagy in Mammalian Cells." *The International Journal of Biochemistry & Cell Biology*, *36* (12). Elsevier, 2445–62.

Meric-Bernstam, Funda. & Ana, Maria Gonzalez-Angulo. (2009). "Targeting the MTOR Signaling Network for Cancer Therapy." *Journal of Clinical Oncology*, *27* (13). American Society of Clinical Oncology: 2278.

Michaud, Mickaël., Isabelle, Martins., Abdul, Qader Sukkurwala., Sandy, Adjemian., Yuting, Ma., Patrizia, Pellegatti., Shensi, Shen., Oliver, Kepp., Marie, Scoazec., Grégoire, Mignot., Santiago, Rello-Varona., Maximilien, Tailler., Laurie, Menger., Erika, Vacchelli., Lorenzo, Galluzzi., François, Ghiringhelli., Francesco, di Virgilio., Laurence, Zitvogel., & Guido, Kroemer. (2011). "Autophagy-Dependent Anticancer Immune Responses Induced by Chemotherapeutic Agents in Mice." *Science*, *334* (6062), 1573 LP-1577. http://science.sciencemag.org/content/334/6062/1573.abstract.

Mizushima, Noboru. (2007). "Autophagy: Process and Function." *Genes & Development*, *21* (22). Cold Spring Harbor Lab, 2861–73.

Mizushima, Noboru, Beth, Levine, Ana, Maria Cuervo. & Daniel, J Klionsky. (2008). "Autophagy Fights Disease through Cellular Self-Digestion." *Nature*, *451* (7182). Nature Publishing Group, 1069.

Mortensen, Monika., Elizabeth, J Soilleux., Gordana, Djordjevic., Rebecca, Tripp., Michael, Lutteropp., Elham, Sadighi-Akha., Amanda, J. Stranks., Julie, Glanville., Samantha, Knight. & Sten-Eirik, W Jacobsen. (2011). "The Autophagy Protein Atg7 Is Essential for Hematopoietic Stem Cell Maintenance." *Journal of Experimental Medicine*, *208* (3). Rockefeller University Press, 455–67.

Nakagawa, Ichiro., Atsuo, Amano., Noboru, Mizushima., Akitsugu, Yamamoto., Hitomi, Yamaguchi., Takahiro, Kamimoto., Atsuki, Nara., Junko, Funao., Masanobu, Nakata. & Kayoko, Tsuda. (2004). "Autophagy Defends Cells against Invading Group A Streptococcus." *Science*, *306* (5698). American Association for the Advancement of Science, 1037–40.

Nakahira, Kiichi., Jeffrey Adam, Haspel., Vijay, A. K. Rathinam., Seon-Jin, Lee., Tamas, Dolinay., Hilaire, C. Lam., Joshua, A. Englert., Marlene, Rabinovitch., Manuela, Cernadas. & Hong, Pyo Kim. (2011). "Autophagy Proteins Regulate Innate Immune Responses by Inhibiting the Release of Mitochondrial DNA Mediated by the NALP3 Inflammasome." *Nature Immunology*, *12* (3). Nature Publishing Group, 222.O'Reilly, Kathryn E., Fredi, Rojo., Qing-Bai, She., David, Solit., Gordon, B. Mills., Debra, Smith., Heidi, Lane., Francesco, Hofmann., Daniel, J. Hicklin. & Dale, L. Ludwig. (2006). "MTOR Inhibition Induces Upstream Receptor Tyrosine Kinase Signaling and Activates Akt." *Cancer Research*, *66* (3). AACR, 1500–1508.

Paul, Suman, Anuj, K Kashyap, Wei. Jia, You-Wen. He. & Brian, C Schaefer. (2012). "Selective Autophagy of the Adaptor Protein Bcl10 Modulates T Cell Receptor Activation of NF-KB." *Immunity*, *36* (6). Elsevier, 947–58.

Proikas-Cezanne, Tassula, Zsuzsanna, Takacs, Pierre, Dönnes. & Oliver, Kohlbacher. (2015). "WIPI Proteins: Essential PtdIns3P Effectors at the Nascent Autophagosome." *J Cell Sci*. The Company of Biologists Ltd, jcs-146258.

Rao, Shuan., Luigi, Tortola., Thomas, Perlot., Gerald, Wirnsberger., Maria, Novatchkova., Roberto, Nitsch., Peter, Sykacek., Lukas, Frank., Daniel, Schramek. & Vukoslav, Komnenovic. (2014). "A Dual Role for Autophagy in a Murine Model of Lung Cancer." *Nature Communications*, *5*. Nature Publishing Group, 3056.

Rodriguez, Olga Catalina., Sujatra, Choudhury., Vamsi, Kolukula., Eveline, E. Vietsch., Jason, Catania., Anju, Preet., Katherine, Reynoso., Jill, Bargonetti., Anton, Wellstein. & Chris, Albanese. (2012). "Dietary Downregulation of Mutant P53 Levels via Glucose Restriction: Mechanisms and Implications for Tumor Therapy." *Cell Cycle*, *11* (23). Taylor & Francis, 4436–46.

Rosenfeldt, Mathias T., Jim, O'Prey., Jennifer, P. Morton., Colin, Nixon., Gillian, MacKay., Agata, Mrowinska., Amy, Au., Taranjit, Singh Rai., Liang, Zheng. & Rachel, Ridgway. (2013). "P53 Status Determines the Role of Autophagy in Pancreatic Tumour Development." *Nature*, *504* (7479). Nature Publishing Group, 296.

Ruiz-Irastorza, Guillermo, Manuel, Ramos-Casals., Pilar, Brito-Zeron. & Munther, A. Khamashta. (2008). "Clinical Efficacy and Side Effects of Antimalarials in Systemic Lupus Erythematosus: A Systematic Review." *Annals of the Rheumatic Diseases*, December.

Sahra, Issam Ben., Kathiane, Laurent., Sandy, Giuliano., Frédéric, Larbret., Gilles, Ponzio., Pierre, Gounon., Yannick, Le Marchand-Brustel., Sophie, Giorgetti-Peraldi., Mireille, Cormont. & Corine, Bertolotto. (2010). "Targeting Cancer Cell Metabolism: The Combination of Metformin and 2-Deoxyglucose Induces P53-Dependent Apoptosis in Prostate Cancer Cells." *Cancer Research*. AACR, 8–5472.

Saitoh, Tatsuya, Naonobu, Fujita, Takuya, Hayashi, Keigo, Takahara, Takashi, Satoh, Hanna, Lee, Kohichi, Matsunaga, Shun, Kageyama, Hiroko, Omori. & Takeshi, Noda. (2009). "Atg9a Controls DsDNA-Driven Dynamic Translocation of STING and the Innate Immune Response." *Proceedings of the National Academy of Sciences*, *106* (49). National Acad Sciences, 20842–46.

Salemi, Souzan, Shida, Yousefi, Mihai, A. Constantinescu., Martin, F. Fey. & Hans-Uwe, Simon. (2012). "Autophagy Is Required for Self-Renewal and Differentiation of Adult Human Stem Cells." *Cell Research*, *22* (2). Nature Publishing Group, 432.

Shacka, John J., Barbara, J Klocke. & Kevin, A. Roth. (2006). "Autophagy, Bafilomycin and Cell Death: The 'AB-Cs' of Plecomacrolide-Induced Neuroprotection." *Autophagy*, *2* (3). Taylor & Francis, 228–30.

Simonsen, Anne. & Sharon, A. Tooze. (2009). "Coordination of Membrane Events during Autophagy by Multiple Class III PI3-Kinase Complexes." *The Journal of Cell Biology*, *186* (6). Rockefeller University Press, 773–82.

Sinha, S. & Levine, B. (2009). "The Autophagy Effector Beclin 1: A Novel BH3-Only Protein." *Oncogene*, *27* (S1). Nature Publishing Group, S137.

Strohecker, Anne Marie., Jessie, Yanxiang Guo., Gizem, Karsli-Uzunbas., Sandy, M Price, Guanghua, Jim Chen., Robin, Mathew, Martin, McMahon. & Eileen, White. (2013). "Autophagy Sustains Mitochondrial Glutamine Metabolism and Growth of BrafV600E-Driven Lung Tumors." *Cancer Discovery*. AACR, CD-13.

Takamura, Akito., Masaaki, Komatsu., Taichi, Hara., Ayako, Sakamoto., Chieko, Kishi., Satoshi, Waguri., Yoshinobu, Eishi., Okio, Hino., Keiji, Tanaka. & Noboru, Mizushima. (2011). "Autophagy-Deficient Mice Develop Multiple Liver Tumors." *Genes & Development*, *25* (8). Cold Spring Harbor Lab, 795–800.

Tasdemir, Ezgi, Chiara Maiuri, M., Lorenzo, Galluzzi, Ilio, Vitale, Mojgan, Djavaheri-Mergny., Marcello, D'amelio, Alfredo, Criollo, Eugenia, Morselli, Changlian, Zhu. & Francis, Harper. (2008). "Regulation of Autophagy by Cytoplasmic P53." *Nature Cell Biology*, *10* (6). Nature Publishing Group, 676.

Thoreen, Carson C., Seong, A. Kang., Jae, Won Chang., Qingsong, Liu., Jianming, Zhang., Yi, Gao., Laurie, J. Reichling., Taebo, Sim., David, M. Sabatini. & Nathanael, S. Gray. (2009). "An ATP-Competitive MTOR Inhibitor Reveals Rapamycin-Insensitive Functions of MTORC1." *Journal of Biological Chemistry*. ASBMB.

Towers, Christina G. & Andrew, Thorburn. (2016). "Therapeutic Targeting of Autophagy." *EBioMedicine*, *14*. Elsevier, 15–23.

Turcotte, Sandra, Denise, A. Chan., Patrick, D. Sutphin, Michael, P. Hay., William, A. Denny. & Amato, J. Giaccia. (2008). "A Molecule Targeting VHL-Deficient Renal Cell Carcinoma That Induces Autophagy." *Cancer Cell*, *14* (1). Elsevier, 90–102.

Wang, Chenran, Chun-Chi, Liang, Christine Bian, Z., Yuan, Zhu. & Jun-Lin, Guan. (2013). "FIP200 Is Required for Maintenance and Differentiation of Postnatal Neural Stem Cells." *Nature Neuroscience*, *16* (5). Nature Publishing Group, 532.

Wang, Hua-Yi, Gui-Fang, Yang, Yan-Hua, Huang, Qi-Wen, Huang, Jun, Gao, Xian, -Da Zhao., Li-Ming, Huang. & Hong-Lei, Chen. (2014). "Reduced Expression of Autophagy Markers Correlates with High-risk Human Papillomavirus Infection in Human Cervical Squamous Cell Carcinoma." *Oncology Letters*, *8* (4). Spandidos Publications, 1492–98.

White, Eileen. (2012). "Deconvoluting the Context-Dependent Role for Autophagy in Cancer." *Nature Reviews Cancer*, *12* (6). Nature Publishing Group, 401.

Wu, Zhaoju., Pei-Ching, Chang., Joy, C. Yang., Cheng-Ying, Chu., Ling-Yu, Wang., Nien-Tsu, Chen., Ai-Hong, Ma., Sonal, J. Desai., Su, Hao Lo. & Christopher, P. Evans. (2010). "Autophagy Blockade Sensitizes Prostate Cancer Cells towards Src Family Kinase Inhibitors." *Genes & Cancer*, *1* (1). SAGE Publications Sage CA: Los Angeles, CA, 40–49.

Yang, Annan., Rajeshkumar, N. V., Xiaoxu, Wang., Shinichi, Yabuuchi., Brian, M. Alexander., Gerald, C. Chu., Daniel, D. Von Hoff., Anirban, Maitra. & Alec, C. Kimmelman. (2014). "Autophagy Is Critical for Pancreatic Tumor Growth and Progression in Tumors with P53 Alterations." *Cancer Discovery*. AACR.

Yang, Shenghong., Xiaoxu, Wang., Gianmarco, Contino., Marc, Liesa., Ergun, Sahin., Haoqiang, Ying., Alexandra, Bause., Yinghua, Li., Jayne, M. Stommel. & Giacomo, Dell'Antonio. (2011). "Pancreatic Cancers Require Autophagy for Tumor Growth." *Genes & Development*. Cold Spring Harbor Lab.

Yasir, Muhammad, Niseema, D. Pachikara, Xiaofeng, Bao, Zui, Pan. & Huizhou, Fan. (2011). "Regulation of Chlamydial Infection by Host Autophagy and Vacuolar ATPase-Bearing Organelles." *Infection and Immunity. Am Soc Microbiol*, IAI-05308.

Yue, Zhenyu, Shengkan, Jin, Chingwen, Yang, Arnold, J. Levine. & Nathaniel, Heintz. (2003). "Beclin 1, an Autophagy Gene Essential for Early Embryonic Development, Is a Haploinsufficient Tumor Suppressor." *Proceedings of the National Academy of Sciences*, *100* (25). National Acad Sciences, 15077–82.

Zhang, Lin., Joseph, J. Y. Sung., Jun, Yu., Siew, C. Ng., Sunny, H. Wong., Chi, H. Cho., Simon, S. M. Ng., Francis, K. L. Chan. & William, K. K. Wu. (2014). "Xenophagy in Helicobacter Pylori-nd Epstein–Barr Virus-induced Gastric Cancer." *The Journal of Pathology*, *233* (2). Wiley Online Library, 103–12.

Zitvogel, Laurence, Oliver, Kepp, Lorenzo, Galluzzi. & Guido, Kroemer. (2012). "Inflammasomes in Carcinogenesis and Anticancer Immune Responses." *Nature Immunology*, *13* (4). Nature Publishing Group, 343.

In: Trends in Biochemistry and Molecular Biology
Editors: Hossain Uddin Shekhar et al.
ISBN: 978-1-53616-434-3

Chapter 11

INSIGHTS INTO THE BIFIDOBACTERIUM: INTEGRAL MEMBERS OF THE GUT MICROBIAL COMMUNITY

Parag Palit and Farhana Tasnim Chowdhury*

Department of Biochemistry and Molecular Biology,
Faculty of Biological Sciences, University of Dhaka, Dhaka, Bangladesh

ABSTRACT

Bifidobacterium refers to members of the gut microbiota that are classified as lactic acid producing bacteria (LAB) and are among the primary members inhabiting the human gut. In recent years, much focus has been placed on various aspects of this bacterium, illuminating insights into the potential beneficial role it may confer to human health and immunity. Distinct members of the gut microflora belonging to the genus *Bifidobacterium* have been shown to be equipped with innate machinery enabling more efficient usage of oligosaccharides in human breast milk, which in turn offers them with a competitive advantage for colonization in the infant's gut ahead of other potentially virulent microorganisms. Over the years, indiscriminate use of antibiotics along with the introduction of formula feeding to infants has seen a noticeable decline in the bifidobacterial colonization of the human gut. This reduction in the overall abundance of *Bifidobacterium* in the human gut has been strongly associated with a multitude of enteric maladies, including necrotizing enterocolitis, shigellosis and even with severe forms of malnutrition. In more recent time, several species of *Bifidobacterium* are increasingly being considered as newer prospects for supplementation in the form of probiotics. This chapter discusses in details about the roles of Bifidobacterium species on human health with insights into some state of the art research in unveiling deeper knowledge into the role of this particular microflora in the host system.

* Corresponding Author's E-mail: farhanatasnim@du.ac.bd.

1. INTRODUCTION

The gastrointestinal tract constitutes a major natural reservoir harboring and extensive and widespread community of microorganisms. The colon, in particular, is the primary site for microbial colonization with more than 500 species of bacteria residing at the site. The human gut microbiome is dominated by facultative and obligate anaerobes, which include members of the genus of *Bifidobacterium, Eubacterium, Clostridium, Peptococcus, Peptostreptococcus, Ruminococcus, Escherichia, Enterobacter, Enterococcus, Klebsiella, Lactobacillus*, and *Proteus* (Bäckhed et al., 2005). The neonatal GI tract is colonized immediately following the birth and is completed within the next few days. The pattern of colonization has been found to be strongly linked with factors such as the mode of delivery (natural vaginal delivery or caesarian section) (Azad et al., 2013; Cabrera-Rubio et al., 2012; Dominguez-Bello et al., 2010), infant diet (exclusive breastfeeding or infant formula feeding) (Penders, 2017; Underwood et al., 2015), geographical location of the birth (developed country or less developed country) (Fallani et al., 2010), order of birth of the infant and so on.

An integral metabolic role displayed by the gut microbial community involves the fermentation of endogenous mucus produced by the epithelia and of substances that would otherwise be categorized as non-digestible dietary carbohydrates. The host gut microbial community also mediates the salvage process of energy in the form of short-chain fatty acids (SCFA), production of nutrients and metabolites substrates that are readily absorbed while the residing gut flora is supported with energy and required nutrition that enables their continual growth and proliferation (Thursby and Juge, 2017).

Recent works also provide key evidence for the human gut microflora to be associated with host immunomodulation whereby the Toll-like receptors (TLR) have been identified as integral signaling devices of these resident microorganisms (Rooks and Garrett, 2016). Subsequently, aberrations in the composition of the gut microbiota have shown significant links to a number of gastrointestinal diseases including, necrotizing enterocolitis and inflammatory bowel disease (Carding et al., 2015; Lynch and Pedersen, 2016; Walters et al., 2014).

Distinct interest has been focused on species of the *Bifidobacterium* genus, which have been found to be linked to conferring positive effects of host health and immunity. Due to their prospective beneficial roles in the host system and their "GRAS (generally regarded as safe)" status, members of the *Bifidobacterium* genus have been incorporated as live components as supplements to the diet, i.e., - probiotics (Ouwehand et al., 2018). According to the FAO/WHO 2011 definition, probiotics refer to "live microorganisms that when administered in adequate amounts, confer a health benefit to the host" (Hill et al., 2014). Research regarding these health beneficiary bacteria has rocketed during the course of the past decade with many probiotic agents being investigated in many animal-model based and clinical studies.

2. Morphology and Abundance

Bifidobacterium sp. constitutes a distinctively diverse group of bacteria that are among the very first natural colonizers of the gastrointestinal tract (particularly of the colon), mouth, and vagina (Tannock, 1999). The first report on the isolation of *Bifidobacterium* was made by Tissier in 1899 who isolated a colony of branched rod-shaped microorganisms from the stool samples of healthy breastfed infants (Biavati et al., 2000). The colony morphology resembled a "Y-like" or "bifid" structure (Biavati et al., 2000; MITSUOKA, 1984). The 'human' group of *Bifidobacterium* comprises of mainly those that had been isolated from the intestine or from the stool samples of adults or infants. This group includes *Bifidobacterium pseudocatenulatum*, *Bifidobacterium catenulatum*, *Bifidobacterium adolescentis*, *Bifidobacterium longum*, *Bifidobacterium infantis Bifidobacterium breve, Bifidobacterium angulatum* and *Bifidobacterium dentium* (O'Callaghan and van Sinderen, 2016).

Bifidobacteria are Gram-positive, non-motile, carry out the anaerobic mode of respiration. They belong to the family of Actinobacteria but are distinct from other lactic acid producing bacteria including the Firmicutes, a group that comprises of *Lactobacilli, Leuconostoc, Pedicoccus or Lactococcus.* The abundance of *Bifidobacteria* in the colon is about hundred times than that of *Lactobacilli* (Sgorbati et al., 1995). Both the *Firmicutes* and *Bifidobacteria* comprise of a group called the lactic acid bacteria (LAB) (Kaplan and Hutkins, 2000), owing to both the groups sharing similar metabolic functions. The abundance of *Bifidobacteria* is the highest among all genus of microflora in the gut of healthy infants, decreasing to a stable count of 2-14% relative abundance by adulthood, before further decreasing in old age (Tiihonen et al., 2010). Gestational age has been associated with the difference in the inhabiting microbial colony. Pre-term infants have been linked with a higher abundance of *Proteobacteria* and an increase in the count of Staphylococcus and Clostridium with a subsequent lower abundance of *Actinobacter* (Khodayar-Pardo et al., 2014). In contrast, full-term infants have been linked with a higher abundance of *Bifidobacteria* and *Firmicutes* (Khodayar-Pardo et al., 2014).

Several other studies exploring the outcome of breastfeeding with formula feeding found significant association with the differential abundance of *Bifidobaterium* species. While *B. breve, B. bifidum, B. longum* ssp. *longum, and B. longum* spp. *infantis* were to be present in both breast-and formula-fed infants as confirmed by both culture-dependent and molecular techniques, greater prevalence of *B. longum* ssp. *infantis* was found in breastfed infants. On the contrary, the abundance of B. *longum* ssp. *longum* was higher amongst bottle-fed infants (Jost et al., 2014; Laursen et al., 2017; Tannock et al., 2013; Ventura et al., 2012). Consequently, one of the more abundant members of the adult gut microflora, *B. adolescentis*, had only been present in formula-fed infants (Jalanka-Tuovinen et al., 2011; Rajilić-Stojanović et al., 2009).

Bifidobacteria have been found to exhibit a crucial role in the degradation of human milk oligosaccharides (HMOs) (Turroni et al., 2018), creating a clear evolutionary link between them and the other members of the gut microbiota. In particular, *B. longum* ssp. *infantis* has been extensively studied for its ability to degrade and utilize various HMOs.

3. Bifidobacterial Genome

The first publication on a complete *Bifidobacterium* genome was made in 2002 when the genome of *Bifidobacterium animalis* strain DN-173 010 was sequenced. Since then, there has been a steady rise in the number of full *Bifidobacterium* genome sequences, made publicly available. Recent reports suggest that the NCBI database currently possesses a total of 254 *Bifidobacterium* genome sequences, among which there are 61 complete genome sequences. Several species of *Bifidobacteria* including *B. adolescentis, B. animalis, B. breve, B. bifidum, B. longum,* and *B. angulatum,* account for more than one distinct genome sequences (Milani et al., 2016; Schell et al., 2002).

The bifidobacterial genome is circular and spans to a length 2.2 Mb on an average, with considerable degree of variation between the different species. For instance, the genome of *B. indicum* LMG11587 encompasses 1.73 Mb, whereas *B. scardovii* JCM12489 harbors a genome of 3.16 Mb. All *Bifidobacterium* genomes typically code for 52-58 tRNA genes, although the genome of *B. longum* subsp. *infantis* ATCC15697 encodes a total of 79 tRNA genes. A crucial index of genomic stability, the G-C content, ranges from 59.2% in *B. adolescentis* to 64.6% in *B. scardovii.* Though the average number of genes harbored by the genome of *Bifidobacterium* is 1825, the genome of *B. infantis*, the predominant species reported in the gut of healthy breastfed infants, *B. infantis,* the genome is 2.8 Mb in length and consists of 2423 annotated genes, along with a further 91 non-coding RNA genes (Lee and O'Sullivan, 2010; Ventura et al., 2009; Ventura et al., 2004).

Sgorbati et al. in 1982, first reported of the presence of plasmids in the *Bifidobacteria*, where they found 900 out of a total of 1461 *Bifidobacterium* strains to harbor extrachromosomal DNA. These included strains of included *B. longum*, *B. globosum*, *B. asteroides* and *B. indicum* species. Other reports independently made by Iwata and Morishita in 1989 and by Bourget et al. in 1993 found plasmids to be harbored by strains of *B. breve* (Ventura et al., 2004).

4. Carbohydrate Metabolism by Bifidobacterial Species

Only eight glycosyl hydrolases (GHs) directly associated with carbohydrate metabolism are predicted to be encoded by the human genome. As a result, a large and significant portion of the complex dietary carbohydrates remain undigested and end up for their eventual degradation by the microflora in the colon. The human gut microbial community is predicted to possess at least 100 times more genes required for carbohydrate metabolism in comparison to that encoded by the host genome. Thereby, a plethora of non-digestible carbohydrates of both plant and host origin such as: e.g., pectin, hemicellulose, xylans, mucin, and glycosphingolipids; are broken down in the colon (Macfarlane and Gibson, 1997).

In the *Bifidobacterium* genus, hexoses are degraded solely by the fructose-6-phosphate pathway (Pokusaeva et al., 2011; Turroni et al., 2010), which involves the enzyme fructose-6-phosphate phosphoketolase. This enzyme serves as a taxonomic tool for identification of the *Bifidobacterium* genus. Lactic acid and acetic acid are the end products of this hexose degradation. A recent study involving comparative genomics performed on the different *Bifidobacterium* genome sequences unraveled that 5.5% of the core *Bifidobacterium* genomic coding sequences (BifCOGs) was indeed involved with the metabolism of carbohydrates and was conserved in the genus.

The utilization of sucrose is mediated by a gene cluster consisting of three genes, namely: ScrP, which encodes a sucrose phosphorylase; ScR, which is a GalR-LacI-type transcriptional regulator and the sucrose transporter, ScrT. Sucrose is catabolized by sucrose-phosphorylases, *Scr*P yielding glucose-1-phosphate, and fructose in the process. Subsequently, glucose-1-phosphate is converted to glucose-6-phosphate by glucose-6-phosphate isomerase for eventual entry into the fructose-6-phosphate (F6P) shunt. The expression of these genes: *scrP*, *scr*R, *scr*T is induced by sucrose and raffinose and on the other hand, repressed by glucose. Bifidobacteria also possess extracellular enzymes that catalyze the degradation of polysaccharides like amylopectin, amylose, and xylan. Another key enzyme in sugar fermentation is galactosidase, which was identified recently (Khoroshkin et al., 2016; Milani et al., 2015; Turroni et al., 2016).

5. Human Milk Oligosaccharides and Its Importance to *Bifidobacterium*

Human milk contains 5 to 23 g/L of oligosaccharides (HMOs), which are among the most abundant components in human milk, following lactose and lipids. The structures of these oligosaccharides consist of lactose-reducing end that is conjugated with a fucosylated and/or sialylated N-acetyllactosamine residue. This results in human milk containing 200 different oligosaccharide structures of varying size, charge, and sequence.

Although it was presumed that these oligosaccharides had little biochemical importance, the beneficial role of these milk oligosaccharides in the proper growth of infants has been well-established and recognized. Particular HMO structures are suited to exhibit the role of "molecular decoys" to enable the binding of pathogens. In the process, a defensive strategy in inhibiting the binding of pathogens to epithelial cells is executed (Aakko et al., 2017; Bode, 2015; Kobata, 2017; Kulinich and Liu, 2016.

HMOs are built up of both neutral and anionic moieties with these building blocks being any five of the monosaccharide units, being: D-glucose, D-galactose, N-acetylglucos-amine, L-fucose, and N-acetylneuraminic acid.

The first report on HMOs dates back to about five decades back when Gyorgy et al. identified N-acetyl-glucosamine containing oligosaccharides, which they termed as the "Bifidus factor." More recent studies indicate *Bifidobacterium* and *Bacteroides* species as being distinguished by their capability in degradation and subsequent consumption of HMOs as the sole source of carbon. These findings were a result of studies conducted among members of the gut microbiota including *Lactobacillus, Clostridium, Eubacterium, E. coli*, *Veillonella,* and *Enterococcus* isolates (Plaza-Díaz et al., 2018).

However, the ability to consume and grow on solely HMOs as the carbon source is not intrinsic to all members of the *Bifidobacterium* genus. This is evident from reports made Xiao et al. which showed/that an integral HMO component in type I glycans, Lacto-N-biose (LNB), only facilitated the growth of *B. bifidum, B. breve, and B. longum (subsp. infantis and longum)* but did not support the growth of *B. adolescentis, B. animalis, B. catenulatum, B. dentium, B. angulatum, and B. pseudolongum* (Odamaki et al., 2015).

6. The Genetic Blueprint for Milk Utilization

B. infantis is distinctly idiosyncratic in comparison to other members of the gut microflora through its remarkable capability to degrade the human milk oligosaccharides (HMOs) and subsequently utilize the by-products of the HMO catabolism. This unique ability is acquired through the possession of a wide array of genes encoding a unique and distinct synchronized assemblage of glycosidases and oligosaccharide transporters. Fucose or sialic acid is common to most HMO structures. Amid all members of the *Bifidobacterium* genus, the ability to generate fucosidases and sialidases is only intrinsic to *B. infantis*, *B. breve*, and *B. bifidum*, with only *B. infantis* being to digest capable of consuming all HMO structures (Underwood et al., 2015).

Preliminary findings made by Garrido et al. showed a distinct group of solute binding proteins *B. infantis* that displayed a marked affinity for mammalian glycan oligosaccharides (Garrido et al., 2016; Garrido et al., 2015). Subsequent studies done by Ewaschuk et al. illustrated that these "soluble factors" produced by *B. infantis* led to a rise in the expression of occludin, a primary tight junction protein in the human intestinal epithelium, thus indicating a potential improvement in gut barrier function (Ewaschuk et al., 2008).

These reports were validated by the discovery of a novel 43 kbp gene cluster in *B. infantis*, which was found to be dedicated towards the import and processing of HMO. This 30-gene cluster, not found in any other *Bifidobacterium* species constitutes of 4-glycosyl hydrolases that are predicted to cleave the HMO structures and its component monosaccharides (Sela et al., 2008).

Additionally, there are two ABC transport permeases along with the associated ATPase in the 5'-terminal of the gene orientation in the cluster. Two more permease pairs occupy the central part of this gene cluster. A total of seven extracellular solute binding proteins (SBP1 family) are prognostic of binding to oligosaccharides to enable interactions with the outer cell wall permeases, components of the ABC transporter complex. Six of these HMO cluster lipoproteins were found to display an exceptional evolutionary divergence in relation to the other SBP1 family proteins from the three completely sequenced bifidobacteria (Sela et al., 2008)

In brief, the entire process of HMO utilization in *B. infantis* starts with transporter proteins binding specifically to particular linkages in the HMO structures, followed by the transport of intact HMOs into the cytoplasm. This phenomenon suggests that the entire HMO utilization gene cluster is "switched on" in the presence of HMOs. Subsequently, glycosidases specific for all possible linkages in the HMO structures are up-regulated in *B. infantis* (Sela et al., 2008).

B. infantis also distinguishes itself from *B. bifidum* and *Bacteroides* in the context of explicitness in the affinity for other glycans. *B. infantis* is unable to degrade the O-glycans in human mucus, unlike *B. bifidum* and *Bacteroides* which are able to deconstruct the glycan linkages in both HMOs and mucus. Through the expression of extracellular glycosyl hydrolases, *B. bifidum* and *Bacteroides* species degrade these complex glycans in the extracellular region, enabling the entry of certain glycan products into the cell and their subsequent consumption while other moieties of the degraded glycans are left outside the cell (Sela et al., 2008; Underwood et al., 2015).

7. Bifidobacterium as Probiotics

Alarming rise in the number of pathogens resistant to conventionally used antibiotics has led to the quest for alternate agents for countering pathogenic infestation. Photo-pharmacotherapy and probiotics in conjugation with prebiotics have become a means of primordial replacement for the indiscriminate conventional use of antibiotics. Probiotics refer to dietary supplements consisting of non-pathogenic or rather beneficial microorganisms to the health of the host. A wide range of commercially available probiotics is available, containing various bacterial species and strains. Majority of these bacteria are selected not on the known basis upon which they confer benefits to the host but rather on the factors related to the ease of large scale production. Henceforth, there is a marked necessity to conduct well-designed clinical studies on the commercially available probiotics to assess their safety and efficacy, particularly in high-risk groups such as the immune-compromised individuals and neonates.

Bifidobacteria are reported to exhibit immunomodulatory roles at both species and strain levels. Isolates of *B. adolescentis* and *B. longum* subsp. *infantis* obtained from human breast milk have been found to exert overwhelming immunomodulatory effects via mechanisms involving Th1/Th2 associated cytokine regulation. This, in turn, raises the scope for their incorporation in probiotic formulation for infants (Sarkar and Mandal, 2016).

Bifidobacteria associated products are increasingly being used in challenging settings such as the neonatal intensive care units, owing to their status as being safe and tolerable upon administration. One particular study, designed as a partially randomized, controlled 2 months clinical trial aiming to assess the safety and tolerability of *B. infantis* supplementation in healthy full-term infants, found no adverse effects of the supplementation along with increased colonization of *B. infantis* in the guts of the infants receiving the supplementation compared to the control group (Smilowitz et al., 2017).

The presence of *Bifidobacterium* in the gut of healthy breast-fed infants has been shown to confer a plethora of health-related benefits to the infant. In one study conducted among Bangladeshi infants, *Bifidobacterium* dominated gastrointestinal tracts were found to be strongly associated with improved vaccine responses (Huda et al., 2014). Moreover, other studies have shown that a higher *Bifidobacterium* load was linked to greater resistance to infestation by a multitude of pathogens and enhanced gut barrier function (Bron et al., 2017; Fukuda et al., 2011; Ling et al., 2016). A study comparing the efficacy of different *Bifidobacterium* species demonstrated that *B. infantis* was more efficient in the colonization of the premature gut when compared to *B. lactis*, particularly in the presence of human milk. Moreover, *B. infantis* was also associated with reduced intestinal permeability and anti-inflammatory effects in the premature gut (Bron et al., 2017; Fukuda et al., 2011; Gareau et al., 2010; Huda et al., 2014; Ling et al., 2016; Sarkar and Mandal, 2016; Smilowitz et al., 2017; Underwood et al., 2015). In more recent times, a clinical trial involving *B. infantis* supplementation in healthy breast-fed infants showed that the colonization of *B. infantis* persisted over a period of 30 days post supplementation. Higher fecal concentrations of lactate and acetate and a consequently lowered fecal pH were observed in the infants receiving the supplementation, thus indicating alterations in intestinal fermentation (Henrick et al., 2018).

CONCLUSION

Bifidobacterium species confer beneficial effects on the health and immunity of the host through its colonization in the host GI tract and subsequent metabolism. Successful attempts in sequencing and annotation of the complete *Bifidobacterium* genomes have enabled the identification of the precise mechanisms underlying their unique metabolic activities. While the knowledge on carbohydrate utilization allows the identification of novel and efficacious prebiotic compounds, complete annotation and characterization of the genes in the carbohydrate utilization pathway would require attempts of site-directed mutagenesis of candidate genes. However, most of the identified *Bifidobacterium* species have been recalcitrant to genetic modification. Hence, future studies would require addressing the development of new and effective molecular tools in unraveling the exact nature of the interaction of *Bifidobacterium* with the host system.

REFERENCES

Aakko, J., Kumar, H., Rautava, S., Wise, A., Autran, C., Bode, L., Isolauri, E., and Salminen, S. (2017). Human milk oligosaccharide categories define the microbiota composition in human colostrum. *Beneficial Microbes 8*, 563-567.

Azad, M.B., Konya, T., Maughan, H., Guttman, D.S., Field, C.J., Chari, R.S., Sears, M.R., Becker, A.B., Scott, J.A., and Kozyrskyj, A.L. (2013). Gut microbiota of healthy Canadian infants: profiles by mode of delivery and infant diet at 4 months. *CMAJ 185*, 385-394.

Bäckhed, F., Ley, R.E., Sonnenburg, J.L., Peterson, D.A., and Gordon, J.I. (2005). Host-bacterial mutualism in the human intestine. *Science 307*, 1915-1920.

Biavati, B., Vescovo, M., Torriani, S., and Bottazzi, V. (2000). Bifidobacteria: history, ecology, physiology and applications. *Annals of Microbiology 50*, 117-132.

Bode, L. (2015). The functional biology of human milk oligosaccharides. *Early human development 91*, 619-622.

Bron, P.A., Kleerebezem, M., Brummer, R.-J., Cani, P.D., Mercenier, A., MacDonald, T.T., Garcia-Ródenas, C.L., and Wells, J.M. (2017). Can probiotics modulate human disease by impacting intestinal barrier function? *British Journal of Nutrition 117*, 93-107.

Cabrera-Rubio, R., Collado, M.C., Laitinen, K., Salminen, S., Isolauri, E., and Mira, A. (2012). The human milk microbiome changes over lactation and is shaped by maternal weight and mode of delivery. *The American journal of clinical nutrition 96*, 544-551.

Carding, S., Verbeke, K., Vipond, D.T., Corfe, B.M., and Owen, L.J. (2015). Dysbiosis of the gut microbiota in disease. Microbial ecology in health and disease *26*, 26191.

Dominguez-Bello, M.G., Costello, E.K., Contreras, M., Magris, M., Hidalgo, G., Fierer, N., and Knight, R. (2010). Delivery mode shapes the acquisition and structure of the initial microbiota across multiple body habitats in newborns. *Proceedings of the National Academy of Sciences 107*, 11971-11975.

Ewaschuk, J.B., Diaz, H., Meddings, L., Diederichs, B., Dmytrash, A., Backer, J., Looijer-van Langen, M., and Madsen, K.L. (2008). Secreted bioactive factors from Bifidobacterium infantis enhance epithelial cell barrier function. *American Journal of Physiology-Gastrointestinal and Liver Physiology 295*, G1025-G1034.

Fallani, M., Young, D., Scott, J., Norin, E., Amarri, S., Adam, R., Aguilera, M., Khanna, S., Gil, A., and Edwards, C.A. (2010). Intestinal microbiota of 6-week-old infants across Europe: geographic influence beyond delivery mode, breast-feeding, and antibiotics. *Journal of pediatric gastroenterology and nutrition 51*, 77-84.

Fukuda, S., Toh, H., Hase, K., Oshima, K., Nakanishi, Y., Yoshimura, K., Tobe, T., Clarke, J.M., Topping, D.L., and Suzuki, T. (2011). Bifidobacteria can protect from enteropathogenic infection through production of acetate. *Nature 469*, 543.

Gareau, M.G., Sherman, P.M., and Walker, W.A. (2010). Probiotics and the gut microbiota in intestinal health and disease. *Nature reviews Gastroenterology & hepatology 7*, 503.

Garrido, D., Ruiz-Moyano, S., Kirmiz, N., Davis, J.C., Totten, S.M., Lemay, D.G., Ugalde, J.A., German, J.B., Lebrilla, C.B., and Mills, D.A. (2016). A novel gene cluster allows preferential utilization of fucosylated milk oligosaccharides in Bifidobacterium longum subsp. longum SC596. *Scientific reports 6*, 35045.

Garrido, D., Ruiz-Moyano, S., Lemay, D.G., Sela, D.A., German, J.B., and Mills, D.A. (2015). Comparative transcriptomics reveals key differences in the response to milk oligosaccharides of infant gut-associated bifidobacteria. *Scientific reports 5*, 13517.

Henrick, B.M., Hutton, A.A., Palumbo, M.C., Casaburi, G., Mitchell, R.D., Underwood, M.A., Smilowitz, J.T., and Frese, S.A. (2018). Elevated fecal pH indicates a profound change in the breastfed infant gut microbiome due to reduction of Bifidobacterium over the past century. *MSphere 3*, e00041-00018.

Hill, C., Guarner, F., Reid, G., Gibson, G.R., Merenstein, D.J., Pot, B., Morelli, L., Canani, R.B., Flint, H.J., and Salminen, S. (2014). Expert consensus document: The International Scientific Association for Probiotics and Prebiotics consensus statement on the scope and appropriate use of the term probiotic. *Nature reviews Gastroenterology & hepatology 11*, 506.

Huda, M.N., Lewis, Z., Kalanetra, K.M., Rashid, M., Ahmad, S.M., Raqib, R., Qadri, F., Underwood, M.A., Mills, D.A., and Stephensen, C.B. (2014). Stool microbiota and vaccine responses of infants. *Pediatrics 134*, e362-e372.

Jalanka-Tuovinen, J., Salonen, A., Nikkilä, J., Immonen, O., Kekkonen, R., Lahti, L., Palva, A., and de Vos, W.M. (2011). Intestinal microbiota in healthy adults: temporal analysis reveals individual and common core and relation to intestinal symptoms. *PloS one 6*, e23035.

Jost, T., Lacroix, C., Braegger, C.P., Rochat, F., and Chassard, C. (2014). Vertical mother–neonate transfer of maternal gut bacteria via breastfeeding. Environmental *Microbiology 16*, 2891-2904.

Kaplan, H., and Hutkins, R.W. (2000). Fermentation of fructooligosaccharides by lactic acid bacteria and bifidobacteria. *Appl Environ Microbiol 66*, 2682-2684.

Khodayar-Pardo, P., Mira-Pascual, L., Collado, M., and Martinez-Costa, C. (2014). Impact of lactation stage, gestational age and mode of delivery on breast milk microbiota. *Journal of Perinatology 34*, 599.

Khoroshkin, M.S., Leyn, S.A., Van Sinderen, D., and Rodionov, D.A. (2016). Transcriptional regulation of carbohydrate utilization pathways in the Bifidobacterium genus. *Frontiers in microbiology 7*, 120.

Kobata, A. (2017). Structures, classification, and biosynthesis of human milk oligosaccharides. *Prebiotics and Probiotics in Human Milk*, 17-44.

Kulinich, A., and Liu, L. (2016). Human milk oligosaccharides: The role in the fine-tuning of innate immune responses. *Carbohydrate research 432*, 62-70.

Laursen, M.F., Bahl, M.I., Michaelsen, K.F., and Licht, T.R. (2017). First foods and gut microbes. *Frontiers in microbiology 8*, 356.

Lee, J.-H., and O'Sullivan, D.J. (2010). Genomic insights into bifidobacteria. *Microbiol Mol Biol Rev 74*, 378-416.

Ling, X., Linglong, P., Weixia, D., and Hong, W. (2016). Protective effects of bifidobacterium on intestinal barrier function in LPS-induced enterocyte barrier injury of Caco-2 monolayers and in a rat NEC model. *PloS one 11*, e0161635.

Lynch, S.V., and Pedersen, O. (2016). The human intestinal microbiome in health and disease. *New England Journal of Medicine 375*, 2369-2379.

Macfarlane, G.T., and Gibson, G.R. (1997). Carbohydrate fermentation, energy transduction and gas metabolism in the human large intestine. In *Gastrointestinal microbiology* (Springer), pp. 269-318.

Milani, C., Lugli, G.A., Duranti, S., Turroni, F., Mancabelli, L., Ferrario, C., Mangifesta, M., Hevia, A., Viappiani, A., and Scholz, M. (2015). Bifidobacteria exhibit social behavior through carbohydrate resource sharing in the gut. *Scientific reports 5*, 15782.

Milani, C., Turroni, F., Duranti, S., Lugli, G.A., Mancabelli, L., Ferrario, C., van Sinderen, D., and Ventura, M. (2016). Genomics of the genus Bifidobacterium reveals species-specific adaptation to the glycan-rich gut environment. *Appl Environ Microbiol 82*, 980-991.

Mitsuoka, T. (1984). Taxonomy and ecology of bifidobacteria. *Bifidobacteria and Microflora 3*, 11-28.

O'Callaghan, A., and van Sinderen, D. (2016). Bifidobacteria and their role as members of the human gut microbiota. *Frontiers in microbiology 7*, 925.

Odamaki, T., Horigome, A., Sugahara, H., Hashikura, N., Minami, J., Xiao, J.-Z., and Abe, F. (2015). Comparative genomics revealed genetic diversity and species/strain-level differences in carbohydrate metabolism of three probiotic bifidobacterial species. *International journal of genomics 2015*.

Ouwehand, A.C., Sherwin, S., Sindelar, C., Smith, A.B., and Stahl, B. (2018). Production of Probiotic Bifidobacteria. In *The Bifidobacteria and Related Organisms* (Elsevier), pp. 261-269.

Penders, J. (2017). Chapter 5 Early diet and the infant gut microbiome: how breastfeeding and solid foods shape the microbiome. In *Microbiota in health and disease: from pregnancy to childhood* (Wageningen Academic Publishers), pp. 1281-1292.

Plaza-Díaz, J., Fontana, L., and Gil, A. (2018). Human Milk Oligosaccharides and Immune System Development. *Nutrients 10*, 1038.

Pokusaeva, K., Fitzgerald, G.F., and van Sinderen, D. (2011). Carbohydrate metabolism in Bifidobacteria. *Genes & Nutrition 6*, 285.

Rajilić-Stojanović, M., Heilig, H.G., Molenaar, D., Kajander, K., Surakka, A., Smidt, H., and De Vos, W.M. (2009). Development and application of the human intestinal tract chip, a phylogenetic microarray: analysis of universally conserved phylotypes in the abundant microbiota of young and elderly adults. *Environmental Microbiology 11*, 1736-1751.

Rooks, M.G., and Garrett, W.S. (2016). Gut microbiota, metabolites and host immunity. *Nature Reviews Immunology 16*, 341.

Sarkar, A., and Mandal, S. (2016). Bifidobacteria—Insight into clinical outcomes and mechanisms of its probiotic action. *Microbiological Research 192*, 159-171.

Schell, M.A., Karmirantzou, M., Snel, B., Vilanova, D., Berger, B., Pessi, G., Zwahlen, M.-C., Desiere, F., Bork, P., and Delley, M. (2002). The genome sequence of Bifidobacterium longum reflects its adaptation to the human gastrointestinal tract. *Proceedings of the National Academy of Sciences 99*, 14422-14427.

Sela, D., Chapman, J., Adeuya, A., Kim, J., Chen, F., Whitehead, T., Lapidus, A., Rokhsar, D., Lebrilla, C.B., and German, J. (2008). The genome sequence of Bifidobacterium longum subsp. infantis reveals adaptations for milk utilization within the infant microbiome. *Proceedings of the National Academy of Sciences 105*, 18964-18969.

Sgorbati, B., Biavati, B., and Palenzona, D. (1995). The genus Bifidobacterium. In *The genera of lactic acid bacteria* (Springer), pp. 279-306.

Smilowitz, J.T., Moya, J., Breck, M.A., Cook, C., Fineberg, A., Angkustsiri, K., and Underwood, M.A. (2017). Safety and tolerability of Bifidobacterium longum subspecies infantis EVC001 supplementation in healthy term breastfed infants: a phase I clinical trial. *BMC pediatrics 17*, 133.

Tannock, G.W. (1999). Identification of lactobacilli and bifidobacteria. *Curr Issues Mol Biol 1*, 53-64.

Tannock, G.W., Lawley, B., Munro, K., Pathmanathan, S.G., Zhou, S.J., Makrides, M., Gibson, R.A., Sullivan, T., Prosser, C.G., and Lowry, D. (2013). Comparison of the compositions of the stool microbiotas of infants fed goat milk formula, cow milk-based formula, or breast milk. *Appl Environ Microbiol 79*, 3040-3048.

Thursby, E., and Juge, N. (2017). Introduction to the human gut microbiota. Biochemical Journal *474*, 1823-1836.

Tiihonen, K., Ouwehand, A.C., and Rautonen, N. (2010). Human intestinal microbiota and healthy ageing. *Ageing research reviews 9*, 107-116.

Turroni, F., Bottacini, F., Foroni, E., Mulder, I., Kim, J.-H., Zomer, A., Sánchez, B., Bidossi, A., Ferrarini, A., and Giubellini, V. (2010). Genome analysis of Bifidobacterium bifidum PRL2010 reveals metabolic pathways for host-derived glycan foraging. *Proceedings of the National Academy of Sciences 107*, 19514-19519.

Turroni, F., Milani, C., Duranti, S., Mahony, J., van Sinderen, D., and Ventura, M. (2018). Glycan utilization and cross-feeding activities by bifidobacteria. *Trends in microbiology 26*, 339-350.

Turroni, F., Milani, C., Duranti, S., Mancabelli, L., Mangifesta, M., Viappiani, A., Lugli, G.A., Ferrario, C., Gioiosa, L., and Ferrarini, A. (2016). Deciphering bifidobacterial-mediated metabolic interactions and their impact on gut microbiota by a multi-omics approach. *The ISME journal 10*, 1656.

Underwood, M.A., German, J.B., Lebrilla, C.B., and Mills, D.A. (2015). Bifidobacterium longum subspecies infantis: champion colonizer of the infant gut. *Pediatric research 77*, 229.

Ventura, M., O'flaherty, S., Claesson, M.J., Turroni, F., Klaenhammer, T.R., Van Sinderen, D., and O'toole, P.W. (2009). Genome-scale analyses of health-promoting bacteria: probiogenomics. *Nature Reviews Microbiology 7*, 61.

Ventura, M., Turroni, F., Motherway, M.O.C., MacSharry, J., and van Sinderen, D. (2012). Host–microbe interactions that facilitate gut colonization by commensal bifidobacteria. *Trends in microbiology 20*, 467-476.

Ventura, M., van Sinderen, D., Fitzgerald, G.F., and Zink, R. (2004). Insights into the taxonomy, genetics and physiology of bifidobacteria. *Antonie van Leeuwenhoek 86*, 205-223.

Walters, W.A., Xu, Z., and Knight, R. (2014). Meta-analyses of human gut microbes associated with obesity and IBD. *FEBS letters 588*, 4223-4233.

In: Trends in Biochemistry and Molecular Biology
Editors: Hossain Uddin Shekhar et al.
ISBN: 978-1-53616-434-3

Chapter 12

COORDINATION OF MOLECULAR TECHNIQUES AND ADVANCED BIOINFORMATICS TOOLS FOR ANALYZING VIRAL EVOLUTIONARY DISTANCE WITH AN EMPHASIS ON FOOT-AND-MOUTH DISEASE VIRUS (FMDV) SEROTYPE O

***Salma Akter*[1,2], *Mohammad Anwar Siddique*[1], *A. S. M. Rubayet Ul Alam*[1,2], *Munawar Sultana*[1,*] *and M. Anwar Hossain*[1,3,*]**

[1]Department of Microbiology, University of Dhaka, Dhaka, Bangladesh
[2]Department of Microbiology, Jahangirnagar University, Savar, Bangladesh
[3]Jashore University of Science and Technology, Jashore, Bangladesh

ABSTRACT

Detection of evolutionary changes in the genome of pathogenic viruses plays a crucial role in the development of preventive vaccines and drugs. Modern bioinformatics study enables *in silico* detection of sequence diversity withinsurface antigens and the effect of mutational changes on translated structural, enzymatic and regulatory proteins. These coordinated analyses can reveal the prominent target sites on the capsid regions of virus for designing vaccines; that can accelerate antiviral immune responses in the host. Therefore, modern approaches integrate high throughput functional genomics and bioinformatics tools for the detection of mutations, recombination, evolutionary distance as well as identification of novel strains and lineages of dominant serotypes of the particular viral pathogen. The resultant data can be implemented to combat the

[*] Corresponding Author's E-mail: hossaina@du.ac.bd and munawar@du.ac.bd.

devastating animal diseases. In this chapter, a sequence and bioinformatics based coordinated study of Foot and Mouth Disease Virus (FMDV) serotype O will be focused on discovering the emergence of two novel sub-lineages. This platform will serve as an example for coordinating molecular techniques with advanced bioinformatics for analyzing viral evolutionary distances necessary for vaccine development.

Keywords: evolutionary distance, bioinformatics, emergence, Foot and Mouth Disease Virus, sub-lineage

1. Introduction

Many viral diseases of both humans and animals are emerging and reemerging in different parts of the world. The world has become a global village and the rapid changes in population dynamics and migration may introduce a new or a rare disease to a new population. For example, Severe Acute Respiratory Syndrome (SARS) virus was introduced in South Asia through a single infected human coming from Mainland China to Hongkong (Chowell, Fenimore et al. 2003). Also, there is prominent evidence of the emergence of new strains of influenza viruses due to genetic re-assortment during the infection cycle within the host body.

Several animal and vector-borne viruses are prone to mutations that may result in severe damage within the host. Avian influenza virus, Foot and Mouth Disease Virus (FMDV), Chikungunya virus, Dengue virus, Nipah virus, Zika virus, etc. are the most dangerous viral pathogens affecting livestock and human.

Viruses with an RNA genome, both positive sense and negative sense, are more prone to acquiring mutations in their genome during replication within host cells. The lack of proofreading activity of RNA dependent RNA polymerase, responsible for replication of viral RNA, causes the synthesis of new sequences of the viral genome. Among these sequences, some escape mutations take place, enabling the new progeny virus to spoof the protective immune system and cause severe damage to the host body. Thus, new strains, lineages and sub-lineages can be developed within a single serotype of the virus. The serotype determinants with surface antigens can remain unchanged, while the other regions of protective antigens can be subjected to antigenic drift through a positive selection of escape mutations. Sometimes, multiple frequent occurrences of pathogenic strains within the same area or same population can cause the emergence of a new lineage, sub-lineage, even strains of the virus as a result of a blending of multiple strains during coinfection within the same host.

The viral pathogens, which are likely to be transmitted across the country within an animal population, can be mutated and dominated by the new strain when available vaccines fail to develop immunity against the new strains. In 2018, a report (Siddique, Ali et al. 2018) from Bangladesh revealed the emergence of two novel sub-lineages in

Foot and Mouth Disease Virus (FMDV) serotype O, which resulted in the high incidence of Foot and Mouth Disease (FMD) among the cattle population of the country. The report was generated from exclusive field isolates of virus from infected animals followed by cultivation in cell culture and detection of genotypic characteristics through molecular and bioinformatics tools. Due to error-prone replication of RNA and the continuous evolution potential of FMDV, one of the dominant FMDV serotypes - type O - will be highlighted in this chapter. The workflow will involve both molecular and bioinformatics techniques for understanding the recent mechanisms for studying viral evolutionary genetics.

2. Classification of Foot and Mouth Disease Virus Types

Understanding the evolution of divergent viruses, especially RNA viruses, by accurate classification into genotypes is critical. Classifying these virus strains into genotypes or subtypes based on their sequence similarities from the multiple sequence alignment (MSA) has become the state-of-the-art technique in understanding their evolution, epidemiology and developing antiviral therapies or vaccines (Kim, Ahn et al. 2010). The conventional classification methods include the following: (1) the nearest neighbor methods that search for the best match of the query to the representatives of each genotype, so-called references; and (2) the phylogenetic methods that look for the monophyletic group to which the query branches. However, with the increasing numbers of sequences, a few outliers are observed that cannot be clearly classified(Kim, Ahn et al. 2010).

Foot-and-Mouth Disease (FMD) is the first reported case of the tie between a virus and an animal disease; and is caused by a single-stranded positive sense RNA genome, FMD Virus (FMDV). This devastating disease infects the domesticated and wild, cloven-hoofed animals which cause substantial economic losses worldwide in terms of production and trades. This virus exists as seven serotypes (A, O, C, Asia1 and SAT 1-3), each divided into topotypes, genetic lineages, sub-lineages, and strains; and also distributed worldwide differentially. Mainly, serotypes O, A and Asia1 are circulating in Asian territory with the dominance of serotype O along with co-circulation of serotype A and Asia1. The O/ME-SA/Ind2001 lineage was the most prominent lineage for serotype O in Asia during 2015-2018, with its first intrusions into Saudi Arabia-UAE-Bahrain (2013-2015), Vietnam-Laos-Myanmar (2015-2016) and China (2017), along with its continued dominance in the Indian subcontinent.

3. SEQUENCE BASED EVOLUTION OF FMDV SEROTYPE O

Serotype O is the most prevalent of the seven serotypes of FMDV and the genetic diversity of serotype O is much greater, allowing the classification of many distinct topotypes and lineages (Samuel and Knowles 2001). The cut-off value of 15% nucleotide difference was used as a standard to identify the topotypes that fell within different geographical boundaries (Vosloo, Knowles et al. 1992, Samuel and Knowles 2001) and if the difference among the sequences was at least 7.5%, then the isolates were considered as a distinct lineage_(Mohapatra, Sanyal et al. 2002).

Globally, in serotype O, 11 geographically restricted topotypes namely Europe-South America (E-SA), Middle East-South Asia (ME-SA), South East Asia (SEA), Cathay (China and East Tartary), Indonesia (ISA)- 1, ISA-2, East Africa (EA)-1, EA-2, EA-3, EA-4, and West Africa have been identified so far (Samuel and Knowles 2001, Ayelet, Mahapatra et al. 2009).

The serotype O isolates collected in the Indian sub-continent were found to be of various lineages within the Middle East-South Asia (ME-SA) topotype (Hemadri, Tosh et al. 2002). So far, there were four lineages under ME-SA topotype namely Ind2001, Ind2011 PanAsia, and PanAsia2 (Knowles, Samuel et al. 2001, Hemadri, Tosh et al. 2002, Subramaniam, Sanyal et al. 2013). Hemadri, Tosh, Sanyal and Venkataramanan reported the first appearance of Ind2001 lineage in 2001(Hemadri, Tosh et al. 2002). The pandemic PanAsia strain was detected as early as 1982 in India and was responsible for most of the outbreaks between 1996 and 2008 being established in South Asia as the most dominant lineage (Knowles, Samuel et al. 2001, Hemadri, Tosh et al. 2002, Subramaniam, Pattnaik et al. 2013). On the other hand, the Ind2001 lineage after causing sporadic cases during 2003–2005, re-surged in 2008 and outcompeted the dominant PanAsia lineage in 2009(Das, Sanyal et al. 2012). The study led by Subramaniam, Sanyal, Mohapatra, Sharma, Biswal, Ranjan, Rout, Das, Bisht, Mathapati, Dash and Pattnaik (Subramaniam, Sanyal et al. 2013) revealed emergence of a new genetic group in 2011, henceforth named as Ind2011 after a gap of 10 years since the identification of Ind2001 lineage. Ind2011 lineage is so far restricted to the southern region in the states of Karnataka, Tamilnadu, Andhra Pradesh and Kerala (Subramaniam, Sanyal et al. 2013).

FMDV viruses of Ind2001 lineage accounted for most of the current type O outbreaks within Indian sub-continent (Suroowan, Javeed et al. 2017). This lineage was found in Nepal, Bangladesh, Bhutan and India on a regular basis but has never been reported from Pakistan (Jamal, Ferrari et al. 2011). In spite of initially identified in the year of 1997 (Valdazo-González, Knowles et al. 2014), Ind2001 lineage was thought to get emerged from variants of PanAsia strains in 2001 (Hemadri, Tosh et al. 2002) and has later diversified into four sub-lineages (Ind2001a, b, c and d) (Subramaniam, Mohapatra et al. 2015). A distinct sub-lineage circulating before 2001 was designated as Ind2001a in compliance with the WRL report. The sub-lineages Ind2001b and Ind2001d include isolates collected during 2001-2002 and 2008-2013, respectively. Interestingly, only a few isolates collected from UAE in 2008 (e.g., UAE/4/2008) represented Ind2001c sub-lineage (Subramaniam, Mohapatra et al. 2015).

4. Genetic Analysis of Antigenic Sites on Capsid: Basis for FMDV Serotype O Identification

In total, there are five reported, experimentally proved antigenic sites on the capsid of FMDV serotype O. The antigenic site 1 consists of both linear and conformational epitopes formed by G-H loop (residues 140-160 or 137-155 or 130-160) and C terminus (residues 200-213 or 190-203) of VP1 (Xie, McCahon et al. 1987, Kitson, McCAHON et al. 1990, Bai, Bao et al. 2010). Within antigenic site 1, positions 144, 146, 147, 148, 154, 206 and 208 have been reported to be antigenically critical (Xie, McCahon et al. 1987, Parry, Barnett et al. 1989, Kitson, McCAHON et al. 1990, Aktas and Samuel 2000). The antigenic site 2 covers the B-C loop (31, 70-80) and the adjacent E-F loop (132-135 or 131-134) of VP2 (Mateu, Hernández et al. 1994, Bai, Bao et al. 2010). Residues of B-C loop (43-59) of VP1 forms the antigenic site 3 where residues 43 and 44 are the most crucial for antigenicity (Aggarwal and Barnett 2002, Bai, Bao et al. 2010). B-B knob (58-61) of VP3 has been reported to be critical for site 4 wherein position 58 is thought to be most important in antigenicity (Xie, McCahon et al. 1987, Kitson, McCAHON et al. 1990, Mateu, Hernández et al. 1994, Aktas and Samuel 2000). Antigenic site 5 is formed by position 149 of VP1 along with other surface located amino acids of G-H loop (Crowther, Farias et al. 1993). All the other identified sites are conformational and trypsin resistant except the antigenic site 1 which is linear and trypsin sensitive. Other as yet undefined sites might also be important in the induction of a protective immune response in addition to the known antigenic sites (Aggarwal and Barnett 2002).

For analyzing the genetic evolution of diverse strains within a serotype of FMDV along with their route of transmission throughout the continent, several bioinformatics tools have been used with an exclusive prediction of the emergence of new lineages of any particular topotype. Besides bioinformatics analysis with a phylogenetic study, it is important to reveal the phylodynamics of newly emerged viral strains and their phylogeography for proper understanding of the viral evolution. Recent emergence of two sub-lineages of FMDV serotype O, namely Ind2001BD1 and Ind2001BD2, under Ind2001 lineage has been reported from Bangladesh (Siddique, Ali et al. 2018) which exclusively revealed the phylodynamic and phylogeographic distribution of newly emerged FMDV strains throughout South Asian territory.

5. Pipeline for FMDV Identification and Bioinformatics Analyses

To identify circulatory FMDV strains, tongue epithelial tissue from FMD suspected (having signature signs and symptoms) cattle are collected. After transportation of the tissue sample to the laboratory having proper biosafety containment, the RNA extraction is executed by using a variety of approaches like an automated extraction system, column-based kit, etc. Immediately after the RNA extraction, the next step of essence is Reverse Transcription for the conversion of the RNA to cDNA. A portion of the extracted RNA and cDNA might be stored at -80°C for further use.

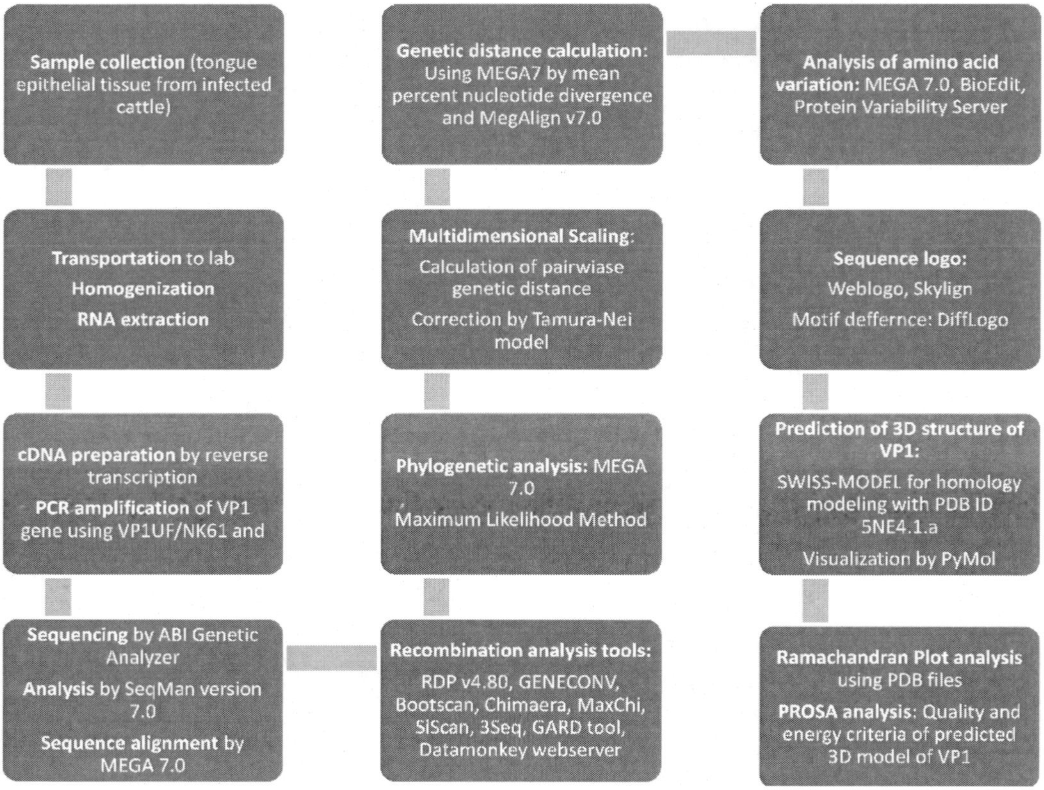

Figure 1. Schematic diagram of workflow from sample collection to Ramachandran plot analysis for coordinated detection of sequence variability within VP1 gene of FMDV.

The presence of the FMDV in the sample of interest, Polymerase Chain Reaction (PCR) targeting the VP portion (preferably the VP1- the most antigenically important site and flanking regions) is performed. The presence of signature amplicon after PCR implies the presence of FMDV in the sample of interest or not.

For sequence-based analyses of the FMDV, the automated cycle sequencing of the amplicon is performed. After successful sequencing of the amplicon, the very first approach is to do BLAST- Basic Local Alignment Search Tool for the serotype level identification of the FMDV. After that, it is recommended to clean the raw sequences for further use based on the alignment pattern as well as raw sequence file.

To uncover the lineage or even sub-lineage level identification of the FMDV, the most important step is to generate a rational dataset covering the significantly enough representative sequences from each lineage and even sub-lineages. After that, phylogenetic analyses using the dataset having the representative sequences from each lineage and even sub-lineages and query sequences are done either Distance-based methods (Neighbor-Joining) or Character-based methods (Maximum likelihood). The Character-based methods are considered powerful to infer than that of Distance-based methods. For sub-lineage level identity confirmation and coining of new sub-lineage within a lineage, the Multi-dimensional scaling is performed.

Since FMDV is an RNA virus it is necessary to check the pattern of mutations throughout the genome or at least the genomic hotspots considering the antigenic profile. Based on the purpose of the study, it is recommended to compare the amino acid sequence of the FMDV in a study with respective vaccine strain in use to map the change in the antigenic site over time and coin the tentative reason for vaccine failure. To further confirm the changes in the gene or genome, the homology modeling of the protein of interest is performed. The overall process is illustrated in Figure 1.

6. Functional Genomics and Computational Analysis of FMDV Serotype O in Bangladesh: An Overview of Sequential Experiment

Throughout the following segments, FMDV serotype O-sequence and computer-based assay performed by our group of Microbial genetics and bioinformatics laboratory in the Department of Microbiology, University of Dhaka is highlighted as a consistent example of the coordinated assay.

6.1. Phylogenetic Analyses Based on VP1 Coding Region

All the FMDV sequences of Bangladesh, sequenced by our group of Microbial genetics and bioinformatics laboratory in the Department of Microbiology, University of Dhaka, were taken into account. From other FMDV sequences of Bangladesh, the representative sequences were taken after screening to reconstruc the phylogenetic tree. Maximum Likelihood (Felsenstein 1981) method was implied for inferring the evolutionary history of each serotype taken in this study. Evolutionary analyses conducted in MEGA7 (Kumar, Stecher et al. 2016) using the Maximum Likelihood method based on the Tamura-Nei model (Tamura and Nei 1993) respectively. For FMDV serotype O, a discrete Gamma distribution of the value of 1.13 was used to model evolutionary rate differences among the sites. It should be noted that the substitution model and parameters of the method can be changed based on variation in the dataset.

6.2. Multidimensional Scaling (MDS) for Clade Visualization

Multidimensional scaling (often abbreviated to MDS) can be defined as the search for a low dimensional space, usually Euclidean, in which points in the space represent the objects wherein one point representing one object. Multidimensional scaling is designed to construct a diagram showing the relationships between groups of sequences, given a matrix of distances between the sequences. Classical multidimensional scaling (MDS) maps the sequences to a high-dimensional principal coordinate space while trying to preserve the distance relationships among them as much as possible (Cox and Reid 2000). Distance matrix of 135 FMDV serotype O sequences was created using MEGA7 by calculating pairwise genetic distances after correcting it by Tamura-Nei (1993) model (Tamura and Nei 1993). Using the classical multidimensional scaling (MDS) method (cmd scale in R version 3.3.2), pairwise genetic distance was converted into Euclidean distances in two-dimensional space which was plotted using ggplot2 (Lemey, Rambaut et al. 2009) (Team, R. C. R. et el 2013).

6.3. Determination of Genetic Distances among Groups of FMDV O

Based on the reference sequence; the topotypes, lineages and sub-lineages need to be chosen and then based on the genetic variation among those groups calculated by MEGA 7 (Kumar, Stecher et al. 2016), the genetic distance between different groups were measured. On a broad spectrum, sequences from Ind2001 and Ind2011 lineages along with sub-lineage Ind2001BD1/Ind2001BD2 were taken for finding out the credibility of Ind2001BD1/BD2 to be a lineage itself. Afterward, sequences of all the established sub-lineages of Ind2001 along with two newly emerged sub-lineages, Ind2001BD1 and Ind2001BD2, were taken into consideration for measuring the genetic distance among those isolates.

Using MegAlign version 7.0 (DNASTAR, Inc., Madison, WI, USA), mean percent nucleotide identity and divergence were calculated taken one representative isolate having average diversity within cluster from the three sub-lineages (Ind2001d, Ind2001BD1 and Ind2001BD2) circulating in Bangladesh. The mean nucleotide divergence within each sub-lineages present in Bangladesh was calculated with the implementation of Seaview v4.6 (Gouy, Guindon et al. 2009).

6.4. Selection Analysis

The synonymous (dS) and non-synonymous (dN) changes at every codon of VP1 coding sequences were estimated using four different selective pressure analyses implemented in DataMonkey (Pond and Frost 2005, Delport, Poon et al. 2010). The analyses were Single Likelihood Ancestor Counting (SLAC), Fixed Effects Likelihood (FEL), Internal Fixed Effects Likelihood (IFEL) and Mixed Effects Model of Evolution (MEME). For nucleotide substitution, the best fit model with the least AIC number was taken for datasets of each serotype. The significance level (p-value) for every analysis implied here was 0.1. The synonymous rate exceeding the non-synonymous rate was considered as negative selection (dS>dN), while the positive selection was defined as when the non-synonymous rate exceeds the synonymous rate (dN>dS) (Kimura 1977, Yang and Bielawski 2000). Neutral selection was defined when the non-synonymous rate equals to the synonymous rate (dN = dS) (Kimura 1977).

6.5. Amino Acid Variation of Bangladeshi Serotype O Isolates

Most variation of the amino acids leading to antigenic changes occurs within the capsid coding regions of the genome of which the VP1 region was the most variable. This helps in providing FMDV with a selective advantage for evading host immune system and helping in survive (Carrillo, Tulman et al. 2005). Viral protein 1 (VP1) contains three of the five neutralizing antigenic sites of the capsid region (site I, III and V) in serotype O (Xie, McCahon et al. 1987, Kitson, McCAHON et al. 1990, Aktas and Samuel 2000). Therefore, constant monitoring of FMDV field strain capsid coding regions especially focused on VP1 to identify the variants must be maintained regularly in endemic regions.

Two types of amino acid variations were checked for the Bangladeshi isolates. An overall variation was checked in the protein level for all the isolates of serotype O. The second type was more precise – the amino acid variation found from the Bangladeshi serotype O isolates under each sub-lineages of Ind2001 lineage.

6.6. Protein Variability Analysis

The coding sequences of the circulatory FMDVs serotype O in Bangladesh, spanning more than 50% of the VP1 region genome were taken to check protein variability. The amino acid was translated based on the standard genetic code after codon based alignment performed with MEGA7 software (Kumar, Stecher et al. 2016). Protein Variability Server (PVS) (Garcia-Boronat, Diez-Rivero et al. 2008) was used for finding out protein variability index using the Wu-Kabat variability coefficient (Kabat, Wu et al. 1977) of the aligned protein sequences. The protein variability index computed using the defined formula was used to determine whether the predicted epitopes were positioned in the least variable, moderately variable, or hypervariable regions. Using the PVS server, a consensus sequence was also derived from this alignment that also finds out the variable site with a threshold level of less than or equal to one. The consensus sequence utilized a VP1 protein sequence that can be derived from the most commonly identified amino acid residues situated at each position.

6.7. Sub-Lineage Specific Variations

Taken BAN/1/2009 (Accession no. HQ630676) as the reference sequence, the variations among the sub-lineages (Ind2001d, Ind2001BD1 and Ind2001BD2) along with the inter sub-lineage virus isolate - BAN/DH/Dh-216/2015 (KY077608) were checked using MEGA7. For a better view of the mutations present among the three sub-lineages, DiffLogo (Nettling, Treutler et al. 2015) were used for visualizing the motif differences at particular antigenic region (C-terminal) of VP1.

6.8. Prediction of 3D Structures of VP1 Coding Region

On the basis of major mutation in VP1 isolated from Bangladesh, the protein modeling was performed taken PDB id 5NE4.1.A (Kotecha, Wang et al. 2017) as template. Four targets having complete VP1 coding region- three strains from each sub-lineage were chosen for homology modeling. From three sub-lineages namely Ind2001BD1, Ind2001BD2 and Ind2001d, isolates BAN/GO/Ka-236(pig)/2015, BAN/BO/Na-161/2013 and BAN/JA/Me-180/2013 were taken.

SWISS-MODEL server (Biasini, Bienert et al. 2014) was implied for generating the protein model and then PyMol (Schrodinger, L, 2016) was used for visualization of the PDB file. Using PDB files of targets, Ramachandran plot (Ramachandran 1963) was generated to find out whether the residues were in one of the three regions-favored, allowed and outlier using RAMPAGE (Lovell, Davis et al. 2003). To evaluate the quality and energy criteria of 3D structures, PROSA (Wiederstein and Sippl 2007) was used. Using PyMol, the predicted VP1 structures of each selected Bangladesh isolates within two novel sub-lineages (Ind2001BD1 and Ind2001BD2) were superimposed on the VP1 structure of the isolate previously identified sub-lineage (Ind2001d).

6.9. Viral Phylodynamics with an Emphasis on Foot-and-Mouth Disease Virus (FMDV)

Viral phylodynamics is the study of how epidemiological, immunological, and evolutionary processes act and potentially interact to shape viral phylogenies to shed light on how these dynamics impact viral genetic variation (Volz, Koelle et al. 2013). So far, phylodynamic studies have tended to focus on a limited number of viral phenotypes in spite of having difference concerning many phenotypes. Phenotypes associated with virulence, viral transmissibility, cell or tissue tropism, and antigenic variability that can facilitate escape from host immunity are the chosen ones. Transmission dynamics and selection can have an impact on viral genetic variation, therefore, viral phylogenies can be used to investigate important epidemiological, immunological and evolutionary processes including epidemic spread, spatio-temporal dynamics, tissue tropism and antigenic drift. The quantitative investigation of these processes is the central aim of viral phylodynamics (Volz, Koelle et al. 2013).

Phylodynamic approaches are mainly used to better understand viral transmission dynamics and spread within infected hosts in different locations and time. Volz, Koelle and Bedford (Volz, Koelle et al. 2013) described some crucial applications of viral phylodynamics, and these include dating epidemic and pandemic origins, tracing the viral spread and ascertaining the effectiveness of viral control efforts. With the rate of evolution measured in real units of time, it is possible to infer the date of the most recent common ancestor (MRCA) for a set of viral sequences. The age of the MRCA of these isolates is a lower bound; the common ancestor of the entire virus population must have existed earlier than the MRCA of the virus sample. Phylogeographic models have the possibility of more directly revealing the hidden transmission patterns as it is very difficult to assess the transmission between diverse groups or classes of a virus, be they geographic-, age, or risk-related from surveillance data alone. Viral control efforts can also impact the rate at which virus populations evolve, thereby influencing phylogenetic patterns. Phylodynamic approaches quantify how evolutionary rates change over time, and on the basis of the information, the effectiveness of control strategies can be checked (49). A better understanding of RNA virus phylodynamics will allow more engaged attempts at pathogen surveillance, facilitate more precise predictions of the epidemiological impact of newly emerged viruses and assist in the control efforts (Holmes and Grenfell 2009).

The outbreak of the foot-and-mouth disease in the UK in 2001 resulted in an ideal and well-documented database in terms of epidemiological scale (Cottam, Haydon et al. 2006). It is one of the most well-documented large outbreaks in terms of the accessibility of spatio-temporal incidence data in parallel with contact tracing and the underlying spatial pattern of the susceptible farms as a measure of the contact network. Additionally, analyses of viral sequences from relatively small samples of farms have drawn important conclusions about the epidemic spread and allowed the testing of new methods to reveal the spatiotemporal patterns written into sequence data (Cottam, Haydon et al. 2006).

Cottam, Haydon, Paton, Gloster, Wilesmith, Ferris, Hutchings and King (Cottam, Haydon et al. 2006) estimated the rate of nucleotide substitution to be 2.26×10^{-5} per site per day and 1.5 substitutions per farm infection as well as the day at which FMDV first infected livestock in the United Kingdom was 7 February 2001. This sufficiently high rate shows that detailed histories of the transmission pathways can be reliably reconstructed (Cottam, Haydon et al. 2006). There were several other reports on phylodynamics analyses of FMDV in different countries as in the Philippines, Vietnam and India (Di Nardo, Knowles et al. 2014, Subramaniam, Mohapatra et al. 2015, Brito, Pauszek et al. 2017). The substitution rate calculated for VP1 of FMDV O/ME-SA/PanAsia circulating in Vietnam between 2009 and 2014 was 1.66×10^{-2} [52] that was similar to a previous estimate of serotype O Cathay topotype (1.06×10^{-2}) (Di Nardo, Knowles et al. 2014).

6.10. Selection of Substitution Model

Using local FMDV serotype O sequences of Bangladesh, our group has performed a selection of substitution model using jModelTest (version 2.1.10) package (Darriba, Taboada et al. 2012) by computing likelihood score out of 88 models. The best-fitted model was selected using the lowest corrected Akaike's Information Criterion (cAIC) (Hurvich and Tsai 1993) and Bayesian Information Criterion (BIC) (Schwarz 1978) values. The default substitution rate was 1.0 with no estimation and four categories of gamma were implied.

6.11. Selection of Tree and Clock Model

Since Bayesian phylogenetic analysis requires appropriate tree and clock model as prior, stepping stone (SS) method (Xie, Lewis et al. 2010) was employed to calculate the marginal log-likelihood estimation (MLE) (Baele, Lemey et al. 2012). Bayesian phylodynamics analysis was performed within a Bayesian coalescent framework by a Markov Chain Monte Carlo (MCMC) approach (Yang and Rannala 1997, Drummond, Nicholls et al. 2002) available in the Bayesian Evolutionary Analysis by Sampling Tree (BEAST) v 2.4.6 software platform (Drummond and Rambaut 2007, Bouckaert, Heled et al. 2014).

A combination of two coalescent and two relaxed clock models (Drummond, Ho et al. 2006) was prepared using BEAUTi (Drummond, Suchard et al. 2012) and the models were compared by their MLE score. Under both relaxed clock model, the specified clock rate was estimated based on previous studies (Jenkins, Rambaut et al. 2002, Tully and Fares 2008, Yoon, Lee et al. 2011, Subramaniam, Sanyal et al. 2013). Stepping Stone (SS) as the state-of-the-art method was employed to construct an overall ranking of competing models, by estimating the (log) marginal likelihood for each model, while accommodating phylogenetic uncertainty. Path Sampler from BEAST 2.4.5 application launcher was implemented for the stepping stone method (alpha value-0.3) with 20 power posteriors. The model with the lowest MLE score and the maximum Bayes Factor (BF) (Kass and Raftery 1995) as compared to other models was selected as the best model.

6.12. MCMC Computation in BEAST

The Markov Chain Monte Carlo (MCMC) analysis was set-up with the model for 10^8 steps for FMDV serotype O using BEAST where no burning was used. Sampling from every 10000 and 50000 samples was done for using as tracing effective sample size (ESS) values (Drummond, Ho et al. 2006) and generating tree wherein auto-detect mode was on for each case.

6.13. Tracing the Log Files

The log files generated in these analyses were visualized in Tracer v1.6 (http://tree.bio.ed.ac.uk/software/tracer/) after 10% burn-in and the statistical uncertainties were summarized in the 95% highest probability density (HPD) intervals (Rambaut, Suchard et al. 2015). Based on the ESS values (>200 at least) of log files checked by Tracer v1.6 after a 10% burn-in, the chain length was 10,000,000. The convergence of sampling was assessed by ESS and the BEAST analyses were run for a sufficiently long time to confirm that ESS was above 200 in all parameters. Estimation of substitution rate and time to a most recent common ancestor (tMRCA) of all nodes were obtained from the tracer by analyzing the values of tree height and ucld mean. The year of the latest isolated sequence was used in case of determining tMRCA.

6.14. Bayesian Skyline Plot Generation

Bayesian Skyline Plot (Drummond, Rambaut et al. 2005) was generated in Tracer (Rambaut, Suchard et al. 2015) to infer the changes in effective population size over time. In this case, the tree prior was coalescent Bayesian skyline with the same parameters set used in phylodynamics analysis. After producing log file with 10^8 runs, one sample among every 20000 samples was taken. Popsizes (Population sizes) and

Groupsizes were fixed as a dimension based on major groups (sub-lineages/groups) that were investigated.

6.15. Phylogeography of RNA Virus

Understanding the geographic context of evolutionary histories is burgeoning across biological disciplines since evolutionary change is only fully comprehended by considering its geographic context. In the light of increasingly detailed geographical and environmental observations, recent endeavors attempt to interpret contemporaneous genetic variation. The development of phylogeographic inference techniques has explicitly aimed to integrate such heterogeneous data. Phylogeography has been defined as the 'field of study concerned with the principles and processes governing the geographical distributions of geographical lineages, especially those within and among closely related species' (Avise 2000).

Viruses represent ideal organisms for the study of phylogeography because of genetic malleability. Understanding the evolutionary processes that give rise to the basic phylogeographical patterns observed in viruses can also be of great epidemiological and clinical importance (Homes, E. C., 2004). Phylogenetic inference from molecular sequences has developed into an increasingly popular tool to trace the patterns of virus dispersal (Lemey et al. 2009). The time-scale of epidemic spread usually provides sufficient time for rapidly evolving viruses, especially RNA viruses to accumulate analytically informative mutations in their genomes. Consequently, spatial diffusion can leave a measurable footprint in gene sequences from these viruses. It is impossible to comprehend and analyze the spread of RNA viruses without considering how this takes place in both time and space. Indeed, the patterns of distribution depend largely on the number and density of susceptible hosts, quantities that will themselves change in time and space (Homes, E. C., 2004). Developed analytical tools are utilized to uncover the footprint of spatio-temporal reconstructed history in contemporaneous molecular sequences providing insights into the origin and epidemic spread (Lemey, Rambaut et al. 2010).

Analytical techniques based on Bayesian MCMC methods (Drummond, Nicholls et al. 2002) have been developed to reconstruct ancestral states of viruses and how viruses migrated in the past after testing hypotheses about the spatial diffusion patterns of viruses. The phylogeographic analysis would benefit even more from fully integrating spatial, temporal and demographic inference. Lemey, Rambaut, Drummond and Suchard (Lemey, Rambaut et al. 2009) implemented the ancestral reconstruction of discrete states in a Bayesian statistical framework for evolutionary hypothesis testing that was geared towards rooted, time-measured phylogenies. Their methods allow character mapping in natural time scales, calibrated under a strict or relaxed molecular clock, in combination with several models of population size change (Lemey, Rambaut et al. 2009). The Bayesian stochastic search variable selection (BSSVS) enables to construct a Bayes factor (BF) test (Kass and Raftery 1995) that identifies the most parsimonious description of the phylogeographic diffusion process (Lemey, Rambaut et al. 2009).

7. Ancestral State Reconstruction

7.1. States with Values of Latitude and Longitude

Sampling location was used as a discrete character for ancestral state reconstruction (Lemey, Rambaut et al. 2010). To infer the ancestral state for the geographical locations, each sequence has been categorized into one of thirteen regions of South Asia. For each state, the latitude and longitude had been set based on the center region under the whole area.

7.2. Setting up of Parameters

BEAST_CLASSIC package was installed and used for discrete phylogeography analysis. After the installation, a discrete trait (location) was added. The clock model and tree prior model were the same as used in phylodynamics analysis.

7.3. Computational Run of MCMC Chain and Tracing the Values

A MCMC analysis was run for 75×10^6 steps that stores log file from every 1000 samples and tree with trait from every 4000 trees. No pre-burning was used and the mode was auto-detect in all the cases. The log file generated was checked in the tracer (Rambaut, Suchard et al. 2015).

7.4. Generation of MCC Tree

The maximum clade credibility (MCC) trees were generated using Tree-Annotator in BEAST2 package (Bouckaert, Heled et al. 2014). The posterior set of trees was used to estimate the maximum clade credibility (MCC) phylogeny of each dataset wherein 10% of the trees were discarded as burn-in.

7.5. Visualization and Editing of Tree

The final tree was visualized using FigTree version 1.4.3 (available at http://tree.bio.ed.ac.uk/software). The estimates of substitution rate and time to most recent common ancestor (tMRCA) of all nodes were obtained based on the MCC trees.

7.6. Discrete Pattern of Phylogeography

Using the parameters described above, the inferred transmission rates were evaluated between character states using an asymmetric model for the discrete traits in BEAST and the significance of the network was estimated with Bayesian stochastic search variable selection (BSSVS), which checks the hypothesis of non-zero transmission rates between discrete characters (Lemey, Rambaut et al. 2009). SPREAD v1.0.6 (Bielejec, Rambaut et al. 2011) was used to produce the kml file format that showed the route along with the states having the time scale presented in Google Earth (https://www.google.com/earth/download/ge/).

7.7. Prediction of the Significant Routes

Statistical support was evaluated using Bayes factor (BF) for discrete traits implemented in SPREAD v1.0.6 to analyze the pattern of geographical spread and significant viral transmission route were assessed by considering BF>3 as significant non-zero transmission cut-off (Lemey, Rambaut et al. 2009). The BFs found from the analysis were divided into three major groups. The range of BFs is from 6 to 10, 11-20 and >20. The Bayes factor based significant routes were plotted on the map found from OpenStreetMap (Haklay and Weber 2008) and the directions of the routes were also shown.

Conclusion

Modern approaches involving molecular and bioinformatics analyses on Foot-and-mouth disease virus (FMDV) starting from serotype O detection until a vaccine candidate isolation have been discussed in this chapter. FMD has been considered as a devastating animal plague since its diagnosis, and FMDVs are transboundary animal pathogens affecting all the major non-avian livestock species, where cattle are the most susceptible. Notably, FMDV serotypes show some regionality, and the O serotype is most common with the worldwide spreading. Reconstructing the evolutionary history, dispersal processes and spatio-temporal patterns from viral sequences contributes to our understanding of the epidemiological dynamics underlying epizootic events. Within the projected investigation, viral VP1 coding region sequence based analysis of FMD samples revealed that FMDV serotype O was circulating in Bangladesh with 82% abundance. Importantly, the VP1 phylogeny revealed the emergence of two novel sub-lineages of serotype O, named as Ind2001BD1 and Ind2001BD2, within the Ind2001 lineage besides the circulation of Ind2001d sub-lineage in Bangladesh. Moreover, the Bayesian phylogenetic framework was used to explore the phylodynamics and spatio-temporal dispersion of the FMDV serotype O ME-SA topotype that caused epidemics in the South Asian region between 2000 and 2016. Interestingly, the phylogeographic study revealed Myanmar as the probable origin country of circulatory FMDV serotype O in 1969, which gradually spread to Pakistan, India, Nepal, Bhutan and Bangladesh. For FMDV serotype O, the evolutionary rate was estimated to be 6.03×10^{-3} substitutions per site, per year. Significant mutations were detected within BC and GH loop of FMDV serotype O, where several antigenically critical codons within the VP1 region were detected to be under positive selection. That investigation conclusively demonstrated the emergence of two novel sub-lineages of FMDV type O Ind2001BD1 and Ind2001BD2 under O/ME-SA/Ind2001 lineage for the first time ever. On the basis of the study, it is highly recommended to select the vaccine strain of FMDV serotype O considering the current circulatory pool. This particular example emphasizes the necessity of coordinated approaches for gradual selection and analyses of candidate animal vaccine strain. Such approaches are also applicable for screening of other pathogenic viruses to combat diseases. The applicability of rational drug and vaccine designing using sequence and computational tools can accelerate the arena of novel therapeutic approaches.

ACKNOWLEDGMENTS

We acknowledge HEQEP (Higher Education, Quality Enhancement Project), World Bank and University Grants Commission (UGC), Bangladesh for funding the experimental part described here.

REFERENCES

Aggarwal, N. and P. Barnett (2002). "Antigenic sites of foot-and-mouth disease virus (FMDV): an analysis of the specificities of anti-FMDV antibodies after vaccination of naturally susceptible host species." *Journal of General Virology* **83**(4): 775-782.

Aktas, S. and A. Samuel (2000). "Identification of antigenic epitopes on the foot and mouth disease virus isolate O1/Manisa/Turkey/69 using monoclonal antibodies." *Revue Scientifique et Technique-Office International des Epizooties* **19**(3): 744-753.

Avise, J. C. (2000). *Phylogeography: the history and formation of species*, Harvard University press.

Ayelet, G., et al. (2009). "Genetic characterization of foot-and-mouth disease viruses, Ethiopia, 1981–2007." *Emerging infectious diseases* **15**(9): 1409.

Baele, G., et al. (2012). "Improving the accuracy of demographic and molecular clock model comparison while accommodating phylogenetic uncertainty." *Molecular Biology and Evolution* **29**(9): 2157-2167.

Bai, X., et al. (2010). "Genetic characterization of the cell-adapted PanAsia strain of foot-and-mouth disease virus O/Fujian/CHA/5/99 isolated from swine." *Virology Journal* **7**(1): 208.

Biasini, M., et al. (2014). "SWISS-MODEL: modelling protein tertiary and quaternary structure using evolutionary information." *Nucleic acids research* **42**(W1): W252-W258.

Bielejec, F., et al. (2011). "SPREAD: spatial phylogenetic reconstruction of evolutionary dynamics." *Bioinformatics* **27**(20): 2910-2912.

Bouckaert, R., et al. (2014). "BEAST 2: a software platform for Bayesian evolutionary analysis." *PLoS computational biology* **10**(4): e1003537.

Brito, B., et al. (2017). "Phylodynamics of foot-and-mouth disease virus O/PanAsia in Vietnam 2010–2014." *Veterinary research* **48**(1): 24.

Carrillo, C., et al. (2005). "Comparative genomics of foot-and-mouth disease virus." *Journal of virology* **79**(10): 6487-6504.

Chowell, G., et al. (2003). "SARS outbreaks in Ontario, Hong Kong and Singapore: the role of diagnosis and isolation as a control mechanism." *Journal of theoretical biology* **224**(1): 1-8.

Cottam, E. M., et al. (2006). "Molecular epidemiology of the foot-and-mouth disease virus outbreak in the United Kingdom in 2001." *Journal of virology* **80**(22): 11274-11282.

Cox, D. R. and N. Reid (2000). *The theory of the design of experiments*, Chapman and Hall/CRC.

Crowther, J., et al. (1993). "Identification of a fifth neutralizable site on type O foot-and-mouth disease virus following characterization of single and quintuple monoclonal antibody escape mutants." *Journal of General Virology* **74**(8): 1547-1553.

Darriba, D., et al. (2012). "jModelTest 2: more models, new heuristics and parallel computing." *Nature methods* **9**(8): 772.

Das, B., et al. (2012). "Field outbreak strains of serotype O foot-and-mouth disease virus from India with a deletion in the immunodominant βG-βH loop of the VP1 protein." *Archives of virology* **157**(10): 1967-1970.

Delport, W., et al. (2010). "Datamonkey 2010: a suite of phylogenetic analysis tools for evolutionary biology." *Bioinformatics* **26**(19): 2455-2457.

Di Nardo, A., et al. (2014). "Phylodynamic reconstruction of O CATHAY topotype foot-and-mouth disease virus epidemics in the Philippines." *Veterinary research* **45**(1): 90.

Drummond, A. J., et al. (2006). "Relaxed phylogenetics and dating with confidence." *PLoS Biology* **4**(5): e88.

Drummond, A. J., et al. (2002). "Estimating mutation parameters, population history and genealogy simultaneously from temporally spaced sequence data." *Genetics* **161**(3): 1307-1320.

Drummond, A. J. and A. Rambaut (2007). "BEAST: Bayesian evolutionary analysis by sampling trees." *BMC Evolutionary Biology* **7**(1): 214.

Drummond, A. J., et al. (2005). "Bayesian coalescent inference of past population dynamics from molecular sequences." *Molecular Biology and Evolution* **22**(5): 1185-1192.

Drummond, A. J., et al. (2012). "Bayesian phylogenetics with BEAUti and the BEAST 1.7." *Molecular Biology and Evolution* **29**(8): 1969-1973.

Felsenstein, J. (1981). "Evolutionary trees from DNA sequences: a maximum likelihood approach." *Journal of molecular evolution* **17**(6): 368-376.

Garcia-Boronat, M., et al. (2008). "PVS: a web server for protein sequence variability analysis tuned to facilitate conserved epitope discovery." *Nucleic acids research* **36**(suppl_2): W35-W41.

Gouy, M., et al. (2009). "SeaView version 4: a multiplatform graphical user interface for sequence alignment and phylogenetic tree building." *Molecular biology and evolution* **27**(2): 221-224.

Haklay, M. and P. Weber (2008). "Openstreetmap: User-generated street maps." *IEEE Pervasive Computing* **7**(4): 12-18.

Hemadri, D., et al. (2002). "Emergence of a new strain of type O foot-and-mouth disease virus: its phylogenetic and evolutionary relationship with the PanAsia pandemic strain." *Virus genes* **25**(1): 23-34.

Holmes, E. C. and B. T. Grenfell (2009). "Discovering the phylodynamics of RNA viruses." *PLoS computational biology* **5**(10): e1000505.

Hurvich, C. M. and C. L. Tsai (1993). "A corrected Akaike information criterion for vector autoregressive model selection." *Journal of time series analysis* **14**(3): 271-279.

Jamal, S. M., et al. (2011). "Genetic diversity of foot-and-mouth disease virus serotype O in Pakistan and Afghanistan, 1997–2009." *Infection, Genetics and Evolution* **11**(6): 1229-1238.

Jenkins, G. M., et al. (2002). "Rates of molecular evolution in RNA viruses: a quantitative phylogenetic analysis." *Journal of molecular evolution* **54**(2): 156-165.

Kabat, E., et al. (1977). "Unusual distributions of amino acids in complementarity determining (hypervariable) segments of heavy and light chains of immunoglobulins and their possible roles in specificity of antibody-combining sites." *Journal of Biological Chemistry* **252**(19): 6609-6616.

Kass, R. E. and A. E. Raftery (1995). "Bayes factors." *Journal of the American statistical association* **90**(430): 773-795.

Kim, J., et al. (2010). "A classification approach for genotyping viral sequences based on multidimensional scaling and linear discriminant analysis." *BMC bioinformatics* **11**(1): 434.

Kimura, M. (1977). "Preponderance of synonymous changes as evidence for the neutral theory of molecular evolution." *Nature* **267**(5608): 275.

Kitson, J. D., et al. (1990). "Sequence analysis of monoclonal antibody resistant mutants of type O foot and mouth disease virus: evidence for the involvement of the three surface exposed capsid proteins in four antigenic sites." *Virology* **179**(1): 26-34.

Knowles, N., et al. (2001). "Outbreak of foot-and-mouth disease virus serotype O in the UK caused by a pandemic strain." *The Veterinary Record* **148**(9): 258-259.

Kotecha, A., et al. (2017). "Rules of engagement between αvβ6 integrin and foot-and-mouth disease virus." *Nature communications* **8**: 15408.

Kumar, S., et al. (2016). "MEGA7: molecular evolutionary genetics analysis version 7.0 for bigger datasets." *Molecular biology and evolution* **33**(7): 1870-1874.

Lemey, P., et al. (2009). "Bayesian phylogeography finds its roots." *PLoS computational biology* **5**(9): e1000520.

Lemey, P., et al. (2010). "Phylogeography takes a relaxed random walk in continuous space and time." *Molecular biology and evolution* **27**(8): 1877-1885.

Lovell, S. C., et al. (2003). "Structure validation by Cα geometry: ϕ, ψ and Cβ deviation." *Proteins: Structure, Function, and Bioinformatics* **50**(3): 437-450.

Mateu, M., et al. (1994). "Antigenic heterogeneity of a foot-and-mouth disease virus serotype in the field is mediated by very limited sequence variation at several antigenic sites." *Journal of virology* **68**(3): 1407-1417.

Mohapatra, J. K., et al. (2002). "Sequence and phylogenetic analysis of the L and VP1 genes of foot-and-mouth disease virus serotype Asia1." *Virus research* **87**(2): 107-118.

Nettling, M., et al. (2015). "DiffLogo: a comparative visualization of sequence motifs." *BMC bioinformatics* **16**(1): 387.

Parry, N., et al. (1989). "Neutralizing epitopes of type O foot-and-mouth disease virus. II. Mapping three conformational sites with synthetic peptide reagents." *Journal of General Virology* **70**(6): 1493-1503.

Pond, S. L. K. and S. D. Frost (2005). "Datamonkey: rapid detection of selective pressure on individual sites of codon alignments." *Bioinformatics* **21**(10): 2531-2533.

Ramachandran, G. N. (1963). "Stereochemistry of polypeptide chain configurations." *J. Mol. Biol.* **7**: 95-99.

Rambaut, A., et al. (2015). *Tracer* v1. 6. 2014.

Samuel, A. and N. Knowles (2001). "Foot-and-mouth disease type O viruses exhibit genetically and geographically distinct evolutionary lineages (topotypes)." *Journal of General Virology* **82**(3): 609-621.

Schwarz, G. (1978). "Estimating the dimension of a model." *The annals of statistics* **6**(2): 461-464.

Siddique, M., et al. (2018). "Emergence of two novel sublineages Ind2001 BD 1 and Ind2001 BD 2 of foot-and-mouth disease virus serotype O in Bangladesh." *Transboundary and emerging diseases* **65**(4): 1009-1023.

Subramaniam, S., et al. (2015). "Evolutionary dynamics of foot-and-mouth disease virus O/ME-SA/Ind2001 lineage." *Veterinary microbiology* **178**(3-4): 181-189.

Subramaniam, S., et al. (2013). "Status of Foot-and-mouth Disease in India." *Transboundary and emerging diseases* **60**(3): 197-203.

Subramaniam, S., et al. (2013). "Emergence of a novel lineage genetically divergent from the predominant Ind2001 lineage of serotype O foot-and-mouth disease virus in India." *Infection, Genetics and Evolution* **18**: 1-7.

Suroowan, S., et al. (2017). "Ethnoveterinary health management practices using medicinal plants in South Asia–a review." *Veterinary research communications* **41**(2): 147-168.

Tamura, K. and M. Nei (1993). "Estimation of the number of nucleotide substitutions in the control region of mitochondrial DNA in humans and chimpanzees." *Molecular biology and evolution* **10**(3): 512-526.

Tully, D. C. and M. A. Fares (2008). "The tale of a modern animal plague: tracing the evolutionary history and determining the time-scale for foot and mouth disease virus." *Virology* **382**(2): 250-256.

Valdazo-González, B., et al. (2014). "Genome sequences of foot-and-mouth disease virus O/ME-SA/Ind-2001 lineage from outbreaks in Libya, Saudi Arabia, and Bhutan during 2013." *Genome Announc*. **2**(2): e00242-00214.

Volz, E. M., et al. (2013). "Viral phylodynamics." *PLoS computational biology* **9**(3): e1002947.

Vosloo, W., et al. (1992). "Genetic relationships between southern African SAT-2 isolates of foot-and-mouth-disease virus." *Epidemiology & Infection* **109**(3): 547-558.

Wiederstein, M. and M. J. Sippl (2007). "ProSA-web: interactive web service for the recognition of errors in three-dimensional structures of proteins." *Nucleic acids research* **35**(suppl_2): W407-W410.

Xie, Q., et al. (1987). "Neutralization of foot-and-mouth disease virus can be mediated through any of at least three separate antigenic sites." *Journal of General Virology* **68**(6): 1637-1647.

Xie, W., et al. (2010). "Improving marginal likelihood estimation for Bayesian phylogenetic model selection." *Systematic biology* **60**(2): 150-160.

Yang, Z. and J. P. Bielawski (2000). "Statistical methods for detecting molecular adaptation." *Trends in ecology & evolution* **15**(12): 496-503.

Yang, Z. and B. Rannala (1997). "Bayesian phylogenetic inference using DNA sequences: a Markov Chain Monte Carlo method." *Molecular biology and evolution* **14**(7): 717-724.

Yoon, S. H., et al. (2011). "Molecular epidemiology of foot-and-mouth disease virus serotypes A and O with emphasis on Korean isolates: temporal and spatial dynamics." *Archives of virology* **156**(5): 817-826.

In: Trends in Biochemistry and Molecular Biology
Editors: Hossain Uddin Shekhar et al.
ISBN: 978-1-53616-434-3

Chapter 13

ENDOPHYTES: A DIVERSE WORLD OF MICROORGANISMS

Farhana Tasnim Chowdhury [*] ***and Tonny Tabassum***
Department of Biochemistry and Molecular Biology,
Faculty of Biological Sciences, University of Dhaka, Dhaka, Bangladesh

ABSTRACT

Microorganisms provide a versatile source of unique and potential bioactive components. A special group of microorganisms that inhabit in plants, during a variable period of their life without causing any detectable disease or symptoms is classified as endophytes. Bioactive natural products and enzymes from these endophytes are of considerable importance to chemists and biologists alike. Endophytes play a beneficial role in various aspects, such as plant growth promotion, enhancement of biotic and abiotic stress tolerance of plant and production of compounds with antimicrobial and anticancer properties which could be exploited in pharmaceutical industries. Many endophytes also emit a diverse class of volatile organic compounds (VOCs), many of which may be biologically active and play an important role in the development of a symbiotic relationship with host plant and inter- and intra-kingdom signaling. The volatile organic compounds emitted by endophytes play a huge role in the combat against plant pathogens, enhancement of host survival in stress condition, enhancement of plant growth, prevention of post-harvest infection in plants and lastly overall crop welfare. Recent studies exposed that endophytic microorganisms can also produce complex hydrocarbon molecules with fuel potential by degrading plant biomass. Endophytes, a relatively under-explored group of microorganisms, are currently being studied by biologists and natural product researchers and a more interdisciplinary systematic exploration will likely uncover their novel use for their future utilization and new opportunities to develop bio-based commercial products to combat global crop and human problems.

---

[*] Corresponding Author's E-mail: farhanatasnim@du.ac.bd.

1. INTRODUCTION

Plants and microscopic organisms maintain a close relationship as all plants in nature accommodate a diverse community of microbes. These microbes can colonize in plant rhizospheres and other tissues, for example, flowers, fruits, leaves, stems, roots, and seeds. "Endophytes" are inter- or intra-cellular microbes living in healthy plant tissues that do not interrupt host metabolism or show any obvious sign of infection or disease (Strobel et al., 2004). This definition does not include pathogens and nodule-producing microbes. Endophytes are mostly fungi and bacteria which resides within a plant for whole or part of their lifecycle. Though endophytes are magniloquent and found in almost all species of plant, the endophyte plant association is still a great conundrum (Hardoim et al., 2015). To define endophytes, the spectrum of plant-microbe co-existence including the entire spectrum of symbiotic interactions such as parasitism, commensalism, and mutualism should be perceived. Each of the 300,000 species of higher plants existing on earth contains a diversity of endophytes creating enormous biodiversity (Strobel and Daisy, 2003).

Multiple recent studies have reported that these intrinsic microbes have beneficial effects on their hosts (Hardoim et al., 2008; Zinniel et al., 2002). Endophytes can promote plant growth by producing certain crucial enzymes e.g., ACC Deaminase or by increasing the availability of nutrients to the plant. Endophytes can indirectly contribute to promote plant growth, for example by suppressing plant diseases or by regulating the toxic effect of environmental pollutants or by neutralizing stresses of the plant caused by an excess of the hormone ethylene, by heavy metals, by drought, and by saline soil. This symbiotic mutualism relationship differs between species and environments (Gao et al., 2010; Hardoim et al., 2008; Rosenblueth and Martínez-Romero, 2006). Another remarkable contribution of endophytes is in the production of bioactive secondary metabolites, materials with possible use in modern medicine, novel antibiotics, agriculture, and industry, antimycotics, immunosuppressants and anticancer compounds (Strobel and Daisy, 2003).

The word 'endophyte' was created by a German scientist, Heinrich Anton De Bary in 1884 (Wilson, 1995). The first-ever discovered plant endophyte was a fungus that was identified in *Persian darnel* (annual grass) by Freeman during 1904 in Germany and grasses with a dense population of endophytes showed to be more resistant to pest attack (Kulkarni et al., 2014). The first report of bacteria living in healthy plant tissues was made around 1926 (Mohanta et al., 2010).

All endophytic microorganisms are divided into two categories primarily: systemic (true) endophytes which live entirely within plant tissues throughout its life without causing any disease and non-systemic (transient) endophytes which often become morbific to their host plants under taxing or resource-limited growing conditions. They were categorized based on the endophyte's genetics, biology, and mechanism of transmission from host to host (Wani et al., 2015).

Fungal taxa of *Ascomycota*, *Basidiomycota*, and *Zygomycota* are the most reported endophytes in different plant species (Strobel and Daisy, 2003). But vast studies need to be conducted to define the ecology and habitat of most groups of endophytic fungi. Reports by forest pathologists had commonly regarded many of the fungi endophytes as minor or secondary pathogens (Mueller et al., 2011). The blurred boundaries categorizing endophytes, facultative pathogens, and latent pathogens can be underscored by the occurrence of endophytic fungi in both healthy and infected plant tissues. Certainly, there are minor dissimilarities between many fungi considered as "endophytic" and those considered to be "latent pathogens" (Carroll, 1988; Rodriguez and Redman, 2008).

Primarily, endophytes are entered to plant tissue through the root zone; sometimes aerial parts of plants, such as flowers, stems, and cotyledons, may also be used for entry. Germinating radicles, secondary stomata can be a special entry point of bacteria into a tissue, as a result of the follicular fissure (Zinniel et al., 2002). Only small fractions of rhizosphere and phyllosphere bacteria are capable of living inside the plant (Compant et al., 2005).

1.1. Endophyte-Plant Communication

It is assumed that the communication between endophyte and the respective host plant may have arisen because of co-evolutionary processes during the appearance of higher plants hundreds of millions of years ago. The mechanism of transmission, the way of infection, and genetic background are the major factors responsible for defining endophyte-plant communication or interaction. The amount of endophytes that will colonize in the host plant is also determined both by host plant defense system and endophytes enzyme machinery. Endophytic microorganisms colonize in the plant parts by secreting several enzymes like cellulase, xylanase, lignin peroxidase, protease, amylase, lipase, etc. which degrade plant external surface permitting intracellular or intercellular colonization (Witzell et al., 2014). In response, host plant produces defense enzymes which cause colonization of endophytes in plant tissue. Plant-microbe interaction can have various outcomes like positive or negative or neutral symbiosis. The relationship between plant and its associated microbes is also reliant on the genotype of both plant and microbe (Jia et al., 2016; Moricca and Ragazzi, 2008).

2. Overview of Potential Roles of Endophytic Microorganisms

Endophytes are proficient of producing various components that are utilized by host plants for protection against parasites or pathogens. Also, many of these components or compounds are demonstrated to be novel and unique drug and are now being used for the treatment of various human diseases like cancer, infection, etc. Many of these bioactive compounds secreted by endophytic microbes are also responsible for promoting the growth of host plants by different mechanisms. Even many recent studies have reported endophytes as a source complex hydrocarbons with fuel potential. The volatiles emitted by endophytic microorganisms has been seen as a new wonder in microbial research and their flavor and fragrance are of great importance in industrial application. Some of the potential endophytes isolated from plants and their activity are mentioned in **Table 1**.

Table 1. List of endophytic microorganisms and their function

Endophyte	Host	Function	References
Fusarium redolens	*Dioscorea zingiberensis*	Antibacterial activity	(Xu et al., 2008)
Chloridium sp.	*Azadirachta indica*	Antibacterial activity	(Kharwar et al., 2009)
Periconia sp.	*Taxus cuspidate*	Antibacterial activity	(Kim et al., 2004)
Fusarium sp.	*Quercus variabilis*	Antibacterial activity	(Shu et al., 2004)
Phoma sp.	*Larrea tridentata*	Biofuel	(Strobel et al., 2011)
Muscodor albus	*Cinnamomum zeylanicum*	Antimicrobial activity	(Strobel et al., 2001)
Burkholderia phytofirmans	*Panicum virgatum*	Plant growth promotion activity	(Kim et al., 2012)
Aspergillus niger	*Cyndon dactylon*	Cytotoxic to the cancer cell	(Song et al., 2004)
Pestalotiopsis microspora	*Terminalia morobensis*	Antioxidant activity	(Harper et al., 2003)
Fusarium subglutinans	*Tripterygium wilfordii*	Immunosuppressive activity	(Lee et al., 1995)
Williopsis saturnus	*Zea mays* L.	Plant growth promotion activity	(Nassar et al., 2005)
Aspergillus fumigatus	*Juniperus communis*	Anticancer activity	(Kusari et al., 2009)
Hypoxylon sp.	*Persea indica*	Fuel	(Tomsheck et al., 2010)
Micromonospora coriariae	*Coriaria myrtifolia*	Antibacterial activity	(Trujillo et al., 2006)
Pseudonocardia acaciae	*Acacia auriculiformis*	Antibacterial activity	(Duangmal et al., 2009)
Streptomyces albidoflavus	*Bruguiera gymnorrhiza*	Antifungal activity	(Yan et al., 2010)
Micromonospora lupini	*Lupinus angustifolius*	Antitumor activity	(Igarashi et al., 2007)
Pseudomonas, Rhizobia	*Arabidopsis*	Plant growth promotion activity	(Van Loon, 2007)
Pseudomonas fluorescens	Tomato	Plant growth promotion activity	(Wang et al., 2005)
Trichoderma	Plam	Plant growth promotion activity	(Lee et al., 2016)

2.1. Plant Growth Promotion

In the recent years, endophytic microbes have attained significant concerns from scientists around the world as a novel resource that can control various devastating plants infections or diseases and can also stimulate of growth of respective host plant (Compant et al., 2005). The benefits of using endophytes as biocontrol agents are that they can adapt easily to live inside the host plants and thus can reliably suppress vascular disease of plants. Moreover, they do not cause environmental and ecological pollution and sometimes they carry out an important part in the reduction of different environmental pollutants (Backman and Sikora, 2008). Almost all deterministic information on beneficial microbes of plants are acquired from rhizosphere bacteria as known about endophytes is very little and it is assumed that the mode of action of endophytes for promotion of plant growth is similar to the mechanism followed by bacteria that reside in the rhizosphere. Bacteria generally stimulate plant development in 2 distinct ways: directly & indirectly (Glick, 1995). Direct pathway of plant development is instigated by making nutrients convenient for use and/or producing phytohormones. The presence of pathogens, pollutants or other stress conditions is often the cause for indirect plant development. The beneficial microbe inactivates or kills the pathogen, usually a fungus. In the case of bioremediation or phytoremediation, the bacteria inactivate pollutant responsible for the prevention of germination of the seed or the growth of the plant. And in the case of pathogens, the beneficial microbe inactivates or kills the pathogen. For the case of stresses caused due to heavy metals, drought, excess ethylene hormone and salinated soil, 1-aminocyclopropane-1 carboxylate (ACC) deaminase enzyme, as well as some other factors secreted by bacteria, has the ability to increase plants leniency towards several biotic or abiotic stresses (Glick, 1995).

2.1.1. Direct Plant Growth Promotion

2.1.1.1. Increased Phytohormones

Phytohormones play an effective role in promoting plant development. Hormones normally synthesized by microbes to promote plant development are auxins, ethylene, cytokinins, gibberellins, and abscisic acid. Auxin or Gibberellins directly promote plant growth whereas ethylene shows the growth inhibitory effect. Indole 3 acetic acid (IAA), which is an auxin hormone, is being able to increase cell growth, division, and differentiation in plants (Glick, 2012). Root-associated bacteria like *Enterobacter* sp. and *Pseudomonas* sp. are found to produce Indole 3 acetic acid (IAA) in a good amount (El-Khawas et al., 1999). Likewise, plant growth-promoting bacteria (PGPB) has displayed its capability to produce enzyme ACC deaminase, that degrades ACC (1-aminocyclopropane-1-carboxylic acid), which is the precursor of ethylene. Thus it decreases the growth inhibitory effects of ethylene (Glick, 2005). Hence endophytes can ease stress instigated by ethylene due to the result of salinity, flood, drought, heavy metals toxicity, and pathogens.

2.1.1.2. Biofertilizers

As biofertilizers, endophytes can upsurge plant development and growth by providing the plant with essential nutrients such as nitrogen, phosphorous and ferric ions when plant growth is stunted due to low concentrations of nutrients (Glick, 2012). An endophyte *Bacillus* sp. SLS18 which is known for its plant growth promotion activity was investigated for its role in biomass production (Luo et al., 2012). Endophytic nitrogen-fixing bacteria enhance the growth of biofuel plants. Some bacteria are capable of binding to nitrogen gas that is present in the atmosphere and transform it into more soluble ammonia which can be absorbed and used by the plant (Loiret et al., 2004).

Deficiency of phosphorus is a common restrictive issue for plant development and growth and surprisingly some endophytic bacteria can produce enzymes and organic acids which can convert the attached phosphorous of different organic or inorganic compounds into solubilized form and thereby making it available for the plant (Vessey, 2003).

Iron is an element that despite its abundance is largely unavailable because of its poor solubility. But iron ions are necessary for all living organisms. Therefore, endophytes scavenge ferric ions under iron limiting conditions. This scavenging strategy implies the biosynthesis and emission of particles called siderophores, which display an increased affinity towards ferric ions. Siderophores are commonly produced by endophytic bacteria. Since the production of siderophores is a way to survive in an iron-deficient situation, it can be assumed that some microorganisms that have the capacity to adapt into an endophytic regime may produce siderophores to overcome the adverse condition of the root interior which is extremely depleted of iron. Recent studies show that production of endophytic siderophores may have a great impact on plant nutrition (Ma et al., 2011; Vessey, 2003).

2.1.2. Indirect Plant Growth Promotion

2.1.2.1. Disease Suppression

Organisms that can restrict infection, development, and persistence of pathogens are recognized as antagonists. The mechanisms used by rhizosphere bacteria to protect plants against pathogenic fungi have been well studied (Mercado-Blanco and Bakker, 2007). Apparently, endophytic bacteria regulate plant pathogens by using analogs mechanism but it is challenging to discover the actual mechanism responsible behind the total system as endophytic microbes live within the plant tissue in a hidden manner. This is due to the facts that it is not possible to discriminate between effects caused by bacteria which are inside the plant and bacteria which are outside, and studies with endophytes often lack the genetic approaches of complementation and mutation needed to draw a fruitful conclusion. Antibiosis and induced systemic resistance are the best-studied mechanisms in endophytic microbes responsible for biocontrol. Many microbes that reside in plant roots are able to secrete antibiotics which kill the surrounding pathogens. So, it can be said that endophytes may exploit the mechanism of antibiosis to control pathogenic microbes and thus suppress disease (Sturz et al., 1999). Some rhizosphere bacteria can induce resistance towards pathogenic fungi, bacteria, and viruses at sites in the plant where the beneficial bacterium is not present. Apparently, these bacteria activate a response in plant tissue that travels through the host plant and improves the protection ability of other plant parts to the successive disease. This special mechanism is titled as Induced Systemic Resistance (ISR) (De Vleesschauwer and Höfte, 2009).

2.1.2.2. Stress Tolerance

Phytohormone ethylene is an important regulator of plant development and growth and a vital element in the response of plants in various stress condition (Glick, 2004). Biosynthesis process of ethylene and its levels in plants are tightly regulated by many environmental factors, such as biotic stresses (Abeles et al., 2012). Ethylene is biosynthesized by the transformation of S-adenosyl methionine (S-AdoMet) into ACC, which is the immediate precursor of ethylene. The enzyme that carries outs this reaction is 1- aminocyclopropane-1-carboxylate synthase (ACS) and this enzyme majorly control the production of ethylene (Wang et al., 2002). The high level of plant ethylene is harmful to plant growth and development and initiates leaf chlorosis, senescence, and abscission. This stress ethylene considerably aggravates the effects of the stress, which was responsible for the generated ethylene response. Therefore, decreasing the amount of stress ethylene by any mechanism would altogether reduce the damage towards the plant tissue that happens due to the stress condition (Glick, 2004). Endophytes use ACC deaminase enzyme to promote plant development as this enzyme degrades ACC, the precursor of ethylene (Honma and Shimomura, 1978). By reducing the amount of ACC in plants, ACC deaminase-synthesizing endophytes lessen the amount of ethylene in plants and thus protects plants from subsequent damage (Glick, 2004). Application of ACC deaminase-producing endophytes may reduce ethylene levels in the plant in stress condition and can help plants become tolerant to stresses caused by salinity, flood, drought, heavy metals contamination, poisonous compounds, and pathogens. Endophyte can also control ethylene levels in plants by inhibiting the ethylene biosynthetic enzymes ACS and/or β –cystathionase (Hardoim et al., 2008). Regulation of ethylene content in the plant by any of these two mechanisms involves that ethylene producing plant cells and endophytes that are able to degrade/inhibit are in close vicinity which is a situation seen in plant-endophyte interaction. It is plausible to think that plants might have favored colonization of microorganisms, having elevated ACC-deaminase activity which eventually become endophytic and contributed to ameliorate stresses caused by high ethylene levels.

2.1.2.3. Bioremediation

In recent times the pollution of soil and water due to components such as herbicides, explosives, polyaromatic hydrocarbons, and heavy metals has become a huge concern. Bioremediation, a process of using microbial metabolism to remove these pollutants is a new technique that can be used for proper treatment and removal of these harmful waste and pollutants from our environment. The term rhizoremediation is used for the degradation of environmental pollutants by microbes residing in the rhizosphere (Kuiper et al., 2004). It is sometimes also called phytoremediation; in this case, it is suggested that the degradation is carried out by the plant but the process is enhanced by the endophytes (Ma et al., 2011). For most pollutants, bacterial strains can be selected which degrade the pollutant.

2.2. Bioactive Secondary Metabolites

Endophytes are a well-known source of biologically active metabolites having various potential activities like antimicrobial, anticancer and immunosuppressive effects besides their diverse application in agriculture and food industry. An endophytic microorganism may persist in a plant as a symbiotic organism, by providing defensive materials that possibly hinder or kill attacking pathogens (Jia et al., 2016; Ludwig-Müller, 2015; Tejesvi et al., 2011). Numerous studies confirmed that many endophytic microorganisms have the capability to synthesize several potent compounds that were previously known to be produced by their host plant (Kaul et al., 2012). Taxol, camptothecin, and vincristine are some of the anticancer drugs that can be synthesized by both plants and endophytic fungi (Jia et al., 2016).

Many fungi, bacteria, plants, and marine organisms are known to synthesize polyketides. They have extensive use in the pharmaceutical industry. Large multimodular enzyme complexes designated as polyketide synthases (PKSs) are responsible for the biosynthesis of Polyketides. These large multimodular enzyme complexes catalyze the polymerization of acyl-CoA thioester building blocks (Harvey, 2008; Horinouchi, 2009).

2.2.1. Antibacterial Compounds

Many endophytes possess the ability to inhibit or kill a broad range of pathogenic bacteria, fungi, and viruses. Therefore, there is a prodigious value of these endophytes to develop novel antimicrobial drugs. A large number of novel antibiotics have already been isolated from endophytes, like munumbicins, javanicin, beauvericin, saadamycin, rapamycin, clarithromycin, cyclododecane, petalostemumol and many more (Gouda et al., 2016). Potential antibiotic, berberine was sequestered from endophyte *Alternaria* sp., sequestered from *Phellodendron amurense* (Duan et al., 2009). An endophytic bacteria named *Streptomyces* NRRL 30562 also produces strong antibiotics named munumbicins A-D and they act strongly against both gram-positive and gram-negative bacteria (Castillo et al., 2002).

2.2.2. Antifungal Compounds

From the endophyte *Streptomyces* sp. TP-A0595, 6-prenylindole was sequestered. This compound showed prominent antifungal activity against devastating pathogen *Fusarium oxysporum* (Sasaki et al., 2002). Cedarmycins A and B which were found from *Streptomyces* sp., showed strong antifungal activity against *Candida glabrata* (Golinska et al., 2015). Another endophytic *Streptomyces* sp. Tc022 strongly inhibited *Colletotrichum musae* and *Candida albicans* (Taechowisan et al., 2003). Moreover, cultures of endophytic fungi *Hypoxylon* sp. has great bio-activity against *Botrytis cinerea*, *Phytophthora cinnamomi*, *Cercospora beticola*, and *Sclerotinia sclerotiorum* proposing that the compounds produced by endophytes may play an important role as antifungal medicine (Tomsheck et al., 2010).

2.2.3. Antitumor Agents

Complex diterpene alkaloid, paclitaxel was known to be a plant-derived compound. But due to horizontal gene transfer (HGT) between microbes and plants, genes associated with paclitaxel production may have been transferred into the plant-associated endophytes. Now many endophytes such as *Metarhizium anisopliae*, isolated from the bark of Taxus tree, are found to be a potent producer of anticancer compound-paclitaxel (Gond et al., 2014).

Three endophytic fungi, *Fomitopsis* sp., *Alternaria alternate*, and *Phomposis* sp. which were also isolated from *Miquelia dentata*, were found to produce camptothecine, a potent anticancer agent (Shweta et al., 2013). Three novel azaphilone alkaloids, chaetomugilides A–C, were sequestered from a new endophytic fungi *Chaetomium globosum* TY1, isolated from *Ginkgo biloba*. These compounds were found to show significant cytotoxicity against HepG2 cell line (Chen et al., 2016).

2.3. Bioactive Volatile Organic Compounds (VOCs)

It has been recognized recently that a group of metabolites which play a vital contribution in microbial communications and interactions are VOCs. These compounds are low molecular weight carbon-containing compound. They evaporate and diffuse under room temperature and air pressures. The physicochemical properties of VOCs facilitate their diffusion and evaporation through gas-filled and water pores in rhizosphere and soil (Chandra and Sharma, 2017; Tumlinson, 2014).

Endophytes are now being regarded as an unfamiliar source of volatile hydrocarbons. VOCs can help to shape the interactions in model and agricultural system. Though not explored thoroughly, VOCs emitted by fungus are getting recognition for their important contribution in many aspects of fungal ecology. VOCs can be of several classes of bioactive chemicals including alcohols, aldehydes, acids, esters, ketones, thiols, terpenes, etc. these bioactive chemicals can act as biofumigant to control the postharvest disease (Junker and Tholl, 2013).

The first well-studied case of isolating fungal volatile organic compound was from endophytic fungi *Muscodor albus*. VOCs produced by *M. albus* are proficient of killing a wide variety of fungi and bacteria which are pathogenic to human and plant community (Ezra et al., 2004).

Endophytic fungi named *Phoma* sp. isolated from *Larrea tridentata* was seen to yield many derivatives of naphthalene. The VOCs of *Phoma* sp. own antifungal activity. Its VOCs inhibit the growth of *Phytophthora palmivora*, *Verticillium dahliae*, *Ceratocycstis ulmi*, and *Cercospora beticola* in a strong manner (Strobel et al., 2011). Recently many other endophytes are also found to secrete VOCs, naturally active and support in persistence inside the host along with benefiting the host (Strobel and Daisy, 2003). These volatile compounds emitted by endophytes can assist their host plant in numerous aspect like, destroy the pathogen, enhancing survival in a desert habitat, attracting or repelling insects and supporting its host in competition with other plants by inhibiting seed germination of other plants (Bitas et al., 2013). The volatome profile and ability of volatiles to inhibit the growth of different pathogenic microbes, postharvest preservation of dried crops and plant development suggest endophytes as a source of volatiles for biocontrol purpose.

Volatile aroma compounds used in the food industries are of great commercial value (Krings and Berger, 1998). While maximum existing aroma compounds are mostly obtained from medicinal plants, many endophytic microorganisms are also proven to be able to yield a high amount of flavors and aroma compounds by *de novo* fermentation mechanism (Krings and Berger, 1998). Recent studies propose that fragrant compounds are emitted not only by aromatic plants but also by their endophytes (Abrahão et al., 2013; Strobel et al., 2017). For example, two recently isolated endophyte of *Rosa damascena* named *Alternaria* sp. and *Aspergillus niger* was found to produce methyl eugenol and 2-phenyl ethanol (2-PE) which are two vital elements of rose oil and responsible for flavor and aroma (Kaul et al., 2008; Wani et al., 2010). The widespread investigation has confirmed that many endophytic fungi produce numerous aroma compounds (Pimentel et al., 2011; Teles et al., 2006). The massive potential of volatile organic compounds has paved the way further screening of endophytic microbes with virtuous features like high growth rate, high waste decomposing enzymes activity for aroma and flavor compound production.

2.4. Bio-Fuel

Due to the increasing global population and rapid urbanization fuels are being exploited excessively. As a result, fossil fuel which is an important necessity to human civilization is nearly on the verge of depletion. So search for alternative fuel has become one of our major goals (Kalscheuer et al., 2006). Biofuel is a type of fuel whose energy is derived from biological carbon fixation (Sims et al., 2010). Concerning biofuels, there are many types of biofuel among which two major categories which are of great importance are: ethanol and biodiesel (fatty acid alkyl ester, FAAE) (Junchen et al., 2012; Uthoff et al., 2009). Biodiesel is fuel comprised of mono-alkyl esters of long-chain fatty acids. It is a mixture of fatty acid methyl esters (FAMEs) or fatty acid ethyl esters (FAEEs) (Knothe et al., 2015).

2.4.1. Biofuel from Microorganisms

Production of biofuel utilizing microorganisms is getting attention as a cost-effective, sustainable alternative (Ahmad et al., 2011). It is well-known knowledge that many oleaginous microbes, such as microalgae, bacteria, fungi or yeast, often accumulate intracellular lipids as their biomass. These oleaginous microbes oil could represent a promising raw material for biofuel production (Allen et al., 2014; Meng et al., 2009). This makes microbes, which are naturally capable of producing biofuel a great economic bioenergy source. These microorganisms can convert its cellular lipids or plant biomass into biofuel or biodiesel by their cellular machinery.

2.4.2. Origin of Crude Oil and Historical Role of Endophytes in Fossil Fuel Production

Endophytes are considered as an exceptional entity among all the microbial organisms present in our atmosphere. Endophytes reside within their plant host with a slower growth pattern and help the plant during its lifetime. After the death of the host plant, endophytes of that particular host plant show their potentiality in disintegrating the plant material under conditions of limited oxygen and producing various hydrocarbons with fuel potential using diverse enzyme machinery which they harbor. Their competency in converting their host plant materials into fuel components supports the role of endophytic microorganisms in the production of ancient shale based crude oil suggested by many researchers (Strobel, 2014; Ul-Hassan et al., 2012). They proposed that endophytic fungi associated with oscillated plant materials may have a role in the production of shale-related oil in the shale beds of Montana states. Those fungi were seen to be present in the intracellular space of host plant which proves their identity as endophytes of the petrified plant. In view of this study, possibly it is not irrational to speculate that many hydrocarbon compounds in the world's upper shroud may have risen through the breakdown and fermentation of plant materials by their endophytic fungi since endophytic microbes are the first ones to perform on plant material as they are positioned in the tissues of plant (Strobel, 2014; Strobel et al., 2008). These findings suggest that other endophytes, which are identified recently can also be seen as a possible contestant for the production of biofuels by utilizing lignocellulosic source as well as other carbon sources since endophytes may have fascinating implications in concern to the processes that subscribe to the production of fuel products.

2.4.3. Biofuel from Endophytes

Endophytes are currently being seen as a new source of alternative biofuels. But these biofuels need to be formed directly from microorganisms by a more proficient biosynthetic process. Fungal endophytes have also been seen to synthesize biofuel or biodiesel directly which have similarities with conventional diesel fuel (Combet et al., 2006; Stadler and Schulz, 2009). More recently, many endophytic fungi have been found to synthesize myco-diesel (Bhagobaty and Joshi, 2015; Gianoulis et al., 2012). A fungal endophyte can produce hydrocarbons, which are major constituents of diesel fuel was demonstrated for the first time in the paper entitled "The production of myco-diesel hydrocarbons and their derivatives by the endophytic fungus *Gliocladium roseum* (NRRL 50072)" (Strobel, 2014). Endophytic fungi that under the *Gliocladium* genus are found to be capable of producing various complex hydrocarbon molecules with fuel potential by degrading plant cellulosic material. Gas chromatography–mass spectrometry analysis of *Gliocladium* volatile organic compounds confirmed the production of numerous C_6–C_{19} hydrocarbon compounds comprising hexane, benzene, heptane, 3,4-dimethyl hexane, 1-octene, m-xylene, 3-methyl nonane, dodecane, tridecane, hexadecane and nonadecane (Ahamed and Ahring, 2011). Another endophytic fungus named *Hypoxylon* sp isolated from *Persea indica* is able to produce a striking spectrum of organic compounds, consisting of 1,8-cineole, 1-methyl-1,4-cyclohexadiene, and (+)-.alpha.-methylene alpha.-fenchocamphorone (Tomsheck et al., 2010). Among them, 1,8-cineole is a monoterpene. This monoterpene is an octane derivative and has potential use as a fuel alternative. Thus, compounds produced by *Hypoxylon* sp. significantly increases their possible uses in energy production (Tomsheck et al., 2010). Another endophytic fungi named *Phoma* sp. isolated from *Larrea tridentata* (creosote bush) produces an exclusive mixture of organic compounds. Its volatome profile includes a chain of sesquiterpenoids and organic compounds like 1-butanol, 2-methyl ethanol, 2-Butanone ethanol, 7-Methanoazulene, β-Chamigrene, trans-Caryophyllene, α-Longipinene, β-Ylangene, butanoic acid, 2-methyl-ethyl ester; butanoic acid, 3-methyl-ethyl ester, etc. which can be used as fuel additives (Strobel et al., 2011).

Conclusion

The global and economic interest in endophytes is increasing day by day and the major reason is their potential use for biotechnological applications. These endophytic microorganisms return to the endophytic stage after application in soil or plant and are protected from biotic and abiotic stress and threats, which is, in fact, their major advantageous point in the biological application. In addition, their genomes harbor enormous plasticity to overcome defense reactions and can ecologically be adapted to the target niche. Due to excessive use of synthetic biocontrol agents like fertilizer, pesticides and food preservatives, nature has been subjected to severe pollution leading to the severe toxification of the earth ecosystem. Use of endophyte-plant communications can result in the promotion of plant health and can play a major role in low-input sustainable agriculture application for both food and nonfood crops. Moreover, if we focus on biological control of plant diseases and biologically improve the crop production by using endophytes, it is both environment-friendly and is also considered as a suitable alternative to chemical fertilizers or fungicides or bactericides. There are also the crises of the emergence of new and resistant pathogenic microorganisms due to the exploitation of antimicrobial drugs. With these knocking at the door, finding eco-friendly natural sources of biocontrol agents and antimicrobial compounds is another main concern. Plant endophytic community represents a promising source of new generation natural bioactive agents. Isolation of novel bioactive compounds from endophytes will pave a new way for a native and cost-effective way of treatment of various diseases and infections. Moreover, the identification of potential endophytes that has the capacity to produce sustainable bioenergy can establish a promising approach to replace the use of conventional fuel. Numerous classes of endophytic microorganisms have been lately recognized that convert plant cellulosic material to a range of complex hydrocarbons, ketones and other oxygenated compounds, which are possibly feasible as biofuels.

References

Abeles, F.B., P.W. Morgan, and M.E. Saltveit Jr, *Ethylene in plant biology.* 2012: Academic press.

Abrahão, M.R., G. Molina, and G.M. Pastore, "Endophytes: Recent developments in biotechnology and the potential for flavor production." *Food research international*, 2013. 52(1): 367-372.

Ahmad, A., N.M. Yasin, C. Derek, and J. Lim, "Microalgae as a sustainable energy source for biodiesel production: a review." *Renewable and Sustainable Energy Reviews*, 2011. 15(1): 584-593.

Ahamed, A. and B.K. Ahring, "Production of hydrocarbon compounds by endophytic fungi Gliocladium species grown on cellulose." *Bioresource technology*, 2011. 102(20): 9718-9722.

Allen, J.W., A.M. Scheer, C.W. Gao, S.S. Merchant, S.S. Vasu, O. Welz, J.D. Savee, D.L. Osborn, C. Lee, and S. Vranckx, "A coordinated investigation of the combustion chemistry of diisopropyl ketone, a prototype for biofuels produced by endophytic fungi." *Combustion and Flame*, 2014. 161(3): 711-724.

Backman, P.A. and R.A. Sikora, "Endophytes: an emerging tool for biological control." *Biological control*, 2008. 46(1): 1-3.

Bhagobaty, R.K., "Endophytic Fungi: Prospects in Biofuel Production." *Proceedings of the National Academy of Sciences, India Section B: Biological Sciences*, 2015. 85(1): 21-25.

Bitas, V., H.-S. Kim, J.W. Bennett, and S. Kang, "Sniffing on microbes: diverse roles of microbial volatile organic compounds in plant health." *Molecular plant-microbe interactions*, 2013. 26(8): 835-843.

Carroll, G., "Fungal endophytes in stems and leaves: from latent pathogen to mutualistic symbiont." *Ecology*, 1988. 69(1): 2-9.

Castillo, U.F., G.A. Strobel, E.J. Ford, W.M. Hess, H. Porter, J.B. Jensen, H. Albert, R. Robison, M.A. Condron, and D.B. Teplow, "Munumbicins, wide-spectrum antibiotics produced by Streptomyces NRRL 30562," endophytic on *Kennedia nigriscansa. Microbiology*, 2002. 148(9): 2675-2685.

Chandra, D. and A. Sharma, "Harnessing Endophytic Microbial Volatile Organic Compound (MVOC) for Sustainable Agroecosystem," in *Endophytes: Crop Productivity and Protection.* 2017, Springer. 239-259.

Chen, L., Q.-Y. Zhang, M. Jia, Q.-L. Ming, W. Yue, K. Rahman, L.-P. Qin, and T. Han, "Endophytic fungi with antitumor activities: Their occurrence and anticancer compounds." *Critical reviews in microbiology*, 2016. 42(3): 454-473.

Combet, E., J. Henderson, D.C. Eastwood, and K.S. Burton, "Eight-carbon volatiles in mushrooms and fungi: properties, analysis, and biosynthesis." *Mycoscience*, 2006. 47(6): 317-326.

Compant, S., B. Duffy, J. Nowak, C. Clément, and E.A. Barka, "Use of plant growth-promoting bacteria for biocontrol of plant diseases: principles, mechanisms of action, and future prospects." *Applied and environmental microbiology*, 2005. 71(9): 4951-4959.

De Vleesschauwer, D. and M. Höfte, "Rhizobacteria-induced systemic resistance." *Advances in botanical research*, 2009. 51: 223-281.

Duangmal, K., A. Thamchaipenet, A. Matsumoto, and Y. Takahashi, "Pseudonocardia acaciae sp. nov., isolated from roots of Acacia auriculiformis A. Cunn. ex Benth." *International journal of systematic and evolutionary microbiology*, 2009. 59(6): 1487-1491.

Duan, L., G. Liwei, and Y. Hong, "Isolation and identification of producing endophytic fungi of berberine from the plant Phellodendron amurense." *J. Anhui. Agric. Sci,* 2009. 22(007): 10.3969.

El-Khawas, H. and K. Adachi, "Identification and quantification of auxins in culture media of Azospirillum and Klebsiella and their effect on rice roots." *Biology and Fertility of Soils*, 1999. 28(4): 377-381.

Ezra, D., W. Hess, and G.A. Strobel, "New endophytic isolates of Muscodor albus, a volatile-antibiotic-producing fungus." *Microbiology*, 2004. 150(12): 4023-4031.

Gao, F.-k., C.-c. Dai, and X.-z. Liu, "Mechanisms of fungal endophytes in plant protection against pathogens." *African Journal of Microbiology Research*, 2010. 4(13): 1346-1351.

Gianoulis, T.A., M.A. Griffin, D.J. Spakowicz, B.F. Dunican, A. Sboner, A.M. Sismour, C. Kodira, M. Egholm, G.M. Church, and M.B. Gerstein, "Genomic analysis of the hydrocarbon-producing, cellulolytic, endophytic fungus Ascocoryne sarcoides." *PLoS Genet*, 2012. 8(3): e1002558.

Glick, B.R., "Bacteria with ACC deaminase can promote plant growth and help to feed the world." *Microbiological research*, 2014. 169(1): 30-39.

Glick, B.R., "Plant growth-promoting bacteria: mechanisms and applications." *Scientifica*, 2012. 2012.

Glick, B.R., "Modulation of plant ethylene levels by the bacterial enzyme ACC deaminase." *FEMS microbiology letters*, 2005. 251(1): 1-7.

Glick, B.R., "The enhancement of plant growth by free-living bacteria". *Canadian journal of microbiology*, 1995. 41(2): 109-117.

Golinska, P., M. Wypij, G. Agarkar, D. Rathod, H. Dahm, and M. Rai, "Endophytic actinobacteria of medicinal plants: diversity and bioactivity." *Antonie Van Leeuwenhoek*, 2015. 108(2): 267-289.

Gond, S., R. Kharwar, and J. White Jr, "Will fungi be the new source of the blockbuster drug taxol?" *Fungal Biology Reviews*, 2014. 28(4): 77-84.

Gouda, S., G. Das, S.K. Sen, H.-S. Shin, and J.K. Patra, "Endophytes: a treasure house of bioactive compounds of medicinal importance." *Frontiers in microbiology*, 2016. 7: 1538.

Hardoim, P.R., L.S. Van Overbeek, G. Berg, A.M. Pirttilä, S. Compant, A. Campisano, M. Döring, and A. Sessitsch, "The hidden world within plants: ecological and evolutionary considerations for defining functioning of microbial endophytes." *Microbiology and Molecular Biology Reviews*, 2015. 79(3): 293-320.

Hardoim, P.R., L.S. van Overbeek, and J.D. van Elsas, "Properties of bacterial endophytes and their proposed role in plant growth." *Trends in microbiology*, 2008. 16(10): 463-471.

Harper, J.K., A.M. Arif, E.J. Ford, G.A. Strobel, J.A. Porco, D.P. Tomer, K.L. Oneill, E.M. Heider, and D.M. Grant, "Pestacin: a 1, 3-dihydro isobenzofuran from Pestalotiopsis microspora possessing antioxidant and antimycotic activities." *Tetrahedron*, 2003. 59(14): 2471-2476.

Harvey, A.L., "Natural products in drug discovery." *Drug discovery today*, 2008. 13(19-20): 894-901.

Honma, M. and T. Shimomura, "Metabolism of 1-aminocyclopropane-1-carboxylic acid." *Agricultural and Biological Chemistry*, 1978. 42(10): 1825-1831.

Horinouchi, S., "Combinatorial biosynthesis of plant medicinal polyketides by microorganisms." *Current opinion in chemical biology*, 2009. 13(2): 197-204.

Igarashi, Y., M.E. Trujillo, E. Martínez-Molina, S. Yanase, S. Miyanaga, T. Obata, H. Sakurai, I. Saiki, T. Fujita, and T. Furumai, "Antitumor anthraquinones from an endophytic actinomycete *Micromonospora lupini sp. nov. Bioorganic & medicinal chemistry letters*, 2007. 17(13): 3702-3705.

Jia, M., L. Chen, H.-L. Xin, C.-J. Zheng, K. Rahman, T. Han, and L.-P. Qin, "A friendly relationship between *endophytic fungi and medicinal plants: a systematic review." Frontiers* in microbiology, 2016. 7: 906.

Junchen, L., M. Irfan, and F. Lin, "Bioconversion of agricultural waste to ethanol: A potential source of energy." *Archives Des Sciences*, 2012. 65(12): 626-642.

Junker, R.R. and D. Tholl, "Volatile organic compound mediated interactions at the plant-microbe interface." *Journal of chemical ecology*, 2013. 39(7): 810-825.

Kalscheuer, R., T. Stölting, and A. Steinbüchel, "Microdiesel: Escherichia coli engineered for fuel production." *Microbiology*, 2006. 152(9): 2529-2536.

Kaul, S., S. Gupta, M. Ahmed, and M.K. Dhar, "Endophytic fungi from medicinal plants: a treasure hunt for bioactive metabolites." *Phytochemistry Reviews*, 2012. 11(4): 487-505.

Kaul, S., M. Wani, K.L. Dhar, and M.K. Dhar, "Production and GC-MS trace analysis of methyl eugenol from endophytic isolate of Alternaria from rose." *Annals of Microbiology*, 2008. 58(3): 443.

Kharwar, R.N., V.C. Verma, A. Kumar, S.K. Gond, J.K. Harper, W.M. Hess, E. Lobkovosky, C. Ma, Y. Ren, and G.A. Strobel, "Javanicin, an antibacterial naphthaquinone from an endophytic fungus of neem, *Chloridium sp.*" *Current microbiology*, 2009. 58(3): 233-238.

Kim, S., S. Lowman, G. Hou, J. Nowak, B. Flinn, and C. Mei, "Growth promotion and colonization of switchgrass (*Panicum virgatum*) *cv. Alamo* by bacterial endophyte Burkholderia phytofirmans strain PsJN." *Biotechnology for Biofuels*, 2012. 5(1): 37.

Kim, S., D.-S. Shin, T. Lee, and K.-B. Oh, "Periconicins, two new fusicoccane diterpenes produced by an endophytic fungus *Periconia sp.* with antibacterial activity." *Journal of Natural Products*, 2004. 67(3): 448-450.

Knothe, G., J. Krahl, and J. Van Gerpen, *The biodiesel handbook.* 2015: Elsevier.

Krings, U. and R. Berger, "Biotechnological production of flavours and fragrances." *Applied microbiology and biotechnology*, 1998. 49(1): 1-8.

Kulkarni, N., J. Dalal, and M. Bodhankar, "Microbial Endophytes: Ecology and Biological Interactions." *International Journal of Research in Agricultural Sciences*, 2014. 1(5): 2348-3997.

Kuiper, I., E.L. Lagendijk, G.V. Bloemberg, and B.J. Lugtenberg, "Rhizoremediation: a beneficial plant-microbe interaction." *Molecular plant-microbe interactions*, 2004. 17(1): 6-15.

Kusari, S., M. Lamshöft, and M. Spiteller, "*Aspergillus fumigatus Fresenius,* an endophytic fungus from *Juniperus communis L. Horstmann* as a novel source of the anticancer pro-drug deoxypodophyllotoxin." *Journal of applied microbiology*, 2009. 107(3): 1019-1030.

Lee, S., M. Yap, G. Behringer, R. Hung, and J.W. Bennett, "Volatile organic compounds emitted by Trichoderma species mediate plant growth." *Fungal biology and biotechnology*, 2016. 3(1): 7.

Lee, J.C., E. Lobkovsky, N.B. Pliam, G. Strobel, and J. Clardy, "Subglutinols A and B: immunosuppressive compounds from the endophytic fungus Fusarium subglutinans." *The Journal of Organic Chemistry*, 1995. 60(22): 7076-7077.

Loiret, F., E. Ortega, D. Kleiner, P. Ortega-Rodés, R. Rodes, and Z. Dong, "A putative new endophytic nitrogen-fixing bacterium Pantoea sp. from sugarcane." *Journal of Applied Microbiology*, 2004. 97(3): 504-511.

Ludwig-Müller, J., "Plants and endophytes: equal partners in secondary metabolite production?" *Biotechnology Letters*, 2015. 37(7): 1325-1334.

Luo, S., T. Xu, L. Chen, J. Chen, C. Rao, X. Xiao, Y. Wan, G. Zeng, F. Long, and C. Liu, "Endophyte-assisted promotion of biomass production and metal-uptake of energy crop sweet sorghum by plant-growth-promoting endophyte Bacillus sp. SLS18." *Applied Microbiology and Biotechnology*, 2012. 93(4): 1745-1753.

Ma, Y., M. Prasad, M. Rajkumar, and H. Freitas, "Plant growth promoting rhizobacteria and endophytes accelerate phytoremediation of metalliferous soils." *Biotechnology advances*, 2011. 29(2): 248-258.

Meng, X., J. Yang, X. Xu, L. Zhang, Q. Nie, and M. Xian, "Biodiesel production from oleaginous microorganisms." *Renewable Energy*, 2009. 34(1): 1-5.

Mercado-Blanco, J. and P.A. Bakker, "Interactions between plants and beneficial Pseudomonas spp.: exploiting bacterial traits for crop protection." *Antonie van Leeuwenhoek*, 2007. 92(4): 367-389.

Mohanta, S., G. Sharma, and B. Deb, "Diversity of endophytic diazotrophs in non-leguminous crops-A Review." *Assam University Journal of Science and Technology*, 2010. 6(1): 109-122.

Moricca, S. and A. Ragazzi, "Fungal endophytes in Mediterranean oak forests: a lesson from Discula quercina." *Phytopathology*, 2008. 98(4): 380-386.

Mueller, G.M., *Biodiversity of fungi: Inventory and monitoring methods.* 2011: Elsevier.

Nassar, A.H., K.A. El-Tarabily, and K. Sivasithamparam, "Promotion of plant growth by an auxin-producing isolate of the yeast *Williopsis saturnus* endophytic in maize *(Zea mays L.)* roots." Biology and Fertility of soils, 2005. 42(2): 97-108.

Pimentel, M.R., G. Molina, A.P. Dionísio, M.R. Maróstica Junior, and G.M. Pastore, "The use of endophytes to obtain bioactive compounds and their application in biotransformation process." *Biotechnology Research International*, 2011. 2011.

Rodriguez, R. and R. Redman, "More than 400 million years of evolution and some plants still can't make it on their own: plant stress tolerance via fungal symbiosis." *Journal of Experimental Botany*, 2008. 59(5): 1109-1114.

Rosenblueth, M. and E. Martínez-Romero, "Bacterial endophytes and their interactions with hosts." *Molecular Plant-Microbe Interactions*, 2006. 19(8): 827-837.

Shu, R., F. Wang, Y. Yang, Y. Liu, and R. Tan, "Antibacterial and xanthine oxidase inhibitory cerebrosides from *Fusarium sp. IFB-121*, and endophytic fungus in *Quercus variabilis*." *Lipids*, 2004. 39(7): 667-673.

Sasaki, T., Y. Igarashi, M. Ogawa, and T. Furumai, "Identification of 6-prenylindole as an antifungal metabolite of Streptomyces sp. TP-A0595 and synthesis and bioactivity of 6-substituted indoles." *The Journal of Antibiotics*, 2002. 55(11): 1009-1012.

Shweta, S., J.H. Bindu, J. Raghu, H. Suma, B. Manjunatha, P.M. Kumara, G. Ravikanth, K. Nataraja, K. Ganeshaiah, and R.U. Shaanker, "Isolation of endophytic bacteria producing the anti-cancer alkaloid camptothecine from *Miquelia dentata Bedd.(Icacinaceae)*." *Phytomedicine*, 2013. 20(10): 913-917.

Sims, R.E., W. Mabee, J.N. Saddler, and M. Taylor, "An overview of second generation biofuel technologies." *Bioresource Technology*, 2010. 101(6): 1570-1580.

Song, Y., H. Li, Y. Ye, C. Shan, Y. Yang, and R. Tan, "Endophytic naphthopyrone metabolites are co-inhibitors of xanthine oxidase, SW1116 cell and some microbial growths. " *FEMS Microbiology Letters*, 2004. 241(1): 67-72.

Stadler, M. and B. Schulz, "High energy biofuel from endophytic fungi?" *Trends in Plant Science*, 2009. 14(7): 353-355.

Strobel, G., A. Ericksen, J. Sears, J. Xie, B. Geary, and B. Blatt, "*Urnula sp.*, an endophyte of *Dicksonia antarctica*, making a fragrant mixture of biologically active volatile organic compounds." *Microbial Ecology*, 2017. 74(2): 312-321.

Strobel, G.A., "Methods of discovery and techniques to study endophytic fungi producing fuel-related hydrocarbons." *Natural Product Reports*, 2014. 31(2): 259-272.

Strobel, G., S.K. Singh, S. Riyaz-Ul-Hassan, A.M. Mitchell, B. Geary, and J. Sears, "An endophytic/pathogenic *Phoma sp.* from creosote bush producing biologically active volatile compounds having fuel potential." *FEMS Microbiology Letters*, 2011. 320(2): 87-94.

Strobel, G.A., B. Knighton, K. Kluck, Y. Ren, T. Livinghouse, M. Griffin, D. Spakowicz, and J. Sears, "The production of myco-diesel hydrocarbons and their derivatives by the endophytic fungus *Gliocladium roseum* (NRRL 50072)." *Microbiology*, 2008. 154(11): 3319-3328.

Strobel, G., B. Daisy, U. Castillo, and J. Harper, "Natural products from endophytic microorganisms." *Journal of Natural Products*, 2004. 67(2): 257-268.

Strobel, G. and B. Daisy, "Bioprospecting for microbial endophytes and their natural products." *Microbiology and Molecular Biology Reviews*, 2003. 67(4): 491-502.

Strobel, G.A., E. Dirkse, J. Sears, and C. Markworth, "Volatile antimicrobials from Muscodor albus, a novel endophytic fungus." *Microbiology*, 2001. 147(11): 2943-2950.

Sturz, A., B. Christie, B. Matheson, W. Arsenault, and N. Buchanan, "Endophytic bacterial communities in the periderm of potato tubers and their potential to improve resistance to soil-borne plant pathogens." *Plant Pathology*, 1999. 48(3): 360-369.

Taechowisan, T., J.F. Peberdy, and S. Lumyong, "Isolation of endophytic actinomycetes from selected plants and their antifungal activity." *World Journal of Microbiology and Biotechnology*, 2003. 19(4): 381-385.

Teles, H.L., R. Sordi, G.H. Silva, I. Castro-Gamboa, V. da Silva Bolzani, L.H. Pfenning, L.M. de Abreu, C.M. Costa-Neto, M.C.M. Young, and Â.R. Araújo, "Aromatic compounds produced by *Periconia atropurpurea*, an endophytic fungus associated with Xylopia aromatica." *Phytochemistry*, 2006. 67(24): 2686-2690.

Tejesvi, M.V., M. Kajula, S. Mattila, and A.M. Pirttilä, "Bioactivity and genetic diversity of endophytic fungi in *Rhododendron tomentosum Harmaja.*" *Fungal Diversity*, 2011. 47(1): 97-107.

Tumlinson, J.H., "The importance of volatile organic compounds in ecosystem functioning." *Journal of chemical ecology*, 2014. 40(3): 212.

Tomsheck, A.R., G.A. Strobel, E. Booth, B. Geary, D. Spakowicz, B. Knighton, C. Floerchinger, J. Sears, O. Liarzi, and D. Ezra, "*Hypoxylon sp.*, an endophyte of *Persea indica*, producing 1, 8-cineole and other bioactive volatiles with fuel potential." *Microbial Ecology*, 2010. 60(4): 903-914.

Trujillo, M.E., R.M. Kroppenstedt, P. Schumann, L. Carro, and E. Martínez-Molina, "*Micromonospora coriariae sp. nov.*, isolated from root nodules of *Coriaria myrtifolia.*" *International Journal of Systematic and Evolutionary Microbiology*, 2006. 56(10): 2381-2385.

Ul-Hassan, S.R., G.A. Strobel, E. Booth, B. Knighton, C. Floerchinger, and J. Sears, "Modulation of volatile organic compound formation in the Mycodiesel-producing endophyte *Hypoxylon sp.* CI-4." *Microbiology*, 2012. 158(2): 465-473.

Uthoff, S., D. Bröker, and A. Steinbüchel, "Current state and perspectives of producing biodiesel-like compounds by biotechnology." *Microbial Biotechnology*, 2009. 2(5): 551-565.

Van Loon, L., "Plant responses to plant growth-promoting rhizobacteria," in *New Perspectives and Approaches in Plant Growth-Promoting Rhizobacteria Research.* 2007, Springer. 243-254.

Vessey, J.K., "Plant growth promoting rhizobacteria as biofertilizers." *Plant and Soil*, 2003. 255(2): 571-586.

Wang, Y., Y. Ohara, H. Nakayashiki, Y. Tosa, and S. Mayama, "*Microarray analysis of the gene expression profile induced by the endophytic plant growth-promoting rhizobacteria, Pseudomonas fluorescens FPT9601-T5 in Arabidopsis.* Molecular Plant-Microbe Interactions, 2005. 18(5): 385-396.

Wang, K.L.-C., H. Li, and J.R. Ecker, "Ethylene biosynthesis and signaling networks." *The plant cell*, 2002. 14(suppl 1): S131-S151.

Wani, Z.A., N. Ashraf, T. Mohiuddin, and S. Riyaz-Ul-Hassan, "Plant-endophyte symbiosis, an ecological perspective." *Applied microbiology and biotechnology*, 2015. 99(7): 2955-2965.

Wani, M.A., K. Sanjana, D.M. Kumar, and D.K. Lal, "GC–MS analysis reveals production of 2–Phenylethanol from Aspergillus niger endophytic in rose." *Journal of basic microbiology*, 2010. 50(1): 110-114.

Wilson, D., *Endophyte: the evolution of a term, and clarification of its use and definition.* Oikos, 1995: 274-276.

Witzell, J., J.A. Martín, and K. Blumenstein, "Ecological aspects of endophyte-based biocontrol of forest diseases," in *Advances in Endophytic Research* 2014, Springer. 321-333.

Xu, L., L. Zhou, J. Zhao, J. Li, X. Li, and J. Wang, "Fungal endophytes from Dioscorea zingiberensis rhizomes and their antibacterial activity." *Letters in Applied Microbiology*, 2008. 46(1): 68-72.

Yan, L.-L., N.-N. Han, Y.-Q. Zhang, L.-Y. Yu, J. Chen, Y.-Z. Wei, Q.-P. Li, L. Tao, G.-H. Zheng, and S.-E. Yang, "Antimycin A 18 produced by an endophytic Streptomyces albidoflavus isolated from a mangrove plant." *The Journal of Antibiotics*, 2010. 63(5): 259.

Zinniel, D.K., P. Lambrecht, N.B. Harris, Z. Feng, D. Kuczmarski, P. Higley, C.A. Ishimaru, A. Arunakumari, R.G. Barletta, and A.K. Vidaver, "Isolation and characterization of endophytic colonizing bacteria from agronomic crops and prairie plants." *Applied and Environmental Microbiology*, 2002. 68(5): 2198-2208.

In: Trends in Biochemistry and Molecular Biology
Editors: Hossain Uddin Shekhar et al.
ISBN: 978-1-53616-434-3

Chapter 14

MORPHOLOGICAL AND MOLECULAR DETECTION OF FUNGI IN BANGLADESH

Shamim Shamsi*
Department of Botany, University of Dhaka University, Dhaka, Bangladesh

ABSTRACT

Bangladesh is a land of natural resources with diversified flora and fauna. Fungi are one of the important components of biodiversity and especially, the mycoflora because of its potential impact onhuman economy and food security. During the tenure of 1952 to 2018, the mycoflorareported from Bangladesh were studied and morphological and molecular detection of fungi has been presented in this account.

Keywords: morphological, molecular detection fungi, Bangladesh

1. INTRODUCTION

Bangladesh is low-lying, mainly riverine country located in South Asia with a coastline of 580km (360mi) on the northern littoral of the Bay of Bengal. The delta plain of the Ganges (Padma), Brahmaputra (Jamuna), and Meghna Rivers and their tributaries occupy 79 percent of the country. Four uplifted blocks (including the Madhupur and Barind Tracts in the centre and northwest) occupy 9 percent, and steep hill ranges up to ca 1,000 m high occupy 12 percent in the southeast (the Chittagong Hill Tracts) and in the northeast. Straddling the Tropic of Cancer, Bangladesh in general possesses a

*Corresponding Author's E-mail: prof.shamsi@gmail.com.

luxuriant vegetation. However, only a small portion of the country's land surface is covered with forests [1].

Climate of Bangladesh is suitable for growth of microflora on various substratum. Most fungi are microscopic because of the small size of their structures, and their saprophytic lifestyles in soil or on dead decaying matter. Some fungi are symbionts of plants, animals, or other fungi and majority of them are also parasites on plants, animals and humans. They started their life cycle from a single spore or a hyphal fragment and gradually colonize on respective substratum. Individual fungus become noticeable either as mushrooms or as molds, when asexual or sexual fruiting structure are developed. Along with actinomycetes and bacteria some fungi play significant role in the decomposition of organic matter and have fundamental roles in nutrient cycling and exchange them in the environment. Fungi have long been used as a direct source of human food, in the form of mushrooms and truffles. They are also used in food processing and in the fermentation of various food products, such as wine, beer, and soy sauce. Since the 1940s, fungi have been used for the production of antibiotics, and, more recently, various enzymes produced by fungi are used industrially. Fungi are also used as biological pesticides to control weeds, plant diseases and insect pests. Many species produce bioactive compounds called mycotoxins, such as alkaloids and polypeptides, which are toxic to animals including humans. The fruiting structures of a few species contain psychotropic compounds and are consumed recreationally or in traditional spiritual ceremonies. Fungi can break down manufactured materials and buildings, and become destructive pathogens of humans and other animals. Plant pathogenic fungi *Phytophthora infestans*, *Drechslera oryzae*, *Pyricularia oryzae*, *Puccinia graminis-tritici* and speices of *Erisiphae* are causing tremendous losses of crops due to fungal diseases [e.g., Late blight of potato, Brown spot and blast diseaserice of rice, wheat rust and members of cucurbitaceae, legume plants and ornamentals]. Some species of *Alternaria*, *Aspergillus*, *Fusarium* and *Rhizopus*are responsible for detoriation of storage grains. Species of *Fusarium*, *Rhizopus* and yeast also cause food spoilage which have a significant impact on human food supplies and local economies [2,3,4].

Around 120,000 species of fungi have been described by taxonomists, but the global biodiversity of the fungus kingdom is not fully understood. A 2017 estimate suggests there may be between 2.2 and 3.8 million species. Mycology is the study of fungi in various aspects. After the invention of microscope in the 17th century mycology rapidly developed. Pier Antonio Michelii considered as father of Mycology. *Nova plantarum genera* written by Michelii was published in 1729. In 1953 Carl Linnaeus introduced the binomial system of nomenclature in his *Species plantarum*. Christian Hendrik Persoon (1761–1836) a Dutch scientist established the first classification of mushrooms with such skill as to be considered a founder of modern mycology. Anton de Bary, in 20th century extensively worked on life cycle of pathogenic fungi and his research strengthen the field of plant pathology [2,3].

Modernization of Mycology was seen in 20thcentury that has come from advances in biochemistry, genetics, molecular biology, and biotechnology. The use of DNA sequencing technologies and phylogenetic analysis has provided new insights into fungal relationships and biodiversity, and has challenged traditional morphology-based groupings in fungal taxonomy [2].

In mycology, species have historically been distinguished by a variety of methods and concepts. Classification based on morphological characteristics, such as the size and shape of spores or fruiting structures, has traditionally dominated fungal taxonomy. Species may also be distinguished by their biochemical and physiological characteristics, such as their ability to metabolize certain biochemicals, or their reaction to chemical tests. The biological species concept discriminates species based on their ability to mate. The application of molecular tools, such as DNA sequencing and phylo-genetic analysis, to study diversity has greatly enhanced the resolution and added robustness to estimates of genetic diversity within various taxonomic groups [2].

2. Materials and Methods

2.1. Collection of Samples

Micro fungi are mostly members of Chytridiomycetes, Oomyctetes, Zygomycetes, Ascomyctetes and Deuteromycetes and few are Basidiomycetes. Macrofungi are mostly members of the class Basidiomycetes. Anamorphic fungi are mostly members of Deuteromycetes which remain in soil debries as saprophytes, facultative parasites with cereal crops, vegetables, fruits, fiber yielding plants, oil yielding plants and ornamental plants.

Most of the micro fungi were found as pathogens or saprophytes on stem, leaf, woody debris and leaf litter environment. Asexual fruiting structures of these fungi were studied directly from the samples or isolated from the samples. Saprophytic fungi mostly isolated from soil debris following 'serial dilution'. Pathogenic or saprophytic fungi associated with plant parts were isolated on PDA (Potato Dextrose Agar) medium following 'tissue planting method' [5]. Seed borne fungi were isolated following 'blotter' and 'paper towel method' method' [6].

The most frequently collected species of the genera are *Alternaria, Aspergillus, Cercospora,* so farrecorded on crop plants, storage grains, vegetables, ornamentals, forest tree, fibre yielding plants, oil yielding plants and medicinal plants are *Cladosporium, Colletotrichum, Curvularia, Drechslera, Fusarium, Penicillium, Pseudocercospora* and *Rhizopus.*

2.2. Morphological Identification

Identification of an organism is key to its classification and taxonomy. In the past, fungi and other organisms have been identified based on their morphological characteristics. The Saccardo system was primarily based on morphology of sporulation structures as they are known in nature as well as the morphology and pigmentation of conidia and conidiophores [7,8]. Anamorphic fungi are identified following relevant standard literatures [9,10, 11, 13].

2.3. Mushroom Collection and Processing

Mushroom were collected at fruiting stage when beautiful basidiocarp were formed in nature after sexual reproduction. Samples were usually collected during day time and field characteristics of mushrooms were recorded in the data sheet. During collection necessary materials and equipment such as isolation kit, slants, petridishes containing medium, isolation chamber, typed data sheet, digital camera for photography, digging equipment, heat convector card board, chemical reagents for biochemical analysis were arranged. Soft mushrooms were collected carefully by using forceps/free hand while the mushrooms growing on wood were collected along with small part of wood. The photograph was taken in their natural habitat. Each sample was wrapped in the paper envelop along with field notes, date of collection, habitat, locality and specimen number on tag.

Freshly harvested mushroom was washed by water for removing debris. Fleshy mushroom is highly perishable as it is susceptible to deterioration by the enzyme and microorganism. During the analysis period some precautions before processing of mushroom Short term preservation were followed and another is long term preservation on the basis of study purpose and structure of the mushroom. Collected mushrooms were dried by using sun heat (Sundry) when collected mushroom from remote area where electricity was not available. But most of the collected samples were dried by using electrical air flow drier. Samples were stored in Ziploc poly bag during research period with Silica gel at the rate of 10% of dry basis for further study [14].

2.4. Mushroom Identification

The collected specimens were brought to the laboratory. The measurements of various parts of mushrooms were recorded and morphological features were observed. The taxonomy has been done on the basis of macro and microscopic characteristic according to the literatures.

The morphological parameters used for the identification of mushroom specimens such as- cap color, cap surface, cap margin, cap diameter, stipe length, gill attachment, gill spacing and spore dimension. Microscopic features were carried out using standard microscopic methods. The information of the various characters stated was used to identify each specimen by comparison with illustrations in colour field guides and also by the use of descriptions and keys.

The specimens were dried in hot air at 40°-50°C and stored in air tight containers with some silica gel for further microscopic studies. The spores of collected mushrooms were mounted on slide by using glycerine and cotton blue for their size measurement. The spore diameter and the photograph of spores were calculated using the Motic Microscope (Motic images plus 2.0) with the magnification of 40x. Collected mushroom species have been categorized as edible, inedible and medicinal uses based on available literature [15,16,17].

2.5. Molecular Identification

Molecular identification and quantification of several techniques have been developed for organisms in the past two decades. Fungi represent as the primary decomposers in ecosystems. It is conservatively estimated that 1.5 million species of fungi exist [2]. Many fungal species are important plant and human pathogens [3]. Rapid and accurate detection of fungal pathogens to species or strain level is often essential for disease management strategy. Different molecular genotypes/varieties can also exist within species, and may have different pathogenic profiles and virulence levels to the host. In addition, unculturable and non-sporulating fungi remain a major challenge when studying biotrophic, endophytic, and mycorrhizal groups. Therefore, novel techniques are required when attempting to detect fungi in the environment [18].

Mian et al. (2003) detected diversity of the blast pathogen *Pyricularia grisea* from Bangladesh by DNA fingerprinting [19].

Molecular identification of three species viz .*Pythium* sp., *Pythiumrhizo-oryzae* and *Pythium catenulatum* under the class oomycetes were done by Rahman and Sarowar (2016) [20].

Eleven of the isolates were isolated in Potato Dextrose Agar (PDA) plates and were identified using molecular methods that included DNA extraction, PCR amplification and subsequent sequencing of the ITS region of the genomic DNA of the samples. BLAST analysis to GenBank.

3. Results and Discussion

Fungi isolated and identified from 1952 to 2018 in Bangladesh from various substrate and host are compiled in this account as far as possible. This data will be useful in the compilation of fungal biodiversity of Bangladesh as well protection of plants from fungal attack.

Ishaque and Talukdar (1967) and Talukdar (1974) extensively worked on fungal diseases of Bangladesh and reported the fugal pathogen involved with respective diseases [21,22]. Khan et al. (1969) contribute a lot on Agaricales of Bangladesh. Author worked on hyperparacsitic fungi associated with rust pustules of *Justicia ganderusa*. Khan and Shamsi reported hyphomycetous fungi from Bangladesh [23,24,25,26,111]. Bakr et al. (2007) compiled various diseases and fungi and published it. This publication is a remarkable document in the field of Mycology and Plant pathology in Bangladesh [27].

Siddiqui et al. (2007) have been reported 275 fungal species under 125 genera from Bangladesh [28]. Shamsi (2017, 2017) presented check list of forty species of lower fungi and 208 species of anamorphic fungi under 51 genera of the family Dematiaceae from Bangladesh [29,30].

3.1. Morphological Identification of Fungi

In Bangladesh Ahmed (1952) reported fungal diseases of jute plants. Jute plant (*Corchorus oilotrius* and *C. capsularis*) is one of the important cash crop of Bangladesh [31]. Stem rot of jute caused by *Macrophomina phaseolina*, black band of jute caused by *Botryodiplodia theobromae*, anthracnose caused by *Colletotrichum corchori*, Die back caused by *Glumerella singulata*, soft rot caused by *Sclerotium rolfsii* and wilting caused by *Rhizocrona solani* are the destructive diseases of jute. Jute pathogen are detected by Ahmed (1952-1956), Ahmed and Ahmed (1966) [27, 32].

Ahmed and Hossain (1985) survey crop diseases in Bangladesh and establish a herbarium at BARI. Plant Pathology Division, BARI, Joydebpur, Gazipur [33].

Shamsi et al. (2016) isolated and reported 30 species of fungi from different parts of infected jute plants during the tenure of October 2008 to January 2016.The isolated fungi were *Alternaria alternata*, *Aspergillus flavus*, *A. fumigatus*, *A. niger*, *A. terreus*, *Cercospora corchori*, *Cladosprium cladosporioides*, *C. oxysporum*, *Curvularia lunata*, *Colletotrichum corchori*, *C. gloeosporioides*, *Corynespora cassiicola*, *Dendryphiella vinoa*, *Diplodia* sp., *Eurotium* sp., *Fusarium avenaceum*, *Fusarium gramienarum*, *Fusarium* sp., *Gebberella zeae*, *Lasidiplodia theobromae*, *Macrophomina phaseolina*, *Phoma* sp. *Phomopsis* sp., *Rhizopus* sp. *Rhizoctonia solani*, *Sclerotium rolfsii*, *Stachybotrys* sp, *Torula* sp. *Trichothecium roseum* and *Ulocladium* sp. Association of *Stakybotrys* sp. and *Ulocladium* sp. with jute seeds is a new record [34].

Considerable works on fungal diseases of rice plant and the detection of rice seed borne fungi have been done in Bangladesh. From Bangladesh Miah et al. (1985), Mia et al.(1993), Shahjahan et al., (1993), Shamsi (1999) and Shamsi et al. (2003) reported fungal diseases of rice plants and associated fungi [35, 36, 37, 38, 39]. The common fungal pathogen and seed borne fungi of rice are *Drechsleara oryzae*, *Pyricularia oryzae*, *Curvularia lunata*, *C. oryzae*, *Cercospora janseana*, *Epicoccum purpurescens*, *Fusarium moniliformae*, *Microdochium oryzae*, *Rynchosporium oryzae Sarocladium oryzae*, *Aspergillus* spp., *Penicillium* spp., *Nigrospora oryzae*, *Tilletia barclayana*, *Phoma* sp., *Pyrenochaeta oryzae*, *Trichoconis padwickii* and *Ustilaginnoides virens*. *Colletrichum* sp. was detected from asomatic rice [40]. Ora et al, (2015) were identified *Rhizopus stolonifer*, *Aspergillus* spp. *Fusarium moniliforme*, *Phoma* sp. *Bipolaris oryzae*, *Curvularia lunata*, *Penicillium* sp. *Alternaria tenuissima*, *Nigrospora oryzae*, *Chaetomium globosum* and *Tilletia barclyana* on two hybrid rice varietiesHira-1 and Hira-2 [41].

Twenty five species of fungi were isolated from diseased rice grains of four commercially cultivated rice varieties namely BRRI-28, 29, Kalijira and Pajam collected from 14 districts of 7 divisions and 40 rice samples viz. Hybrid2, 3, 4, BR-7, 11, 12, 14,16, 22,23, 25, 26 and BRRI -28 to BRRI-55 were collected from Bangladesh Rice Research Institute at Joydebpur. Amongst twenty five species of fungi nine viz., *A. alternata*, *A. flavus*, *C. lunata*, *D. oryzae*, *F. moniliforme*, *F. solani*, *M. oryzae*, *P. guepinii* and *S. oryzae* were found to be pathogenic to rice seeds Chowdhury et al. (2018) [42].

Fungal diseases of wheat and associated fungi were extensively studied by Fakir et al.(1987), Basak et al. (1987), Alam et al.(1995) and Malaker et al. (2005c). Leaf rust caused by *Puccinia triticina*, Black point, seedling blight and BpLb (Bipotaris leaf blight) caused by *Bipolaris sorokiniana* are the most destructive diseases of wheat [43, 44, 45, 46].

In recent years, blast of wheat caused by *Pyricularia oryzae* is a devastating disease of wheat in Bangladesh (Sadat and Choi 2016) [47].

Momtaz et al. (2018) reported thirty five fungal species, representing 20 genera were associated with Bipolaris leaf blight infected leaves of twenty one wheat varieties. The isolated fungi were *Alternaria alternata*, *A. triticina*, *Arthirinium* sp., *Aspergillus flavus*, *A. fumigatus*, *A. niger*, *A. terreus*, *Aspergillus* sp., *Bipolaris cynodontis*, *B. oryzae*, *B. sorokiniana*, *B. tetramera*, *B. victoriae*, *Bispora antenata*, *Chaetomium globosum*, *Chaetophoma* sp., *Cladosporium cladosporioides*, *Coniothyrium* sp., *Curvulariaaffinis*, *C. lunata*, *C. pallescens*, *Drechslera dematioidea*, *D. hawaiiensis*, *Epicoccum purpurascens*, *Eurotium* sp., *Fusarium moniliforme*, *F. nivale*, *F. semitectum*, *Nigrospora oryzae*, *N. sacchari*, *Penicillium digitatum*, *Pestalotiopsis guepinii*, *Rhizopus stolonifer*, *Syncephalastrum racemosum* and *Trichoderma viride*. *Bispora antenata* is new record for

Bangladesh. The infected samples were collected from eight districts (Dhaka, Gazipur, Dinajpur, Joypurhat, Pabna, Sirajgonj, Kushtia and Chuadanga) of Bangladesh [48].

Maize (*Zea mays* L.) is also a popular cereal crop in Bangladesh. The plant suffer from five major diseases under the agro-ecological conditions of Bangladesh Seed and soil borne fungi associated maize were found involved in the causal complex of seed rot and seedling blight, such as *Pythium aphanidermatum, Rhizoctonia*, Stack rot caused by *Geberella zeae*, *Fusarium moniliformae*, *Diplodia maydes*, *Helminthosporium*, *Colletotrichum graminicola*, *Aspergillus* and *Penicillium.* Leaf blight is caused by *Drechslera turcicum* and *D. maydis* [27]. Ear rot caused by *Nogrospora oryzae* and *Penicillium* sp. Downey mildew was caused by *Sclerospora* spp. Shamsi and Yasmin (2007) reported *Curvularia harveyi* Shipton on mayze [49].

Leaf blight caused by *B. sorokiniana* and foot rot caused by *S. rolfsii* are major diseases of Barley and Millets [27].

Sorghum (*Sorghum vulgare*) is also the cereal crop that are cultivated in Bangladesh. Seeds of sorghum were collected from eight different locations of Bangladesh and fungi were isolated following Blotter method. The isolated fungi were *Curvularia lunata*, *Fusarium moniliforme*, *Alternaria tenuis*, *Bipolaris sorghicola*, *Colletotrichum graminicola*, *Botrytis cinerea*, *Aspergillus niger*, *Penicillium oxalicum* and *A. flavus* [27].

A total of 34 fungi representing 23 genera were recorded from the seed samples of blackgram (*Vigna mungo*), chickpea (*Cicer arietinum*), Lentil, mungbean, fieldpea and soybean. The important pathogenic fungi enlisted were *B. theobromae*, *Botrytis cinerea*, *Cercospora* sp., *Colletotrichum dematium*, *C. lindemuthianum*, *F. moniliformae*, *F. oxysporum*, *F. solani*, *M. Phaseolina* and *R. solani*. Fungal species recorded in started seeds included *Aspergillus*, *Alternaria*, *Colletotrichum*, *Crynespora*, *Drechslera*, *Fusarium*, *Macrophomina*, *Myrothecium*, *Phoma* and *Rhizoctonia*. Out of this *F. oxysporum*, *F. solani*, *F. equiseti*, *Myrothecium roridum*, *Drechslera* spp., *Alternaria tenuis*, *A. flavus*, *A. clavatus* and *M. phaseolina* were found pathogenic to germinating seed and seedling of mung bean [27].

Nine species of fungi were recorded on seeds of nine varieties of chickpea. The isolated fungi were *Alternaria alternata*, *A. flavus*, *A. niger, A. fumigatus*, *A. nidulans*, *Curvularia lunata*, *Penicillium* sp., *Rhizopus stolonifer* and *Trichoderma virid*e [50].

Leaf spot and pod spot of *Brassica* spp. The causal organism were recorded *Alternaria brassicae*, *A. brssicicola* [27,50]. The predominant and pathogenic seedborne fungi detected in mustard seed include *Alternaria brassicae*, *A. brssicicola*, *A. tenuis*, *A. flavus*, *A. niger*, *C. lunata*, *Penicillium* spp. and *R. stolonifer*. Shamsi and Hosen (2016) reported *Monochaetia karstenii* var. *gallica* (Stay.) Sutton on *Brassica napus* L. [51,52]. *Scerlotinia sclerotiarum* caused white spot of *Brassica* spp. [51].

Groundnut (*Arachis hypogaea* L. suffers from 14 fungal diseases which cause 30-40% yield loss. Common diseases of groundnut are Leaf spot, rust and stem rot are important causal agents of the diseases are *Cercospoa arachidicola*, *C. personeta*,

Puccinia arachidis and *Sclerotium rolfsii*. Associated fungi in groundnut seed includes *M. phasesolina* and *S. rolfsii* [27,51].

Shamsi and Sharmin (2012) reported a total of total 48 fungal species on groundnut. The fungi were *Alternaria alternata, Aspergillus flavus, A. fumigatus, A.niger, A. terreus, Aspergillus* sp., *Cercospora arachidicola, Chaetophoma* sp., *Cladosporium cladosporioides, Colletotrichum acutatum, Colletotrichum dematium, Colletotrichum gloeosporioides, Colletotrichum lindemuthianum, Colletotrichum orbiculare, Colletotrichum*sp.1, *Colletotrichum*sp. 2. *Curvularia lunata, Curvularia lunata*var. *aeria, Curvularia ovoidea, Curvularia pallescens, Curvularia pennisetii, Curvularia stapeliae, Curvularia uncinata, Cylindrocladium* sp. *Epicoccum purpurascens, Eurotium* sp., *Fusarium buharicum, F. equiseti, F. heterosporum*, F. *moniliforme* var. *subgluitans, F. semitectum, Fusarium* sp.1, *Fusarium* sp.2, *Fusarium* sp. 3, *Fusarium* sp.4, *Geotrichum*sp., *Lasiodiplodia theobromae* (=*Botryodiplodia theobromae*), *Monilia* sp., *Nigrospora sacchari, Phaeoisariopsis personata, Penicillium* sp.,*Pestalotia* sp., *Pleospora* sp.,*Puccinia arachidis, Rhizopus stolonifer, Robillarda* sp., *Sclerotium rolfsii, Sordaria* sp. and *Trichoderma viride* [53].

Sesame (*Sesamum indicum* L.) attacked by nine diseases (among which stem rot and Cercospora leaf spot are the most important. In sesame seed *Alternaria alternata, A. tenuis, A. flavus, A. niger, F. moniliforme. C. cassicola, M. phaseolina, Penicillium* spp. and *R. stolonifer* [27, 51]. Shamsi and Sultana (2012). Shamsi and Hosen (2016) detected *Fusarium merismoides* Corda. from sesame [54, 55].

In soybean seeds *Colletotricum dematium* var, *truncatum, M. phaseolina, F. oxysporum, Cercospora kikuchi, Aspergillus* spp., *Curvularia* spp. and *R.stolonifer* were recorded [27].

Late blight caused by *Phytophthora infestans* (Mont) de Bary is the most destructive diseases of potato in Bangladesh Twenty races of the fungus have been identified in this country [27].

Very little works on the detection of seedborne fungi are available in the country. In onion seed 10 different seed borne fungi were detected. Twenty six fungal species were detected from stored seeds of chilli. Fungal pathogen associated with different types of fruit rot of chilli were *Alternaria tenuis, B. theobromae, Cercospora capsici, Colletotrichum capsici, F. moniliformae, Phomopsis capsici* and *M. phaseolina* [27].

Among the vegetables tomato seeds attacked by *Alternaria* spp., *Aspergillus* spp., *Botrytis* spp., *Chaetomium* spp., *Curvularia* spp., *Fusarium* spp., *Penicillium* spp. and *Rhizopus* sp. [27].

In brinjal seeds *Phomopsis vexans, Alternaria solani, A. flavus, A. niger, Fusarium* spp., *Penicillium* spp. and *Rhizopus* sp. were recorded [27].

Citrus spp. are infected by *Penicillium digitatum* [27].

Shamsi et al. (2015) isolated *Aspergillus niger*, *Candida krusei, Fusarium* sp., *Rhizopus stolonifer*, *Penicillium digitatum* and *Trichoderma* sp. from lemon fruit in storage [27, 56].

White rust is a destructive disease of *Amrathus* spp. caused by *Albugo candida* [27]. Twelve species of fungi were associated with *Basella alba* and *B. rubra*. The isolated fungi were *Alternaria alternata*, *Ascochyta* sp, *Aspergillus, Aspergillus fumigatus, Aspergillus niger, Aspergillus terreus*, *Colletotrichum dematium*, *Curvularia pallescens*, *Fusarium oxysporum, Macrophomina phaseolina* and two species of *Penicillium.* Pathogenicity test showed that, *A. fumigatus, A. terreus, F. oxysporum* and *C. pallescens* were pathogenic to *Basella* spp. [22, 27, 57].

Sultana et al. (2014) isolated fungi from fruits and powdered coriendar, cucurma and red chili.Shamsi et. al. (2015). Shamsi and Hosen reported fungi associated with black cumin. *Aspergillus* spp and *Rhizopus* sp. was the predominating fungi with spices [58,59,60].

Mycoflora of *Momordica cochinchinensis*, *M. charantia, Trichosanthe sanguina, T. dioica,*and *Annona sqamosa* were studied and enlisted [61,62,63,64].

Cotton suffers from anthracnose, leaf spot, boll rot, rust and leaf spot. Diseases of cotton, its causal agents and associated fungi were extensively studied and recorded from 1974 to 2015 [22,27, 65,66,67,68]. Related publications are remarkable work in the field of cotton research.

Mycoflora of some herbal medicinal plants were examined and related causal agents and associated fungi were identified. *Smilax zeylanica, Datura metel, Senna alata, Clitoria ternatea*, *Oxalis* spp., *Catharanthus roseus, Rouwolfia serpentina, Huttuyania cordata* and *Aloe vera* [69, 70, 71, 72, 73, 74, 75, 76, 77, 78].

Red rot caused by *Colletotrichum falcatum* is a destructive disease of sugarcane [27].

Common members of cucurbitaceae, rose, dahlia and few legume plants are susceptible to species of *Erysiphe* and its anamorph *Oidium* causing powdery mildew disease [27].

Seed pathogens of cucurbits are *A. niger, A. flavus, A. sydawii, Chaetomium* sp., *Epicoccim* sp., *Penicillium* spp. *Doratomycets, Rhizopus* sp. *Ceratocystis* sp. *Curvularia* sp. and *Corynespora* sp. [27].

Bamboo blight are caused by *Conithyrium fucklli* and *Sarocladium oryzae* are the common fungi causing bamboo blight [27].

Jack fruit are infected by *C. gloeosporioides*, *R. atrocarpi* and *Fusarium* sp. [27].

Banana is infected by *Colletotrichum musae* and Fusarium oxysporumf. sp. cubens. Mango, guavaand papaya tree and post harvest fruits areinfected by C.*gloeosporioides* and *Fusarium* spp. [27].

Fungal disease of rare plants *Sechium edule*, *Theobromae cacao*, Breabfruit (*Atrocurpu saltalis*), two species of liliaceae, *Lea macrophylla* were examined and respective fungi were enlisted [79, 80, 81, 82, 83].

Diplocarpon rosea and its anamorph *Marssonina rosea* causing black spot of rose. Rust of rose is caused by *Phragmidium micronatum. Diplodia rosea*. Black spot of rose caused by *Pestalotiopsis guepinii* were recorded for the first time [27, 84, 85].

Gerbera spp. is commercially grown in Bangladesh. Anthracnose and blight is the destructive diseases of the plant. *Alternaria citrii, A. tennuissima, Colletotrichum capcisi, C. dematium, C. coffeanum* and*Curvularia clavata* were found to be pathogenic to *Gerbera* plant [86].

Twenty species of fungi were recorded from *Tagetes erecta* and *Tagetes patula* [27, 87].

Powdwey mildew, Grey mold, blight and white mold caused by species of *Erysiphe*, Botrytis cinerea, *Alternaria alternata*, *Curvularia lunata, Aspergillus fumigates* and *Slerotinia sclerotiorum* are the destructive diseasesof*Tagetes* spp. Top dying of sundary is caused by *Botyosphaeria ribis* [27].

Wilt and dye back of sissoo is caused by *Colletotrichum gloeosporioides* [27]. Shamsi et al. (2008) detected *Phyllactenia dalbergae* Piroz. and its anamorph *Ovulariopsis sissoo*sp. nov. on *Dalbergia sissoo* Roxb. from Bangladesh. In 2012 *Cylidrocladium*, an unidentified rust fungi and other associated fungiwere detected from sisso [27,88,89].

Rubber diseases are powdery mildew caused by *Oidium hevae*, die back caused by *C. gloeosporioides*, root rot caused *Ganoderma pseudoferrum*, foot and root rot caused by *F. oxysporum* [27]. Shamsi and Chowdhury.(2014) isolated *Aspergillus flavus, A. fumigatus, A. niger, Cladosporium cladosporioidis, Corynespora cassiicola, Colletotrichum* sp., *Fusarium* sp, *Mucor* sp., *Penicillium* sp. and *Trichoderma viride* from processed and unprocessed rubber sheets [90].

Shamsi et al. reported the fungi associated with leaves of *Sonneratia aetala*and *S. caseolaris* from Rangabali caostal zone of Bangladesh [91].

Hyperparasitic fungi also detected as as biocontrol agent for rusts [27]. Biocontrol agents also detected on phylloplane in nature [92,93,94].

*Trichoderma herzineum, T.koninzii, T. viride, Gliocladium viride*are common biocontrol agents in soil [27]. In addition to *T. viride, Aspergillus niger, A. flavus* and *A, fumigatus* also showed antifungal activities against various fungi [27,95,96,97].

Shamsi et al. (2015) also detected biocontrol agents from contaminated fungal culture plates [98].

Shamsi et al, (2014) observed fungal colonies growing as patches round the year in 2011 on the distempered indoor damp walls of ground floor of the Department of Botany, University of Dhaka. Fungi were isolated and identified as *Cladosporium oxysporum, Curvularia lunata* and species of *Fusarium* and *Penicillium* [99].

The materials of this investigation was the indoor damp walls of the Department of Botany, Dhaka University and a private house of New DOHS within Dhaka City Samples were collected from the indoor damp walls during the period of May 2012 to May 2013.

A total of nine species of fungi and one sterile fungi were isolated from the samples. The fungi were *Aspergillus flavus, A. Fumigates A. niger A. niger A.terreusA. ustus, Fusarium* sp., *Penicillium* sp. *Ulocladium* sp. and a fungus with sterile mycelia [100].

Aeromycoflora of Karwan bazar vegetable market in relation to occurrence and prevalence, their seasonal variation and distribution of fungal spores in air was studied from October 2010 to September 2011. During the study a total of 31 fungal species belonging to 18 genera of fungi were isolated and identified [101]. Sultana et al. (2015) studied monthly variation in air borne fungal propagules of Dhaka Metropolitan city and recorded that *Aspergillus* spp. were the predominant fungi [102].

Shamsi et al. (2015) detected crystal formation in the culture of *Aspergillus niger, A. fluvus*, *Aspergillus* sp., *Bipolaris sorokiniana, Cladosporium* spp., *Colletotrichum gloeosporioides, Colletotrichum orbiculare, Curvularia* spp., *Fusarium* sp., *Paecilomyces* spp.,*Penicillium* sp., *Pestalotia* sp., *Pestalotiopsis guepinii,Phaeoisariopsis personata, Trichodermaviride* andsevensterile culture offungi. Eighteen types of crystallographic structures were detected from different fungal isolates. This is the first report of formation of crystal in fungal cultures from Bangladesh [103].

Ascobolus sp. *Chaetomim globosum*, *C. magnum*, *Botryosphaeria ribis,* G*lumeralla cingulata*, *Nurospora crassa*, *Botryotrichum* sp. and *Gibberella fujikuroi* are common asocomycetous fungi in Bangladesh [28]. *Sordaria* sp. [27], *Microasacu*s sp. *Emerecella* n*idulans*, *Gibberella zeae*, *Euritium rubrum* and G*lumeralla*, are also reported from Bangladesh [27,103,104,105,106,107].

Shamsi and Mamun (2016) noticed diversity of fungi associated with fallen deadspadix of *Cocos nucifera* [108].

Mazid and Khan, Mazid and Nahar, and Khan et al. (1968 and 1969 reported agarics of Bangladesh (East Pakistan) [27, 109,110,111].

Wood decaying fungi are identified as follows: *Dedalia flavida*, *D. coniferogosa*, *D. quersina*, *Fomes albomarginatus, F. sublinteus*, *Ganoderma lucidum*, *Hexagonia tenuis*, *Irpex flavus*, *Lentinus badius*, *L. palisort*, *Polyporus versatalis*, *P. ostresformis*, *Polystictus xanthopus*, *P. hirsutus* and *Sterium petaloides* [27].

Fifty five fungi have been detected in seeds of 17 different tree species. Seed borne fungi responsible for causing seed rots and germination failure. Most common seed borne fungi are *Botryodiolodia theobromae*, *Fusarium* spp., *Macrophomina phaseolina*., and species of *Aspergillus* and *Penicillium*. Out of f 74 seedling diseases known to occur on 38 different species in the plantation nurseries, most damaging diseases are damping off (*Fusarium* spp*.,* *Phytophthora* sp. and *Rhizoctonia solani*), seedling blight (*Colletotrichum* sp., *Coniclla* sp., F*usarium* sp., and *Guignardia calami*) collar and root rots (*Fusarium solani*, *Phytophthora* sp. and *R. solani*), die-back (*B. theobromae*) and leaf spots (*Colletotrichum capsici*, *Pestalotia* sp. and *Thanatephorus cucumeris*) [27].

Two hundred forty nine decay fungi representing 44 different genera cause 24 types of wood rot in 122 wood species in the country. Among the decay fungi, species of

Daedalea, *Fomes*, *Ganoderma*, *Irpex*, *Lentinus*, *Polyporus, Polystictus, Porea* and *Trametes* are of major importance [27].

A survey was conducted to study the biodiversity, habitat and morphology of mushroom grown in leaved, deciduous and mixed forest of Bangladesh. A total of 117 samples were collected from nine selected districts of Bangladesh viz. Barisal, Borguna, Patuakhali, Perojpur, Jhalokathi, Bandorban, Dhaka, Gazipur and Tangail. About 12 different species were found under 10 families viz. Amanitaceae (*Amanita bisporigera*), Pyronemataceae (*Aleuria aurantia*), Boletaceae (*Boletus subvelutipes*), Agaricaceae (*Agaricus* sp.), Tricholomataceae (*Callistosporium* sp.), Marasmiaceae (*Gymnopus* sp.), Cortinariaceae (*Cortinarius corrugatus*) Mycenaceae (*Mycena epipterygia*) Entolomataceae (*Nolanea Strictia*), Ganodermataceae (*Ganoderma lucidum*, *Ganoderma applanatum*, *Ganoderma tsuage*) [112].

Rubina et al. (2017) identified macro fungi of National Botanical Garden, Dhaka. The fungi were *Ganoderma lucidum, G. tropicum, G.boninense*, *Ganoderma* sp., *G, tsugae*, *G. lipsiense*, *G. applanatum*, *G. lobatum, G. pfeifferi, Daedaleopsis confragosa*, *Russula. nobilis*, *Psathyrellacandolleana, Lycoperdonpyriforme, Flammulinavelutipes, Psilocybecubensis, Cantharelluscinereus,* C*repidotusvariabilis, Lepiotacristata, L. procera* and*Lepiota* sp. [113].

3.2. Molecular Detection of Fungi

In Bangladesh molecular detection of fungi is inadequate. From available literature few information on molecular identification of fungi are incorporated in this account.

Mian et al. (2003). Studied the blast pathogen *Pyricularia grisea* from Babgladesh by DNA fingerprinting [114].

Hussain et al. (2012) reported wilt of guava plants in Bangladesh. Sixteen isolates of *Fusarium oxysporum* was identified based on morphologicalcharacters. *Eleven* isolates were confirmed as *F. oxysporum* through polymerase chain reaction (PCR) using species specific primers designed from the conserved regions of 18S rRNA gene [115].

Four morphological mutants (albino, ropy, conidial band and buff) of *Neurospora crassa* were characterized based on morphological features and molecular markers. Most of the mutants viz; albino, conidial band and buff showed characteristic RAPD banding profile. However, no band was found in wild Ema, Ema and mutant ropy. Highest number of RAPD bands were found in albino. The mutant albino showed very different morphological and molecular features from the rest specimens [116]. Islam et al. (2016) was studied morphological and molecular characteristics ofwheat blast in Bangladeshcaused by a South American lineage of *Magnaporthe oryzae* [117].

Molecular identification of three species viz.*Pythium* sp., *Pythium rhizo-oryzae* and *Pythium catenulatum* under the class oomycetes were doneby Rahman and Sarowar (2016) [20].

Das et al. (2017) isolated and identified an endophytic Basidiomycete, Grammothele lineata *Corchorus olitorius* [118].

Naznin et al. (2017) detected the presence of plant pathogenic *Pythium* spp. around Bangladesh Agricultural University campus, Mymensingh, Bangladesh [119].

Tasnim et al. (2018) detected antimicrobial activity and Identification of bioactive volatile metabolites of Jute endophyticfungus *A spergillus flavus* [120].

A total of 24 fungal isolates was morphologically identified from twenty (BRRI dhan 56 to BRRI dhan 76] BRRI rice varieties, of which 13 fungal isolates could be identified by molecular technique during the tenure of January 2015 to July 2016. Among the isolated fungi, association of *Fusarium proliferatum, Microdochium fisheri, Pestalotiopsis oxyanthi* and *Phanerochaete chrysosporium* are first time recorded from Bangladesh [121].

Present document will be helpful for understanding the morphological and habitat diversity of fungi in Bangladesh as well as for management of fungal diseases in agriculture and forestry.

REFERENCES

[1] Anonymous, (2018). *Bangladesh*. From Wikipedia, the free encyclopedia. https:// en.wikipedia.org/wiki/Bangladesh.

[2] Anonymous, (2018). *Fungus*. From Wikipedia,the free encyclopedia. https:// en.wikipedia.org/wiki/Fungus.

[3] Agrios G.N.(2002). *Plant Pathology* (5th ed.). Academic Press Inc., N.Y.

[4] Alexopoulos C.J., Mims C.W., Blackwell M. (1996*). Introductory Mycology*. John Wiley and Sons.

[5] Anonymous, 1968. *Plant Pathologist's Pocket Book.* 1st ed. The Commonwealth Mycological Institute, England.pp. 267.

[6] Anonymous. 2014. *International Rules for Seed Testing*. International Seed Testing Association, Switzerl and. pp. 10.

[7] Barnett, H.L. and B.B. Hunter. 1972. *Illustrated Genera of Imperfect Fungi*. Burgess Publishing Company, U.S.A. Third Edition. pp. 241.

[8] Barnett, H.L.and Hunter, B.B. 1998. *Illustrated genera of imperfect fungi.* 4th ed. BurgessPublishing Company: Minneapolis, 241 p.

[9] Ellis, M.B. 1971. *Dematiaceous Hyphomycetes.* The Commonwealth Mycological Institute, England pp. 608.

[10] Ellis, M.B. 1976. *More Dematiaceous Hyphomycetes.* The Commonwealth Mycological Institute, England. pp. 507.

[11] Thom, C., and Raper, K.B. 1945. *A Manual of the Aspergilli.* The Williams & Wilkins Company. Baltimore. pp.373.

[12] Booth, C. 1971. *The Genus Fusarium.* Commonwealth Mycological Institute, Kew, Surrey, England. pp. 237.

[13] Sutton, B.C. 1980. The Coelomycetes. *Fungi Imperfecti with Pycnidia, Acervuli and Stromata.* Commonwealth Mycological Institute, England, pp. 696.

[14] Rubina, H., Aminuzzaman, F.M., Chowdhury, M.S.M. and Das, K. (2017). Morphological Characterization of Macro Fungi Associated with Forest Tree of National Botanical Garden, Dhaka. *Journal of Advances in Biology & Biotechnology* 11(4): 1-36.

[15] Purakasthya R.P., Chandra A. *Manual of Indian Edible Mushrooms.* Today and Tomorrow's Publication, New Delhi; 1985.

[16] Singer R. *The Agaricales in Morden Taxonomy.* J. Cramer, Weinheim, 4th ed. 1986; 912.

[17] Roy A, De A.B. Polyporaceae of India. *International Book Distributors.* Dehradun; 1998.

[18] Clement K. M. Tsui, J.Woodhall, W. Chen, C. A. Lévesque, A. Lau, Cor D. Schoen, C. Baschien, M. dJ. Najafzadeh, and G.S. de Hoog. 2011. Molecular techniques for pathogen identification and fungus detection in the environment. *IMA Fungus.* 2011 2(2): 177189.

[19] Mian, S.M., Stevens, C., Mia, M.A.T. 2003. Diversity of the blast pathogen *Pyricularia grisea* from Bangladesh by DNA fingerprinting. *Bangladesh J. Plant Pathol.* 19(1&2):81-85.

[20] Rahman K.M.M. and M.N. Sarowar. (2016). Molecular characterization of oomycetes from fish farm located in Mymensingh sadar during summer. *Asian J. Med. Biol. Res.* 2016, 2 (2), 236-246.

[21] Ishaque, M.J. Talukdar 1967. Survey of fungal flora of East Pakistan. *Agril. Pakistan.* **18**:17-26.

[22] Talukder, M.J. 1789086. Plant diseases in Bangladesh. *Bangladesh J. Agril. Res.* **1**(1):61-86.

[23] Khan, A.Z.M. Nowsher A. and S. Shamsi. 1983. Cercosporae from Bangladesh I.*Bangladesh J. Bot.* **12**(1): 66-80.

[24] Khan, A.Z.M. Nowsher A. and S. Shamsi. 1983. Cercosporae from Bangladesh II.*Bangladesh J. Bot.* **12**(2): 105-118.

[25] Khan, A.Z.M. Nowsher A. and S. Shamsi. 1986. Hyphomycetes from Bangladesh. *Bangladesh J. Bot.* **15**(2): 111-121.

[26] Khan, A.Z.M. Nowsher A., S. Shamsi. and R. Akhter. 2003. Hyphomycetes from Bangladesh II. *Bangladesh J. Bot.* **32**(1): 47-48.

[27] Bakr., M.A., Ahmed H.U. and M.A. Wadud Mian. 2007. Research on crop disease management at Bangladesh Agricultural University. *Advances in Plant Pathological Research in Bangladesh.* Plant Pathology Division. BAR. Gazipur. Bangladesh. pp. 344.

[28] Siddiqui, K.U., Islam, M.A., Begum, Z.N.A., Hassan, M.A., Khandker, M., Rahman, M.M., Kabir, S.M.H., Ahmad, M., Ahmed, A.T.A., Rahman, A.K.A. and Haque, E.U. (eds.), 2007. *Encyclopedia of Flora and Fauna of Bangladesh.* Vol.2. Cyanobacteria, Bacteria and Fungi. Asiatic Society of Bangladesh, Dhaka. 415 pp.

[29] Shamsi, S. 2017. Check list of fungi in Bangladesh: (lower Fungi). *Plant Environ. Dev.* **6**(1):1-9.

[30] Shamsi, S. 2017. Checklist of deuteromycetous fungi of Bangladesh. *I. J. Bangladesh Acad. Sci.* **41**(2):115-126.

[31] Ahmed, Q.A. 1952. Diseases of jute in East Pakistan. *Jute and Jute Fabrics.***7**: 147-151

[32] Ahmed, Q.A. and N. Ahmed. 1966. Fungi of East Pakistan. Supliment 1. *Pakistan Journal of Forestry.* **16**(4):402-411.

[33] Ahmed, H.U. and M.M. Hossain. 1985. *Crop disease survey and establishment of a herbarium at BARI.* Plant Pathology Division, BARI, Joydebpur, Gazipur. pp.107.

[34] Shamsi, S., R. Sultana and P. Chowdhury. 2016. Prevalence of fungi associated with different parts of jute plant. *Plant Environment Development.* **5**(1):19-25.

[35] Miah, S.A., A.K.M. Shahjahan, M.A. Hossain and N.R. Sharma. 1985. Survey of rice disease in Bangladesh. *Tropical Pest Management* 31(3): 208-213.

[36] Mia, M.A.T. 1993. Status of research on seed in Bangladesh and future need. Progress in Plant Pathology. *Procedings of the Fifth Biennial conference of the Phytopathological Society* held during 27-28 June 1993. Pp. 108.

[37] Shahjahan, A.K.M. 1993. *Practical approaches to crop pest and disease management in Bangladesh.* BARC. Dhaka, Bangladesh. pp.168.

[38] Shamsi, S. 1999. *Investigations into the sheath rot disease of rice (Oryza sativa L.) in Bangladesh.* Ph. D. thesis. Department of Botany University of Dhaka. Bangladesh.pp. xii + 132.

[39] Shamsi, S., A.Z.M. Nowsher A. Khan, A.K.M. Shahjahan and Siddique Ali Miah. 2003. Fungal species associated with sheaths and grains of sheath rot affected rice varieties from Bangladesh. *Bangladesh J. Bot.* **32**(1): 17-22.

[40] Shamsi, S., N. Nahar, P, Chawdhuri and S. Momtaz. 2010. Fungal diseases of three aromatic rice (*Oryza sativa* L.) *Journal of Bangladesh Academy of Sciences.* **34**(2):63-70.

[41] Qra, N., A.N. Faruq, M. Toihidul Islam and M Rahman M.M.2015. Detection and Identification of Seed orne Pathogens from Some Cultivated Hybrid Rice Varieties in Bangladesh. *Middle East Journal of Scientific Research* **10**(4):482-488.

[42] Chowdhury, P.S. Shamim and M. Abul Bashar (2019). Mycoflora associated with diseased rice grains and their pathogenic potentiality. Department of Botany, University of Dhaka. Dhaka-1000. Bangladesh. *Abstract presented in 18th annual conference* held in 16 Mach 2019, at Cox's Bazar, Chittagong.

[43] Fakir, G.M. 1987. *An annotated list of seed-borne diseases in Bangladesh Agril.* Inf. Service, Ministry of Agriculture and Forests. Dhaka. pp. 17.

[44] Basak, A.B., Karim M.B., Hoque, M.N., and Biawas A.P.1987. Studies on the fungi associated with different varieties of wheat seeds grown in Bangladesh. *Seed Research.* **15**(1):17-75.

[45] Alam, K.B., P.K. Malaker, M.A. Shaheed, M.U. Ahmed, F. Ahmed and M.S. Haque. 1995. Yield loss assessment of wheat due to Bipolaris leaf blight in Bangladesh. *Bangladesh J. Pl. Pathol.* **11**(1&2): 35-38.

[46] Malaker, P.K., S.K. Nayar, M. Prashad, S.C. Bhardwaj and M.M.A. Reza 2005c. Analysis of pathotypes and potulation of genes for resistance to leaf rust in wheat. *Bangladesh J. Agric.* 29&39:33-42.

[47] Sadat M.A. and J.Choi 2016. Wheat Blast: A New Fungal Inhabitant to Bangladesh Threatening World Wheat Production. *Bangladesh J. Pl. Pathol.* **33**(2): 103-108.

[48] Momtaz, M.S., S. Shamsi and T.K. Dey. 2018. Prevalence of fungi associated with Bipolaris leaf blight (BpLB) of different wheat varieties in Bangladesh. *Bioresearch Communications.* **4**(2):530-540.

[49] Shamsi, S. and A. Yasmin. 2007. *Curvularia Harveyi* Shipton: A new hyphomycetes record for Bangladesh. *Bangladesh J. Plant Taxon.***14** (1): 67-69.

[50] Shamsi, S. and A. Khatun. 2016. Prevalence of fungi in different varieties of chickpea (*Cicerarietinum* L.)seeds in storage. *Bangladesh Acad. Sci.* **40**(1):37-44.

[51] Bakr, M.A., M.A. Hossain and M.M. Karim. 2009. Gradient of oil seed crop diseasemanagement, fungal associations and mycotoxin contamination. *Advances in Oilseed Researches in Bangladesh.* Oilseed Research Centr. BARI., Gazipur, Bangladesh. pp180.

[52] Shamsi, S. and S. Hosen. 2016. New records of *Monochaetia karstenii* var. *gallica*(Stay.) Sutton on *Brassica napus* L. from Bangladesh. *Asiat. Soc. Bangladesh. Sci.* **42**(1):127-128.

[53] Shamsi, S. and S. Sharmin. 2012. *Fungal diseases of Groundnut from Bangladesh.* Lambert Publishers. Germany. pp. 54.

[54] Shamsi, S. and R. Sultana. 2012. New records of two species of *Corynespora* on sesame (*Sesamum indicum* L.) from Bangladesh. *Bangladesh Journal of Plant Pathol.*27(1&2):75-76.

[55] Shamsi and M. D. Hosen. 2016. *Fuarium merismoids* Corda. – A new record of anamorphic fungus for Bangladesh. *Bangladesh Acad. Sci.* **40**(2):207-209.

[56] Shamsi, S., N. Naher and T. Saha. 2015. Mycoflora associated with lemon (*Citrus lemon*) fruits in storage. *Bangladesh J. Plant Pathol.* **31**(1&2): 27-30.(published in 2016).

[57] Akter, S and S. Shamsi. 2013. Mycoflora associated with *Basella* spp. and their pathogenicpotentiality. *Bangladesh J. Sci. Res.* **26**(1&2):83-87.

[58] Sultana, T., S. Shamsi and M.A. Bashar. 2014. Fungi associated with common spices in Bangladesh. *J.Asiat. Soc. Bangladesh. Sci.* **40**(2):179-186.

[59] Shamsi, S, T. Sultana and P. Chowdhury. 2015. Mycoflora associated with paanch phoron and its management with common salt. *Bangladesh J. Sci. and Res.* **28**(1):79-83.

[60] Shamsi, S. and S. Hosen,2017. Report on Mycoflora associated with infected stems and capsules of *Nigella sativa* L. (Black Cumin). *Bioresearch Communications.* **3**(1): 298-301.

[61] Shamsi, S. S. Hosen, Md. Al-Mamun and M. Begum. 2016. Mycoflora associated withinfected fruits of*Momordica cochinchinensis* (Lour.) Spreng. *Bangladesh Journal of Plant Taxonomy.* **23** (2):181-188.

[62] Miya, M.D. and S. Shamsi. 2016. Mycoflora associated with *Momordica charantia* and theirpathogenic potentiality. *Asiat. Soc. Bangladesh. Sci.* **42**(2):129-144.

[63] Islam, A.A. and S. Shamsi. 2016. Eco Friendly Management of Mycoflora Associated with *Trichosanthes Anguina* L. and *Trichosanthes Dioica* Roxb. *International Journal of Research Studies in Biosciences (IJRSB).* **4** (1): 52-56.

[64] Shamsi, S. andS. Hosen, 2017. Mycoflora associated with fruit of custard apple (*Annona squamosa* L,). *Bangladesh J. Sci. and Res.***2 9**(1):173-176.

[65] Shamsi, S., A. Yasmin and R.J.F. Lutfunnessa. 2008. Association of*Sclerotium rolfsii* Sacc. With boll rot disease of cotton (*Gossypium hirsutum* L.)var. CB3 in Bangladesh. *Dhaka. Univ. J. Biol. Sci.* **17**(2): 155-158.

[66] Lutfunnessa, R.J.F. and S. Shamsi. 2011. Fungal diseases of cotton plant-(*Gossypium hirsutum* L.) in Bangladesh. *Dhaka Univ. J. Biol. Sci.* **20**:(2):139-146.

[67] Shamsi S. and N. Naher. 2014. Boll rot of cotton (*Gossypiumhirsutum* L.) caused by *Rhizopus oryzae* Went &Prins. Geerl. – A new record in Bangladesh. *J. Agril. Res.* **39**(3):547-551.

[68] Shamsi, S, N. Naher and R.Azad. 2015. Mycoflora of cotton plant (*Gossypium hirsutum* L.) - with three new records of Deuteromycets from Bangladesh. *J. Bangladesh Acad. Sci.* **39**(2): 213 221.

[69] Shamsi, S. and N. Naher and M. Hossain. 2008. fungi associated with *Smilax zeylanica L.* and their pathogenicpotentiality. *Bangladesh J. Plant Pathol.* **24**(1&2):83-89.

[70] Aktar, M. and S. Shamsi. 2010. Fungi associated with *Datura metel* L. *Dhaka Univ. J. Biol. Sci.* **19**(1): 83-89.

[71] Shamsi, S.P. Chowdhury and N. Naher. 2013. Mycoflora associated with the leaves of *Senna alata* (L.) Roxb. *Journal of Bangladesh Academy of Sciences.* **37**(2):249-252.

[72] Shamsi, S., P. Chawdhury and T. Sultana. 2014. Report on mycoflora associated with *Clitoria ternatea* L. – an herbal medicinal plant in Bangladesh. *MAPRJ.* **2**(2), pp. 28-32.

[73] Fatema, Y. and S. Shamsi. 2012. Fungi associated with two species of *Oxalis. Bangladesh J. Sci. Res.* **25**(1):53-60.

[74] Shamsi and R. Sultana. 2014. Report on phylloplane mycoflora associated *with Catharanthus roseus* (L.) G. Don. An herbal medicinal plant in Bangladesh. *Bangladesh J. Sci. Res.* **27**(2):201- 207.

[75] Yasmin, Z. and S. Shamsi.2015. Report on anthracnose of *Rouwolfia serpentina* (L.) Benth exKurz. caused by *Colletotrichum gloeosporioides* (Penz.)Sacc. from Bangladesh. *J. Asiat. Soc. Bangladesh. Sci.* **41**(2): 184-193.

[76] Yasmin, Z. and S. Shamsi. 2016. Report on endophytic fungi associated with *Rouolfia serpentine* (L.) Benth ex Kurz - an herbal medicinal plant. *Bioresearch Communications.* **2**(1):280-283.

[77] Azad, R. and S.Shamsi. 2011. Identification and pathogenic potentiality of fungi associated with *Huttuyania cordata* Thunb. *Dhaka Univ. J. Biol. Sci.* **20**:(2):131-138.

[78] Shutrodhar, A.R. and S. Shamsi. 2013. Anthracnose and leaf spot diseases of *Aloe vera* L. from Bangladesh. *Dhaka Univ. J. Biol. Sci.* **22**(2):104-8209.

[79] Shamsi, S. and M.R. Khan. 2009. Microbes associated with Chayote – *Sechium edule* (Jacq.)Sw. in Bangladesh. *Dhaka Univ. J. Biol. Sci.* **18**(1): 79-82.

[80] Shamsi, S., N. Nahar and S. Momtaz. 2010. First report of Lasiodiplodia pod rot disease of cacao- (*Theobromae cacao.* L.) form Bangladesh. *Bangladesh J. Pl. Pathol.* **26**(1&2):81-82.

[81] Shamsi, S.R. Sultana and Y. Fatema. 2012. Association of fungi with Breabfruit (*Atrocurpus Atalis* Fosb.). *Journal of Bangladesh Academy of Sciences.* **36**(1): 143-146.

[82] Shamsi, S. and N. Naher. 2013. Microflora associated with two species of Liliaceae in Bangladesh. *Dhaka Univ. J. Biol. Sci.* **22**(1):39-46.

[83] Hoq, R. and S. Shamsi. 2012. Report of Association of Mycoflora with Leaves and Stem of *Lea macrophylla* Roxb. Ex Harnem from Bangladesh. *Journal of Bangladesh Academy of Sciences.* **36**(2):257-262.

[84] Ghosh, A and S. Shamsi. 2014. Fungal diseases of rose plant in Bangladesh. *Journal of Bangladesh Academy of Sciences.* **38**(2):335-233.

[85] Shamsi, S. and A. Ghosh. 2013. *Pestalotiopsis guepinii* (Desm.) Stay. - a new pathogen of Black spot disease of rose in Bangladesh. *Bangladesh J. Plant Pathol.* **29**(1&2): 11-14. (published in 2015).

[86] Yasmin, F. and S. Shamsi. 2013. Phylloplane mycoflora of *Gerbera* spp. and their pathogenic potentiality. *Journal of Bangladesh Academy of Sciences.* **37**(2):211-217.

[87] Aktar, M and S. Shamsi. 2018. Incidence and Severity of Blight Disease of *Tagetes erecta* and *T. patula. Bioresearch Communications.* (1):464-469.

[88] Shamsi, S., Razia Sultana and Rumana Azad. 2008. New record of *Phyllactenia dalbergae* Piroz. and its anamorph *Ovulariopsiss issoo* sp. nov. on *Dalbergia sissoo* Roxb. from Bangladesh. *Bangladeh J. of Plant Pathol.* **24** (1&2): 87-89.

[89] Shamsi, S.R. Sultana and R. Azad. 2012. Occurrence of leaf and pod diseases of sissoo (*Dalbergia sissoo* Roxb.)in Bangladesh. *Bangladesh J. Plant Pathol.* **28**(1&2):45-52.

[90] Shamsi, S.and P. Chowdhury. 2014. Microflora associated with rubber sheets and itsmanagement by common salt (Sodium chloride*).J. Asiat. Soc. Bangladesh. Sci.* **40**(1):79-87.

[91] Shamsi,S., S. Hosen and A. Ahmed.2018. Fungi associated with leaves of *Sonneratia aetala* BuchHam and *Sonneratia caseolaris* (L.) Englar from Rangabalic ostal zone of Bangladesh. *Dhaka Univ. J. Biol. Sci.* **37**(2):155-162.

[92] Shamsi, S. and N. Naher. 2010. Hyperparasitic fungus *Tuberculina persicina* (Ditm. Ex. Fr. Sac., as biocontrol agent for rusts. *Dhaka Univ. J. Biol. Sci.***19** (1): 91-94.

[93] Shamsi, S. and N. Naher. 2010. Phylloplane mycoflora on *Vigna sinensis* L. *Dhaka Univ. J. Biol. Sci.***19** (1): 203-206.

[94] Shamsi, S., N. Naher and R. Hoq. 2012. Fungi as biocontrol agents in nature: I – Interaction offungi on phylloplane of *Daturametel* L. and *Vignacatjang*. L. *Bangladesh J. Agril. Res.* **37**(3):537-541.

[95] Hosen, S., S. Shamsi and M.A. Bashar. 2016. *In vitro* biological control of *Colletotrichum gloeosporioides* (Penz.) Sacc. and *Sclerotiumrolfsii*Sacc., causal agent of anthracnose and softrot of *Corchorus capsularis* L. *Bangladesh J. Bot.***45**(1):175-180.

[96] Ameen, Md., S. Shamsi, S. Hosen and M.A. Bashar. 2017. Antagonistic potential of soil fungi against post-harvest pathogenic fungi of *Musa sapientum* L. *Bangladesh J. Bot.* **46**(2):733-738.

[97] Helal, R.B. and S. Shamsi. 2019. Antagonistic potential of some soil fungi against three post- harvest pathogenic fungi of *Carica papaya* L. *Dhaka Univ. J. Biol. Sci.* **38**(1):1-7.

[98] Shamsi, S., P. Chawdhury and N. Naher. 2015. Detection of biocontrol agents from contaminated fungal culture plates. *Plant Environment Development.* **4**(2):21-25.

[99] Shamsi, S., M.A. Bashar and A. Aziz. 2014. Fungi on indoor walls and their management by fungicides. *Bangladesh J. Bot.* **43**(2):239-242.

[100] Shafa, S., S. Shamsi and M.A. Bashar. 2014. Indoor fungi on damp walls of buildings and their management. *Dhaka Univ. J. Biol. Sci.* **23**(1):9-16.

[101] Shamsi, S., N. Naher, M.T.I. Chowdhury and A.K.M. Wahiduzzaman. 2014. Seasonal variation in vegetable market of Karwan Bazar, Dhaka, Bangladesh. *Journal of Bangladesh Academy of Sciences.* **38**(1):49-59.

[102] Sultana, T., M.A. Bashar and S. Shamsi 2015. Monthly variation in air borne fungal propagules of Dhaka Metropolitan city. *Dhaka Univ. J. Biol. Sci.* 24(1):25-33.

[103] Shamsi, S., N. Naher and Y. Fatema. 2015. Crystal formation in culture of fungi: New records from Bangladesh. *J. Asiat. Soc. Bangladesh, Sci.* **41**(1):59-66.

[104] Shamsi, S. and Razia Sultana. 2008. *Gibberella zeae* (Schw.) Petch – A new record of ascomyceteous fungus for Bangladesh. *Bangladesh J. Plant Taxon.* **15**(2): 163-165.

[105] Shamsi, S., N. Nahar, S. Momtaz and P. Chawdhury. 2010. New records of ascomyceteson aromatic rice variety-kataribhog. *Bangladesh J. Pl. Pathol.* **26**(1&2):77-78.

[106] Shamsi, S. and F. Yasmin. 2014. *Eurotium rubrum* Bremer an ascomycetous fungus isolated from *Gerbera aurantiaca* L. in Bangladesh. *Plant Enviro. Dev.* **3**(2):40-42.

[107] Shamsi, S. andN. Naher. 2015. Disease severity and mycoflora associated with anthracnose on leaves of five angiosperms. Bangladesh *J. Sci. and Res.* **28**(2):103-111.

[108] Shamsi, S. and M.A. Mamun. 2016. Diversity of fungi associated with fallen dead spadix of*Cocosnucif era* L.J.Biodivers. Conserv. *Bioresour. Manag.* **2**(1):61-68.

[109] Mazid, F.Z. and L. Nahar. 1968. Studies on the Agarics of Dacca, East Pakistan. Part-I. *Sci. Res. East Reg. Lab.***1**:50-57.

[110] Mazid, F.Z.and Khan N.A. 1968. Studies on the Agarics of Dacca, East Pakistan. Part- II. *Sci. Res. East Reg. Lab.***1**:58-65.

[111] Khan, AZMNA, FZ Majid and I. Nahar. 1969. Studies on the agarics of Dhaka, East Pakistan. III. White spored agarics. *Pakistan Journal of Botany* **1**:35-37.

[112] Rumainul MI, Aminuzzaman FM. 2016. Macro fungi biodiversity at the central and northern biosphere reserved areas of tropical moist deciduous forest region of Bangladesh. *Journal of Agriculture and Ecology Research International.* **5**(4):1-11.

[113] Rubina, H., F. M. Aminuzzaman,M.S.M. Chowdhury and K. Das. 2017. Morphological Characterization of macro fungi associated with forest tree of National Botanical Garden, Dhaka. *Journal of Advances in Biology and Biotechnology.* **11**(4):1-18.

[114] Mian, S.M., Stevens, C., Mia, M.A.T. 2003. Diversity of the blast pathogen *Pyricularia grisea* from Babgladesh by DNA fingerprinting. *Bangladesh J. PlantPathol.* **19**(1&2):81-85.

[115] Hussain, M.Z., M.A. Rahman, M.N. Islam, M.A. LatifandM.A. Bashar. 2012. Morphological and Molecular Identification of *Fusarium oxysporum* Sch. Isolated From Guava Wilt in Bangladesh. *Bangladesh J. Botany.* **41**(1):49-54.

[116] Akter, N.A. Habib, S.P. Beauty, T. Rahim, M.N. Islam.2014. Molecular characterization of four morphological mutants of *Neurosporacrassa.* Dhaka Univ. *Journal of Bio. Sci.***23** (1): 85-91.

[117] Islam,M.T.D. Croll, P. Gladieux,D.M. Soanes, A. Persoons, P. Bhattacharjee,M.S. Hossain,D.R. Gupta, M.M. RahmanM.G. Mahboob, N. Cook,M.U. Salam,M.Z. SurovyV. B. Sancho,J.L.N. Maciel, A. NhaniJúnior,V.L. Castroagudín,J.T. de A. Reges,P.C. Ceresini, S. Ravel, R. Kellner, E.Fournier, D. Tharreau, M.H. Lebrun,B.A. McDonald, T. Stitt,D. Swan,N.J. Talbot, D G.O. Saunders, J. Win and S. Kamoun. 2016. Emergence of wheat blast in Bangladesh was caused by a South American lineage of *Magnaportheoryzae.BMC Biol.*2016; 14: 84.

[118] Das, A., M.I. Rahman, A.S. Ferdous, Al-Amin, M.M. Rahman, N. Nahar, M.A. Uddin, M.R.Islam,H. Khan. 2017. An endophyticBasidiomycete, *Grammothele-lineata*, isolated from *Corchorusolitorius*, produces paclitaxel that shows cytotoxicity. *PLoS ONE* **12**(6):17.

[119] Naznin, T., Hossain, M.J., Nasrin, T., Hossain, Z. and Sarowar, M.N. (2017). Molecular characterization reveals the presence of plant pathogenic*Pythium* spp. around Bangladesh Agricultural University campus, Mymensingh, Bangladesh. *International Journal of Agricultural Research,* **12**(4), 199-205.

[120] Tasnim,F., M. Chowdhury, M. Sarker, M. Saiful Islam, H.P. Nur, M.R. Islam, H. Khan. 2018. Investigation of Antimicrobial Activity and Identification of Bioactive Volatile Metabolites of Jute Endophytic Fungus *Aspergillus flavus. Bioresearch Communications.* **4**(1):476-482.

[121] Sultana, T.S. Shamsi and Md. A.l Bashar. 2018. Morphological and molecular identificationof fungi associated with seeds of selected BRRI rice varieties. *Poster presented in Biotechnical fair* 8-9, September, Nov theatre, Dhaka. Bangladesh.

In: Trends in Biochemistry and Molecular Biology
Editors: Hossain Uddin Shekhar et al.
ISBN: 978-1-53616-434-3

Chapter 15

APPLICATION OF MOLECULAR BIOLOGY TECHNIQUES IN AQUATIC ANIMAL HEALTH RESEARCH

Mohammad Shamsur Rahman*, Nusrat Jahan Punom, Md. Mostavi Enan Eshik and Mst. Khadiza Begum
Aquatic Animal Health Group, Department of Fisheries,
Faculty of Biological Sciences, University of Dhaka, Dhaka, Bangladesh

ABSTRACT

Three major pillars of aquatic animal health management are diagnostics, prevention and control of classical and emerging diseases. There is a huge impact of molecular biology techniques to apply in any of these three important fields of health management. Classical and modern molecular techniques, viz. ELISA, Immunohistochemistry, DNA hybridization, qualitative and quantitative PCR, MLST can use to immediately and accurately diagnose various bacterial, viral, fungal and parasitic aquatic diseases. To prevent and control the diseases after properly diagnose; probiotics, vaccination, drug design could be important steps using molecular biology techniques.

1. INTRODUCTION

Aquatic animal protein has become an important source for food protein. In the developing countries, fish is one of the common and essential food items of the daily diet for most of the people. Globally, the fish consumption is higher than any other type of

* Corresponding Author's E-mail: shamsur@du.ac.bd.

animal meat as a source of protein. The increasing demands for fish and other aquatic products can be met through aquaculture. The losses due to diseases is one of the major constraints to aquaculture production. The sector has confronted serious problems with disease epidemics which caused substantial economic losses over the decades. One of the key problems is the emergence and spread of infectious pathogens. Intensive aquaculture practices enhance the emergence of pathogens. Worldwide trade in fish, other aquatic animals and their products facilitates the pathway for trans-boundary spread of pathogens as well.

Fish and shellfish are susceptible to various pathogenic organisms like bacteria, viruses, parasites and fungi. The classical diseases of fish and shellfish include White Spot Syndrome Virus (WSSV) in marine water shrimp, Epizootic Ulcerative Syndrome (EUS), Yellow Head Virus (YHV), Motile *Aeromonas* Septicemia (MAS) and Vibriosis. Some emerging diseases like Tilapia Lake Virus (TiLV) infection, Early Mortality Syndrome (EMS) in shrimp, *Enterocytozoon hepatopenaei* (EHP) infection in Shrimp, *Macrobrachium rosenbergii* Nodavirus (MrNV) infection in prawn and Shrimp Hemocyte Iridescent Virus (SHIV) infection are now threats to world's aquaculture industry. The Office International des Epizooties/World organization for Animal Health (OIE) has listed certain important diseases infecting the aquaculture systems worldwide (OIE, 2018).

Aquatic animals live in a diverse environment which makes the disease control measures insignificant. Therefore, effective management and control of infectious diseases to maintain the health of fish and shellfish is of prime concern. Several molecular methods have been developed which are now being widely practiced for diagnosis, prevention and treatment of infectious pathogens. The general molecular approaches used in aquatic animal disease research are histopathology, Enzyme-linked Immunosorbent Assay (ELISA), allozyme study, immunofluorescence test, immunohistochemistry, nucleic acid hybridization, different types of Polymerase Chain Reaction (PCR) and molecular typing methods. Advanced molecular techniques include Real Time PCR (qPCR), multilocus sequence typing, DNA microarray assay, Next Generation Sequencing (NGS) and Whole Genome Sequencing. Many monoclonal or polyclonal antibodies against fish pathogens are now available commercially, which are different in sensitivity and specificity. Western blotting and immunoblotting techniques can be useful for confirmation of certain pathogens, such as viral pathogens (e.g., white spot syndrome virus and yellow head virus in shrimp), that cannot be grown in cell lines. The techniques which can be routinely used in future are based on nanoparticles and next-generation sequencing technologies, which enables more sensitive, rapid identification, detection and genetic characterization of aquatic pathogens (Kim et al. 2017).

Disease control strategies combine preventive and curative measures. Several molecular techniques have been proved to be useful for prevention and treatment of infectious diseases. Prevention, rather than treatment, should be the foremost task in fish health management. Good management practice of fish farms is of primary importance in avoiding bacterial and viral disease and parasite problems. In addition, the highly effective and economical strategies in protecting the health of fish and other aquatic animals from various infectious pathogens are vaccination and diet supplementation with probiotics, prebiotics and synbiotics. Molecular techniques are equally important for disease prevention to discover new vaccine candidates to combat different types of fish pathogens and to identify effective probiotic isolates which have antagonistic activity against the pathogen. As well as disease treatment also rely greatly on molecular techniques by detecting the pathogen accurately. Whole genome sequencing provides the opportunity to identify the disease causing gene and accurate drug design can be possible to maintain the health status of aquatic animals.

This chapter is focused on different molecular techniques used in aquatic animal diseases research, classical and emerging diseases of fish and shellfish and application of molecular techniques in diagnosis and control of infectious diseases for aquatic animal health management.

2. Classical and Advanced Molecular Biology Techniques

2.1. Histopathology

When a pathogen or a disease occurring organism such as bacteria or viruses find their suitable host, within a short period of time histological changes occurred in host tissues so histopathology has been utilized as a diagnostic tool for detection of pathogens in tissue sections. Observation of histopathological changes can support the results of other diagnostic methods, leading to confirmatory diagnosis. Indeed, several studies have used this technique as the primary method of detecting pathogens, especially parasites, or toxicopathic changes as a target endpoint in fish (Feist and Longshaw, 2008). When a new disease spreads rapidly in a geographical area and there is no established diagnostic tools for the detection of emerging pathogens, for this circumstance histopathology play a vital role to detect pathogens. On the other hand, histopathology are less sensitive and specific than other methods such as culture-based and molecular techniques. For histology technique, sample preparation procedure is complex which are important for subsequent data analysis, and is time-consuming and costly.

2.2. Nucleic Acid Hybridization

Two single-stranded nucleic acid molecules of complementary base sequence form a double stranded hybrid. Nucleic acid can be hybridized because of their basic property to couple with each other, which allows a part of the known sequences to find complementary sequences in an unknown DNA sample. The known sequence is called the probe. Besides, both DNA and RNA show the similar pattern of base-pairing, the formation of hybrid during the nucleic acid hybridization process can occur DNA-DNA, DNA-RNA, RNA-RNA (Felix, 2010). Keep in mind that in hybridization, the two participating population of nucleic acid should be single stranded.

2.3. Southern Blotting and Northern Blotting

Southern and Northern Blots are general techniques for the evaluation of specific sequences from DNA or RNA samples. Southern blotting was named after Edward M. Southern who explained this method in 1970. Southern blotting has been used for the detection of a specific DNA sequence in a mixture of DNA fragments or total cell DNA. For southern blot, firstly the DNA is isolated from the target organism and digested with restriction enzymes and the DNA fragments are separated by the gel electrophoresis method. The DNA fragments in the gel are denatured by gel electrophoresis method because single strand DNA is required for hybridization. DNA denaturation in the gel is done by exposed mild alkali i.e., NaOH. Then DNA fragments are transferred from gel to a suitable membrane. The membrane with the attached DNA is exposed to an appropriate labeled probe for hybridization. This probe may be chemofluoresent or radiolabeled RNA or DNA sequenced which hybridizes to complementary sequences on the blot. Then the blot is washed with salt or detergent in solutions to remove all residual nonbinding probe. Lastly, the blot is visualized using CCD camera or autoradiography film (Overturf, 2009).

Northern blotting was established by James Alwine, George Stark and David Kemp in 1977. Northern blotting helps to detection the expression of a particular gene by estimating corresponding mRNA. The basic difference between Southern and northern blotting is that RNA is being analyzed rather than DNA in the northern blot. The fundamental principle of the Northern blotting procedure is same as the southern blotting except RNA samples are run in the gel and affixed to a membrane. Since most mRNA is shorter in size than the genomic DNA, digestion by restriction enzymes is not required prior to electrophoresis. In Northern blotting method, RNA molecules in the sample are denatured by using formaldehyde as a denaturing agent (Overturf, 2009).

2.4. Enzyme-Linked Immunosorbent Assay (ELISA)

Enzyme-linked immunosorbent assay (ELISA) is an analytical technique involves detection of an antigen–antibody reaction by a change in colour or fluorescence due to by-product of an enzymatic reaction. ELISA is used for the detection and quantification of proteins typically secreted or released from cells. Using this technique one can determine how much antibody is in a sample, or determine how much protein is bound by an antibody. This procedure involves at least one antibody with specific antigen specificity. The sample containing an unknown amount of antigen on a solid substrate (usually a polystyrene microtiter plate) is either unspecified (by adsorption to the surface) or specified (by capturing by another antibody specific to the same antigen, in "sandwich"). After the antigen is immobilized, the antibody is added to the detector, building an antibody-antigen complex. Antibodies can be linked to one of the enzymes, or they can be detected by a secondary antibody linked to the enzyme through biological coupling. Between each step, the plate is usually washed with a mild cleaning solution to remove any specifically unrelated proteins or antibodies. After the final wash step, the plate is developed by adding an enzymatic substrate to produce a visual signal, which indicates the amount of antigen in the sample. The advantages of using ELISA techniques are specific, time- and cost-effective but are less sensitive than polymerase chain reaction (PCR)-based methods and also can produce false-negative results and may exhibit cross-reactivity with closely related antigens (Law et al., 2015).

2.5. Allozyme

Allozyme are variant forms of enzymes encoded by structural genes which differs structurally but not functionally from other allozymes coded by different alleles at the same locus (Bader 1998). Enzymes have a net electric charge, depending on the length of amino acids comprising the protein. When a mutation occur in DNA that change the amino acid sequence, the net electric charge of the protein can be changed. Because changes in electric charge and conformation can affect the migration rate of proteins in an electric field, allelic variation can be detected by gel electrophoresis and subsequent enzyme-specific stains. Each protein in the gel migrates in a direction and at a rate that depends on the proteins net electric charge and molecular size. Allozyme variation provides data on single locus genetic variation and these locus genetic data allow us to answer many basic questions about fish and fish populations. Because allozyme analysis does not require DNA extraction or the availability of sequence information, primers or probes, they are quick and easy to use. The main drawback of allozymes is their relatively low abundance and low level of polymorphism (Kephart 1990).

2.6. Immunofluorescence Test

Immunofluorescence (IF) is a powerful technique which is based on the use of specific antibodies which have been chemically conjugated to fluorescent dyes to detect specific target antigens. The technique has a number of different biological applications including tissue sections, cultured cell lines, or individual cells, and may be used to analyze the distribution of proteins, glycans. Fluorescence signals depend on some factors and these factors are quality and concentration of the antibody, proper handling of the specimen, and detection with the appropriate secondary antibodies (Odell and Cook, 2013).

Direct Immunofluorescence (IF) Primary, or direct immunofluorescence technique uses only a single antibody that is chemically linked to a fluorophore. The antibody recognizes the target molecule (antigens) and binds to it, and the fluorophore it carries can be detected via microscopy. Direct IF is a rapid and specific technique, although less sensitive than other diagnostic methods; moreover, the lack of availability of antibodies is problematic (Odell and Cook, 2013).

Indirect Immunofluorescence Secondary, or indirect immunofluorescence uses two antibodies. The unlabeled primary antibody binds to the target molecule, and the secondary antibody, which carries the fluorophores, identifies and fixes to the primary antibody (Odell and Cook, 2013). Multiple secondary antibodies can be associated with one primary antibody. This provides amplification of the signal by increasing the number of fluorine molecules in each antigen. This process is more complex and time-consuming than the initial (or direct) process, but it offers more flexibility because a variety of different secondary antibodies and detection methods can be applied for a particular antibody.

2.7. Immunohistochemistry

Immunohistochemistry is a technique that uses antibodies conjugated to enzymes that catalyze reactions to form detectable compounds to visualize and localize specific antigens in a tissue sample. Enzymes, such as Horseradish Peroxidase (HRP) or Alkaline Phosphatase (AP), are commonly used to catalyze a color-producing reaction. Immunohistochemistry (IHC) is an extension of traditional histology where formalized, paraffin or wax-embedded tissue is sectioned and incubated with a pathogen-specific antibody (Adams & Marin de Mateo 1994). IHC makes it possible to see where specific antigens are expressed in a tissue sample. To find out exact location of specific proteins is important for diagnostics in fields such as cancer or infectious diseases. The principle of immunohistochemistry is that antibodies bind specifically to an antigen in biological tissues for example in the jaw, gonads or heart. The antigen-antibody interactions can be visualized by a marker that includes fluorescent dyes, enzymatic color reactions, radioactive elements, or colloidal gold. However, IHC tests may produce false positive and false negative results due to many different factors including, fixation, cross-reactivity, antigen retrieval, antibody sensitivity, etc. IHC is less sensitive than IFAT, although amplification methods such as those based on biotin- streptavidin can increase the sensitivity of the reaction (Hsu & Raine 1981) to detect viral haemorrhagic septicaemia virus.

2.8. Polymerase Chain Reaction (PCR)

Polymerase chain reaction is an in-vitro technique in which oligonucleotide primers are used to amplify specific segment of a DNA. In this technique, a segment of DNA amplified into multiple copies of DNA. PCR makes possible to amplify DNA from any types of sample either live or dead. To amplify a desired target sequence of a DNA using Polymerase Chain Reaction technique, these five important components are required such as Primers, buffers, thermostable DNA polymerase enzymes, dNTPs and Template DNA (Felix 2010). There are three majors step in a PCR, which are repeated for 30 or 40 cycles. In denaturation step at 94°C, the double strand DNA break to form single stranded DNA. In annealing at 54°C, ionic bonds are constantly formed and broken between the single stranded primer and the single stranded template. On the other hand polymerase can attach and starts copying the template. Extension at 72°C, this is the suitable temperature for the working of polymerase. The bases (complementary to the template) are coupled to the primer on the 3' side. PCR can detect infectious agents that are frequently impossible to isolate or culture under artificial conditions. PCR is a highly specific and sensitive method of detecting pathogens, given its complexity, an insufficient PCR product yield and occurrence of non-specific binding between primers and other DNAs in the mixture can hamper its utility (Kim et al. 2017).

2.9. Reverse Transcriptase PCR

One of the major disadvantages of conventional PCR is that it cannot amplify single stranded RNA molecules. Detection of mRNA is useful in detecting genes that are actively expressed. Reverse transcriptase PCR is a variation of the PCR technique where cDNA (complementary DNA) is produce from RNA with association of reverse transcriptase enzyme. PCR is based on the Taq polymerase enzyme; RNA is not an effective substrate for this enzyme. For this reason, the target of interest is first transferred to the complementary DNA (cDNA), which can then be amplified. For example, Yellow Head Virus and Tilapia Lake Virus contains RNA genome. To perform

PCR assay for RNA virus, RT-PCR has to be applied which includes a primary steps to obtain cDNA from extracted viral DNA. Therefore RT-PCR has three steps: 1. Isolation of RNA from disease affected organisms; 2. Preparation of cDNA from the isolated RNA with the help of reverse transcriptase enzyme; 3. PCR of cDNA (Felix 2010). Certain RT enzymes do not perform outside the non-stringent hybridization temperature of 42°C. At times, single stranded RNA forms secondary structures that are stable and prevent the conversion of RNA into cDNA. A recombinant DNA polymerase from the *Thermus thermophilus* (Tth pol) contain both polymerase and RT activity in the presence of Mn^{2+}.

2.10. Nested PCR

Nested PCR is another variation of PCR in which two sets of primers are used to amplify specific sequence of gene. In first round of PCR, a large fragment is produced which used as the template for the second PCR assay. The aim of nested PCR is to reduce non-specific binding in products due to the amplification of unexpected primer binding sites (van Pelt-Verkuil et al. 2008). Nested PCR is a variation of standard PCR that enhances the specificity and yield of the desired amplicons (Haff 1994). Since the production of the second amplicon depends on the success of the first amplicon production, the second amplicon production automatically confirms the accuracy of the first amplicon. Nested PCR can be performed in a single tube or two tube method. In single tube mode, both outer and internal primer sets are added at the same time. Sensitivity and specificity of both RNA and DNA amplification can be increased significantly through the nested PCR method. However, the risk of contamination is a weakness to this method.

2.11. Multiplex PCR

Multiplex PCR conditions can detect several pathogens simultaneously in a multiplex reaction that would improve the time and cost effectiveness of this methodology, and confront one of the main arguments against adopting these techniques as a routine measure (Williams et al., 1999). The multiplex PCR can amplify more than one target sequence using more than a pair of primers in the reaction. Multiplex PCRs are described to detect viral or bacterial agents and / or other infectious agents in a single reaction tube. Multiplex PCR has now been applied successfully in many areas of DNA diagnosis, including genetic deletion analysis, quantitative detection and RNA detection. In the field of infectious diseases, this technique has proved to be a valuable tool for identifying viruses, bacteria, fungus and parasites. It may be difficult to standardize the assay because there is a need to establish annealing temperatures for each set of primers, the size of each amplicon should be sufficient to produce distinct bands by gel electrophoresis, and the specificity and sensitivity for all pathogens should be established (Adams and Thompson 2011).

2.12. Molecular Typing Methods

2.12.1. Randomly Amplified Polymorphic DNA (RAPD)

Randomly amplified polymorphic DNA is PCR based multilocus DNA fingerprint technique. The RAPD procedure was first developed in 1990 using PCR to randomly amplify unknown sequences of DNA with one short PCR primer (8-10 bp) in length. Because primers are short and relatively low annealing temperatures (often 36-40°C), the probability of many products being amplified is good, with each product usually representing a different location (Liu 2008). Random amplified polymorphic DNA (RAPD) typing is a reliable, reproducible, accurate and sensitive discriminatory method for epidemiological typing of various micro-organisms (Kim et al., 2017). For RAPD techniques, no knowledge of the DNA sequence of the target genome is required, as the primers will bind somewhere in the sequence. RAPD has been used to discriminate various fish and shellfish pathogens. However, this method is labour intensive and time-consuming.

2.12.2. Restriction Fragment Length Polymorphism (RFLP)

RFLP has been applied to distinguish organisms by analyzing patterns derived from their DNA cleavage. If two organisms differed in the distance between sites of a specific nucleotide division, the length of the fragments produced would vary when the DNA was digested with the restriction enzyme. Similar patterns generated from RFLP can be used to distinguish species (and even strains) from one another. This technique allows for the study of small variations called polymorphisms that occur in the DNA sequence between individuals of the same species. Polymorphism can occur as a result of the deletion, inversion, additions, substitutions and translocation into the DNA sequence. However, this method is not able to differentiate closely related species that have a similar or almost identical 16S rRNA gene sequence because all produce the same RFLP pattern (Figueras et al., 2011).

2.12.3. Amplified Fragment Length Polymorphism (AFLP)

A rapid technique relies on PCR, AFLP can be used to type prokaryotes and eukaryotes. The method is based on selective PCR amplification of the entire genome's restriction fragments, and has been shown to be fast, highly reproducible and highly discriminatory (Vos et al., 1995). The selected markers are amplified in PCR, making AFLP an easy and rapid tool for strain identification in aquaculture pathogens. In AFLP assay, the first step is restriction of genomic DNA and ligation of adaptors or linkers containing the restriction sites to both ends of the DNA fragments. Subsequent PCR involves two amplification steps using primers complementary to the adaptor sequences to amplify a selected subset of the restriction fragments. The number of amplicons generated in individual assays should be in a manageable range (e.g., 50–100 fragments of 50–500 bp). AFLP has been used for molecular typing of bacterial pathogens.

2.12.4. Multilocus Sequencing Typing (MLST)

MLST was first proposed in 1998 and is the "gold standard" of typing that enables an unambiguous characterization of bacterial isolates using the sequence of internal fragments of seven house-keeping genes to provide a universal, portable, and accurate way to type bacteria (Maiden et al. 1998). Approximately 450 to 500 bp internal fragments of each gene can be used, where they can be accurately sequenced to both sides using an automated DNA sequencer. For each housekeeping gene, the different sequences within the bacterial species are assigned as distinct alleles. For each isolate, alleles in each of the seven sites determine the profile of the allele or sequence type (ST) (Maiden et al. 1998). MLST has been applied as a tool for epidemiological analysis and pathogen surveillance, as well as for investigating the population structure and evolution of bacteria and eukaryotic organisms. MLST has also been used in population structure studies of non-pathogenic bacteria. Most bacterial species have enough variability within house-keeping genes to provide multiple alleles, allowing billions of distinct allelic features to be distinguished using seven house-keeping loci.

2.13. Advanced Molecular Techniques

2.13.1. Real Time PCR

Real-time PCR is highly sensitive techniques which used to amplify and quantify of a specific nucleic acid sequence with detection of the PCR product in real time (Higuchi et al., 1993). Real-time also called "quantitative" or "kinetic" is an in-vitro technique that can be used to determine the relative gene expression quantity as well as genotyping by detection of single-nucleotide polymorphisms (SNP) (Bluth and Bluth 2013). In real time PCR, from the affected samples DNA, cDNA, or RNA targets can be easily quantified achieved by determining the cycle when the PCR product can first be detected. The main principle of real time PCR is enable to accurate quantification of nucleic acids which is not possible in conventional PCR. The same procedure followed in real time PCR as like as conventional PCR but real time PCR also uses an additional oligonucleotide probe. This probe is specific to the target and contains a fluorochrome in one end and a quencher molecule at the other end. Differences of individual nucleotides such as SNPs in PCR products can be detected by sequencing the sequence-specific hybridization of the probe.

Because of different color fluorochromes, the probes can be labelled in several ways, allowing both SNPs to be typed in the same tube (Bluth and Bluth 2013). These molecules can be used in a closed system to differentiate the allele of PCR products. Both real-time assays and endpoint formats can be read, using a fluorescent thermocycler or LightCycler. This PCR-based test can be enhanced by combining the amplification primer and the fluorescent detection component in the same molecule to enable real-time gene profiling. Three types of labelled probe are used in real-time PCR: cleavage-based probes, molecular beacons and FRET (Förster resonance energy transfer) probes (Kim et al., 2017). Cleavage-based probes, also known as TaqMan® probes are used most widely and frequently incorporate a high-energy dye termed a reporter at the 5′ end, and a low-energy molecule termed a quencher at the 3′ end. When the probe is cleaved by the 5′ to 3′ exonuclease activity of Taq DNA polymerase, the distance between the reporter and the quencher increases, causing the transfer of energy to stop. Thereby, the fluorescence of the reporter increases and that of the quencher decreases. Few FRET and molecular beacons have been applied to detection of fish and shellfish pathogens. Real-time RT-PCR can take place in a two-step or one-step reaction. With two-step RT-PCR, the RNA is first reverse transcribed into cDNA using oligo-dT primers, random oligomers, or gene-specific primers. An aliquot of the reverse-transcription reaction is then added to the real-time PCR. The use of oligo-dT primers or random oligomers for reverse transcription means that many different transcripts can be analyzed by PCR from a single RT reaction. In addition, precious RNA samples can be copied immediately to a more stable cDNA for later use and long term storage. In single-step RT-PCR - also referred to as one tube RT-PCR - both reverse transcription and real-time PCR occur in the same tube. This is possible due to specialized reaction chemistries and cycling protocols. Real-time PCR is highly suited for a wide range of applications, such as gene expression analysis, determination of viral load, detection of genetically modified organisms (GMOs), SNP genotyping, and allelic discrimination (Logan et al., 2009).

2.13.2. DNA Microarrays

A large number of genes cannot be studied same time using conventional methods. DNA Microarray is such a techniques that enables researchers to investigate and address issues that were once thought to be untraceable. One can analyze the expression of many genes in a single reaction quickly and effectively. For molecular diagnostics, DNA microarrays play a vital role in determining gene expression. In one test, this technique allows for simultaneous assessment of the rate of expression of thousands of genes of a given sample. The two types of micro-arrays of DNA are widely used: cDNA microarrays and oligonucleotide/DNA chips. To determine whether an individual has a mutation of a particular disease, one can first obtains a DNA sample from the infected organisms as well as a control sample - a sample that does not contain a mutation in the gene of interest. Then denatures the DNA in the samples and this denaturation process make single-stranded DNA molecules by separating the two complementary strands of DNA. The next step is to cut long DNA into smaller, more manageable parts and then label each fragment by connecting the fluorescent dye. The DNA of the individual is marked by green dye and the control - or normal - DNA is marked with red dye. Both sets of labeling DNA are then inserted into the chip and allowed to hybridize - or attach - to the synthetic DNA on the chip. If the individual does not have a mutation for the gene, both the red and green samples will be linked to the sequences of the chip that represent the sequence without the mutation. If the individual has a mutation, the individual DNA will not be properly linked to the DNA sequence on the chip that represents the "normal" sequence, but instead the sequence will be linked to the chip representing the mutated DNA. The use of DNA microarrays as a diagnostic tool for aquaculture is in its infancy (Kostic et al. 2008). The advantage of this technology is that a large number of DNA spots from different pathogens can be included on a single slide, allowing multiplexing for deterrent pathogens. DNA microarray can be used in the field of gene discovery, drug discovery, disease diagnosis and toxicological research.

2.13.3. Next Generation Sequencing

In modern age, it is possible to sequence thousands to millions of DNA molecules at a time. Next-generation sequencing (NGS) is a high-throughput methodology that enables the sequencing of thousands to millions of DNA molecules simultaneously. The Sanger method is accepted worldwide as a first-generation technology. This technique was considered to be fairly static with the main change being in the increase in number of samples that could be analyzed during any single run. However, the length of reads per run remains the same. Its limits include a maximum reading length of usually less than 1000 bp, very high cost, long experimental setup times, high DNA concentrations required and some areas are unsequenceable. NGS generates hundreds of gigabases of short reads, 50–500 base pairs (Zhang et al., 2011).

Next generation methods of DNA sequencing have three general steps: Library preparation, amplification and Sequencing. In library preparation step, libraries are prepared using random fragmentation of DNA, followed by ligation with custom linkers. Besides in amplification stage, the prepared library is amplified using clonal amplification methods and PCR. In sequencing steps, DNA is sequenced using one of several different approaches. NGS features are improved accuracy and speed, but also reduces manpower and cost. It can be said that the greatest improvement is the development of parallel analysis, which increased the speed of sequencing.

2.13.4. Whole Genome Sequence

Whole genome sequencing also known as full genome sequencing or complete genome sequencing or entire genome sequencing is apparently the process of determining the complete DNA sequence of an organism's genome (for example human, bacteria, viruses etc.) at a single time. This process sequencing all types of DNA from an organism's such as chromosomal DNA as well as DNA contained in the mitochondria. At present, most genome projects use the short gun sequencing strategy for sequencing the genome. In the first step, DNA is cut into small random fragments. Depending on the technology, this process sequenced independently to a certain length. The powerful computer algorithms are then used to aggregate the resulting sequence together read back to longer contigs, a process known as *de novo* assembly. For proper assembly, it is important to have sufficient overlap between the reading sequence at each position in the genome, requiring high sequencing coverage (or read depth). Of course, for longer sequence read, more overlap can be expected, which reduces the depth of the required raw reads. The longer parts (several hundred base pairs) are usually sequenced from both ends (the paired-end sequencing) to provide additional information about the correct read position in the assembly. After initial assembly, contigs are typically joined to form longer sequence (known as scaffolds). To achieve this, libraries of long fragments of DNA spanning across several kilobases (kb) of sequencing are prepared and their endpoints sequenced. Based on the technology and characteristics of library preparation, these libraries are called, for example paired-end, mate-pair or jump libraries. If the sequence of the end points of many separate fragments lies in two different contigs, they are attached to a scaffold. The expected length of the library section provides information about the actual distance between contigs, and fills the gap created with the uninformative base pair 'N'. Subsequent gap closing techniques, which typically use long reading through repeated sequences, help fill in missing base pair information. In the last step, the resulting scaffolds are often joined into linkage groups or placed on the chromosomes (Ellegren et al. 2012).

2.13.5. Transgenesis of Fish

Transgenesis is a process by which foreign gene transferred into new hosts. Transgenic are organisms into which transgene has been artificially introduced and the transgene stably integrated into their genomes (Pinkert et al. 1995). Transgenic fish are produced by the artificial transfer of rearranged genes into newly fertilized eggs. The main purposes of producing transgenic fish to enhancing fish quality, growth, disease resistance and productivity. The genes of interest in transgenic fish production are growth hormone genes, metallothionein, crystalline, antifreeze protein, esterase, and regulatory gene sequences.

Microinjection, electroporation, retrovirus infection, transposable elements, and particle gun bombardment techniques have been used to introduce foreign genetic material (DNA) into vertebrate cells (Rembold et al. 2006). Microinjection is a commonly used technique for the production of transgenic fish and due to its simplicity and reliability (Zbikowska 2003). Transgenic technology through DNA microinjection into zebra fish embryos has made great gain in the last decade. It is shown that the DNA injected into the cytoplasm of fertilized zebra fish eggs could integrate into the fish genome and be inherited in the germ line (Stuart et al. 1990). The transposable elements, such as Tol2, has been useful for transgenesis in most fishes, including zebrafish, medaka and tilapia (Fujimura and Kocher 2011).

3. Major Aquatic Animal Diseases

3.1. White Spot Disease (WSD)

During the past three decades in tropical and subtropical areas of the world, shrimp industry has faced a huge impairment due to the emergence of WSD or White Spot Disease. A very large, enveloped, double-stranded DNA (dsDNA) virus of Nimaviridae which hold its own new genus, Whispovirus and familiar as White Spot Syndrome Virus (WSSV) is the causative agent of WSD (Sánchez-Paz, 2010). The large virions of WSSV are ovoid or elliptical to bacilliform in shape, and with a tail-like flagellum carries at one end of the virion (Durand et al., 1997). Lo et al. confirmed WSSV have different geographical isolates with genotypic variability but this double-stranded DNA virus contains around 305 kbp sized genome (Lo et al., 2011). All the variety of this virus is classified as a single species (white spot syndrome virus) within the genus Whispovirus (Lo et al., 2011).

18 penaeid shrimp species are highly susceptible to infection with WSSV, among them *P. indicus, P. japonicas, Pmerguiensis, P. monodon, P. penicillatus, and P. vannamei* are mostly infected. Crabs, crayfish, freshwater prawns, spiny lobsters and clawed lobsters are prone to be infected with WSSV (Lo & Kou, 1998). WSD infected shrimp has been identified in China, Japan, Korea, South-East Asia, South Asia, the Indian Continent, the Mediterranean (Stentiford & Lightner, 2011), the Middle East, and the Americas. According to the guidelines of OIE Aquatic Animal Health Code, Australia, Africa and some specific zones are free from infection with WSSV (Lo et al., 1996).

In the shrimp farming industry, WSD remains the most significant disease in penaeid shrimp where WSSV causes huge economic loss. After the first outbreak in 1992, China faced a great threat in the production over 70% and within three years the loss was over US$ 2 billion. In Thailand, before the hit of WSSV, annual production rate was roughly 34,000 tonnes per year, but in 1994 their production reduced to 2.65 thousand tonnes, or US$ 1.6 billion by value. In Ecuador, outbreak occurred at 1999 and from that time production decreased over 60% in two years, resulting in losses of over US$ 1 billion from 1998-2001 (Chakraborty and Ghosh, 2014). In 1994, the disease spread to southwest region of Bangladesh, affecting approximately 90% of traditional shrimp farms and causing a 20% drop in national shrimp production. Therefore, Shrimp trades fell from 25,742 tonnes to 18,630 tonnes in 1997–1998 (Debnath et al., 2014).

3.2. Yellow Head Disease (YHD)

In the early 1990s, a significant mortality and massive losses in Thailand introduced the whole world with a new viral disease of *Penaeus monodon*, named by yellow head disease, its gross sign included yellowish cephalothorax in the infected shrimp (Chantanachookin et al. 1993). A rod-shaped, enveloped, nidovirus which contains single-stranded RNA (positive-sense) genome is the causative agent of this pandemic disease (Chantanachookin et al. 1993; Tang and Lightner 1999). Different geographical types of this RNA virus explained as yellow head complex which represented their genomic mutations. Among the eight genetic variants, genotype 1 of yellow head virus (YHV) has the most devastating nature to form an outbreak in different farmed shrimp sp. of Southeast and East Asia (Walker et al., 2001).

YHV is enough to infect diverse species of penaeid shrimp like as *Penaeus aztecus, Penaeus duorarum, Penaeus merguiensis, Penaeus monodon, Penaeus setiferus, Penaeus stylirostris and Litopenaeus vannamei*. Some other small shrimp *Palaemon styliferus, Metapenaeus ensis, Euphasia superha* and krill (*Acetes* spp.) and crabs (e.g., *Scylla serrata*) are also get infected (Flegel et al., 1995). Yellow head disease caused by YHV1 has already been visited in a number of Asian countries including Thailand, India, Sri Lanka, Malaysia, Vietnam, Indonesia, the Philippines and in Chinese Taipei (Walker et al., 2001). Western hemisphere has also been affected by YHV, where *P. vannamei* of Mexico has come to the front light (Sanchez-Barajas et al., 2009).

Shrimp farming industry of Thailand faces approximately 30 million US dollars of losses after the severe attack of YHV (Flegel et al. 1995). In India 1994, mass mortalities of 70-100% was occurred in tiger shrimp (*Penaeus monodon*) within 2 to 3 days which was reported by Mohan et al. (1998).

3.3. White Tail Disease (WTD)

During 1997, an outbreak of farm-reared freshwater prawn *Macrobrachium rosenbergii* was reported in Guadeloupe Island (French West Indies) (Arcier et al., 1999). This newly emerged viral disease was described as white tail disease (WTD) or white muscle disease due to whitish or milky appearance of tails and abdominal muscle of the infected larvae (Qian et al. 2003). *M. rosenbergii* nodavirus (*Mr*NV) (Arcier et al., 1999) and extra small virus (XSV) (Sri Widada and Bonami, 2004) are the causative agents of white tail disease. Larval, post larval and early juvenile stages of the giant freshwater prawn are vulnerable to *Mr*NV threat and adult stage of *M. rosenbergii* are act as the vertical transmitter of this virus (Arcier et al., 1999). *Mr*NV is an icosahedral, non-enveloped, small sized virus carrying distinct spikes on the outer surface of them. *Mr*NV belongs to the family *Nodaviridae.* Due to the new characteristics of this variant positioned them into a unique genus of *Gammanodavirus*. An extra small satellite virus (XSV) of *Mr*NV is interconnected to each other to create a devastating outbreak of the white leg disease in prawn (Lin et al. 2018). This naked, icosahedral virus contains a linear single-stranded positive-sense RNA which encoded a capsid protein, cp-17 but they are in lacking of gene-encoding enzymes necessary for their replication. It is also established that *Mr*NV is the major cause in the accompanied infection of WTD (Zhang et al., 2006).

According to the Aquatic Code of OIE (2018), mainly giant river prawn (*Macrobrachium rosenbergii*) but also white leg shrimp (*Penaeus vannamei*) infected by *Mr*NV. In addition, pathogen-specific positive PCR results have been reported in the following species: kuruma prawn (*Penaeus japonicus*), Indian white prawn (*Penaeus indicus*), giant tiger prawn (*Penaeus monodon*), dragonfly (*Aeshna* sp.), giant water bug (*Belostoma* sp.), beetle (*Cybister* sp.), backswimmer (*Notonecta* sp.), hairy river prawn (*Macrobrachium rude*), monsoon river prawn (*Macrobrachium malcolmsonii*), brine shrimps (*Artemia* sp.) and red claw crayfish (*Cherax quadricarinatus*).

WTD was first reported in the French West Indies (Arcier et al., 1999), after that in China (Qian et al., 2003), India (Hameed et al., 2004), Thailand (Yoganandhan et al., 2006) and Australia (Owens et al., 2009).

3.4. Shrimp Hemocyte Iridescent Virus Disease (SHIVD)

A newly emerging virus which tentatively named by the shrimp hemocyte iridescent virus (SHIV), attacked *Litopenaeus vannamei* (white leg shrimp) in China, 2014 (Qiu et al. 2017). The outbreak of China due to this newly discovered virus could express as shrimp hemocyte iridescent virus disease (SHIVD). After the analysis of amino acid sequences of the major capsid protein (MCP) and ATPase of this novel virus, it is also proposed by Qiu et al. 2017 SHIV could belong to the new genus *Xiairidovirus* and in 2018 Qiu et al. separated this virus into another subfamily *Betairidovirinae* under the family *Iridoviridae* (Qiu et al. 2017; Qiu et al. 2018). Evidence confirmed that insects and crustaceans are mainly infected by the members of the iridescent virus of the subfamily *Betairidovirinae* whereas the subfamily *Alphairidovirinae* commonly attacked ectothermic vertebrates namely bony fish, amphibians and reptiles (Chinchar et al. 2017).

Besides *L. vannamei* (white leg shrimp), SHIV is also detected in *Fenneropenaeus chinensis* and *Macrobrachium rosenbergii* in China (Qiu et al. 2017). Other reports confirmed the presence of six different crustacean iridoscent virus including an irido-like virus detected in marine crab *Macropipus depurator* (Montanie et al., 1993), a putative iridovirus infected in penaeid shrimp *Protrachypene precipua* (Lightner and Redman, 1993), Sergestid iridovirus (SIV) affected in sergestid shrimp *Acetes erythraeus* (Tang et al., 2007), invertebrate iridovirus 31 (IIV-31) found in pill bug *Armadillidium vulgare* (Piegu et al., 2014), CQIV identified in freshwater lobster *Cherax quadricarinatus* (Xu et al., 2016), and SHIV isolated in *L. vannamei* (Qiu et al., 2017). Shrimp hemocyte iridescent virus (SHIV) caused severe infection and high mortality in farmed *L. vannamei* in December 2014 in China that was identified in 2017 (Qiu et al. 2017).

3.5. Early Mortality Syndrome (EMS)/ Acute Hepatopancreatic Necrosis Disease (AHPND)

In the year 2009, early mortality syndrome (EMS) introduce a new misery to shrimp farmers because of a novel disease in shrimp farms of Asia. EMS emerges within 20 to 30 days of postlarvae where mortalities can reach 100% in severe conditions. In the mid of 2011, the EMS diseased shrimp described with the unusual histopathological profile where these shrimp has characterized by acute, massive sloughing of hepatopancreatic tubule epithelial cells. The EMS infected shrimp were attacked by unique strains of *Vibrio parahaemolyticus* and the disease was defined as acute hepatopancreatic necrosis disease (AHPND) (Lightner et al. 2012). The AHPND positive *Vibrio parahaemolyticus* isolates carried a unique plasmid (pVA) that have a number of potential toxin genes (Gomez-Gil et al. 2014). Among them Pir-like toxin genes *Pir*A and *Pir*B (Han et al. 2015) alone can regulate the ultimate cause of AHPND in the affected shrimp. But more recent studies reported that pVA plasmid and variants appear in many *V. parahaemolyticus* serotypes and also in other *Vibrio* species viz. *Vibrio campbellii*, *Vibrio harveyi*, and *Vibrio owensii* also harbor this plasmid (Han et al. 2017; Xiao et al., 2017).

According to the Aquatic Code of OIE 2018, the most cultivated shrimp of Southeast Asia black tiger shrimp, *Penaeus monodon*, and American white leg shrimp, *Litopenaeus vannamei*, are affected by EMS/AHPND. In fleshy prawn, *P. chinensis* this disease also reported but other shrimp or prawn species are impervious or less vulnerable to EMS/AHPND (OIE, 2018). The mass mortality due to EMS/AHPND has been reported from China (2010), Vietnam (2010), Malaysia (2011), Thailand (2012) (Flegel, 2012; Lightner et al., 2012), Mexico (2013) (Nunan et al., 2014), the Philippines (2014) (de la Pena et al., 2015), and Bangladesh (2016 and 2017) (Eshik et al. 2017 and 2018).

Due to this catastrophic, acute disease affected area in China had lost 80% of their products. Whereas in Vietnam the economic loss was estimated 5.7 lakh till 7.2 million USD on 2011 and 2012. During 2012 approximately 7% of total shrimp production deficit in Thailand (Lightner et al., 2013). In the Western Hemisphere and Mexico during

2013 and made about 118 million USD economic lost in a part of this country (Schryver et al., 2014).

3.6. Epizootic Ulcerative Syndrome (EUS)

A serious disease of fresh and brackish water fish has faced due to the attack of oomycete *Aphanomyces invadans* and the disease known as Epizootic ulcerative syndrome (EUS). This disease was first described in Japan 1971. This waterborne disease is most frequent at colder temperatures (18 to 22°C) during the winter and rainy seasons (OIE, 2018). More than 125 fish species have been located to be exposed to this pathogen (Kamilya and Baruah, 2014) where juveniles are more vulnerable than older fish. Acute skin and muscle ulceration with heavy mortality explained the clinical signs of EUS. *Aphanomyces invadans* disperses a proteolytic enzyme which promoting invasion into fish muscle and triggering slight to deep lesions, dominating to the severe die-offs in the fish population (Kamilya and Baruah, 2014).

Almost all freshwater and estuarine fish species are vulnerable to EUS but among them certain species viz. striped mullet (*M. cephalus*), snakehead (*Channa* spp), Indian major carps (*Catla catla*, *Labeo rohita* and *Cirrhinus cirrhosus*) and some other important food fishes are particularly open to getting the infection (Roberts et al. 1994). After the first record of outbreak in Japan a number of countries consequently infected through EUS. Outbreaks of EUS have visited more frequently and dispersed worldwide including Asia, Australia, North America and more recently Africa.

EUS is an OIE listed disease and its economic impact is huge. It has caused an estimated loss of US $110 million only in few countries of Asia-Pacific region during late 80's and early 90's (Lilley et al. 1998). Recently severe economic losses due to EUS is not very common.

3.7. Motile *Aeromonas* septicaemia (MAS)

Aeromonas hydrophila, a gram-negative, oxidase-positive, motile, facultative anaerobic bacterium which is the ascendant of motile *Aeromonas* septicaemia. This opportunistic aquatic pathogen causing disease in fish under stress. The genus *Aeromonas* consists of 31 motile bacterial species (Chen et al. 2016). Besides *A. hydrophila*, some other aeromonads like as *A. sobria*, *A. caviae*, *A. allosaccharophila*, *A. veronii* biogroups *sobria* and *veronii, A. encheleia* could influence disease in fish. A number of pathogenic factors like haemolysin and aerolysin which produced by this bacterium for stimulating MAS. Both freshwater and marine water fish species are susceptible to MAS. Haemorrhages, ulcerations, abscesses, ascitic fluid and anaemia are clinical signs of this devastating disease where fatality rates are high, and this induce massive economic losses.

According to Aoki (1999), brown trout, rainbow trout, chinook salmon, ayu, carp, channel catfish, clariid catfish, Japanese eel, American eel, gizzard shad, goldfish, snakehead fish and tilapia are infected by MAS. Besides fish, *A. hydrophila* is pathogenic for amphibians, reptiles and mammals, including men (Lallier and Higgins, 1988; Aoki, 1999).

Globally, the annual economic losses of aquaculture facilities due to the diseases are millions of dollars. In China, losses attributed to *A. hydrophila*, *Y. ruckeri* and *V. fluvialis* were exceeded 120 million dollars between 1990 and 1992 (Pridgeon and Klesius 2012). In west Alabama, USA, MAS outbreaks in channel catfish, *Ictalurus punctatus* have led to an estimated loss of more than USD 3 million in 2009 (Pridgeon et al., 2014).

3.8. Tilapia Lake Virus Disease (TiLVD)

A newly emerging segmented RNA virus called Tilapia Lake Virus (TiLV), has been discovered as the causative virus of Syncytial Hepatitis of Tilapia (SHT) or Tilapia Lake Virus Disease (TiLVD). This novel virus constituted with a high cumulative mortalities (80–90%) in farmed tilapia of Israel, Ecuador, Thailand and Colombia (Bacharach et al. 2016; Eyngor et al. 2014; Dong et al. 2017; Tsofack et al. 2017). Due to the massive die-offs which impact on global food security and nutrition as well as brought a huge economic losses, the UN Food and Agriculture Organization (FAO) and other organizations- NACA, OIE have issued a special alert to all tilapia culturing countries and especially those translocating live tilapia from affected areas like Bangladesh. On May 2017, Dong and his colleagues have published an urgent update with a list of 43 tilapia imported countries that may have been infected with TiLV. According to that, Bangladesh listed at high risk of TiLV disease outbreaks. Currently, tilapia are the second most important group of farmed fish worldwide after carp, wherein the foremost cultured species is Nile tilapia (*Oreochromis niloticus*) (FAO 2005).

TiLV is a segmented, negative-sense ssRNA virus of 10kb in length (Bacharach et al. 2016). TiLV contains 10 genome segments, each with an open reading frame (ORF). Nine of the segments have no recognizable homology to other known sequences but have conserved, complementary sequences at their 5` and 3` termini; one segment predicts a protein with weak homology to the PB1 subunit of influenza C virus, an orthomyxovirus (Bacharach et al. 2016). In 2014, Eyngor and his colleagues investigated outbreaks of tilapia in Israel and reported a syndrome comprising lethargy, endophthalmitis, skin erosions, renal congestion, and encephalitis. Fathi et al. (2017) observed liver and central nervous system lesions, hemorrhagic patches, detached scales, open wounds, dark discoloration and fin rot as sign of the infected fish in Egypt.

Susceptible host of TiLV are Nile tilapia (*O. niloticus*) (Fathi et al. 2017), hybrid tilapia (*O. niloticus* × *O. aureus hybrids*) (Eyngor et al. 2014), Red tilapia (*Oreochromis* sp.) (Dong et al. 2017), Wild tilapines (*Sarotherodon galilaeus, Tilapiazilli, O. aureus* and *Tristamellasimonis intermedia*) (Eyngor et al. 2014). Global tilapia production was estimated at 4.5 million tons with a current value in excess of U.S.$7.5 billion and it is expected to surge to 7.3 million tons by 2030 (FAO, 2014). After the attack of summer mortality syndrome in Egypt 2015 the estimated production loss was 98 000 metric tons, at a value of around USD 100 million (Fathi et al. 2017).

4. Application of Molecular Techniques in Aquatic Animal Health Management

4.1. Applied Molecular Techniques for Fish and Shellfish Disease Diagnosis

World Organisation for Animal Health (OIE) has published 'Manual of Diagnostic Tests for Aquatic Animals (Aquatic Manual)' that provide standardized reference approaches of diseases diagnosis listed by OIE (OIE 2019). This section describe the molecular diagnostic methods documented in Aquatic Manual and described by other authors as well.

4.1.1. Diagnostic Methods for WSD

4.1.1.1. Conventional PCR

PCR has been extensively used to assess the prevalence and geographic distribution of WSD among cultured shrimps and several methods have been developed for the diagnosis of WSSV (Lo et al., 1996; Hossain et al., 2004). Currently, commercial kits are available for the detection of shrimp viruses including WSSV using PCR-based techniques.

4.1.1.2. Real-Time PCR and Multiplex RT-PCR

Methods for quantification of WSSV (Tang and Lightner, 2000), real-time PCR (Dhar et al., 2001) and isothermal DNA amplification (Kono et al., 2004) have also been described. Taqman real-time PCR, a highly specific detection method for WSSV developed by Durand and Lightner (2002). Detection of WSSV has been also tested in multiplex RT-PCR experiments with other shrimp virus (Khawsak et al., 2008).

4.1.1.3. Other Molecular Methods

In situ hybridization (ISH) uses WSSV-specific DNA probes with histological tissue sections to confirm the presence of WSSV nucleic acid in infected cells (Nunan and

Lightner, 1997; Wang et al., 1998). A mini-array method allows one-step multiple detection of WSSV by hybridization of a PCR product, greatly increases the pathogen detection (Quere et al., 2002).

Among the available methods, due to availability, utility, and diagnostic specificity/ sensitivity, PCR is the best method according to OIE (2019) for the detection and to monitor WSSV in post-larvae (PL), juvenile and adult shrimp.

4.1.1.4. Loop-Mediated Isothermal Amplification (LAMP)

The LAMP method is simple, sensitive and rapid diagnostic tool which has been used for the detection of the white spot syndrome virus (WSSV), a serious viral pathogen affecting all life stages of penaeid shrimps. This assay was first described by Kono et al. (2004).

4.1.1.5. DNA Sequencing

For the confirmatory diagnosis of WSSV, DNA sequencing has been described and used within several years (Claydon et al., 2004).

4.1.1.6. Whole Genome Sequencing

Complete genome sequencing was performed by Oakey and Smith (2018) but they cannot identify the source of the virus and suggested to develop another alternative genotyping method for this purpose (Oakey and Smith, 2018).

4.1.2. Diagnostic Methods for YHD

4.1.2.1. Real-Time PCR and Multiplex RT-PCR

A semi-nested RT-PCR can be assayed for YHV (Kiatpathomchai et al., 2004). OIE recommended three RT-PCR methods for YHD (2019). The first one-step protocol, adapted from Wongteerasupaya et al. (1997) will detect only YHV genotype and can be used to confirm YHV in shrimp suspected with YHD. For the differential detection of different genotypes of YHV, mRT-PCR was proven its sensitivity and efficiency (Khawsak et al., 2008).

4.1.2.2. Loop-Mediated Isothermal Amplification (LAMP)

A rapid, cost-effective, sensitive and specific procedure RT-LAMP (reverse transcription loop-mediated isothermal amplification) assay has been used to detect YHV in the heart and gill from infected shrimp and for detecting the structural glycoprotein gene of YHV (Mekata et al., 2006). The assay has a potential usefulness for rapid diagnosis of YHD.

Sequencing: For the confirmatory diagnosis of YHV, sequencing has been performed and the resulting sequences compared with corresponding YHV and GAV sequences reported in GenBank (Castro-Longoria et al. 2008).

4.1.2.3. Whole Genome Sequencing

To reveal the unique gene organization of GAV among nidoviruses complete genome sequencing of GAV has performed by Cowley and Walker (2002). Sittidilokratna et al. (2008) reported completed YHV genome.

4.1.3. Diagnostic Methods for WTD

4.1.3.1. Real-Time PCR and Multiplex RT-PCR

For the detection of MrNV/XSV, protocol for the RT-PCR developed by Sri Widada et al. (2003), Hameed et al. (2004). Nested RT-PCR (nRT-PCR) is also available and recommended for screening broodstock and seed (Sudhakaran et al., 2007a). In 2005, Yoganandhan et al. described that MrNV and XSV can be detected by RT-PCR separately using a specific set of primers or both viruses can be detected simultaneously using a single-tube one-step multiplex RT-PCR (Yoganandhan et al., 2005).

4.1.3.2. Quantitative RT-PCR

To quantify the MrNV/XSV in the infected samples quantitative RT-PCR (RT-qPCR) has been performed using the SYBR Green dye (Zhang et al., 2006).

4.1.3.3. Loop-Mediated Isothermal Amplification (LAMP)

Haridas et al. (2010) has applied loop-mediated isothermal amplification (LAMP) for rapid diagnosis of MrNV and XSV in the freshwater prawn. Most recently Lin et al. (2018) proposed a duplex reverse transcription loop-mediated isothermal amplification assay combined with a lateral flow dipstick method for establishing a rapid, convenient, sensitive and selective detection of MrNV-chin and XSV-chin simultaneously.

4.1.3.4. Sequencing

Yoganandhan et al. (2006) sequenced the DNA fragment amplified from the PCR to confirm the suspected new hosts of MrNV/XSV.

4.1.3.5. Whole Genome Sequencing

High-throughput sequencing has been used to investigate host-pathogen interactions by investigating transcriptomic responses of both the host and virus. A complete genomic RNAs of *Macrobrachium rosenbergii* nodavirus was synthesized from cDNA clones by in vitro transcription (Jariyapong et al. 2018).

4.1.3.6. Cell Culture/Artificial Media

MrNV/XSV can be easily propagated in the C6/36 mosquito *Aedes albopictus* cell line (Sudhakaran et al., 2007b).

4.1.4. Diagnostic Methods for SHIVD

4.1.4.1. Conventional Nested PCR

Qiu et al. (2017 and 2018) has been applied nested PCR method for the detection of SHIV-positive shrimp samples where DNA sample were extracted from the cephalothorax tissue of infected shrimp.

4.1.4.2. TaqMan Probe Based Real-Time PCR

To detect and quantify SHIV in the affected shrimp, a high sensitive and reliable TaqMan probe based qPCR has developed very recently by Qiu et al. (2018). They also concluded that this method could be applied to investigate the prevalence and distribution of SHIV and can be measured its content of different tissues in experimental challenged *L. vannamei*.

4.1.4.3. Sequencing

To confirm the identity of the potential iridescent virus in the affected shrimp samples, high-throughput sequencing was conducted by Novogene (Beijing, China) with the Hiseq. 2000 platform (Qiu et al. 2017). In this study, phylogenetic analyses indicated that SHIV positive sample did not belong to any of the five classified genera of *Iridoviridae*.

4.1.4.4. Whole Genome Sequencing

The complete genome sequence of SHIV was determined and analyzed by Qiu et al. (2018). This study suggested that SHIV should be considered a member of the proposed new genus "*Xiairidovirus.*"

4.1.5. Diagnostic Methods for EMS/AHPND

4.1.5.1. Conventional and Nested PCR

Rapid diagnosis of AHPND disease through PCR kit was developed by Lightner et al. (2013). Later, a new and improved PCR method (AP3 method) for detection of AHPND bacteria has been developed from the results of laboratory test in screening AHPND isolates (NACA, 2014). After AP3 primer method, another efficient and sensitive AP4 nested PCR method was developed by Sritunyalucksana, et al. (2015). AP4 nested PCR method was 100 times higher detection sensitivity (Dangtip, et al., 2015).

4.1.5.2. Real-Time Florescence-Dependent RPA Assay

Early and rapid diagnosis of AHPND, Liu et al. (2017) developed a highly sensitive and specific real-time florescence-dependent RPA assay targeting pirA-like gene for AHPND detection in shrimp.

4.1.5.3. Quantitative Real-Time PCR

In 2015, Han et al. (2015) developed a quantitative polymerase chain reaction (qPCR) technique, based on a TaqMan probe, to detect and quantify a virulence plasmid contained by Vibrio parahaemolyticus causing acute hepatopancreatic necrosis disease (AHPND). This test was specific with high sensitivity (10 copies of virulence plasmid).

4.1.5.4. Loop-Mediated Isothermal Amplification (LAMP)

Arunrut et al. (2016) developed a simpler but equally sensitive approach for detection of VPAHPND based on LAMP combined with unaided visual reading of positive amplification products using a DNA-functionalized, ssDNA-labeled nanogold probe (AuNP). They reported that the detection limit (100 CFU) was comparable to that of other commonly-used methods for nested PCR detection of VPAHPND and 100-times more sensitive than 1-step PCR (10^4 CFU). In 2016, Koiwai et al. also published their research on LAMP as diagnostic tool for the detection of AHPND infection. Their method can detect toxin PirAB-like faster than the conventional PCR without compromising the quality and result.

4.1.5.5. PCR-DNA Chromatography

Koiwai et al. (2018) proposed a diagnostic system for four shrimp diseases (WSD, IHHN, AHPND and EHP infections) using multiplex PCR and STH chromatographic PAS, named PCR-DNA chromatography. This could detect as few as 10 copies of WSD, AHPND. They concluded that PCR-DNA chromatography will be useful for diagnosing not only shrimp diseases, but also other fishery pathogens in laboratory or diagnostic centre due to its ease of use, speed and flexibility.

4.1.5.6. Whole Genome Sequencing

Yang et al. (2014) were sequenced four *Vibrio parahaemolyticus* strains, three of which caused serious acute hepatopancreatic necrosis disease in China and Thailand by an Illumina MiSeq sequencer where Kondo et al. (2014) were sequenced three AHPND and three non-AHPND strains of Thailand using the Illumina MiSeq. After that, Fu et al. (2017) performed the complete genome sequencing of different AHPND positive Vibrio parahaemolyticus strains.

4.1.6. Diagnostic Methods for EUS

4.1.6.1. Conventional and Species-Specific PCR

To disclose the geographic range of *A. invadans* in USA PCR was performed by Blazer et al. (2002) where samples were collected from naturally infected, formalin preserved fish of several epidemics. Lilley et al. (2003) worked on species specific PCR to detect *A. invadans* isolates and 12 other oomycetes. After that, species specific PCR have been utilized extensively for rapid the identification and screening of *A. invadans* from infected fish samples. Commercial DNA extraction kits have been used successfully (Vandersea et al., 2006).

4.1.6.2. Quantitative RT-PCR (qRT-PCR)

To evaluate differential expression levels of nine up-regulated genes and three down-regulated genes of EUS infected samples with control samples, qRT-PCR has been explored by Kumaresan et al. (2018).

4.1.6.3. Sequencing

All PCR products of *A. invadans* isolates sequencing has performed and the results has compared with the sequence deposited in the public gene databanks (Vandersea et al., 2006).

4.1.6.4. Advanced Molecular Approaches

Most recently, Kumaresan et al. (2018) studied on the molecular mechanism of immune response in murrel against this infection through transcriptome technique where the control (CF) and infected fish (IF) groups were sequenced using Illumina Hi-seq sequencing technology. Majeed et al. tested the role of extracellular proteases produced by the *Aphanomyces invadans,* oomycete responsible for the EUS virulence through CRISPR/Cas9 system in many fish species. After the successful and effective mutation in the target gene *A. invadans* zoospores did not produce EUS clinical signs in the fish. Therefore, this method establish a promising approach for functional genomics studies in *A. invadans* and provide novel avenues to develop effective strategies to control this pathogen (Majeed et al. 2018).

4.1.7. Diagnostic Methods for MAS

4.1.7.1. Conventional and Species-Specific PCR

For the identification *A. hydrophila/ A. dhakensis* from the diseased fish samples, conventional PCR techniques were most commonly used where analysis were performed by targeting different genes like as 16S rRNA, *gyr*B and *rpo*D genes (Carriero et al.

2016). Besides these three genes *rec*A, *dna*J and *gyr*A also tested to reclassification of *A. hydrophila* subsp. *dhakensis* (Beaz-Hidalgo et al. 2013).

4.1.7.2. Quantitative RT-PCR (qRT-PCR)

Carriero et al. (2018) studied on the influence of iNOS during *A. dhakensis* infection, mRNA expression on experimentally infected pacus (*Piaractus mesopotamicus*) was determined by quantitative real-time PCR using specific primers. In that work qPCR specific primers were designed based on the obtained pacu iNOS cDNA sequence.

4.1.7.3. Quartz Crystal Microbalance (QCM) Immunosensor System

Hong et al. (2017) has been reported a quartz crystal microbalance (QCM) immunosensor system for the rapid and simple detection of human and fish pathogen *A. hydrophila*. In that experiment, a QCM immunosensor system was able to detect *A. hydrophila* cells. The sensitivity and specificity were enough to use as detection system of field sample through in vivo test.

4.1.7.4. Gold Nanoparticle Probe-Based Assay

A rapid, specific and sensitive assay for detection of *A. hydrophila* DNA without prior amplification utilizing a specific probe conjugated with gold nanoparticles (Elsheshtawy et al. 2018). The modified *Aeromonas*-AuNPs-probe assay could enable the visual detection of *A. hydrophila* DNA within 30 min without any prior amplification, at a detection limit of 10.9 ng.

4.1.7.5. Sequencing

According to Tazumi et al. (2009) sequencing of the 16S–23S rDNA intergenic spacer region (ISR) is now considered a robust and sensitive taxonomic tool which is widely used in bacterial taxonomy. To identify *Aeromonas* species, gene sequences of 60 kDa chaperonin (cpn60), DNA gyrase B subunit (*gyr*B), and RNA polymerase sigma factor (*rpo*D) have also been used (Minana-Galbis et al. 2010).

4.1.7.6. Whole Genome Sequencing

Several studies were conducted on different *A. hydrophila* strains where complete genome sequencing were performed by some researchers viz. Wang et al. (2018); Teng et al. (2017).

4.1.8. Diagnostic Methods for TiLVD

4.1.8.1. Reverse Transcriptase-PCR (RT-PCR)

At the very beginning, Eyngor et al. (2014) developed a reverse transcriptase-PCR (RT-PCR) method for TiLV detection using TiLV-specific primers targeting segment 3.

After using RT-PCR method, a more sensitive nested RT-PCR assay was used to detect TiLV in both fresh and preserved (RNAlater; QIAGEN) samples from diseased fish (Tsofack et al. 2017). Later, a SYBR green-based qPCR method using nested primers (ME1 & 7450/150R/ME2) achieved a detection limit of 70 copies.

An alternative semi-nested RT-PCR method has been developed where the primer Nested ext-2 was omitted to reduce the risk of false-positive detections (Dong et al. 2017) and this was proved to detect TiLV from clinically healthy fish (Senapin et al. 2018). Most recently, Tattiyapong et al. (2018) developed a new SYBR green-based reverse transcription quantitative PCR (RT-qPCR) method targeting the same genome segment 3 for detection of TiLV from clinical samples with sensitivity of two copies/L.

4.1.8.2. Cell Culture

Experiments confirmed that multiple cell lines to be suitable for TiLV cell culture (Eyngor et al. 2014; Tsofack et al. 2017).

4.2. Applied Molecular Technique for Fish and Shellfish Disease Prevention

At present day, molecular biology plays a significant role for the rapid development of new molecular methods for preventing and controlling fish and shellfish diseases. In contrast with traditional method (culture, serology and histology), molecular techniques are potentially faster or more sensitive for preventing and controlling fish diseases.

In the field of biomedical as well as aquatic health, biotechnological and molecular techniques have shown their tremendous success. To design specific and sensitive diagnostic tools, study the immune systems of fish and shellfish, and look at the relationship between pathogens and their hosts, molecular techniques can be used. The creation of disease-resistant strains with gene transfer and the development of effective vaccines and treatments and new ways to deliver them are also made possible with biotechnology. Vaccination is one of the most essential and probably the priority, approaches to prevention and control of infectious disease of fish (Assefa and Abunna, 2018). DNA vaccines made of one or more genes of a pathogen which is suspected as an infectious disease causing organism. To confers immediate and a durable protection from diseases in farmed salmonids against economically important diseases such as infectious hematopoietic necrosis virus (Ballesteros et al. 2015) and viral hemorrhagic septicemia virus (Cho et al. 2017), DNA vaccination through intramuscular injection is one of the best approach. Genetically Modified Vaccines which include in vitro passaging of organism's results in a build-up of genome mutations that make the organism weaken that is also used for fish diseases control (Assefa and Abunna, 2018). *Aeromonas salmonicida* in salmon can be prevented by using genetically modified vaccines. Synthetic peptide vaccines are made of short sequences of amino acids prepared synthetically which work as antigens. These vaccines have been in use as prevention of infectious disease like nodavirus, viral hemorrhagic septicaemia, rhabdovirus, and birnavirus (Dadar et al., 2016). Hitra disease appeared in salmonid aquaculture since 1980s in Norway caused by pathogenic bacterium, *Vibrio salmonicida* (Egidius et al. 1986). Since 1988 most Atlantic salmon and rainbow trout in Norway have been vaccinated (initially via immersion) against this disease. *V. anguillarum, V. salmonicida and A. salmonicida* produce bacterins with antigens which were added with mineral oil adjuvants contributed to effective control of diseases, which without immunoprophylaxis would have caused great losses to the industry (Gudding and Van Muiswinkel, 2013). Now-a-days, vaccines are available against many serious infectious bacterial diseases of fish farming industries (Hastein et al., 2005). Vaccines containing *E. ictaluri* and *F. columnare* bacterins have low efficacy and the use of live attenuated isolates is a new and promising approach (Klesius et al. 1999). A live attenuated *E. ictaluri* vaccine, licensed in the USA a few years ago, has proven efficacious by immersion of fish as early as 7–10 days post hatching (Shoemaker et al., 1999). Czechoslovakian company (Bioveta) in 1982 first produce viral vaccine for fish. The vaccine was mainly used against a carp rhabdovirus responsible for occurring spring viremia disease of carp (SVC). An inactivated viral vaccine against pancreas disease (PD, caused by an aquatic alphavirus) is available in Ireland under a provisional license (Christie et al. 1998, McLoughlin et al. 2003), and a vaccine against infectious salmon anemia (ISA, caused by an orthomyxovirus) is available in Canada and the USA.

Probiotics are cultured products or live microbial feed supplements, which mainly improving the intestinal (microbial) balance of the host (Fuller, 1989). Probiotics have noticed a great concern in aquaculture because probiotics provide benefits to the hosts viz., improving the host growth (Silva et al. 2013), reducing the incidence of diseases (Newaj-Fyzul et al. 2007; Silva et al. 2013), and requiring less chemotherapy (Hai et al. 2009). The wide applications belong to endospore-forming members of *Bacillus* genera (Hong et al. 2005), in which *Bacillus subtilis* is commonly used in aquaculture. *Streptococcus phocae* which is known as a fish pathogen (Austin and Austin, 2012), but they enhanced the growth of black tiger prawn post larvae and protected the animals against challenge with *V. harveyi* (Swain et al. 2009). *Aeromonas hydrophila* and *Aer. sobria* are proved as fish pathogens (Austin and Austin 2012), while they reduced infections of *Aeromonas salmonicida* (Irianto and Austin, 2002). *V. alginolyticus* and *V. proteolyticus* are used as a probiotics for Atlantic salmon (Salmo salar) (Austin et al. 1995) and turbot (*Scophthalmus maximus*) (De Schrijver and Ollevier, 2000) respectively. *Vibrio fluvialis* is a probiotic for *Penaeus monodon* (Alavandi et al. 2004) and *Penaeus japonicus* (El-Sersy et al. 2006). Probiotic *L. lactis* RQ516 that is being used in tilapia (*Oreochromis niloticus*) has shown inhibitory activity against *Aeromonas hydrophila* (Zhou et al. 2010). Balcazar et al. (2007) presented that probiotic *L. lactis* had antibacterial activity towards two fish pathogens namely, *Aeromonas salmonicida* and

Yersinia rukeri. Zapata and Lara-Flores (2013) found that *Leuconostoc mesenteroides* was able to inhibit the growth of fish pathogenic bacteria in Nile tilapia (*O. niloticus*). Ghosh et al. (2008) found that *Bacillus subtilis* significantly reduced the number of motile *Aeromonads*, presumptive *Pseudomonads* and total Coliforms in ornamental fishes (Newaj-Fyzul and Austin, 2014). *Pseudomonas, Vibrio, Aeromonas* spp. and *Coryneforms* had antiviral activity against infectious hematopoietic necrosis virus (IHNV) (Kamei et al. 1988). Li et al. (2009) demonstrated that feeding with a *Bacillus megaterium* strain increased the resistance to white spot syndrome virus (WSSV) in the Vannamei shrimp. It was documented that probiotics such as *Bacillus* and *Vibrio* sp. positively protect shrimp *L. vannamei* against WSSV (Balcazar, 2003). Application of *Lactobacillus* probiotics as a single strain or mixed with Sporolac improved disease resistance against lymphocystis viral disease in olive flounder (Harikrishnan et al. 2010). Lategan et al. (2004) isolated Aeromonas media (strain A199) from eel (*Anguilla australis*) culture water and was observed to have a strong inhibitory activity against *Saprolegnia* sp. In a study with juvenile tiger shrimp (*Penaeus monodon*), Lb. acidophilus 04 (10^5 CFU g−1) was administered for 1month and increased resistance (80% survival) was observed after exposure to *V. alginolyticus* (Sivakumar et al. 2012).

4.3. Applied Molecular Technique for Fish and Shellfish Disease Control

The rapid spread of infection in fish farm have been increasing due to high stocking density, lack of sanitary barriers between farming sites, and separation of fish farms from infected animals (Cabello, 2006; Naylor et al. 2000; Naylor and Burke, 2005). The control of bacterial infections with commonly used antibiotics has become the only solutions for aqua farmers. With the increasing demand of intensive culture, antibiotics application with feed has become essential in aquacultural activities (Xu et al. 2013). Wei (2002) in one of his studies estimated that bacterial infections responsible for 15–20% loss of annual production, and more than 200 diseases have been identified in cultured aquatic species in China. Penicillins, macrolides and quinolones which are classified as critically important antibiotics for treating humans' diseases, are used for treating infectious agent in fish (WHO, 2012). In recent investigations, different antibiotics, including sulfonamides, fluoroquinolones (e.g., ciprofloxacin), tetracyclines (e.g., oxytetracycline), and macrolides (e.g., erythromycin), were detected in receiving waters or sediment of aquaculture farms (Xu et al. 2013; Xue et al. 2013; Zheng et al. 2012; Zou et al. 2011). Sulfonamides, β-lactams, and macrolides are often used as growth promotor and disease preventor in the form of feed additives or as aqua drugs for the treatment of infections. It has been reported that *Aeromonus hydrophila* resistance against a wide range of antimicrobial compounds, including ampicillin, chloramphenicol, erythromycin, nitrofurantoin, novobiocin, streptomycin, sulphonamides and tetracycline (DePaola et al,

1988). They estimated that as many as 38% of the *Aeromonas hydrophila* isolates from diseased catfish are resistant to oxytetracycline. More recently, polymyxin B nonapeptide has been found to inhibit *Aeromonas salmonicida*, probably by disrupting the A-layer (McCashion and Lynch, 1987). Oxytetracycline (OTC) is the most important antibacterial drug used to treat fish as it is effective against pathogenic microorganisms in fish, namely *Flexibacter* sp., *Vibrio* sp., *Aeromonas* sp., *Yersinia* sp., and *Edwardsiella* sp (Terech-Majewska et al. 2006). Tetracycline is considered to be an immunosupressant (Terech-Majewska et al. 2006). Oxytetracycline HCL (Terramycin ® 10) used in salmonids for the control of ulcer disease caused by *Hemophilus piscium*, furunculosis caused by *Aeromonas salmonicida*, bacterial hemorrhagic septicemia caused by *A. liquefaciens*, and *pseudomonas* disease and used in cat fish for the control of bacterial hemorrhagic septicemia caused by *A. liquefaciens* and pseudomonas disease (FDA, 2001). Sufamerazine in Fish Grade antibiotic used in trout to control of furunculosis in salmonids caused by *Aeromonas salmonicida*. (FDA, 2001).

CONCLUSION

Since intensive culture can increase the likelihood of disease, monitoring and management of aquatic animal health is important in the aquaculture industry. A disease outbreak can devastate farmed species and severely impact the aquaculture industry. A rapid response is essential to enable accurate diagnosis of the problem and to apply appropriate control measures. The modern molecular techniques are generally more advantageous in terms of higher specificity and sensitivity for quick detection and identification of pathogens. The application of molecular methods are increasing in aquaculture day by day and should be continuously developed. Diagnosing aquatic animals by the clinical signs and symptoms only is not effective. That's why it is important to apply a rapid and accurate diagnostic methods to be important for proper prevention and control of infectious disease. NGS technology has opened the pathway for genomic research by reducing the costs and accelerating the speed of DNA sequencing. The invention of NGS technologies has really caused the revolution in life sciences. Many unique information in aquatic animal health research would have remained unrevealed or it would take much longer time to expose them without using NGS technologies (Kumar and Kocour 2017). Metagenomics based on NGS is a traditional culture-independent sequence-based technique for microbial detection and identification as well as description of the entire microbiota of a representative sample directly from the environment. This novel technology are now being used in veterinary medicine to study genomes, transcriptomes and host–pathogen interactions, with a particular focus on disease control and management.

REFERENCES

Adams, A. *and* Marin de Mateo, M. *(*1994*)* Immunohistochemical detection of fish pathogens. In *Techniques in Fish Immunology*, edited by J. S. Stolen, T. C. Fletcher, S. L. Kaattari and A. F. Rowley, *3:* 133–144. SOS Publications, Fair Haven, NJ, USA.

Adams, A. and Thompson, K. D. (2011). Development of diagnostics for aquaculture: challenges and opportunities. *Aquaculture Research*, 42: 93-102.

Alavandi, S. V., Vijayan, K. K., Santiago, T. C., Poornima, M., Jithendran, K. P., Ali, S. A. and Rajan, J. J. S. (2004). Evaluation of *Pseudomonas* sp. PM 11 and *Vibrio fluvialis* PM 17 on immune indices of tiger shrimp, *Penaeus monodon. Fish & Shellfish Immunology,* 17 (2): 115-120.

Aoki, T. (1999). Motile aeromonads (*Aeromonas hydrophila*). In *Fish Diseases and Disorders: Viral, Bacterial and Fungal Infections*, edited by Woo, P. T. K. and Bruno, D. W. 427–453. CABI Publishing, London.

Arcier, J. M., Herman, F., Lightner, D. V., Redman, R. M., Mari, J. and Bonami, J. R. (1999). A viral disease associated with mortalities in hatchery-reared postlarvae of the giant freshwater prawn *Macrobrachium rosenbergii. Diseases of Aquatic Organisms,* 38 (3): 177-181.

Arunrut, N., Kampeera, J., Sirithammajak, S., Sanguanrut, P., Proespraiwong, P., Suebsing, R. and Kiatpathomchai, W. (2016). Sensitive visual detection of AHPND bacteria using loop-mediated isothermal amplification combined with DNA-functionalized gold nanoparticles as probes. *PloS One,* 11 (3): e0151769.

Assefa, A. and Abunna, F. (2018). Maintenance of Fish Health in Aquaculture: Review of Epidemiological Approaches for Prevention and Control of Infectious Disease of Fish. *Veterinary Medicine International,* 2018: 5432497

Austin, B., Austin, D. A., Austin, B. and Austin, D. A. (2012). *Bacterial Fish Pathogens.* Heidelberg, Germany: Springer.

Austin, B., Stuckey, L. F., Robertson, P. A. W., Effendi, I. and Griffith, D. R. W. (1995). A probiotic strain of *Vibrio alginolyticus* effective in reducing diseases caused by *Aeromonas salmonicida, Vibrio anguillarum* and *Vibrio ordalii. Journal of Fish Diseases,* 18 (1): 93-96.

Bacharach, E., Mishra, N., Briese, T., Zody, M. C., Tsofack, J. E. K., Zamostiano, R., Berkowitz, A., Ng, J., Nitido, A., Corvelo, A. and Toussaint, N. C. (2016). Characterization of a novel orthomyxo-like virus causing mass die-offs of tilapia. *MBio* 7 (2): e00431-16.

Bader, J. M. (1998). Measuring genetic variability in natural populations by allozyme electrophoresis. In *Tested studies for laboratory teaching*, edited by S. J. Karcher, 19: 25-42. Proceedings of the 19th workshop/conference of the Association for Biology Laboratory Education.

Balcazar, J. L. (2003). *Evaluation of probiotic bacterial strains in Litopenaeus vannamei.* Final Report. National Center for Marine and Aquaculture Research, Guayaquil, Ecuador.

Balcázar, J. L., Vendrell, D., de Blas, I., Ruiz-Zarzuela, I., Gironés, O. and Múzquiz, J. L. (2007). *In vitro* competitive adhesion and production of antagonistic compounds by lactic acid bacteria against fish pathogens. *Veterinary Microbiology,* 122 (3-4): 373-380.

Ballesteros, N. A., Alonso, M., Saint-Jean, S. R. and Perez-Prieto, S. I. (2015). An oral DNA vaccine against infectious haematopoietic necrosis virus (IHNV) encapsulated in alginate microspheres induces dose-dependent immune responses and significant protection in rainbow trout (*Oncorrhynchus mykiss*). *Fish & Shellfish Immunology,* 45 (2): 877-888.

Beaz-Hidalgo, R., Martínez-Murcia, A. and Figueras, M. J. (2013). Reclassification of *Aeromonas hydrophila* subsp. *dhakensis* Huys et al. 2002 and *Aeromonas aquariorum* Martínez-Murcia et al. 2008 as *Aeromonas dhakensis* sp. nov. comb nov. and emendation of the species *Aeromonas hydrophila. Systematic and Applied Microbiology,* 36 (3): 171-176.

Blazer, V. S., Lilley, J. H., Schill, W. B., Kiryu, Y., Densmore, C. L., Panyawachira, V. and Chinabut, S. (2002). *Aphanomyces invadans* in Atlantic menhaden along the east coast of the United States. *Journal of Aquatic Animal Health,* 14 (1): 1-10.

Bluth, M.J. and Bluth, M.H. (2013). Molecular pathology techniques. *Clinics in Laboratory Medicine*, 33(4): 753-772.

Cabello, F.C. (2006). Heavy use of prophylactic antibiotics in aquaculture: a growing problem for human and animal health and for the environment. *Environmental Microbiology,* 8 (7): 1137-1144.

Carriero, M. M., Henrique-Silva, F., Caetano, A. R., Lobo, F. P., Alves, A. L., Varela, E. S., del Collado, M., Moreira, G. S. and Maia, A. A. (2018). Characterization and gene expression analysis of pacu (*Piaractus mesopotamicus*) inducible nitric oxide synthase (iNOS) following *Aeromonas dhakensis infection. Fish and Shellfish Immunology,* 74: 94-100.

Carriero, M. M., Mendes Maia, A. A., Moro Sousa, R. L. and Henrique-Silva, F. (2016). Characterization of a new strain of *Aeromonas dhakensis* isolated from diseased pacu fish (*Piaractus mesopotamicus*) in Brazil. *Journal of Fish Diseases,* 39 (11): 1285-1295.

Castro-Longoria, R., Quintero-Arredondo, N., Grijalva-Chon, J.M. and Ramos-Paredes, J. (2008). Detection of the yellow-head virus (YHV) in wild blue shrimp, *Penaeus stylirostris*, from the Gulf of California and its experimental transmission to the Pacific white shrimp, *Penaeus vannamei. Journal of Fish Diseases,* 31 (12): 953.

Chakraborty, S. and Ghosh, U. (2014). White Spot Syndrome Virus (WSSV) in Crustaceans: An Overview of Host-Pathogen Interaction. *Journal of Marine Biology and Oceanography,* 3 (1), *doi:10.4172/2324-8661.1000121*

Chantanachookin, C., Boonyaratpalin, S., Kasornchandra, J., Direkbusarakom, S., Ekpanithanpong, U., Supamataya, K., Sriurairatana, S. and Flegel, T. W. (1993). Histology and ultrastructure reveal a new granulosis-like virus in Penaeus monodon affected by yellow-head disease. *Diseases of Aquatic Organisms,* 17: 145-145.

Chen, P. L., Lamy, B. and Ko, W. C. (2016). *Aeromonas dhakensis*, an increasingly recognized human pathogen. *Frontiers in Microbiology,* 7: 793.

Chinchar, V. G., Hick, P., Ince, I. A., Jancovich, J. K., Marschang, R., Qin, Q., Subramaniam, K., Waltzek, T. B., Whittington, R., Williams, T. and Zhang, Q. Y. (2017). ICTV virus taxonomy profile: Iridoviridae. *Journal of General Virology,* 98 (5): 890-891.

Cho, S. Y., Kim, H. J., Lan, N. T., Han, H. J., Lee, D. C., Hwang, J. Y., Kwon, M. G., Kang, B. K., Han, S. Y., Moon, H. and Kang, H. A. (2017). Oral vaccination through voluntary consumption of the convict grouper *Epinephelus septemfasciatus* with yeast producing the capsid protein of red-spotted grouper nervous necrosis virus. *Veterinary Microbiology,* 204: 159-164.

Christie, K. E., Mockett, K., Fyrand, K., Goovaerts, D. and Rødseth, O. M. (1998) Vaccination of Atlantic salmon *Salmo salar L.* against pancreas disease. In *4th International Symposium on viruses of Lower Vertebrates.*

Claydon, K., Cullen, B. and Owens, L. (2004). OIE white spot syndrome virus PCR gives false-positive results in *Cherax quadricarinatus. Diseases of Aquatic Organisms,* 62 (3): 265-268.

Cowley, J. A. and Walker, P. J. (2002). The complete genome sequence of gill-associated virus of *Penaeus monodon* prawns indicates a gene organisation unique among nidoviruses. *Archives of Virology,* 147 (10): 1977-1987.

Dadar, M., Dhama, K., Vakharia, V. N., Hoseinifar, S. H., Karthik, K., Tiwari, R., Khandia, R., Munjal, A., Salgado-Miranda, C. and Joshi, S. K. (2017). Advances in aquaculture vaccines against fish pathogens: global status and current trends. *Reviews in Fisheries Science & Aquaculture,* 25 (3): 184-217.

Dangtip, S., Sirikharin, R., Sanguanrut, P., Thitamadee, S., Sritunyalucksana, K., Taengchaiyaphum, S., Mavichak, R., Proespraiwong, P. and Flegel, T. W. (2015). AP4 method for two-tube nested PCR detection of AHPND isolates of *Vibrio parahaemolyticus. Aquaculture Reports,* 2: 158-162.

De Schrijver, R. and Ollevier, F. (2000). Protein digestion in juvenile turbot (*Scophthalmus maximus*) and effects of dietary administration of *Vibrio proteolyticus. Aquaculture,* 186 (1-2): 107-116.

De Schryver, P., Defoirdt, T. and Sorgeloos, P. (2014). Early mortality syndrome outbreaks: a microbial management issue in shrimp farming? *PLoS Pathogens,* 10 (4): e1003919.

Debnath, P., Karim, M. and Belton, B. (2014). Comparative study of the reproductive performance and White Spot Syndrome Virus (WSSV) status of black tiger shrimp (*Penaeus monodon*) collected from the Bay of Bengal. *Aquaculture,* 424: 71-77.

DePaola, A., Flynn, P. A., McPhearson, R. M. and Levy, S. B. (1988). Phenotypic and genotypic characterization of tetracycline-and oxytetracycline-resistant Aeromonas hydrophila from cultured channel catfish (*Ictalurus punctatus*) and their environments. *Applied and Environmental Microbiology*, 54 (7): 1861-1863.

Dhar, A. K., Roux, M. M. and Klimpel, K. R. (2001). Detection and quantification of infectious hypodermal and hematopoietic necrosis virus and white spot virus in shrimp using real-time quantitative PCR and SYBR Green chemistry. *Journal of Clinical Microbiology,* 39 (8): 2835-2845.

Dong, H. T., Siriroob, S., Meemetta, W., Santimanawong, W., Gangnonngiw, W., Pirarat, N., Khunrae, P., Rattanarojpong, T., Vanichviriyakit, R. and Senapin, S. (2017). Emergence of tilapia lake virus in Thailand and an alternative semi-nested RT-PCR for detection. *Aquaculture,* 476: 111-118.

Durand, S. V. and Lightner, D. V. (2002). Quantitative real time PCR for the measurement of white spot syndrome virus in shrimp. *Journal of Fish Diseases,* 25 (7): 381-389.

Durand, S., Lightner, D. V., Redman, R. M. and Bonami, J. R. (1997). Ultrastructure and morphogenesis of white spot syndrome baculovirus (WSSV). *Diseases of Aquatic Organisms,* 29 (3): 205-211.

Egidius, E. (1987). Vibriosis: pathogenicity and pathology. A review. *Aquaculture* 67 (1-2): 15-28.

Ellegren, H., Smeds, L., Burri, R., Olason, P. I., Backström, N., Kawakami, T., Künstner, A., Mäkinen, H., Nadachowska-Brzyska, K., Qvarnström, A. and Uebbing, S. (2012). The genomic landscape of species divergence in *Ficedula flycatchers*. *Nature,* 491: 756.

El-Sersy, N. A., AbdelRazek, F. A. and Taha, S. M. (2006). Evaluation of various probiotic bacteria for the survival of *Penaeus japonicus* larvae. *Fresenius Environmental Bulletin,* 15 (12): 1506.

Elsheshtawy, A., Yehia, N., Elkemary, M. and Soliman, H. (2018). Direct detection of unamplified *Aeromonas hydrophila* DNA in clinical fish samples using gold nanoparticle probe-based assay. *Aquaculture,* 500: 451-457.

Eshik, M. E. E., Punom, N. J., Begum, M. K., Khan, T., Saha, M. L. and Rahman, M. S. (2018). Molecular characterization of Acute Hepatopancreatic Necrosis Disease causing *Vibrio parahaemolyticus* strains in cultured shrimp *Penaeus monodon* in south-west farming region of Bangladesh. *Dhaka University Journal of Biological Sciences,* 27: 57-68.

Eshik, M. M. E., Abedin, M. M., Punom, N. J., Begum, M. K. and Rahman, M. S. (2017). Molecular Identification of AHPND Positive *Vibrio parahaemolyticus* causing an outbreak in South-West Shrimp Farming Regions of Bangladesh. *Journal of Bangladesh Academy of Sciences,* 41 (2): 127-135.

Eyngor, M., Zamostiano, R., Tsofack, J. E. K., Berkowitz, A., Bercovier, H., Tinman, S., Lev, M., Hurvitz, A., Galeotti, M., Bacharach, E. and Eldar, A. (2014). Identification of a novel RNA virus lethal to tilapia. *Journal of Clinical Microbiology*, 52 (12): 4137– 4146.

Fathi, M., Dickson, C., Dickson, M., Leschen, W., Baily, J., Muir, F., Ulrich, K. and Weidmann, M. (2017). Identification of Tilapia Lake Virus in Egypt in Nile tilapia affected by 'summer mortality' syndrome. *Aquaculture,* 473: 430-432.

FDA. (2001). Chapter 22: Aquaculture Drugs. In *Fish and Fishery Products Hazards and Controls Guide.* (Second Edition). FDA (Food and Drug Administration): Washington, available at <http: //seafood. oregonstate. edu/ sites/ agscid7/ files/ snic/ compendium/chapter-22-aquaculture-drugs.pdf>

Feist, S. W. and Longshaw, M. (2008). Histopathology of fish parasite infections– importance for populations. *Journal of Fish Biology,* 73 (9): 2143-2160.

Felix, S. (2010). *Marine and Aquaculture Biotechnology*. Agrobios, India.

Figueras, M. J., Beaz-Hidalgo, R., Collado, L. and Martínez-Murcia, A. J. (2011). Point of view on the recommendations for new bacterial species description and their impact on the genus *Aeromonas* and *Arcobacter. The Bulletin of Bergeys International Society for Microbial Systematics,* 2: 1-16.

Flegel, T. W. (2012). Historic emergence, impact and current status of shrimp pathogens in Asia. *Journal of Invertebrate Pathology,* 110 (2): 166-173.

Flegel, T. W., Sriurairatana, S., Wongteerasupaya, C., Boonsaeng, V., Panyim, S. and Withyachumnarnkul, B. (1995). Progress in characterization and control of yellow-head virus of *Penaeus monodon.* In *Swimming through troubled water; Proceedings of the special session on shrimp farming*, edited by Browdy, C. L. and Hopkins, J. S. 76-83. World Aquaculture Society, Baton Rouge, LA.

Fu, S., Tian, H., Wei, D., Zhang, X. and Liu, Y. (2017). Delineating the origins of *Vibrio parahaemolyticus* isolated from outbreaks of acute hepatopancreatic necrosis disease in Asia by the use of whole genome sequencing. *Frontiers in Microbiology,* 8: 2354.

Fujimura, K. and Kocher, T. D. (2011). Tol2-mediated transgenesis in tilapia (*Oreochromis niloticus*). *Aquaculture,* 319 (3-4): 342-346.

Fuller, R. (1989). Probiotics in man and animals. *Journal of Applied Bacteriology,* 66 (5): 365-378.

Ghosh, S., Sinha, A. and Sahu, C. (2008). Dietary probiotic supplementation in growth and health of live-bearing ornamental fishes. *Aquaculture Nutrition,* 14 (4): 289-299.

Gomez-Gil, B., Soto-Rodríguez, S., Lozano, R. and Betancourt-Lozano, M. (2014). Draft genome sequence of *Vibrio parahaemolyticus* strain M0605, which causes severe mortalities of shrimps in Mexico. *Genome Announcements,* 2 (2): e00055-14.

Gudding, R. and Van Muiswinkel, W.B. (2013). A history of fish vaccination: science-based disease prevention in aquaculture. *Fish & Shellfish Immunology* 35 (6): 1683-1688.

Haff, L. A. (1994). Improved quantitative PCR using nested primers. *Genome Research,* 3 (6): 332-337.

Hai, N. V., Buller, N. and Fotedar, R. (2009). Effects of probiotics (*Pseudomonas synxantha* and *Pseudomonas aeruginosa*) on the growth, survival and immune parameters of juvenile western king prawns (*Penaeus latisulcatus* Kishinouye, 1896). *Aquaculture Research,* 40 (5): 590-602.

Hameed, A. S., Yoganandhan, K., Widada, J. S. and Bonami, J. R. (2004). Studies on the occurrence of *Macrobrachium rosenbergii* nodavirus and extra small virus-like particles associated with white tail disease of *M. rosenbergii* in India by RT-PCR detection. *Aquaculture,* 238 (1-4): 127-133.

Han, J. E., Tang, K. F. J., Aranguren, L. F. and Piamsomboon, P. (2017). Characterization and pathogenicity of acute hepatopancreatic necrosis disease natural mutants, pirABvp (-) *V. parahaemolyticus*, and pirABvp (+) *V. campbellii* strains. *Aquaculture,* 470: 84-90.

Han, J. E., Tang, K. F., Tran, L. H. and Lightner, D. V. (2015). Photorhabdus insect-related (Pir) toxin-like genes in a plasmid of *Vibrio parahaemolyticus*, the causative agent of acute hepatopancreatic necrosis disease (AHPND) of shrimp. *Diseases of Aquatic Organisms,* 113 (1): 33-40.

Haridas, D. V., Pillai, D., Manojkumar, B., Nair, C. M. and Sherief, P. M. (2010). Optimisation of reverse transcriptase loop-mediated isothermal amplification assay for rapid detection of *Macrobrachium rosenbergii* noda virus and extra small virus in *Macrobrachium rosenbergii*. *Journal of Virological Methods,* 167 (1): 61-67.

Harikrishnan, R., Balasundaram, C. and Heo, M. S. (2010). Effect of probiotics enriched diet on *Paralichthys olivaceus* infected with lymphocystis disease virus (LCDV). *Fish & Shellfish Immunology,* 29 (5): 868-874.

Hastefnl, T., Guo'ding, R. and Eve-risen, B. (2005). Bacterial Vaccines for Fish- An Update. *Developments in Biologicals,* 121: 55-74.

Higuchi, R., Fockler, C., Dollinger, G. and Watson, R. (1993). Kinetic PCR: Real time monitoring of DNA amplification reactions. *Biotechnology*, 11: 1026-1030.

Hong, H. A., Duc, L. H. and Cutting, S. M. (2005). The use of bacterial spore formers as probiotics. *FEMS Microbiology Reviews,* 29 (4): 813-835.

Hong, S. R., Kim, M. S., Jeong, H. D. and Hong, S. (2017). Development of real-time and quantitative QCM immunosensor for the rapid diagnosis of *Aeromonas hydrophila* infection. *Aquaculture Research*, 48 (5): 2055-2063.

Hossain, M. S., Otta, S. K., Chakraborty, A., Kumar, H. S., Karunasagar, I. and Karunasagar, I. (2004). Detection of WSSV in cultured shrimps, captured brooders, shrimp postlarvae and water samples in Bangladesh by PCR using different primers. *Aquaculture,* 237: 59-71.

Hsu, S. M. and Raine, L. (1981). Protein A, avidin, and biotin in immunohistochemistry. *Journal of Histochemistry & Cytochemistry*, 29 (11): 1349-1353.

Irianto, A. and Austin, B. (2002). Probiotics in aquaculture. *Journal of Fish Diseases,* 25 (11): 633-642.

Jariyapong, P., Pudgerd, A., Weerachatyanukul, W., Hirono, I., Senapin, S., Dhar, A. K. and Chotwiwatthanakun, C. (2018). Construction of an infectious *Macrobrachium rosenbergii* nodavirus from cDNA clones in Sf9 cells and improved recovery of viral RNA with AZT treatment. *Aquaculture,* 483: 111-119.

Kamei, Y., Yoshimizu, M., Ezura, Y. and Kimura, T. (1988). Screening of bacteria with antiviral activity from fresh water salmonid hatcheries. *Microbiology and Immunology,* 32 (1): 67-73.

Kamilya, D. and Baruah, A. (2014). Epizootic ulcerative syndrome (EUS) in fish: history and current status of understanding. *Reviews in Fish Biology and Fisheries,* 24 (1): 369-380.

Kephart, S. R. (1990). Starch gel electrophoresis of plant isozymes: a comparative analysis of techniques. *American Journal of Botany*, 77 (5): 693-712.

Khawsak, P., Deesukon, W., Chaivisuthangkura, P. and Sukhumsirichart, W. (2008). Multiplex RT-PCR assay for simultaneous detection of six viruses of penaeid shrimp. *Molecular and Cellular Probes,* 22 (3): 177-183.

Kiatpathomchai, W., Jitrapakdee, S., Panyim, S. and Boonsaeng, V. (2004). RT-PCR detection of yellow head virus (YHV) infection in *Penaeus monodon* using dried haemolymph spots. *Journal of Virological Methods,* 119 (1): 1-5.

Kim, A., Nguyen, T. L. and Kim, D. (2017). Modern methods of diagnosis. In *Diagnosis and Control of Diseases of Fish and Shellfish,* edited by Austin B. and Newaj-Fyzul, A. 109-145. John Wiley & Sons Ltd.

Klesius, P. H. and Shoemaker, C. A. (1999). Development and use of modified live *Edwardsiella ictaluri* vaccine against enteric septicemia of catfish. *Advances in Veterinary Medicine,* 41: 523-537.

Koiwai, K., Kodera, T., Thawonsuwan, J., Kawase, M., Kondo, H. and Hirono, I. (2018). A rapid method for simultaneously diagnosing four shrimp diseases using PCR-DNA chromatography method. *Journal of Fish Diseases*, 41 (2): 395-399.

Koiwai, K., Tinwongger, S., Nozaki, R., Kondo, H. and Hirono, I. (2016). Detection of acute hepatopancreatic necrosis disease strain of *Vibrio parahaemolyticus* using loop-mediated isothermal amplification. *Journal of Fish Diseases,* 39 (5): 603-606.

Kondo, H., Tinwongger, S., Proespraiwong, P., Mavichak, R., Unajak, S., Nozaki, R. and Hirono, I. (2014). Draft genome sequences of six strains of *Vibrio parahaemolyticus* isolated from early mortality syndrome/acute hepatopancreatic necrosis disease shrimp in Thailand. *Genome Announcements,* 2 (2): e00221-14.

Kono, T., Savan, R., Sakai, M. and Itami, T. (2004). Detection of white spot syndrome virus in shrimp by loop-mediated isothermal amplification. *Journal of Virological Methods,* 115 (1): 59-65.

Kostic, T., Francois, P., Bodrossy, L. and Schrenzel, J. (2008). Oligonucleotide and DNA microarrays: versatile tools for rapid bacterial diagnostics. In *Principles of Bacterial Detection: Biosensors, Recognition Receptors and Microsystems*, edited by Zourob, M., Elwary, S. and Turner, A., 629-657. Springer, New York.

Kumar, G. and Kocour, M. (2017). Applications of next-generation sequencing in fisheries research: a review. *Fisheries Research,* 186:11-22.

Kumaresan, V., Pasupuleti, M., Arasu, M. V., Al-Dhabi, N. A., Arshad, A., Amin, S. M. N., Yusoff, F. M. and Arockiaraj, J. (2018). A comparative transcriptome approach for identification of molecular changes in *Aphanomyces invadans* infected Channa striatus. *Molecular Biology Reports* 45 (6): 2511-2523.

Lallier, R. and Higgins, R. (1988). Biochemical and toxigenic characteristics of *Aeromonas* spp. isolated from diseased mammals, moribund and healthy fish. *Veterinary Microbiology,* 18 (1): 63-71.

Lategan, M. J., Torpy, F. R. and Gibson, L. F. (2004). Control of saprolegniosis in the eel *Anguilla australis* Richardson, by Aeromonas media strain A199. *Aquaculture,* 240: 19-27.

Law, J. W. F., Ab Mutalib, N. S., Chan, K. G. and Lee, L. H. (2015). Rapid methods for the detection of foodborne bacterial pathogens: principles, applications, advantages and limitations. *Frontiers in Microbiology,* 5: 770.

Li, J., Tan, B. and Mai, K. (2009). Dietary probiotic Bacillus OJ and isomaltooligosaccharides influence the intestine microbial populations, immune responses and resistance to white spot syndrome virus in shrimp (*Litopenaeus vannamei*). *Aquaculture,* 291 (1-2): 35-40.

Lightner, D. V. and Redman, R. M. (1993). A putative iridovirus from the penaeid shrimp Protrachypene precipua Burkenroad (Crustacea: Decapoda). *Journal of Invertebrate Pathology,* 62 (1): 107-109.

Lightner, D. V., Redman, C. R., Pantoja, B. L., Noble, L. M., Nunan, L. T. and Tran, L. (2013). *Documentation of an Emerging Disease (early mortality syndrome) in SE Asia & Mexico*. OIE Reference Laboratory for Shrimp Diseases, Department of Veterinary Science & Microbiology, The University of Arizona, USA.

Lightner, D. V., Redman, R. M., Pantoja, C. R., Noble, B. L. and Tran, L. (2012). Early mortality syndrome affects shrimp in Asia. *Global Aquaculture Advocate,* 15 (1): 40.

Lilley, J. H., Callinan, R. B., Chinabut, S., Kanchanakhan, S., MacRae, I. H. and Phillips, M. J. (1998). *Epizootic ulcerative syndrome (EUS) technical handbook,* The Aquatic Animal Health Research Institute (AAHRI), Bangkok, Thailand.

Lilley, J. H., Hart, D., Panyawachira, V., Kanchanakhan, S., Chinabut, S., Söderhäll, K. and Cerenius, L. (2003). Molecular characterization of the fish-pathogenic fungus *Aphanomyces invadans*. *Journal of Fish Diseases,* 26 (5): 263-275.

Lin, F., Liu, L., Hao, G. J., Sheng, P. C., Cao, Z., Zhou, Y., Lv, P., Xu, T., Shen, J. and Chen, K. (2018). The development and application of a duplex reverse transcription loop-mediated isothermal amplification assay combined with a lateral flow dipstick method for *Macrobrachium rosenbergii* nodavirus and extra small virus isolated in China. *Molecular and Cellular Probes,* 40: 1-7.

Liu, L., Jiang, L., Yu, Y., Xia, X., Pan, Y., Yan, S. and Wang, Y. (2017). Rapid diagnosis of *Vibrio owensii* responsible for shrimp acute hepatopancreatic necrosis disease with isothermal recombinase polymerase amplification assay. *Molecular and Cellular Probes,* 33: 4-7.

Liu, Z.J. (2008). *Aquaculture Genome Technologies*. John Wiley & Sons.

Lo, C. F. and Kou, G. H. (1998). Virus-associated white spot syndrome of shrimp in Taiwan: a review. *Fish Pathology,* 33 (4): 365-371.

Lo, C. F., Ho, C. H., Peng, S. E., Chen, C. H., Hsu, H. C., Chiu, Y. L., Chang, C. F., Liu, K. F., Su, M. S., Wang, C. H. and Kou, G.H. (1996). White spot syndrome baculovirus (WSBV) detected in cultured and captured shrimp, crabs and other arthropods. *Diseases of Aquatic Organisms* 27: 215-225.

Lo, C.F., Aoki, T., Bonami, J. R., Flegel, T., Leu, J. H., Lightner, D. V., Stentiford, G., Söderhäll, K., Walker, P. J., Wang, H. C., Xun, X., Yang, F. and Vlak, J. M. (2011). Nimaviridae. In *Virus Taxonomy, IXth Report of the International Committee on Taxonomy of Viruses*, https://talk.ictvonline.org/ictv-reports/ictv_9th_report/dsdna-viruses-2011/w/dsdna_viruses/119/nimaviridae.

Logan, J., Edwards, K. and Saunders, N. (2009). *Real-Time PCR: Current technology and applications*. Caister Academic Press.

Maiden, M. C. J., Bygraves, J. A., Feil, E., Morelli, G., Russell, J. E., Urwin, R., Zhang, Q., Zhou, J., Zurth, K., Caugant, D. A. and Feavers, I. M. (1998). Multilocus sequence typing: a portable approach to the identification of clones within populations of pathogenic microorganisms. *Proceedings of the National Academy of Sciences,* 95 (6): 3140-3145.

Majeed, M., Soliman, H., Kumar, G., El-Matbouli, M. and Saleh, M. (2018). Editing the genome of *Aphanomyces invadans* using CRISPR/Cas9. *Parasites & Vectors,* 11 (1): 554.

Mccashion, R. N. and Lynch, W. H. (1987). Effects of polymyxin B nonapeptide on *Aeromonas salmonicida. Antimicrobial Agents and Chemotherapy,* 31 (9): 1414-1419.

McLoughlin, M. F., Christie, K. E., Knappskog, D., Koumans, S., Graham, D., Rodger, H. and Turnbull, T. (2003). Field trial experiences with an inactivated monovalent pancreas disease virus vaccine. In *3rd International Symposium on Fish Vaccinology*, 9-11.

Mekata, T., Kono, T., Savan, R., Sakai, M., Kasornchandra, J., Yoshida, T. and Itami, T. (2006). Detection of yellow head virus in shrimp by loop-mediated isothermal amplification (LAMP). *Journal of Virological Methods,* 135 (2): 151-156.

Minana-Galbis, D., Farfan, M., Loren, J. G. and Fuste, M. C. (2010). The reference strain *Aeromonas hydrophicla* CIP 57.50 should be reclassified as *Aeromonas salmonicida* CIP 57.50. *International Journal of Systematic and Evolutionary Microbiology,* 60 (3): 715-717.

Mohan, C. V., Shankar, K.M., Kulkarni, S. and Sudha, P. M. (1998). Histopathology of cultured shrimp showing gross signs of yellow head syndrome and white spot syndrome during 1994 Indian epizootics. *Diseases of Aquatic Organisms,* 34 (1): 9-12.

Montanie, H., Bonami, J. R. and Comps, M. (1993). Irido-like virus infection in the crab *Macropipus depurator* L. (Crustacea, Decapoda). *Journal of Invertebrate Pathology,* 61 (3): 320-322.

NACA. (2014). Acute hepatopancreatic necrosis disease card (updated June 2014). *Network of Aquaculture Centres in Asia-Pacific (NACA).* www.enaca.org.

Naylor, R. and Burke, M. (2005). Aquaculture and ocean resources: raising tigers of the sea. *Annual Review of Environment and Resources,* 30: 185-218.

Naylor, R. L., Goldburg, R. J., Primavera, J. H., Kautsky, N., Beveridge, M. C., Clay, J., Folke, C., Lubchenco, J., Mooney, H. and Troell, M. (2000). Effect of aquaculture on world fish supplies. *Nature,* 405: 1017.

Newaj-Fyzul, A. and Austin, B. (2015). Probiotics, immunostimulants, plant products and oral vaccines, and their role as feed supplements in the control of bacterial fish diseases. *Journal of Fish Diseases,* 38 (11): 937-955.

Newaj-Fyzul, A., Adesiyun, A. A., Mutani, A., Ramsubhag, A., Brunt, J. and Austin, B. (2007). *Bacillus subtilis* AB1 controls Aeromonas infection in rainbow trout (*Oncorhynchus mykiss*, Walbaum). *Journal of Applied Microbiology,* 103 (5): 1699-1706.

Nunan, L., Lightner, D., Pantoja, C. and Gomez-Jimenez, S. (2014). Detection of acute hepatopancreatic necrosis disease (AHPND) in Mexico. *Diseases of Aquatic Organisms,* 111 (1): 81-86.

Nunan, L.M. and Lightner, D.V. (1997). Development of a non-radioactive gene probe by PCR for detection of white spot syndrome virus (WSSV). *Journal of Virological Methods,* 63 (1-2): 193-201.

Oakey, H. J. and Smith, C. S. (2018). Complete genome sequence of a white spot syndrome virus associated with a disease incursion in Australia. *Aquaculture,* 484: 152-159.

Odell, I. D. and Cook, D. (2013). Immunofluorescence techniques. *Journal of Investigative Dermatology,* 133 (1): 1-4.

OIE (2018). *Aquatic Animal Health Code.* World Organisation for Animal Health. http://www.oie.int/standard-setting/aquatic-code/access-online/

OIE (2019). *Manual of Diagnostic Tests for Aquatic Animals.* World Organisation for Animal Health. http://www.oie.int/standard-setting/aquatic-manual/access-online/

Okumus, I. and Ciftci, Y. (2003). Fish population genetics and molecular markers: II-molecular markers and their applications in fisheries and aquaculture. *Turkish Journal of Fisheries and Aquatic Sciences*, 3 (1): 51-79.

Overturf, K. (2009). *Molecular Research in Aquaculture*. John Wiley & Sons.

Owens, L., La Fauce, K., Juntunen, K., Hayakijkosol, O. and Zeng, C. (2009). *Macrobrachium rosenbergii* nodavirus disease (white tail disease) in Australia. *Diseases of Aquatic Organisms,* 85 (3): 175-180.

Piégu, B., Guizard, S., Yeping, T., Cruaud, C., Asgari, S., Bideshi, D.K., Federici, B.A. and Bigot, Y. (2014). Genome sequence of a crustacean iridovirus, IIV31, isolated from the pill bug, *Armadillidium vulgare. Journal of General Virology,* 95 (7): 1585-1590.

Pinkert, C. A., Irwin, M. H. and Moffatt, R. J. (1995). Transgenic animal modeling. In *Molecular Biology and Biotechnology: A Comprehensive Desk Reference*, edited by Meyers, R. A., 901-907. John Wiley & Sons.

Pridgeon, J. W. and Klesius, P. H. (2012). Major bacterial diseases in aquaculture and their vaccine development. *Animal Science Reviews,* 7: 1-16.

Pridgeon, J. W., Zhang, D. and Zhang, L. (2014). Complete genome sequence of the highly virulent *Aeromonas hydrophila* AL09-71 isolated from diseased channel catfish in west Alabama. *Genome Announcements,* 2 (3): e00450-14.

Qi, W. (2002). Social and economic impacts of aquatic animal health problems in aquaculture in China. In *Primary aquatic animal health care in rural, small-scale, aquaculture development*, edited by Arthur, J. R., Phillips, M. J., Subasinghe, R. P., Reantaso, M. B. and MacRae, I. H., 55-61. FAO Fisheries technical paper, 406.

Qian, D., Shi, Z., Zhang, S., Cao, Z., Liu, W., Li, L., Xie, Y., Cambournac, I. and Bonami, J. R. (2003). Extra small virus-like particles (XSV) and nodavirus associated with whitish muscle disease in the giant freshwater prawn, *Macrobrachium rosenbergii. Journal of Fish Diseases,* 26 (9): 521-527.

Qiu, L., Chen, M. M., Wan, X. Y., Li, C., Zhang, Q. L., Wang, R. Y., Cheng, D. Y., Dong, X., Yang, B., Wang, X. H. and Xiang, J. H. (2017). Characterization of a new member of Iridoviridae, Shrimp hemocyte iridescent virus (SHIV), found in white leg shrimp (*Litopenaeus vannamei*). *Scientific Reports,* 7 (1): 11834.

Qiu, L., Chen, M. M., Wan, X. Y., Zhang, Q. L., Li, C., Dong, X., Yang, B. and Huang, J. (2018). Detection and quantification of shrimp hemocyte iridescent virus by TaqMan probe based real-time PCR. *Journal of Invertebrate Pathology,* 154: 95-101.

Quéré, R., Commes, T., Marti, J., Bonami, J. R. and Piquemal, D. (2002). White spot syndrome virus and infectious hypodermal and hematopoietic necrosis virus simultaneous diagnosis by miniarray system with colorimetry detection. *Journal of Virological Methods,* 105 (2): 189-196.

Rembold, M., Lahiri, K., Foulkes, N. S. and Wittbrodt, J. (2006). Transgenesis in fish: efficient selection of transgenic fish by co-injection with a fluorescent reporter construct. *Nature Protocols,* 1 (3): 1133.

Roberts R. J., Frerichs G. N., Tonguthai K. and Chinabut S. (1994). Epizootic ulcerative syndrome of farmed and wild fishes. In *Recent Advances in Aquaculture V,* edited by Muir J. F. and Roberts R. J. 207–239. Blackwell Science.

Sánchez-Barajas, M., Liñán-Cabello, M.A. and Mena-Herrera, A. (2009). Detection of yellow-head disease in intensive freshwater production systems of *Litopenaeus vannamei. Aquaculture International,* 17 (2): 101-112.

Sánchez-Paz, A. (2010). White spot syndrome virus: an overview on an emergent concern. *Veterinary Research,* 41 (6): 43.

Senapin, S., Shyam, K. U., Meemetta, W., Rattanarojpong, T. and Dong, H. T. (2018). Inapparent infection cases of tilapia lake virus (TiLV) in farmed tilapia. *Aquaculture,* 487: 51-55.

Shoemaker, C. A., Klesius, P. H. and Bricker, J. M. (1999). Efficacy of a modified live *Edwardsiella ictaluri* vaccine in channel catfish as young as seven days post hatch. *Aquaculture,* 176 (3-4): 189-193.

Silva, E. F., Soares, M. A., Calazans, N. F., Vogeley, J. L., do Valle, B. C., Soares, R. and Peixoto, S. (2012). Effect of probiotic (B acillus spp.) addition during larvae and postlarvae culture of the white shrimp *Litopenaeus vannamei. Aquaculture Research,* 44 (1): 13-21.

Sittidilokratna, N., Dangtip, S., Cowley, J. A. and Walker, P. J. (2008). RNA transcription analysis and completion of the genome sequence of yellow head nidovirus. *Virus Research,* 136 (1-2): 157-165.

Sivakumar, N., Sundararaman, M. and Selvakumar, G. (2012). Probiotic effect of *Lactobacillus acidophilus* against vibriosis in juvenile shrimp (*Penaeus monodon*). *African Journal of Biotechnology,* 11 (91): 15811-15818.

Sri Widada, J. and Bonami, J. R. (2004). Characteristics of the monocistronic genome of extra small virus, a virus-like particle associated with *Macrobrachium rosenbergii* nodavirus: possible candidate for a new species of satellite virus. *Journal of General Virology,* 85 (3): 643-646.

Sri Widada, J., Durand, S., Cambournac, I., Qian, D., Shi, Z., Dejonghe, E., Richard, V. and Bonami, J. R. (2003). Genome-based detection methods of *Macrobrachium rosenbergii* nodavirus, a pathogen of the giant freshwater prawn, *Macrobrachium rosenbergii*: dot-blot, in situ hybridization and RT-PCR. *Journal of Fish Diseases,* 26 (10): 583-590.

Sritunyalucksana, K., Dangtip, S., Sanguanrut, P., Sirikharin, R., Taengchaiyaphum, R., Thitamadee, S., Mavichak, R., Proespraiwong, P. and Flegel, T. W. (2015). A two-tube, nested PCR detection method for AHPND bacteria. *Network of Aquaculture Centres in Asia-Pacific (NACA).* http://www. enaca. org/modules/news/article. php.

Stentiford, G. D. and Lightner, D. V. (2011). Cases of white spot disease (WSD) in European shrimp farms. *Aquaculture,* 319 (1-2): 302-306.

Stuart, G. W., Vielkind, J. R., McMurray, J. V. and Westerfield, M. (1990). Stable lines of transgenic zebrafish exhibit reproducible patterns of transgene expression. *Development,* 109 (3): 577-584.

Sudhakaran, R., Ishaq Ahmed, V. P., Haribabu, P., Mukherjee, S. C., Sri Widada, J., Bonami, J. R. and Sahul Hameed, A. S. (2007a). Experimental vertical transmission of *Macrobrachium rosenbergii* nodavirus (MrNV) and extra small virus (XSV) from brooders to progeny in *Macrobrachium rosenbergii* and *Artemia. Journal of Fish Diseases,* 30 (1): 27-35.

Sudhakaran, R., Parameswaran, V. and Hameed, A. S. (2007b). In vitro replication of *Macrobrachium rosenbergii* nodavirus and extra small virus in C6/36 mosquito cell line. *Journal of Virological Methods,* 146 (1-2): 112-118.

Swain, S. M., Singh, C. and Arul, V. (2009). Inhibitory activity of probiotics *Streptococcus phocae* PI80 and *Enterococcus faecium* MC13 against vibriosis in shrimp *Penaeus monodon. World Journal of Microbiology and Biotechnology,* 25 (4): 697-703.

Tang, K. F. and Lightner, D. V. (2000). Quantification of white spot syndrome virus DNA through a competitive polymerase chain reaction. *Aquaculture,* 189: 11-21.

Tang, K. F. J. and Lightner, D. V. (1999). A yellow head virus gene probe: nucleotide sequence and application for in situ hybridization. *Diseases of Aquatic Organisms,* 35 (3): 165-173.

Tang, K. F., Redman, R. M., Pantoja, C. R., Le Groumellec, M., Duraisamy, P. and Lightner, D. V. (2007). Identification of an iridovirus in *Acetes erythraeus* (Sergestidae) and the development of in situ hybridization and PCR method for its detection. *Journal of Invertebrate Pathology,* 96 (3): 255-260.

Tattiyapong, P., Sirikanchana, K. and Surachetpong, W. (2018). Development and validation of a reverse transcription quantitative polymerase chain reaction for tilapia lake virus detection in clinical samples and experimentally challenged fish. *Journal of Fish Diseases,* 41 (2): 255-261.

Tazumi, A., Ono, S., Sekizuka, T., Moore, J.E., Millar, B.C. and Matsuda, M. (2009). Molecular characterization of the sequences of the 16S-23S rDNA internal spacer region (ISR) from isolates of Taylorella asinigenitalis. *BMC Research Notes,* 2 (1): 33.

Teng, L., Deng, L., Dong, X., Wei, S., Li, J., Li, N. and Zhou, Y. (2017). Genome Sequence of Hypervirulent *Aeromonas hydrophila* Strain HZAUAH. *Genome Announcements,* 5 (11): e00012-17.

Terech-Majewska, E. and Siwicki, A. K. (2006). Influence of oxytetracycline on the metabolic and phagocyte activity of macrophages and the proliferative response of lymphocytes in Carp and European Catfish. *Medycyna Weterynaryjna,* 62 (12): 1431-1434.

Tsofack, J. E. K., Zamostiano, R., Watted, S., Berkowitz, A., Rosenbluth, E., Mishra, N., Briese, T., Lipkin, W. I., Kabuusu, R. M., Ferguson, H. and del Pozo, J. (2017). Detection of Tilapia Lake Virus in Clinical Samples by Culturing and Nested Reverse Transcription-PCR. *Journal of Clinical Microbiology,* 55 (3): 759-767.

van Pelt-Verkuil, E., van Belkum, A. and Hays, J. P. (2008). *Principles and Technical Aspects of PCR Amplification*. Springer Science & Business Media.

Vandersea, M. W., Litaker, R. W., Yonnish, B., Sosa, E., Landsberg, J. H., Pullinger, C., Moon-Butzin, P., Green, J., Morris, J. A., Kator, H. and Noga, E. J., 2006. Molecular assays for detecting *Aphanomyces invadans* in ulcerative mycotic fish lesions. *Applied and Environmental Microbiology,* 72 (2): 1551-1557.

Vos, P., Hogers, R., Bleeker, M., Reijans, M., Lee, T. V. D., Hornes, M., Friters, A., Pot, J., Paleman, J., Kuiper, M. and Zabeau, M. (1995). AFLP: a new technique for DNA fingerprinting. *Nucleic Acids Research,* 23 (11): 4407-4414.

Walker, P. J., Cowley, J. A., Spann, K. M., Hodgson, R. A. J., Hall, R. M. and Withyachumnarnkul, B. (2001). Yellow head complex viruses: transmission cycles and topographical distribution in the Asia-Pacific region. In *The New Wave: Proceedings of the Special Session on Sustainable Shrimp Farming,* edited by Browdy, C. L. and Jory, D. E. 227-237. World Aquaculture Society, Baton Rouge.

Wang, C. S., Tsai, Y. J. and Chen, S. N. (1998). Detection of white spot disease virus (WSDV) infection in shrimp using in situ hybridization. *Journal of Invertebrate Pathology,* 72 (2): 170-173.

Wang, N., Liu, J., Pang, M., Wu, Y., Awan, F., Liles, M. R., Lu, C. and Liu, Y. (2018). Diverse roles of Hcp family proteins in the environmental fitness and pathogenicity of *Aeromonas hydrophila* Chinese epidemic strain NJ-35. *Applied Microbiology and Biotechnology,* 102 (16): 7083-7095.

Williams, K., Blake, S., Sweeney, A., Singer, J. T. and Nicholson, B. L. (1999). Multiplex reverse transcriptase PCR assay for simultaneous detection of three fish viruses. *Journal of Clinical Microbiology,* 37 (12): 4139-4141.

Wongteerasupaya, C., Tongchuea, W., Boonsaeng, V., Panyim, S., Tassanakajon, A., Withyachumnarnkul, B. and Flegel, T. W. (1997). Detection of yellow-head virus (YHV) of Penaeus monodon by RT-PCR amplification. *Diseases of Aquatic Organisms,* 31 (3): 181-186.

World Health Organization (WHO). (2012). *Critically important antimicrobials for human medicine*. World Health Organization.

Xiao, J., Liu, L., Ke, Y., Li, X., Liu, Y., Pan, Y., Yan, S. and Wang, Y. (2017). Shrimp AHPND-causing plasmids encoding the PirAB toxins as mediated by pirAB-Tn903 are prevalent in various *Vibrio* species. *Scientific Reports,* 7: 42177.

Xu, L., Wang, T., Li, F. and Yang, F. (2016). Isolation and preliminary characterization of a new pathogenic iridovirus from red claw crayfish *Cherax quadricarinatus. Diseases of Aquatic Organisms,* 120 (1): 17-26.

Xu, W., Yan, W., Li, X., Zou, Y., Chen, X., Huang, W., Miao, L., Zhang, R., Zhang, G. and Zou, S. (2013). Antibiotics in riverine runoff of the Pearl River Delta and Pearl River Estuary, China: concentrations, mass loading and ecological risks. *Environmental Pollution,* 182: 402-407.

Xue, B., Zhang, R., Wang, Y., Liu, X., Li, J. and Zhang, G. (2013). Antibiotic contamination in a typical developing city in south China: occurrence and ecological risks in the Yongjiang River impacted by tributary discharge and anthropogenic activities. *Ecotoxicology and Environmental Safety,* 92: 229-236.

Yang, Y. T., Chen, I. T., Lee, C. T., Chen, C. Y., Lin, S. S., Hor, L. I., Tseng, T. C., Huang, Y. T., Sritunyalucksana, K., Thitamadee, S. and Wang, H. C. (2014). Draft genome sequences of four strains of Vibrio parahaemolyticus, three of which cause early mortality syndrome/acute hepatopancreatic necrosis disease in shrimp in China and Thailand. *Genome Announcements,* 2 (5): e00816-14.

Yoganandhan, K., Leartvibhas, M., Sriwongpuk, S. and Limsuwan, C. (2006). White tail disease of the giant freshwater prawn *Macrobrachium rosenbergii* in Thailand. *Diseases of Aquatic Organisms,* 69 (2-3): 255-258.

Yoganandhan, K., Sri Widada, J., Bonami, J. R. and Sahul Hameed, A. S. (2005). Simultaneous detection of *Macrobrachium rosenbergii* nodavirus and extra small virus by a single tube, one-step multiplex RT-PCR assay. *Journal of Fish Diseases,* 28 (2): 65-69.

Zapata, A. A. and Lara-Flores, M. (2012). Antimicrobial activities of lactic acid bacteria strains isolated from Nile Tilapia intestine (*Oreochromis niloticus*). *Journal of Biology and Life Science,* 4 (1): 164-171.

Zbikowska, H. M. (2003). Fish can be first–advances in fish transgenesis for commercial applications. *Transgenic Research,* 12 (4): 379-389.

Zhang, H., Wang, J., Yuan, J., Li, L., Zhang, J., Bonami, J. R. and Shi, Z. (2006). Quantitative relationship of two viruses (MrNV and XSV) in white-tail disease of *Macrobrachium rosenbergii. Diseases of Aquatic Organisms,* 71 (1): 11-17.

Zhang, J., Chiodini, R., Badr, A. and Zhang, G. (2011). The impact of next-generation sequencing on genomics. *Journal of Genetics and Genomics,* 38 (3): 95-109.

Zhou, X., Wang, Y., Yao, J. and Li, W. (2010). Inhibition ability of probiotic, *Lactococcus lactis*, against *A. hydrophila* and study of its immunostimulatory effect in tilapia (*Oreochromis niloticus*). *International Journal of Engineering, Science and Technology,* 2 (7): 73-80.

Zou, S., Xu, W., Zhang, R., Tang, J., Chen, Y. and Zhang, G. (2011). Occurrence and distribution of antibiotics in coastal water of the Bohai Bay, China: impacts of river discharge and aquaculture activities. *Environmental Pollution,* 159 (10): 2913-2920.

In: Trends in Biochemistry and Molecular Biology
Editors: Hossain Uddin Shekhar et al.
ISBN: 978-1-53616-434-3

Chapter 16

UTILITIES OF THE CONCEPT OF STATISTICAL ANALYSIS IN THE BIOLOGICAL SCIENCES

Murshida Khanam*

Department of Statistics, University of Dhaka, Dhaka, Bangladesh

ABSTRACT

To analyze the data related to Biological Sciences, it is very much necessary to have some ideas about statistical analysis. In this chapter, the effectiveness of the concept of statistical analysis in the Biological Sciences has been explained in different ways. Statistics means to collect some data for some predetermined purposes. The collected data will be processed and analyzed to obtain the required results. When the data is related to Biological Sciences, it is called Biostatistics.

In this chapter, firstly, some definitions, say, population, sample, parameter, statistic, variable, mean, median, mode etc. have been presented by using some numerical examples. A key issue whether data follow normal distribution or not, has been focused. Next, the correlation analysis as well as regression analysis have been explored. When two or more variables have been studied, then it is necessary to find out whether the variables are related or not, and which variables are influenced by which variables and with what extent.

In Biological research, the concept of test of hypothesis is very much essential. The present book chapter sheds some lights on test of hypothesis, null hypothesis, alternative hypothesis, test statistic, type-I error, type-II error, level of significance etc. It also discusses about one sample t-test, dependent sample, paired t-test, one-way ANOVA and two –way ANOVA by using some numerical examples.

Keywords: statistics, variable, test of hypothesis, paired t-test, ANOVA

---

* Corresponding Author's E-mail: murshida@du.ac.bd.

1. INTRODUCTION

To analyze the Biological Science related data, it is necessary to have some concept of statistical analysis. According to Steel, Torrie and Dockey (1997) Statistics is the science, pure and applied, creating, developing and applying techniques, by which the uncertainty of inductive inferences may be evaluated. It can be said that Statistics means the collection of data or information for predetermined purposes, after that data should be processed and analyzed by some techniques and something will be interpreted as well as comprehended. Statistics always deals with the aggregation of individuals rather than individual simply. For example, the single measurement of the blood sugar level of a person is 6.11, is not a statistical statement but the statement that the average measurement of blood sugar level of some persons is 6.11, is a statistical statement and this measurement is a statistics. Now we shed some light on Biostatistics. Biostatistics is the term that can be used when the statistical tools or techniques are applied on statistical data related to Biological Sciences. Most of the things in Biological Sciences need to be well-researched. For example, a researcher wants to find out some factors which are responsible for blood pressure of the patients. In that case a multiple regression analysis will be applied to find out some important factors which are responsible for blood pressure of the patients. On the other hand, a researcher wants to find out the relationship between weight and blood sugar level of the patients. In that case it is necessary to perform correlation analysis.

Before going into the in-depth study regarding the necessitates of Statistics in Biological Sciences, it is necessary to focus on some terminologies.

1.1. Population

An aggregate of all individuals is called a population. The population is always determined on the basis of the purpose of the study or research. For example, the whole blood of a patient is the population if the researcher or doctor wants to test whether the patient is attacked with Typhoid or not. Similarly the total respondents in a community will be a population if someone investigates his or her research question regarding the community.

1.2. Sample

A part or portion of the population is called a sample. A sample must be a representative part of the population. For example, to test whether the patient is attacked with typhoid or not then a syringe of blood will be taken. A syringe of blood is called a sample. On the other hand, to test whether the average income of the people of the certain community is BDT 3500 or not then some people will be taken from the whole community. This portion of the people is called a sample.

1.3. Variable

A variable is a characteristic that can be varied from person to person, individual to individual. Usually a variable is measured quantitatively. For example,

1) Gender is a variable which has two categories, male and female.
2) Household size, such as 1, 2, 3, 4 etc.
3) Blood pressure of different persons.
4) Blood sugar levels of different persons.
5) Eye color, such as black, brown, etc.

1.4. Parameter

Population characteristics are known as parameter.

1.5. Statistic

Sample characteristics are known as statistic.

When the data is large enough, it is difficult to comprehend the data. Then it is necessary to summarize the data to understand, to comprehend and to interpret.

1.6. Measures of Central Tendency

Measures of central tendency means a single value that summarizes the whole data set. We use different types of averages as measures of central tendency.

Here we consider only three averages: say I) the arithmetic mean, II) the median, and III) the mode.

Examples:

- The average height of people of a community is 5 feet 2 inch.
- The average income of a community is BDT 25,635, the median income of the community is BDT 23,223 and the modal income value of that community is BDT 19,817.

1.7. Arithmetic Mean

The measure which is obtained by the sum of all observations, divided by the total number of observations is called the Arithmetic mean.

1.8. Median

Median is the middle most observations of the whole set of data.

1.9. Mode

The value that appears most frequently in a data is called a modal value.

So, to get an idea of the whole data set we can use these three measures and summarize the whole data set.

In the following, we have used a data set, from this data set we have calculated mean, median and mode. And finally interpretations have been given.

We have given the following data. The data source is as follows:http://staff.bath.ac.uk/pssiw/stats2/page16/page16.html

The data is named as exam revision.sav. This data got information regarding the students' performance related to examination.

The above data gives information about student's exam score and how many hours they spent for revising their reading. In addition the data has other information. But we have used only student's exam score and how many hours they spent for their revision.

By using SPSS (version 20) we calculated some measures which are important measures of central tendency and these are as follows:

2. Interpretation

It has been found that the average of the exam score of the students is 61. On the other hand, the median is 62. That means the 50% students got score below 62 and the

50% of the students got score more than 62. The model value is 58. That means most of the students got score 58 in their examination.

Table 1. Student's exam score and student's spending hours for their revision

Score	Hours
62.00	40.00
58.00	31.00
52.00	35.00
55.00	26.00
75.00	51.00
82.00	48.00
38.00	25.00
55.00	37.00
48.00	30.00
68.00	44.00
62.00	32.00
62.00	40.00
72.00	61.00
58.00	35.00
65.00	45.00
42.00	30.00
68.00	39.00
68.00	47.00
58.00	41.00
72.00	46.00

Table 2. Measures of central tendency (mean, median and mode) of the above data set

Measures	exam score	hours spent revising
Mean	61	39.15
Median	62	39.50
Mode	58*	30*

*Multiple modes exist, the smallest value is shown.

The above similar interpretation is applicable for hours spent revising (the students spent hours for their revision). The students spent on an average 39.15 hours for their revision.

The median is 39.15. That means 50% students spent below 39.5hours and 50% students spent more than 39.5 hours for their revision.

The model value is 30 hours. That means most of the students spent 30 hours for their revision.

In Biological research, it is also necessary to know the measures of variability of a data set. That means we need to know the concept of measures of dispersion. According to Jalil and Ferdous (1999), the measurement of the scatter of the values of a data set among themselves is called a measure of dispersion. There are so many measures of dispersion, among them standard deviation is the best measure among all measures of dispersion.

If x be the variate value and $\bar{x}$ be the arithmetic mean of the variate values, then the s.d. σ is given by

$$\sigma = \sqrt{\frac{\sum(x-\bar{x})^2}{n}} = \sqrt{\frac{\sum x^2}{n} - \left(\frac{\sum x}{n}\right)^2}$$

Where n =number of observations.

2.1. Continuous Series

The standard deviation in the continuous series is defined by

$$\sigma = \sqrt{\frac{\sum f(x-\bar{x})^2}{n}}$$

n = total frequency and x is the mid values of the classes and f is the frequencies of the classes.

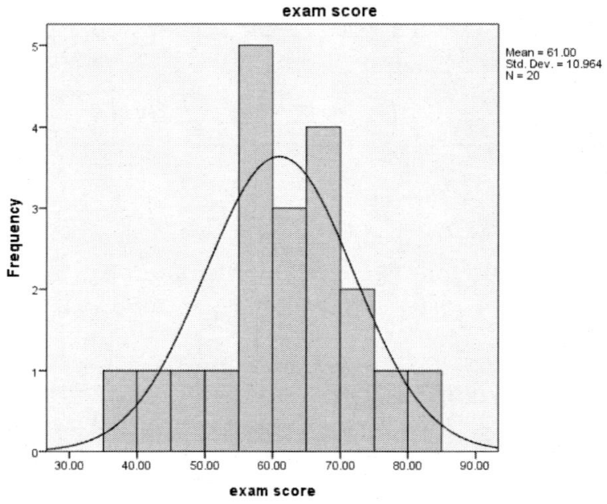

Figure 1. Variable exam score follows normal distribution.

In Figure 1, we have a data set named, exam revision.sav. In that case we can calculate the standard deviation of the variables named scores and hours respectively is 10.96 and 9.07 (by using SPSS 20 version).

In Biological research, it is needed to perform some statistical tests. In that case it is necessary to know whether the data is drawn from normal distribution or not.

To check the normality assumption of a variable we can make a histogram and a normal curve like above, we can say that the variable is normally distributed.

In Biological Sciences it is frequent that simultaneously two or more variables should be studied. In that case it is necessary to find out the relationship between the variables. On the other hand, finding correlation is not enough to comprehend a data, in that case it is necessary to investigate the strength of the relationship between variables.

Correlation analysis means to investigate whether there is linear relationship between the variables or not. On the other hand, the strength of the relationship between two variables is studied with the regression analysis. In the following a data has been downloaded from the website: http://www.principlesofeconometrics.com/excel.htm

The name of the data set in the website is food. The data is as follows:

The correlation and regression analyses have been performed and the interpretations have been given as follows.

Table 3. Food expenditure and income

food_exp	income	food_exp	income
115.22	3.69	406.34	20.13
135.98	4.39	171.92	20.33
119.34	4.75	303.23	20.37
114.96	6.03	377.04	20.43
187.05	12.47	194.35	21.45
243.92	12.98	213.48	22.52
267.43	14.2	293.87	22.55
238.71	14.76	259.61	22.86
295.94	15.32	323.71	24.2
317.78	16.39	275.02	24.39
216	17.35	109.71	24.42
240.35	17.77	359.19	25.2
386.57	17.93	201.51	25.5
261.53	18.43	460.36	26.61
249.34	18.55	447.76	26.7
309.87	18.8	482.55	27.14
345.89	18.81	438.29	27.16
165.54	19.04	587.66	28.62
196.98	19.22	257.95	29.4
395.26	19.93	375.73	33.4

2.2. Interpretation of Correlation Analysis

After analyzing the data by SPSS (Version 20), it has been found that there exists moderate positive correlation between food expenditure and income as the value of the Pearson's product moment correlation coefficient is 0.620. Even this correlation is highly significant at 1% level of significance as the p-value is 0.0000.

2.3. Interpretation of Regression Analysis

From the regression analysis it has been found that the overall model is highly significant at 5% level of significance. The F-statistic value is 23.789 and the p-value is 0.000. The estimated model is:

$$\hat{y} = \underset{(0.062)}{83.416} + \underset{(0.000)}{10.21x}$$

Here y indicates food expenditure and x indicates income. We can say that a one unit increase in income will increase the food expenditure with the rate of 10.21. Here both of the p-values are significant at 10% level of significance. The adjusted R^2 value is 0.369. That means. 36.9% variation of y can be explained by x.

The concept of probability is very much necessary for Biological research because in our daily lives, we are always faced with decision-making situation with uncertainty. Suppose a medical researcher wants to know whether a new medicine will be more effective than the old one to recover fever or not. The probability of an event is calculated by dividing the favorable outcome by the total outcome.

The distribution of a random variable is called probability distribution. Briefly it can be said that the normal distribution is very much important in case of probability distribution. In Biological research it is necessary to consider that the sample will come from a normal population. According to M.N. Islam (2012), in the first place, many scientists have observed that the random variables studied in various physical experiments follow a pattern of variation that is similar to the normal distribution.

The most important concept is the test of hypothesis.

Test of hypothesis is a rule or procedure which if applied to sample data leads to a decision of accepting or rejecting the hypothesis under consideration. Null hypothesis means the hypothesis which is chosen for test and is denoted by H_0. The research question is placed under the head of alternative hypothesis. Test statistic is a formula which is used to test the hypothesis. To perform test of hypothesis there are two types of error that can be taken into account: type I error and type II error. Type I error means that rejecting H_0 when H_0 is true and type II error means accepting H_0 when H_0 is false. The probability of type I error is called the level of significance and is denoted by α. Generally we use level of significance, α = 0.05 or 0.01.

In Biological research when a researcher wants to verify his or her research question then it is necessary to perform a statistical test.

In the following we discuss one sample t-test in a short.

Test of hypothesis of a single mean: (Using the above data set: http://www.principlesofeconometrics.com/excel.htm)

Hypothesis to be tested:

H_0: The average food expenditure is equal to BDT 500.

H_1: The average food expenditure is not equal to BDT 500.

The above one sample t-test has been performed through SPSS and the output is given as follows.

From the table it has been found that the null hypothesis is rejected at 5% level of significance as the P-value is 0.000. That means the average food expenditure is not equal to BDT 500.

Table 4. One-sample t- test

	Test Value = 500					
	t	df	Sig. (2-tailed)	Mean Difference	95% Confidence Interval of the Difference	
					Lower	Upper
Food expenditure	-13.558	19	.000	-259.817	-299.9251	-219.7089

Now we discuss the concept about dependent sample. An example will be helpful to understand about the dependent sample. Suppose a doctor suggests a group of diabetic patients to walk regularly. Before starting walk it has been taken their blood sugar level measurements. Subsequent to one month, after walking regularly, their blood sugar level measurements have been taken. Then it has been investigated that their average blood sugar level (before and after) is same or not. Such type of sample is called a dependent sample because the individuals are same.

In the above situation the paired t-test will be applied. Hypothesis to be tested:

H_0: $\mu_d = 0$
H_1: $\mu_d \neq 0$

We use a hypothetical example.

20 diabetic patients have been considered and their blood sugar level has been recorded. A researcher wants to know whether the walking is effective or not. The data is as follows.

Hypothesis to be tested:

H_0: There is no difference between before and after blood sugar level measurements.

H_1: There is some difference between before and after blood sugar level measurements.

After performing paired t-test, it has been performed that the test statistic value is 3.336 and the corresponding p-value is 0.003. Comparing with the 5% level of significance it can be said that the null hypothesis is rejected. That means there is some difference between before and after blood sugar level measurements. It has been observed that the walking is effective to control the excessive blood sugar level.

In Biological research such type of research is frequently performed.

Table 5. Blood sugar level measurements of 20 diabetic patients before and after walking

Before	After
13.6	12
7.32	12.2
7.8	8.1
9.8	6.61
10.2	11.1
12.3	10.1
20.1	17.1
22.2	18.1
20	19.3
19.5	9.1
17.1	10.1
16.5	10.3
14.2	12.1
13.2	7.8
14.1	8.7
10.1	8.8
9.6	9.1
8.9	10.4
10.3	6.5
21.0	6.9

At last we can focus the concept on the Analysis of variance (ANOVA). ANOVA is a technique which is used to test the equality of several means. In the following we discuss an example. Suppose five medicines were used to recover fever and medicines were used for several patients. The researcher wants to know whether the average recovery times are same or not for five medicines to recover fever. The hypothesis to be tested:

H_0 : All the treatment means are equal. Or, H_0: $\mu_1=\mu_2=\mu_3=\mu_4=\mu_5$
H_1 : At least two of them are unequal.

The hypothetical data is as follows.

In this example, medicines were considered as the treatment. The recovery days are given as observations which are considered as yield.

The output of one way ANOVA has been placed here.

Table 6. Recovery days from fever corresponding to five medicines

Medicine 1	Medicine 2	Medicine 3	Medicine 4	Medicine 5
3	6	3	6	4
4	3	2	7	8
2	3	3	3	3
3	4	4	5	9
8	2	3	2	4

Table 7. One-way ANOVA output

ANOVA					
Yield					
	Sum of Squares	df	Mean Square	F	Sig.
Between Groups	19.760	4	4.940	1.241	.325
Within Groups	79.600	20	3.980		
Total	99.360	24			

3. COMMENT

Since p-value is greater than 0.05, so null hypothesis is accepted. That means, the effectiveness of five medicines to recover fever were equal.

In the following we consider a two way ANOVA. Consider an example, suppose someone wants to know is there any difference among five medicines to recover an infection of the patients and also wants to know whether the effect of five medicines varied among five laboratories or not, then it becomes two way ANOVA. Two-way ANOVA is also called two way classification because data is classified on the basis of two factors. One is treatment and the other one is block.

A hypothetical example of two way ANOVA (two-way classification) is as follows:

Five different medicines have been used for the treatment of a certain infection. These medicines were tested on patients in five different laboratories. The following table gives the number of cases of recovery from the infections of the patients who have used the medicines.

The researcher wants to know, whether the average effects of medicines are equal or not. Simultaneously, effects of medicines were varied among different laboratories or not.

Here laboratories can be considered as blocks and the medicines are considered as treatments.Hypothesis to be tested:

For treatment,

H_0 : All the treatment means are equal. Or, H_0: $\mu_1=\mu_2=\mu_3=\mu_4=\mu_5$
H_1 : At least two of them are unequal.

For block,

And, H_0: All the block means are equal. Or, H_0: $\mu_1=\mu_2=\mu_3=\mu_4=\mu_5$
H_1 : At least two of them are unequal.

The output is as follows:

Table 8. Recovery cases from an infection corresponding to five medicines and five labs

	Medicine 1	Medicine 2	Medicine 3	Medicine 4	Medicine 5
Lab 1	23	17	28	40	41
Lab 2	25	19	22	36	32
Lab 3	19	29	26	26	25
Lab 4	30	15	17	25	26
Lab 5	12	20	11	19	20

From Table 9 it has been found that the block effects and treatment effects are significant at 5% level of significance because the p-values are .023 and .048 respectively. That means the average effects of medicines are not equal to recover a certain infection and even then the average effects of medicines varied among different laboratories to recover a certain infection of the patients.

Table 9. Two-way ANOVA output

Source	**Type III Sum of Squares**	**df**	**Mean Square**	**F**	**Sig.**
block	510.64	4	127.66	3.835	.023
treat	405.44	4	101.36	3.045	.048
Error	532.56	16	33.285		
Total	15993	25			
Corrected Total	1448.64	24			

a. R Squared = .632 (Adjusted R Squared = .449).

From this chapter, we can get some ideas and concepts about the utilities of statistical knowledge and application which are very wide and vast in Biological Sciences.

REFERENCES

Islam, M.N. 2012. *An Introduction to Statistics and Probability.* Mullick & Brothers.160-161, Dhaka New Market, Dhaka-1205.

Jalil, M.A. Ferdous, R. 1999. *Basic Statistics.* Methods and Applications. Robi Publications: Dhaka, Bangladesh.

Steel, Robert G.D., James, H. Torrie and David A. Dickey (1997). *Principle and Procedures of Statistics: A Biometrical Approach.* McGrow-Hill.

EDITORS' CONTACT INFORMATION

Professor Dr. Hossain Uddin Shekhar
Department of Biochemistry and Molecular Biology
Faculty of Biological Sciences
University of Dhaka, Dhaka, Bangladesh
Email: hossainshekhar@du.ac.bd, shekhardu@hotmail.com

Dr. M. M. Towhidul Islam
Assistant Professor, Department of Biochemistry and Molecular Biology
Faculty of Biological Sciences
University of Dhaka, Dkaha, Bangladesh
Email: towhidbmb@du.ac.bd

INDEX

#

A

B

C

D

E

F

G

H

I

L

M

N

O

P

Q

R

S

T

V

W

Y

Related Nova Publications

Horseradish Peroxidase: Structure, Functions and Applications

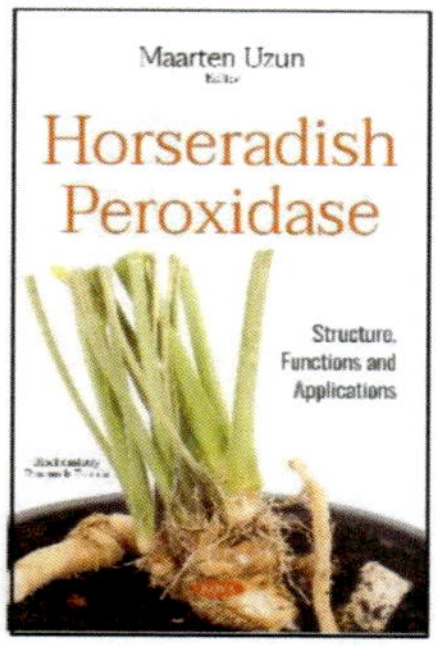

Editor: Maarten Uzun

Series: Biochemistry Research Trends

Book Description: In this compilation, the authors discuss the commercial source for the enzyme horseradish peroxidase, the tuberous roots of the horseradish plant which is native to the temperate regions of the world. Horseradish peroxidase is an oxidoreductase belonging to the highly ubiquitous group of peroxidases, indicating that this enzyme came into existence in the early stages of evolution and has been conserved thereafter.

Softcover ISBN: 978-1-53615-912-7
Retail Price: $95

Hemagglutinins: Structures, Functions and Mechanisms

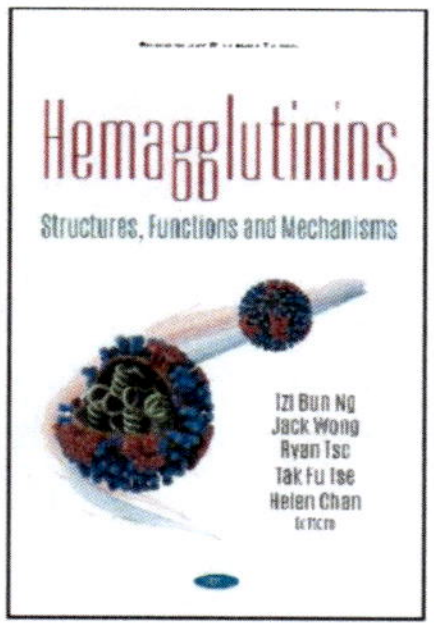

Editors: Tzi Bun Ng, Jack Wong, Ryan Tse, Tak Fu Tse, and

Series: Biochemistry Research Trends

Book Description: Hemagglutinins refers to glycoproteins which bring about agglutination of erythrocytes or hemagglutination. Hemagglutination can be used to identify surface antigens on erythrocytes (with known antibodies) and, hence, the blood type of an individual.

Hardcover ISBN: 978-1-53615-708-6
Retail Price: $230

To see a complete list of Nova publications, please visit our website at www.novapublishers.com

Related Nova Publications

Biochemistry Laboratory Manual for Undergraduate Students

Authors: Buthainah Al Bulushi, Raya Al-maliki, and Musthafa Mohamed Essa, Ph.D.

Series: Biochemistry Research Trends

Book Description: This laboratory manual has been designed for nutrition students for a better understanding of the lab assessments including biochemistry and food chemistry lab assessments. This manual includes both qualitative and quantitative analyses of some of the macro and micronutrients.

Online Book ISBN: 978-1-53614-967-8
Retail Price: $0

Amylases: Properties, Functions and Uses

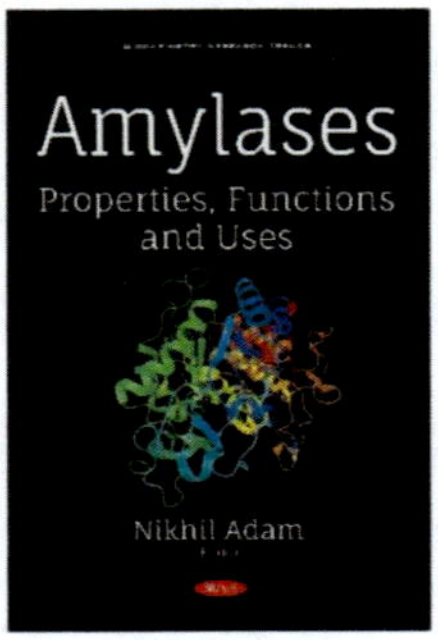

Editor: Nikhil Adam

Series: Biochemistry Research Trends

Book Description: *Amylases: Properties, Functions and Uses* opens with an analysis of the methods commonly used for the immobilization of amylase on particles, the effect that the processes of adsorption and covalent immobilization have on the activity and stability of the enzyme, as well as on its stability and reusability.

Softcover ISBN: 978-1-53614-993-7
Retail Price: $82